VARIETIES OF FLUVIAL FORM

INTERNATIONAL ASSOCIATION OF GEOMORPHOLOGISTS

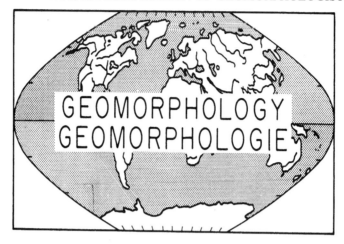

Publication No. 1

THE EVOLUTION OF GEOMORPHOLOGY
A Nation-by-Nation Summary of Development

Edited by H.J. Walker *and* W.E. Grabau

Publication No. 2

RIVER GEOMORPHOLOGY

Edited by Edward J. Hickin

Publication No. 3

STEEPLAND GEOMORPHOLOGY

Edited by Olav Slaymaker

Publication No. 4

GEOMORPHIC HAZARDS

Edited by Olav Slaymaker

Publication No. 5

LANDSLIDE RECOGNITION

Edited by Richard Dikau, Denys Brunsden, Lothar Schrott *and* Maïa-Laura Ibsen

Publication No. 6

GEOMORPHOLOGY SANS FRONTIÈRES

Edited by S. Brian McCann *and* Derek C. Ford

Publication No. 7

VARIETIES OF FLUVIAL FORM

Edited by Andrew J. Miller *and* Avijit Gupta

VARIETIES OF FLUVIAL FORM

Edited by

ANDREW J. MILLER
Department of Geography and Environmental Systems, University of Maryland
Baltimore County, Baltimore, Maryland, USA
and
AVIJIT GUPTA
School of Geography, University of Leeds, Leeds, UK

JOHN WILEY & SONS
Chichester · New York · Weinheim · Brisbane · Singapore · Toronto

Other Wiley Editorial Offices

John Wiley & Sons, Inc., 605 Third Avenue,
New York, NY 10158-0012, USA

WILEY-VCH Verlag GmbH, Pappelallee 3,
D-69469 Weinheim, Germany

Jacaranda Wiley Ltd, 33 Park Road, Milton,
Queensland 4064, Australia

John Wiley & Sons (Asia) Pte ltd, 2 Clementi Loop #02-01,
Jin Xing Distripark, Singapore 129809

John Wiley & Sons (Canada) Ltd, 22 Worcester Road,
Rexdale, Ontario M9W 1L1, Canada

Library of Congress Cataloging-in-Publication Data
Varieties of fluvial form / edited by Andrew J. Miller and Avijit
Gupta.
        p.   cm. — (International Association of Geomorphologists;
  no. 7)
      Includes bibliographical references and index.
      ISBN 0-471-97351-3 (alk. paper)
      1. Rivers—Congresses. 2. Geomorphology—Congresses. I. Miller,
Andrew J. II. Gupta, Avijit. III. Series: Publication
(International Association of Geomorphologists); no. 7.
GB1201.2.V37 1999
551.48′3—dc21   98-36750
                                                                CIP

British Library Cataloguing in Publication Data

A catalogue record for this book is available from the British Library

ISBN 0-471-97351-3

Typeset in 10/12pt Times from editor's disk by MHL Typesetting Ltd, Coventry
Printed and bound in Great Britain by Bookcraft (Bath) Ltd, Midsomer Norton, Somerset
This book is printed on acid-free paper responsibly manufactured from sustainable forestry,
in which at least two trees are planted for each one used for paper production.

To Sandra and Anthea

y

Contents

Contributors

V.R. Baker   Department of Hydrology and Water Resources, University of Arizona, Tucson, AZ 85721, USA

Mary C. Bourke   Center for Earth and Planetary Studies, MRC 315, National Air and Space Museum, Smithsonian Institution, Washington, DC 20560, USA

Gary J. Brierley   School of Earth Sciences, Macquarie University, North Ryde, NSW 2109, Australia

L.J. Broadhurst   Centre for Water in the Environment, University of Witwatersrand, Private Bag 3, WITS 2050, Johannesburg, South Africa

Armando Chávez   Faculty of Geography, University of Guadalajara, Guadalajara, Mexico

Zhongyuan Chen   Estuarine and Coastal State Key Laboratory, East China Normal University, Shanghai 200062, People's Republic of China

Geoff Day   Department of Geology and Geophysics, University of California, Berkeley, CA 94720, USA

Leena A. Deodhar   Department of Geography, S.P. College, Pune 411 030, India

William E. Dietrich   Department of Geology and Geophysics, University of California, Berkeley, CA 94720, USA

Wayne D. Erskine   Forest Research and Development Division, State Forests of New South Wales, PO Box 100, Beecroft, NSW 2119, Australia

Rob J. Ferguson   School of Earth Sciences, Macquarie University, North Ryde, NSW 2109, Australia

Christopher R. Fielding   Department of Earth Sciences, University of Queensland, Queensland 4072, Australia

Paul E. Grams   Department of Geology, Utah State University, Logan, UT 84322-5240, USA

Avijit Gupta   School of Geography, University of Leeds, Leeds, LS2 9JT, UK

Johan Hattingh   Trans Hex Mining Ltd, Private Bag X3, Alexander Bay 8290, South Africa

George L. Heritage   Telford Institute of Environmental Systems, Department of Geography, University of Salford, Manchester, M5 4WT

H.Q. Huang   School of Geosciences, University of Wollongong, Northfields Avenue, Wollongong, NSW 2522, Australia

Vishwas S. Kale   Department of Geography, University of Poona, Pune 411 007, India

Barbara A. Kennedy   School of Geography, Oxford University, Mansfield Road, Oxford, OX1 3TB, UK

M.J. Kirkby   School of Geography, University of Leeds, Leeds, LS2 9JT, UK

G. Komatsu   Lunar and Planetary Laboratory, University of Arizona, Tucson, AZ 85721, USA

Elizabeth A. Livingstone   School of Geography, University of New South Wales, Sydney, NSW 2052, Australia

Andrew J. Miller   Department of Geography and Environmental Systems, UMBC, 1000 Hilltop Circle, Baltimore, MD 21250, USA

B.P. Moon   Centre for Water in the Environment, University of Witwatersrand, Private Bag 3, WITS 2050, Johannesburg, South Africa

Gerald C. Nanson   School of Geosciences, University of Wollongong, Northfields Avenue, Wollongong, NSW 2522, Australia

David Palacios   Department of Physical Geography, Complutense University, Madrid, Spain

Gary Parker   St Anthony Falls Laboratory, University of Minnesota, Minneapolis, MN 55414, USA

Geoff Pickup   CSIRO, Land and Water, Canberra, ACT 2601, Australia

S.N. Rajaguru   Department of Archaeology, Deccan College, Pune 411 006, India

Izak C. Rust   Department of Geology, University of Port Elizabeth, Port Elizabeth, South Africa

A.C.T. Scheepers Department of Earth Sciences, University of the Western Cape, Private Bag X17, Bellville 7535, South Africa

John C. Schmidt Department of Geography and Earth Resources, Utah State University, Logan, UT 84322-5240, USA

Stephen Tooth Department of Geology, University of the Witwatersrand, Johannesburg, WITS 2050, South Africa

A.W. van Niekerk (deceased) formerly Centre for Water in the Environment, University of Witwatersrand, Private Bag 3, WITS 2050, Johannesburg, South Africa

John A. Webb School of Earth Sciences, La Trobe University, Bundoora, Victoria 3083, Australia

Rainer Wende School of Geosciences, University of Wollongong, Northfields Avenue, Wollongong, NSW 2522, Australia

M. Gordon Wolman Department of Geography and Environmental Engineering, The Johns Hopkins University, Charles and 34th Street, Baltimore, MD 21218, USA

Foreword

M. GORDON WOLMAN

Scientists may convey a sense of joy in personal interviews or reminiscences about why they pursue their trade, or more often in conversation. They are reluctant to convey this same sense in formal sessions. The Introduction to this exciting volume sets forth superbly both the intent of the sessions held at the June 1995 South East Asia Conference of the International Association of Geomorphologists and some themes evident in the papers included here. What can be added is the enthusiasm and excitement that pervaded two full days of presentation and discussion of fluvial geomorphology in the sessions at Singapore. That spirit, I believe, reflects the pleasure of discovering, of seeing and learning new things and the pleasure of skewering and defending founded and unfounded generalizations. The authors of these papers have, of course, seen and studied, some for many years, the rivers they write about. Others, like myself and hopefully many readers of this volume, have not encountered the variety of rivers presented here. The convenors and editors are to be congratulated on bringing together, as they intended, the remarkable observations and analyses of the "broad range of geographic and morphological diversity" of rivers. The collection should do much to stimulate inquiry and comparison of the varied rivers found throughout the globe. The results will enhance our understanding of why rivers look the way they do and how they got that way.

As is often the case, much that is new contains or perhaps returns to the old. While Davis' ideal "model" of landscape evolution assumed a particular tectonic history, a pluvial–fluvial "normal" climate, and distinctive sequential morphology over time, the too simple shorthand version is captured in the phrase: landforms are a function of structure, process and time.

Many of the studies reported here emphasize the importance of structure, reflected in rock type and tectonics, and its dominant control of both macro- and micro-features of river systems and form. This emphasis is associated with a departure from an equilibrium framework emphasizing instead non-equilibrium in many different ways including dominance of bedrock controls, great variability in hydrologic regimen along with emphasis upon the dominance of large, episodic cataclysmic events as the principal moulders of the riverine landscape. Thus, as the Introduction notes, most of the papers in this collection represent a significant departure from the prevailing concepts of adjustment and equilibrium which have perhaps dominated the literature over the past several decades. The remarkable examples brought together here demonstrate the enormous spectrum of river scenes found in nature and the multiple combinations and permutations of variables responsible for this diversity. They also demonstrate the value of observations and descriptions of Earth surface features unburdened by preconceived hypotheses about causes and evolution. The results are a very valuable fresh look contributing to a renewed balance in explaining the elements responsible for fluvial landscapes.

Yet, as the editors suggest, the papers in this volume also present a challenge. Some of the studies suggest not simply significant differences in river channel and valley landforms, but a continuity in landform from those dominated by sediments to those dominated by bedrock and tectonics. Others demonstrate an inextricable mix of depositional and erosional processes influencing river form and evolution. In addition, some imply significant climatic control, particularly in semi-arid regions, but comparison of several different rivers flowing on bedrock demonstrates that they may share many attributes despite very different climatic regimes. Similarities rather than differences dominate.

Collectively these studies might lead one to ask: Is every river different from every other river? Given the enormous range of possible combinations of rocks, tectonic and climatic histories, at some temporal and spatial scales one might make a plausible argument for such a view. At the same time, the editors' skilful groupings suggest that at least some dominant controls are distinguishable among the diverse examples. Perhaps, at the risk of pushing the quest too far, one can ask: Are there generalizations that can usefully be made to classify and characterize the spectrum of fluvial landscapes encountered in nature? To do so, even to test possible speculation, more descriptions of reality are needed including ways of characterizing structural features as well as the rocks themselves. New techniques of dating surfaces used to measure the duration of surface features as well as rates of erosion and incision may help in comparing the characteristics and evolution of rivers and valleys. Temperament, perhaps, determines whether one searches for diversity or commonality. The beauty of this volume is that it should inspire those of either temperament to describe the broadest range of rivers in the field and to search for comprehensive explanations of river phenomena.

Introduction

ANDREW J. MILLER AND AVIJIT GUPTA

This collection of papers is based largely on contributions by participants in a series of special sessions at the June 1995 South East Asia Conference of the International Association of Geomorphologists. The concept that draws these papers together is perhaps best explained in the following excerpt from the call for papers that was issued in advance of the conference:

> The purpose of this special session is to provide an opportunity for discussion of the diversity of fluvial forms that may arise in "unusual" settings... These may include settings where external forcing factors exert a powerful influence precluding or modifying the type of equilibrium form that is expected to develop under quasi-stable conditions in an alluvial channel: for example, patterns of response to tectonic activity... longitudinal variations in channel and floodplain form associated with spatially varying patterns of bedrock lithology or jointing... arid-region rivers with longitudinal discharge trends that are not monotonically increasing. Also of potential interest are morphological features forming under climatic/ hydrologic regimes associated with monsoon, tropical wet, extremely arid or extremely cold environments; or patterns of vegetation distribution that exert a controlling influence on channel form. It is hoped that the papers contributed to this session will encompass a broad range of geographic as well as morphological diversity.

In fact the final roster of papers presented here includes examples from Australia, Papua New Guinea, Sri Lanka, India, China, South Africa, Namibia, Antarctica, the United States and Mexico, as well as the moon, Mars and Venus.

The underlying premise was that the world's rivers encompass a wide range of influences and associated fluvial forms, but that the geomorphic literature traditionally has focused most intensively on examples comprising mainly alluvial rivers from humid temperate, semi-arid, or proglacial environments. Although the range of examples covered in the professional literature has certainly broadened over the last couple of decades, the textbooks through which most students are first exposed to geomorphology are focused increasingly on the mechanics of alluvial rivers but for the most part do not treat a spectrum of morphological types that is proportionally representative of the rivers that actually exist on the Earth's surface. (The diversity of influences on fluvial form is perhaps better illustrated in some of the texts published by authors from the southern hemisphere; see, for example, Selby, 1985.) We believe that a presentation that includes some of the outliers on this spectrum may broaden our scientific vocabulary and lead ultimately to improved understanding of fluvial processes and their role in landscape evolution.

Varieties of Fluvial Form. Edited by A.J. Miller and A. Gupta
© 1999 John Wiley & Sons Ltd

Thus the intention of this volume is, first and foremost, to present an international set of case studies with a strong descriptive component, representing the diversity of fluvial forms. Although these examples are by no means unique and they clearly are all governed by the same physical laws, it is also true that virtually everyone who attended the sessions in Singapore had an opportunity to see something that he or she had not seen before. To quote Baker and Komatsu in the opening paper of this volume, "excitement and pleasure in science derive not so much from achieving the final explanation as from discovering the fascinating range of new phenomena to be explained". It is our hope that readers of this volume will share some of that excitement and pleasure.

In addition to the case studies, this volume includes three additional contributions. The first is a Foreword by M. Gordon Wolman, one of the authors of the "bible" of fluvial geomorphology, *Fluvial Processes in Geomorphology*, published in 1964. In preparing the current volume we were acutely conscious of the legacy of this work and we requested that Dr Wolman, who was keynote speaker at the Singapore conference, offer us his impressions of how our understanding of the range of fluvial forms has changed and how it has stayed the same.

We also include two review papers, each of which raises questions about the potential relevance of the case studies provided here in modelling different aspects of fluvial systems. Dr Christopher Fielding, of the University of Queensland, provides a perspective on implications for facies modelling and for interpretation of stratigraphic sequences. Dr M.J. Kirkby, of Leeds University, offers his perspective on implications for landform modelling.

A reading of the entire set of papers reveals several themes that turn up with some frequency. The volume is not organized strictly according to these themes, for the simple reason that any single paper may have some content related to multiple themes on this list. What follows is an attempt to identify the major themes and the papers in which each theme plays a role.

Most of the papers are concerned in some way with the influence of "extrinsic" factors, most commonly in the form of lithology, geologic structure and tectonics; valley confinement as an influence on flow patterns and fluvial morphology; climatic/ hydrologic regimes; large floods; coastal interactions; and the role of backwater/base level effects.

BEDROCK LITHOLOGY AND TECTONICS

Several papers treat the influence of spatially varying lithology, structure and tectonics. At the grandest scale, the paper by Baker and Komatsu reminds us that interplanetary comparisons provide a unique vantage point on large-scale structures and their formative processes, and that the term "fluvial" may be applied both to catastrophic meltwater floods on Mars and to channels carved in solid rock by flowing lava on Venus. Closer to home, both Heritage et al. and van Niekerk et al. are concerned with classification and with the genesis and evolution of a series of related channel types in incised bedrock channels. The anastomosing bedrock channel type is one particularly interesting example that bears some resemblance to the erosional patterns associated with cataclysmic floods in a variety of settings cited by Baker and Komatsu. Anastomosing channels incised in bedrock are found both in the Sabie River of South Africa and along some reaches of the

Narmada River in India, described by Gupta et al. as a mixed bedrock–alluvial system with some particularly abrupt morphological transitions that are affected both by lithological boundaries and regional uplift.

Transitions between bedrock-bound canyons and more alluvial park-like reaches are also observed in incised rivers of the western USA, although, as Grams and Schmidt point out, the controls on channel morphology in the Green River of Utah may be more directly related to the calibre and amount of debris derived from tributaries and valley walls than to direct interactions between the river and the underlying bedrock surface. Deodhar and Kale note a connection between channel and valley form and underlying lithology in the Deccan Trap region of India, and they explore trends using hydraulic geometry as a tool. Wende, on the other hand, working at a smaller spatial scale and examining the erosion and transport of some enormous tabular boulders in bedrock river systems in Australia, notes the relationship between dip, joint patterns, bed gradient and the mechanics affecting both the mode of transport and the threshold required for transport. Palacios and Chávez discuss the role of Holocene and Pleistocene lava flows, tectonically induced debris avalanches, and regional tilting in their influence on fluvial morphology in the Colima Graben of Mexico; and Hattingh and Rust trace the late Cenozoic development of drainage patterns and the spatial distribution of fluvial deposits affected by stream capture, bedrock lithologies of varying resistance, and regional tilting. Dietrich et al. comment on differences in rates of uplift in source areas and downstream depositional areas along the Fly River in Papua New Guinea, noting also the radical change in channel and floodplain configuration from one to the other.

VALLEY CONFINEMENT

There is obviously a strong connection between bedrock lithology, erosion resistance and valley width. The degree and longitudinal pattern of valley confinement in turn exerts an important influence on flow dynamics and associated channel and floodplain characteristics. This theme is addressed in detail by Ferguson and Brierley in their mapping of the longitudinal distribution of valley widths, locations of bedrock spurs, and spatial patterns of different styles of floodplain development along the Tuross River in New South Wales. Erskine and Livingstone, also working in New South Wales, observe that the degree of valley confinement exerts a strong influence on the development of in-channel benches, the relative importance of vertical accretion as a depositional process, and the recurrence interval of the flow required to inundate the various bench levels or the valley flat. As mentioned in the preceding paragraphs, valley confinement is largely controlled by alternating bedrock and alluvial reaches along the Narmada River in India (Gupta et al.) and by zones of alternating lithology and bedrock resistance along the Green River in Utah (Grams and Schmidt); more-confined and less-confined reaches are associated with radically different suites of fluvial landforms. Bourke and Pickup characterize the different "topographic domains" of the Todd River in central Australia, including constrictions created by bedrock strike ridges and indurated Pleistocene alluvial terraces, as well as broader, relatively unconfined alluvial plains, in some cases with multiple distributary channels. The role of the longitudinal pattern of valley confinement in the development of floodout zones and of "reforming" channels in arid central Australia is also noted by Tooth; confinement in this study may be controlled by terrace deposits bordering the channel but also includes examples controlled by bedrock and by aeolian dunes.

HYDROCLIMATOLOGY AND FLOOD REGIMES

Setting aside the even more extreme examples found on the moon, Mars and Venus, climatic regions represented in this volume range from the hyper-arid Skeleton Coast of Namibia, home to the Uniab system as described by Scheepers and Rust; to the arid interior of Australia, with examples described by Tooth and by Bourke and Pickup; to the frigid, arid locale of the fans in East Antarctica, forming under the influence of processes described by Webb and Fielding; to the monsoon climates affecting the rivers of the Deccan Trap uplands (Deodhar and Kale) and the Narmada River in India (Gupta et al.), as well as the Mahaveli Ganga in Sri Lanka (Kennedy). The range of climatic extremes is rounded out by the inclusion of a wet equatorial example, the Fly River of Papua New Guinea, as described by Dietrich et al. The distinctive imprint of the hydroclimatological regime on fluvial morphology is noteworthy, particularly in very arid climates and in settings with highly variable flows. One notable exception, however, is described in the study by Webb and Fielding, who observe that morphology and sedimentology of small debris fans in coastal East Antarctica are not strongly diagnostic of the climatic region.

The spectrum of hydroclimatic influences has more complexity than can be explained by the simple wet–dry and warm–cold axes. About half of the papers explicitly note the importance of floods as formative events. Baker and Komatsu note the formative role of catastrophic floods occurring at the very largest scales and they draw parallels between resulting landforms even in cases where different flood-generating mechanisms may be responsible. Several papers identify different scales of fluvial landforms dominated by flows of different scales; hence the tripartite sequence of dry-season low flows, "normal" monsoon floods and rare high-magnitude floods is related to a channel-in-channel morphology along the Narmada (Gupta et al.) and along the Deccan Trap rivers (Deodhar and Kale) and is reminiscent of a similar sequence of flood magnitudes (within-channel or small-scale; flow across floodplains or medium-scale; and "superflood" or large-scale) and associated landforms described by Bourke and Pickup from the Todd River of arid central Australia. Erskine and Livingstone point out the geomorphic influence of high flood variability in their study area in New South Wales, and they note that floods of varying scale are associated with inset surfaces or in-channel benches at different relative elevations. In addition, the largest floods are identified as channel-clearing events, whereas smaller floods tend to build in-channel benches during the subsequent recovery period. The imbricated clusters of giant tabular boulders described by Wende indicate the persistent influence of rare superfloods on incised bedrock channels in monsoonal northwestern Australia. Catastrophic changes such as avulsion, incision of new channels, and transport of large boulders are attributed to such superfloods by Scheepers and Rust (Uniab, Namibia). The role of large floods in avulsion and channel abandonment is well known along the lower Yangtze, and is briefly described by Chen.

Downstream trends in flood magnitude are just as important as the temporal sequencing of large floods in controlling the fluvial morphology of some systems. Notable cases documented here include the Deccan Trap upland (Deodhar and Kale) and the arid region of central Australia (Tooth), both of which have runoff source areas feeding into rivers which subsequently traverse areas with declining precipitation. The net result of this may be a decrease in channel capacity which in some cases is detectable primarily as a statistical trend in a hydraulic geometry relationship, but in other cases leads to total breakdown and disappearance of the channel system. A similar trend is noted by Bourke and Pickup, who also point out that "asynchronous tributary activity and differences in flow magnitude from contributing systems" may lead to abrupt changes in fluvial morphology at river confluences. In contradistinction to the arid-region tendency towards longitudinally declining flow patterns is the Narmada River, where occasional superfloods are influenced by the tendency of monsoonal storms to follow a track that heads downstream along the axis of the valley (Gupta et al., Figure 5.2), thus reinforcing the flood peak rather than allowing it to attenuate.

The history of climate change forms yet another axis in the relationship between hydroclimatology, flood regime and fluvial morphology. Several authors cite stratigraphic evidence for a distinctly different hydroclimatology and flood regime in the past (Gupta et al., Hattingh and Rust, Dietrich et al.)

BACKWATER/BASE-LEVEL EFFECTS AND COASTAL INTERACTIONS

Base-level controls on fluvial form are imposed most obviously as rivers approach sea level, and several papers examine interactions between coastal and fluvial processes or between base-level/backwater controls and fluvial processes. These include the contribution by Chen, which focuses primarily on Holocene development of the Yangtze delta plain as affected by "a combination of eustasy, isostasy, tectonic displacement, fluvial transport and coastal hydrodynamics". Dietrich et al. note the possible influence of similar factors (other than coastal dynamics) on the broad, low-gradient alluvial plain of the Fly River in Papua New Guinea, where the reach most dramatically affected by rising base level is also characterized by elongate meanders with very slowly migrating channels, low rates of aggradation, and excellent preservation of relict features. A counterintuitive pattern of coastal or base-level interaction is observed by Scheepers and Rust in their study of the Uniab alluvial fan on the Namibian coast, truncated by shore erosion during the Holocene episode of sea-level rise: rapid truncation has led to channel incision even as base level is rising. A more subtle interaction is observed by Webb and Fielding on a series of small fans bordering a tidally influenced gorge in the Prince Charles Mountains of East Antarctica: in addition to some coastal and wind-driven sediment transport, wetting and drying of sandstone clasts in the intertidal zone causes weathering and disintegration of particles to an extent not observed above high tide. Ferguson and Brierley note that the lower end of their study reach on the Tuross River in New South Wales has been incised to a depth of 42 m below sea level and subsequently filled during the Holocene by cohesive estuarine muds overlain by fluvial deposits; this reach exhibits radically different patterns of fluvial morphology and floodplain development from other parts of the Tuross that are situated well upstream of marine influence.

As several studies indicate, backwater effects may be considerably more dynamic than a simple reduction of base level, and they may be caused by a variety of factors other than rising or falling sea level. Dietrich et al. make note of flow reversals and leakage of floodwaters back and forth between the lower Fly River and its tributaries through a network of tie channels connecting the main channels with swamp areas, oxbows and blocked valley lakes. Kennedy comments on the strong base-level influence of a large trunk river on the lower reaches of its tributaries, and notes the presence of ponding, sediment trapping, flow reversals and sedimentary structures that are reminiscent of estuarine tidal inlets. Bourke and Pickup observe changes in morphostratigraphy at tributary confluences resulting from asynchronous flood peaks on the tributary and the main stem: a pulse of sediment emanating from the tributary may block the main channel, causing an upward step in the channel bed, local ponding and aggradation of fine sediment at low flow, whereas a subsequent flow event on the main stem may trench the channel and leave the tributary mouth exposed as a hanging valley. Both Tooth and Scheepers and Rust note instances where aeolian barriers may act as dams, blocking flow and causing either ponding and deposition or diversion of flow into distributary channels. Grams and Schmidt comment on the role of debris fans as the dominant grade controls along the Green River and other channels of the Colorado Plateau; resistant bedrock highs also act as local base-level controls inducing backwater effects and local alluviation along the bedrock anastomosing reaches of the Sabie River (van Niekerk et al., Heritage et al.). Palacios and Chávez suggest that alternating cycles of channel blockage and lacustrine sedimentation, followed by renewed incision, are controlled by recurring volcanic debris avalanches.

ORGANIZATION OF THIS VOLUME

As the preceding discussion shows, many of the papers in this volume treat overlapping themes. Although there is no simple way to divide the volume into non-overlapping categories, it is useful to have some proximity between papers that have more in common and some classification into major topic headings is obligatory. We have therefore identified the following topic headings and sequence of presentation.

Part 1 *Interplanetary Comparisons*
   1 Baker and Komatsu

Part 2 *Bedrock and Mixed Bedrock/Alluvial Rivers*
   2 van Niekerk et al.
   3 Heritage et al.
   4 Grams and Schmidt
   5 Gupta et al.
   6 Hattingh and Rust
   7 Palacios and Chávez
   8 Wende

Part 3 *Arid-Region Rivers*
   9 Tooth
  10 Bourke and Pickup
  11 Scheepers and Rust
  12 Deodhar and Kale
  13 Webb and Fielding

Part 4 *Patterns of Alluvial Deposition*
  14 Dietrich et al.
  15 Ferguson and Brierley
  16 Kennedy
  17 Chen
  18 Erskine and Livingstone
  19 Nanson and Huang

Part 5 *Review Papers*
  20 Fielding
  21 Kirkby

As will be noted from the preceding discussion, this effort to gather a set of case studies of "unusual" rivers has in fact led to identification of a number of common themes meriting further investigation. We anticipate that the examples collected here will be useful to students of fluvial geomorphology by providing a more expansive view of the range of forms that may be observed, and we hope this volume will help to stimulate additional research exploring these themes. We would like to thank the referees for the papers included in this book, and in alphabetical order they are: Ned Andrews, Victor Baker, Gary Brierley, Paul Carling, Mike Church, Bill Dietrich, Ron Dorn, Chris Fielding, Tom Gardner, Will Graf, Alan Howard, Robb Jacobson, Dick Marston, Gerald Nanson, Waite Osterkamp, Geoff Pickup, John Pitlick, James Pizzuto, Bruce Rhoads, John Ritter, Izak Rust, Jack Schmidt, Michael Sheridan, Greg Tucker, Ellen Wohl and Reds Wolman. It is a distinguished list.

Finally, as editors of this volume we wish to take this opportunity to honour the memory of Andre van Niekerk, senior author and co-author respectively of two of the papers in this volume. Dr van Niekerk was a respected member of the research community in South Africa, and he lost his life in an accident while engaged in field research. We extend our condolences to his family, friends and colleagues.

REFERENCE

Selby, M.J. 1985. *Earth's Changing Surface: an introduction to geomorphology*. Oxford University Press, Oxford, 607 pp.

Part 1

Interplanetary Comparisons

1

Extraterrestrial Fluvial Forms

V.R. BAKER[1,2] AND G. KOMATSU[2]

[1]*Department of Hydrology and Water Resources, University of Arizona, Tucson, Arizona, USA*
[2]*Lunar and Planetary Laboratory, University of Arizona, Tucson, Arizona, USA*

ABSTRACT

Fluvial-like landforms occur on at least four solar system bodies. While the "riverine" features of the Moon and Venus seem to have been produced by lava, those of Earth and Mars are water-generated. Similarities and differences among the various fluvial forms pose a challenge to claims of generalization for any explanatory system (theories, models, hypotheses) posed by fluvial geomorphologists. The following aspects of extraterrestrial fluvial forms need to be explained by fluvial geomorphological theory: (1) the development of broadly similar morphologies (meandering, anastomosis) through erosion into rock by both water and lava; (2) the clear importance of cataclysmic processes in generating fluvial landforms; and (3) the phenomenal range of scaling relationships for fluvial landforms on the various planets.

INTRODUCTION

Extraterrestrial geomorphology affords an opportunity for the discovery of anomalies that lead to the questioning of established theories (Baker, 1993). In a science where controlled experimentation is usually impossible (Baker, 1996a), newly discovered landscapes are central to stimulating and testing geomorphological explanations (Baker, 1984, 1985a,b). More importantly, such discoveries provide real-world phenomena for stimulating the analogical reasoning that is central to geomorphological inferences (Gilbert, 1886, 1896; Baker, 1996a). This paper will briefly review the morphologies, diversity and present state of genetic understanding of various extraterrestrial fluvial forms in the spirit of such reasoning.

LUNAR SINUOUS RILLES

If we exclude the fascinating pseudo-scientific studies of "canals" on Mars (Sheehan, 1988), the first extraterrestrial channelized water flows to be inferred from observed

Varieties of Fluvial Form. Edited by A.J. Miller and A. Gupta
© 1999 John Wiley & Sons Ltd

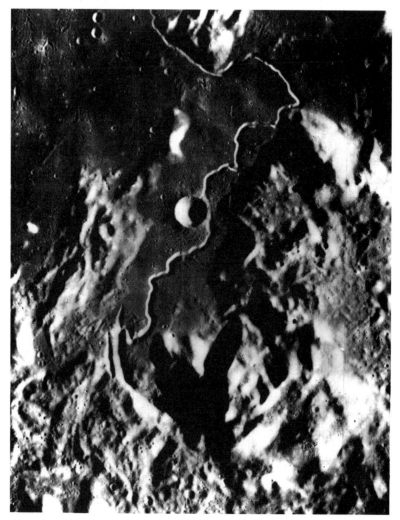

FIGURE 1.1   Lunar sinuous rille. This image from Apollo Metric Camera frames shows the Hadley Rille, which was visited by the Apollo 15 astronauts. The rille is about 1 km wide and extends for about 100 km from the elongate depression at centre, lower left

river-like morphologies are probably the lunar sinuous rilles (Figure 1.1). These meandering channels range from 10 to 340 km in length and are up to 1 km in width (Schubert et al., 1970). The largest are visible telescopically from Earth, and they were early recognized as showing similar morphological attributes to terrestrial river channels (Pickering, 1904; Firsoff, 1960). This observation was consistent with models of a primordial water-rich lunar hydrosphere (Gilvarry, 1960). High-resolution spacecraft images of the Moon became available in 1966, and early interpretation of these reinforced the fluvial hypothesis. The famous Nobel laureate chemist, Harold Urey (1967) concluded that the fluvial-like morphology of lunar sinuous rilles was so overwhelming that it was necessary to postulate some special processes that would have to occur in order for water to flow on the Moon. He even suggested that a large comet impact might have generated a temporary water-rich atmosphere of sufficient density to permit surface-water flow. Gilvarry (1968) built on this proposal to suggest that the lunar maria might be largely underlain by sediments and sedimentary rocks. Theoretical modelling by Lingenfelter et al. (1968) showed that ice-covered rivers required less extreme changes in environment, and that ice-covered water flows were capable of eroding a lunar sinuous rille in about a century. As another consequence of their putative aqueous origin, sinuous rilles were interpreted as indicators of the distribution of volatiles on the Moon (Peale et al., 1968).

The Apollo missions returned samples that provided overwhelming evidence that the Moon had no early hydrosphere (Taylor, 1982). Indeed, the lunar rocks are so deficient in water that this was a major factor in recognizing the Moon's origin by impact of a Mars-sized object with the newly accreted Earth in the very early history of the solar system (Taylor, 1992). Alternative processes were invoked to explain the sinuous rilles, including collapsed lava tubes or lava channels (Baldwin, 1963; Oberbeck et al., 1969), erosion by pyroclastic flows (Cameron, 1964), and a combination of faulting and subsidence (Quaide, 1965). Using an experimental apparatus for venting gas through granular materials, Schumm (1970) produced crater chains and sinuous troughs that mimicked the properties of the lunar sinuous rilles, including meandering and sinuosity. Close-up inspection of the Hadley Rille by the Apollo 15 astronauts and the associated regional geological study indicated that the collapsed lava tube and lava channel hypotheses best accounted for available data (Greeley, 1971a; Howard et al., 1972).

FIGURE 1.2 Lava channels on the southwest rift zone of Mauna Loa Volcano, Hawaii. These channels are up to 10 m wide, and they emanate from linear zones of cinder-and-spatter cones that are aligned along the rift zone

Stimulated in part by the puzzle posed by lunar sinuous rilles, terrestrial lava channels (Figure 1.2) and lava tubes have received considerable study, greatly advancing understanding of their diverse morphologics and processes of formation (Greeley, 1971b; Swanson, 1973; Carr, 1974).

THE CHANNELS ON MARS

The Martian fluvial features are much more diverse and extensive than the fluvial-like forms on the Moon. Immediately following their discovery by the 1972 Mariner 9 imager (Masursky, 1973; Milton, 1973), it was recognized that two general categories were easily distinguishable: (1) outflow channels, morphologically similar to cataclysmic flood channels on Earth (Baker and Milton, 1974; Sharp and Malin, 1975); and (2) valley networks, resembling the tributary patterns of terrestrial drainage basins (Pieri, 1976, 1980). The outflow channels emanate from extensive zones of collapse (Figure 1.3), presumably marking the sites of release of subsurface water (Baker, 1978a, 1982; Carr, 1979; Mars Channel Working Group, 1983). Distally from these source areas, the outflow channels comprise immense complexes of erosional and depositional troughs up to 150 km wide and 2000 km long. At their largest scale, the channels are broadly anastomosing and split by residual uplands, or "islands", of pre-flood-modified terrain (Figure 1.4). The channels have low sinuosity and high width–depth ratios. Pronounced flow expansions and constrictions occur, as do prominent divide crossings, hanging valleys and structural control of erosion. At a finer scale, streamlining of the residual uplands (Figure 1.5) is very well developed (Baker and Kochel, 1978; Komar, 1983, 1984), as are longitudinal grooves, inner channels, cataract complexes, scabland and bar complexes (Figure 1.6) (Baker and Kochel, 1979; Baker, 1982; Baker et al., 1992a). Although a variety of other fluid-flow systems were invoked to explain these features (see Baker, 1982, 1985b, for reviews), the whole assemblage of these landforms seemed most

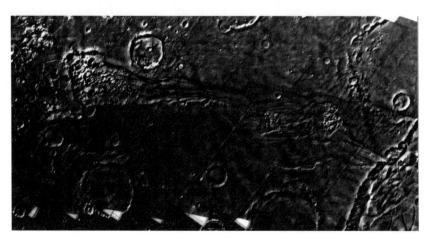

FIGURE 1.3   Ravi Vallis, a relatively small Martian outflow channel, showing the chaotic collapse zone (Aromatum Chaos) at its source (left) and the streamlining and groove terrain along its path. Total channel length is approximately 250 km

FIGURE 1.4   Residual uplands with prominent crater (C), around which channels with grooves (D) are scoured. Note that floodwater encountered an elongate ridge (A) and spilled through multiple gaps (B). Viking Orbiter frame 20AG2

economically explained by cataclysmic flood processes, with particular analogy to the origin of the Channeled Scabland (Baker, 1978b, 1981, 1982; Baker and Nummedal, 1978). However, it was also recognized that some important differences derived from the peculiar Martian environment, notably its lower gravitational acceleration, its much lower atmospheric pressure, and its prevailing subfreezing temperature in comparison to Earth. These special conditions on Mars would influence such factors as sediment transport mechanics (Komar, 1979, 1980), cavitation and ice formation (Baker, 1979), debris flowage (Nummedal and Prior, 1981), and possible large-scale ice processes in the channels (Lucchitta, 1982).

Although they were recognizable on Mariner 6 and 7 imagery from 1969 (Schultz and Ingerson, 1973), the extensive valley networks of Mars were most clearly described from the Mariner 9 and Viking imagery of the 1970s (Milton, 1973; Pieri, 1976; Carr and Clow,

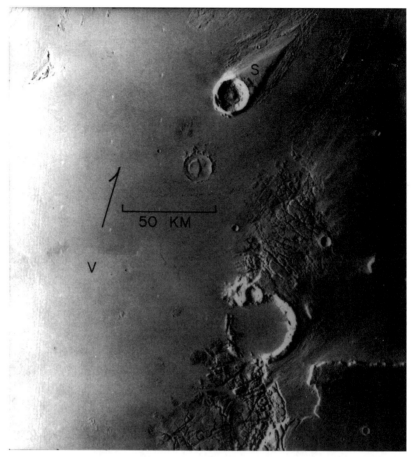

FIGURE 1.5   Large streamlined upland (S) in Kasei Vallis. Chaotic terrain (C) and longitudinal grooves (G) are also visible. Arrow shows north. Viking Orbiter frame 519A01

1981). In contrast to the channels, Martian valley networks do not have obvious bedforms. They comprise integrated networks of tributaries (Figure 1.7) with widths of about 10 km or less, and lengths from <5 km to nearly 1000 km. Drainage densities are generally much lower than for terrestrial valley networks (Baker and Partridge, 1986), although an interesting set of valleys on Martian volcanoes is more similar to terrestrial valleys in their general morphology (Figure 1.8) (Gulick and Baker, 1989, 1990). Among the other important morphological attributes of Martian valley networks are the following: theatre-like valley heads, prominent structural control, low junction angles, quasi-parallel patterns, hanging valleys, irregular widening and narrowing, and indistinct thermal areas. In general, this assemblage of features seems best explained by groundwater sapping processes (Baker et al., 1990), excellent examples of which occur in the sandstone terrains of the Colorado plateau (Figure 1.9) (Laity and Malin, 1985). Recently, Carr (1995) proposed that headward erosion by mass wasting and slow downvalley transport of debris by groundwater-aided debris flowage created the valley networks, but he did not cite terrestrial examples for generating valley networks by such processes.

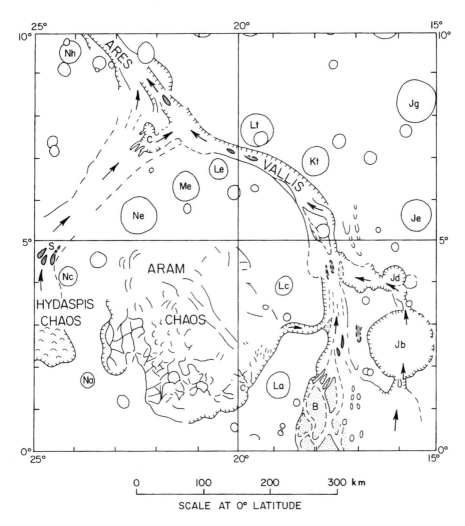

SCALE AT 0° LATITUDE

FIGURE 1.6 Sketch map of upper Ares Vallis showing effects of large-scale flooding. Craters Jb and Jd were breached by floodwater spilling over their rims. Streamlined uplands (S) and an inner channel cataract complex (C) occur, as does a possible area of depositional bars (B)

The general climatic and environmental conditions associated with valley network formation have been the subject of considerable scientific debate (Carr, 1996; Squyres and Kasting, 1994). A prevailing view is that, early in Mars' geological history, conditions favouring the atmospheric cycling of water and liquid surface runoff were made possible through the action of atmospheric warming by radiatively active gases, such as carbon dioxide (Pollack et al., 1987). Alternatively, a globally enhanced geothermal heat flow, early in the planet's history, warmed the ice-rich permafrost to generate surface-water flow and sapping processes (Squyres, 1989). Another group of explanations focuses on the role of more local endogenetic heat sources, including intrusions (Wilhelms and Baldwin, 1989), impact craters (Schultz et al., 1982; Brakenridge et al., 1985) and volcanoes (Gulick, 1993).

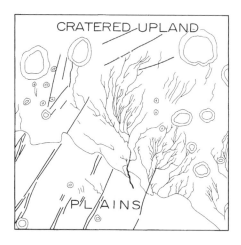

FIGURE 1.7   Sketch map of Warrego Vallis. The area shown is approximately 300×300 km. Note the incomplete dissection of the cratered upland

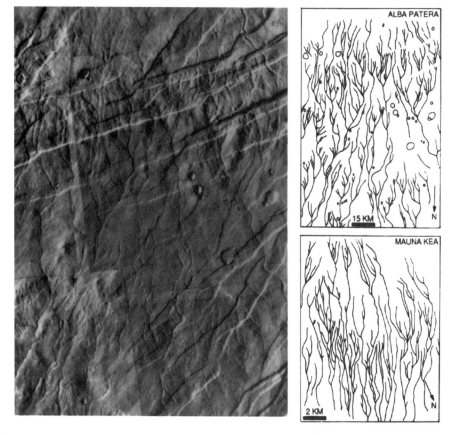

FIGURE 1.8   Fluvial valleys on Alba Patera volcano. At left is the Viking Orbiter mosaic (MTM 45107) and in the upper right is a sketch map of the drainage pattern. Note the more complete dissection of terrain than in Figure 1.7. A map of terrestrial drainage on Mauna Kea Volcano, Hawaii, is shown at lower right for comparison

FIGURE 1.9   Valleys formed by spring sapping and slope retreat in sandstone of southeastern Utah. Note the theatre-like valley heads which retreat headward as they are undermined by the sapping processes

Sinuous ridges with fluvial-like patterns (Figure 1.10) are common in many areas of Mars, particularly at middle and high latitudes. When these ridges were first seen on the Viking Orbiter images of the planet, their resemblance to terrestrial eskers (Figure 1.11) was immediately apparent (Carr et al., 1980, p.136). However, it was also recognized that a glacial origin of eskers has profound implications for the atmospheric cycling of water (needed to form large glaciers), and for the wet-based glacial conditions that favour esker formation. These implications are resisted by many Mars planetologists for reasons reviewed by Carr (1996). Partly motivated by such issues, numerous other hypotheses were offered to explain the fluvial-like patterns of the Martian sinuous ridges. These alternative hypotheses include inverted stream topography (Howard, 1981), shoreline features (Parker et al., 1986), linear dunes (Ruff and Greeley, 1990), exhumed clastic dykes (Ruff and Greeley, 1990) and volcanic landforms (Tanaka and Scott, 1987). Nevertheless, the esker hypothesis for ridge origin (Kargel and Strom, 1992) possesses an attribute that is not a quality of its alternatives. The eskers are part of an assemblage of temporally and spatially related landforms. The assemblage shows a proximal-to-distal transition in the following order: erosional grooves, eskers and drumlins, kames and kettles, morainal ridges, glaciolacustrine plains, and shorelines. Such assemblages occur in terrestrial glaciated terrains and presumably indicate similar origins on Mars.

The Martian sinuous ridges have lengths that can exceed 200 km, and widths of approximately 1 km. They are remarkably continuous, unlike terrestrial eskers which generally contain many gaps in their longitudinal continuity. Nevertheless, the largest terrestrial eskers are comparable in scale to the Martian sinuous ridges. Eskers can have

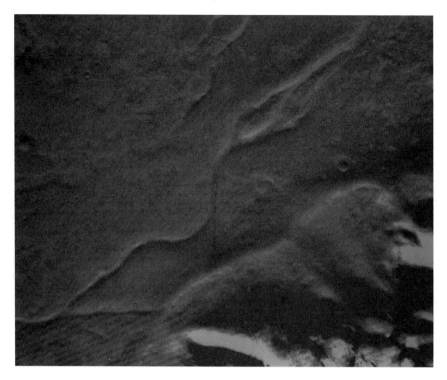

FIGURE 1.10   Sinuous ridges on the floor of the Argyre impact basin, Mars. Note the bifurcation into multiple channels, one of which (right centre) is buried by a lobate debris apron. Scene width is approximately 50 km. Viking Orbiter frame 567B30

lengths of hundreds of kilometres, widths of 400–700 m and heights of 40–50 m (Bennett and Glasser, 1996). Differences between Martian and terrestrial fluvial landforms commonly involve issues of scale, with Martian forms being larger than their terrestrial counterparts.

The abundance and variety of fluvial forms on Mars strongly suggest a complex history of hydrological change on that planet (Baker et al., 1991). Long-term water storage is as ground ice in the permafrost or as subpermafrost groundwater. Episodic, cataclysmic outbursts of water, probably driven by massive geothermal processes associated with mantle plumes, resulted in outflow channels and temporary lakes or "seas" on the lowland northern plains of Mars. A temporary glacial climate accompanied these outburst phases, and the eskers were emplaced during the glacial episodes. The glacial climate was inherently unstable, and Mars readily reverted to its stable conditions of cold, dry surface environment (Baker et al., 1991).

VENUSIAN CHANNELS

Although it has not received the popular media attention of other planetary exploration discoveries, the recognition in 1991 of channels on the planet Venus was both unexpected and important. More than 200 valley and channel complexes are recognized from study

FIGURE 1.11   Oblique aerial photograph of an esker that formed beneath the Okanogan Lobe of the Cordilleran Ice Sheet on the Waterville Plateau of north-central Washington

of the Magellan synthetic aperture radar imager (Baker et al., 1992b; Komatsu et al., 1993). Some of these channels are very similar in morphology to the lunar sinuous rilles. They emanate from distinct, circular or elongated regions of collapse (generally several kilometres in diameter); they are approximately 1 to 2 km wide and several tens of kilometres long; and they become narrower and shallower distally. As on the Moon, most sinuous rilles on Venus are not associated with detectable lava flow margins. The morphological similarity between Venusian and lunar sinuous rilles suggests an origin by the thermal erosion of flowing lava (Hulme, 1973, 1982). However, the exact origin of these channels could have been vastly different given that Venusian sinuous rilles tend to cluster on or near tectonovolcanic features that are not found on the Moon, such as coronae, corona-like features and arachnoids (Komatsu et al., 1993).

Perhaps the most remarkable discovery on Venus was a channel type named "canali" (Italian for "channels"). The name is an attempt to correct a famous error in which this term was applied by the Italian astronomer Giovanni Schiaparelli to linear features he observed on the planet Mars. The American astronomer Percival Lowell applied the English term "canals" to very poorly resolved telescopic interpretation of Mars lineations, leading to decades of misunderstanding of that planet. The Venusian canali have a nearly constant width and depth. They are typically 1 to 3 km in width and are up to 500 km in length, although a few canali are as wide as 10 km, and the length reaches 6800 km in one case (Figure 1.12). Canali may locally exhibit abandoned channel segments, cut-off meander bends, levees, and radar-dark terminal deposits. Sources and termini are generally indistinct. Canali are generally located in Venusian plains regions, which are likely to be the product of extensive and prolonged emplacement of lavas (Head et al., 1991). Canali morphology suggests their probable formation by, and conveyance of, large discharges of low-viscosity lava to distant regions over prolonged periods (Komatsu et al., 1992).

Liquid water is known to be unstable on the surface of Venus, and, somewhat like the Moon, the crust and upper mantle of Venus are known to be unusually dry (Kaula, 1995). Therefore, another very surprising discovery was that of Venusian outflow channels, which display large-scale landforms indicative of cataclysmic fluid flows: streamlined

FIGURE 1.12   Baltis Vallis, Venus, a canali-type channel located at 49–51°N, 165–168°E (F-50N163; left-looking radar illumination). Channel width is approximately 1–4 km; scale bar equals 50 km. Baltis Vallis is the longest channel yet discovered in the solar system with a total length of *c*.6800 km

residual hills, spillovers, bar-like forms and regional anastomosis. Preliminary flow discharge estimates indicate that the responsible "lavas" had fluxes similar to the catastrophic flows of water in terrestrial scabland channelways (Baker et al., 1997). An example outflow channel is Kallistos Vallis (Figure 1.13), which is 1200 km long and up to 30 km in width. Its scale and morphology are comparable to those of the outflow channels of Mars, but the channel-forming fluid was most probably a low-viscosity lava (Baker et al., 1992b).

Venusian lava channels have meander dimensions that relate to their mode of formation. Their meander properties generally follow terrestrial river trends of wavelength (*L*) to width (*W*) ratios, suggesting an equilibrium adjustment of channel form (Komatsu and Baker, 1994). Slightly higher *L/W* for many Venusian channels in comparison to terrestrial rivers may relate to non-aqueous flow processes. The unusually low *L/W* values for some Venusian and lunar sinuous rilles probably indicate modification of original meander patterns by lava-erosional channels widening. Canali-type channels tend to have slightly higher *L/W* ratios than terrestrial rivers, and they often have poorly defined low-wavelength meanders superposed on long-wavelength meanders.

Experimental lava channels, simulated with hot water flowing on polyethylene glycol (Huppert and Sparks, 1985), indicate that meander sinuosity is primarily determined by

FIGURE 1.13 Kallistos Vallis, Venus, an "outflow channel" that probably developed from highly fluid lava. The total channel length is approximately 1200 km

initial flow discharge (the higher the discharge, the lower the sinuosity). Because the hot water thermally widens and deepens the channel while retaining the original meander sinuosity (Huppert and Sparks, 1985), L/W ratios decrease with time. Because both Venusian and lunar sinuous rilles display unusually low L/W ratios, flowing lava may have eroded (through thermal and mechanical processes) both the floors and walls of these channels (Hulme, 1973; Komatsu and Baker, 1994).

Studies of terrestrial rivers (Leopold and Wolman, 1957; Schumm et al., 1972) and flume experiments (Schumm and Khan, 1972) show that meandering occurs within specific ranges of slope and discharge. Theoretical considerations, using energy minimization principles (Chang, 1988), also suggest that channel sinuosity is a function of discharge. The morphologic similarity between Venusian channels and terrestrial fluvial channels (Komatsu et al., 1993) suggests that meanders observed in Venusian channels also developed within a limited range of discharge and slope.

DISCUSSION

Because new planets are first encountered globally at low resolution, extraterrestrial fluvial studies have necessarily taken a global perspective. Moreover, the fluvial features have always been identified in their full geological context, integrated into the long-term planetary history manifest in the global perspective. In contrast, terrestrial fluvial studies are mainly accomplished on restrictive scales, both temporally and spatially. Of course, there are exceptions, including the recognition of geological dominance by "big rivers" (Potter, 1978), but the spatial and temporal narrowness of geomorphological concern has tended to limit its objects of inquiry, to specify its methodologies and to direct its theoretical concerns (Baker and Twidale, 1991; Baker, 1993). Scaling relationships for various geomorphological processes (Baker, 1983) show that nature is not so narrow as modern styles of geomorphological research would seem to imply. Therefore, one of the most important aspects of extraterrestrial fluvial study is to open up the scales of real phenomena that are comparatively studied. Theories applied to phenomena at one scale or on one planet can then be evaluated for their general applicability to phenomena unrecognized in the course of theory formulation.

The restriction of conventional fluvial geomorphology to Earth rivers, particularly at spatially restrictive and temporally recent scales, also has its terrestrial implications. Much in the way that remote sensing has opened up a world of planetary surfaces to discovery, new side-scan sonars have revealed a remarkable array of submarine channel patterns (Mienert et al., 1993). Meandering channels on submarine fans (Flood and Damuth, 1987) have patterns similar to terrestrial rivers (Clark et al., 1992), though their origin is probably through the action of turbidity currents (Normark et al., 1993).

Multichannel patterns are common among the extraterrestrial fluvial and fluvial-like landforms described above. The terrestrial environments for such patterns have mostly been described from alluvial, rather than bedrock channel settings. Thus, Smith and Smith (1980) distinguished anastomosed from braided streams such that the former have relatively stable, cohesive sediment channels separated by relatively large islands that are usually cut from the continuous floodplain. Braided channels have unstable bars within the channel, composed of non-cohesive sediment. Nanson and Knighton (1996) use the term "anabranching" for a whole range of multiple-channel rivers separated by vegetated semipermanent alluvial islands, and anastomosing channels become a subcategory of anabranching channels.

In bedrock settings, which probably correspond to most of the extraterrestrial examples, the bar forms are not relevant to channel stability. Indeed, the high resistance of the bedrock corresponds best to the cohesive banks of alluvial channels. The term "anastomosis" was applied very early to the catastrophic flood channelways cut in rock in the Channeled Scabland (Bretz, 1923). Anastomosing patterns are a distinguishing attribute of cataclysmic flood flow invading relatively small pre-flood valleys, and spilling across divides to create isolated island remnants of the pre-flood topography (Bretz, 1928; Baker, 1978c). Although somewhat neglected in fluvial geomorphological study, anastomosing bedrock patterns are well developed on modern terrestrial rivers, particularly in tropical cratonic areas (Garner, 1967; Kale et al., 1996). Perhaps the importance of these patterns on other planets will stimulate more work on understanding Earth examples, especially given the access to evidence for formative processes that can be obtained with field access.

The recognition of very large cataclysmic flow channels on both Mars and Venus also has important implications for terrestrial studies. Although the Channeled Scabland example has been known for many years, more recently the association of great cataclysmic flooding with deglaciation has been recognized as a general phenomenon of the late Pleistocene in North America and Eurasia (Baker, 1994, 1996b). Large cataclysmic flood features in Siberia are now seen to be similar in formative processes (Baker et al., 1993; Rudoy and Baker, 1993) to features in the Channeled Scabland (Baker, 1973, 1981; Baker and Komar, 1987). The recognition of cataclysmic flooding as an important planetary-scale fluvial process also appears in the controversial hypothesis of massive subglacial flooding as the cause for the very interesting late Pleistocene assemblage in much of Canada of tunnel channels, Rogen moraine and drumlins (Shaw, 1994). While the exact processes responsible for these planetary-scale landscapes are not yet fully resolved, the increased interest in their study reveals the intellectual productivity of changing one's viewpoint from that currently predominating in a discipline (Baker, 1996a).

On 4 July 1997, the terrestrial scientific community received a sudden reminder of the ability of an expanded discovery of reality to stimulate creative scientific thinking. The successful landing on Mars that day of the Pathfinder spacecraft (Golombek, 1997) resulted in the on-site imaging of an immense cataclysmic flood channel, Ares Vallis. This Martian outflow channel shows a great variety of cataclysmic flood landforms at the resolution scales of orbital spacecraft imagery (Golombek et al., 1997; Komatsu and Baker, 1997). Imagery from the Pathfinder Lander reveals the sedimentary and erosional features appropriate to the local scale of ground-based observation (Figure 1.14). This discovery also shows that the lack of ability to perform controlled experiments on cataclysmic floods, large-scale bedrock erosion or similarly interesting phenomena need not relegate geomorphology to a lower scientific status than those sciences able to test their theories experimentally in a laboratory. Geomorphological explanations applicable to fluvial phenomena on Earth should, if they are generalizable theory, also apply to newly discovered fluvial phenomena on other planets. Regardless of such methodological concerns, excitement and pleasure in science derive not so much from achieving the final explanation as from discovering the fascinating range of new phenomena to be explained.

FIGURE 1.14   Photomosaic of images taken at the Mars Pathfinder landing site in July 1997. The smooth slopes for the two hills on the horizon were produced by the erosion of cataclysmic flood flows that also probably emplaced the boulders in the foreground. Some of the boulders also show imbricate arrangements consistent with their fluvial deposition

ACKNOWLEDGEMENTS

Our work in extraterrestrial fluvial geomorphology has been supported by the National Aeronautics and Space Administration Planetary Geology and Geophysics Program, grant NAGW-285. Goro Komatsu is a Research Fellow of the Japan Society for the Promotion of Science by Young Scientists, Japan Ministry of Education.

REFERENCES

Baker, V.R. 1973. *Paleohydrology and sedimentology of Lake Missoula in eastern Washington.* Geological Society of America Special Paper 144, 79 pp.

Baker, V.R. 1978a. A preliminary assessment of the fluid erosional processes that shaped the Martian outflow channels. In *Proceedings of the 9th Lunar and Planetary Science Conference*, Vol. 3. Pergamon Press, New York, 3205–3223.

Baker, V.R. 1978b. The Spokane Flood controversy and the Martian outflow channels. *Science*, **202**, 1249–1256.

Baker, V.R. 1978c. Large-scale erosional and depositional features of the Channeled Scabland. In V.R. Baker and D. Nummedal (Eds), *The Channeled Scabland*. National Aeronautics and Space Administration, Washington, DC, 81–115.

Baker, V.R. 1979. Erosional processes in channelized water flows on Mars. *Journal of Geophysical Research*, **84**, 7985–7993.

Baker, V.R. (Ed.) 1981. *Catastrophic Flooding: The Origin of the Channeled Scabland*. Hutchinson Ross, Stroudsburg, Pennsylvania, 360 pp.

Baker, V.R. (Ed.) 1982. *The Channels of Mars*. University of Texas Press, Austin, Texas 198 pp.

Baker, V.R. 1983. Large-scale fluvial palaeohydrology. In K.J. Gregory (Ed.) *Background to Palaeohydrology: A Perspective*. Wiley, Chichester, 453–478.

Baker, V.R. 1984. Planetary geomorphology. *Journal of Geological Education*, **32**, 236–246.

Baker, V.R. 1985a. Relief forms on planets. In A.F. Pitty (Ed.), *Themes in Geomorphology*. Croom Helm, London, 245–259.

Baker, V.R. 1985b. Models of fluvial activity on Mars. In M. Woldenberg (Ed.), *Models in Geomorphology*. Allen and Unwin, London, 287–312.

Baker, V.R. 1993. Extraterrestrial geomorphology: Science and philosophy of Earthlike planetary landscapes. *Geomorphology*, **7**, 9–35.

Baker, V.R. 1994. Glacial to modern changes in global river fluxes. In *Material Fluxes on the Surface of the Earth*. National Academy Press, Washington, DC, 86–98.

Baker, V.R. 1996a. Hypotheses and geomorphological reasoning. In B.W. Rhoads and C.E. Thorn (Eds), *The Scientific Nature of Geomorphology*. Wiley, New York, 57–85.

Baker, V.R. 1996b. Megafloods and glaciation. In I.P. Martini (Ed.), *Late Glacial and Postglacial Global Changes: Quaternary, Carboniferous–Permian and Proterozoic*. Oxford University Press, Oxford, 98–108.

Baker, V.R. and Kochel, R.C. 1978. Morphometry of streamlined forms in terrestrial and Martian channels. In *Proceedings of the 9th Lunar and Planetary Science Conference*, Vol. 3, Pergamon Press, New York, 3193–3203.

Baker, V.R. and Kochel, R.C. 1979. Martian channel morphology: Maja and Kasei Valles. *Journal of Geophysical Research*, **84**, 7961–7983.

Baker, V.R. and Komar, P.D. 1987. Cataclysmic flood processes and landforms. In W.L. Graf (Ed.), *Geomorphic Systems of North America*. Geological Society of America, Boulder, Colorado, 423–443.

Baker, V.R. and Milton, D.J. 1974. Erosion by catastrophic floods on Mars and Earth. *Icarus*, **23**, 27–41.

Baker, V.R. and Nummedal, D. (Eds) 1978. *The Channeled Scabland*. National Aeronautics and Space Administration, Washington, DC 186 pp.

Baker, V.R. and Partridge, J.B. 1986. Small Martian valleys: Pristine and degraded morphology. *Journal of Geophysical Research*, **91**, 3561–3572.

Baker, V.R. and Twidale, C.R. 1991. The reenchantment of geomorphology. *Geomorphology*, **4**, 73–100.

Baker, V.R., Kochel, R.C., Laity, J.E. and Howard, A.D. 1990. Spring sapping and valley network development. In C.G. Higgins and R.D. Coates (Eds), *Groundwater Geomorphology*. Geological Society of America Special Paper 252, 235–266.

Baker, V.R. Strom, R.G., Gulick, V.C., Kargel, J.S. and Komatsu, G. 1991. Ancient oceans, ice sheets and the hydrological cycle on Mars. *Nature*, **352**, 589–594.

Baker, V.R., Carr, M.H., Gulick, V.C., Williams, C.R. and Marley, M.S. 1992a. Channels and valley networks. In H.H. Kieffer, B. Jakosky and C. Snyder (Eds), *Mars*. University of Arizona Press, Tucson, 493–522.

Baker, V.R., Komatsu, G., Parker, T.J., Gulick, V.C., Kargel, J.S. and Lewis, J.S. 1992b. Channels and valleys on Venus: A preliminary analysis of Magellan data. *Journal of Geophysical Research*, **97**, 13 421–13 444.

Baker, V.R., Benito, G. and Rudoy, A.N. 1993. Paleohydrology of late Pleistocene superflooding, Altay Mountains, Siberia. *Science*, **259**, 348–350.

Baker, V R., Gulick, V.C , Komatsu, G. and Parker, T.J. (1997). Channels and valleys. In S.W. Bougher, D.M. Hunten and R.J. Phillips (Eds), *Venus II*. University of Arizona Press, Tucson, 757–793.

Baldwin, R.B. 1963. *The Measure of the Moon*. University of Chicago Press, Chicago, 488 pp.

Bennett, M.R. and Glasser, N.F. 1996. *Glacial Geology: Ice Sheets and Landforms*. Wiley, Chichester, 364 pp.

Brakenridge, G.R., Newsom, H.E. and Baker, V.R. 1985. Ancient hot springs on Mars: Origins and paleoenvironmental significance of small Martian valleys. *Geology*, **13**, 859–862.

Bretz, J H. 1923. The Channeled Scabland of the Columbia Plateau. *Journal of Geology*, **31**, 617–649.

Bretz, J H. 1928. The Channeled Scabland of eastern Washington. *Geographical Review*, **18**, 446–477.

Cameron, W.S. 1964. An interpretation of Schroter's Valley and other lunar sinuous rilles. *Journal of Geophysical Research*, **69**, 2423–2430.

Carr, M.H. 1974. The role of lava erosion in the formation of lunar rilles and Martian channels. *Icarus*, **22**, 1–23.

Carr, M.H. 1979. Formation of Martian flood features by release of water from confined aquifers. *Journal of Geophysical Research*, **84**, 2995–3007.

Carr, M.H. 1995. The Martian drainage system and the origin of networks and fretted channels. *Journal of Geophysical Research*, **100**, 7479–7507.

Carr, M.H. 1996. *Water on Mars*. Oxford University Press, New York, 229 pp.

Carr, M.H. and Clow, G.D. 1981. Martian channels and valleys: Their characteristics, distribution and age. *Icarus*, **48**, 91–117.

Carr, M.H., Baum, W.A., Blasius, K.R., Briggs, G.A., Cutts, J.A., Duxbury, T.C., Greeley, R., Guest, J., Masursky, H., Smith, B.A., Soderblom, L.A., Veverka, J. and Wellman, J.B. 1980. *Viking Orbiter Views of Mars*. National Aeronautics and Space Administration, Special Paper 441, Washington, DC, 182 pp.

Chang, H.H. 1988. *Fluvial Processes in River Engineering*. Wiley, New York, 432 pp.

Clark, J.D., Kenyon, N.H. and Pickering, K.T. 1992. Quantitative analysis of the geometry of submarine channels: Implications for the classification of submarine fans. *Geology*, **20**, 633–636.

Firsoff, V.A. 1960. *Strange World of the Moon*. Basic Books, New York, 189 pp.

Flood, R.D. and Damuth, J.E. 1987. Quantitative characteristics of sinuous distributary channels on the Amazon deep-sea fan. *Geological Society of America, Bulletin*, **98**, 728–738.

Garner, H.F. 1967. Rivers in the making. *Scientific American*, **216**, 84–94.

Gilbert, G.K. 1886. The inculcation of scientific method by example. *American Journal of Science*, **31**, 284–299.

Gilbert, G.K. 1896. The origin of hypotheses illustrated by the discussion of a topographic problem. *Science*, **3**, 1–13.

Gilvarry, J.J. 1960. Origin and nature of lunar features. *Nature*, **188**, 886–891.

Gilvarry, J.J. 1968. Observational evidence for sedimentary rocks on the Moon. *Nature*, **218**, 336–341.

Golombek, M.P. 1997. The Mars Pathfinder Mission. *Journal of Geophysical Research*, **102**, 3953–3965.

Golombek, M.P., Cook, R.A., Moore, H.J. and Parker, T.J. 1997. Selection of the Mars Pathfinder landing site. *Journal of Geophysical Research*, **102**, 3967–3988.

Greeley, R. 1971a. Lunar Hadley Rille: Considerations of its origin. *Science*, **172**, 722–725.

Greeley, R. 1971b. Observations of actively forming lava tubes and associated structures, Hawaii. *Modern Geology*, **2**, 207–223.

Gulick, V.C. 1993. *Magmatic Intrusions and Hydrothermal Systems: Implications for the Formation of Small Martian Valleys*. Unpublished PhD Dissertation, University of Arizona, Tucson.

Gulick, V.C. and Baker, V.R. 1989. Fluvial valleys and Martian paleoclimates. *Nature*, **341**, 514–516.

Gulick, V.C. and Baker, V.R. 1990. Origin and evolution of valleys on Martian volcanoes. *Journal of Geophysical Research*, **95**, 14 325–14 344.

Head, J.W., Campbell, D.B., Elachi, C., Guest, J.E., McKenzie, D.P., Saunders, R.S., Schaber, G.G. and Schubert, G. 1991. Venus volcanism: Initial analysis from Magellan data. *Science*, **252**, 276–299.

Howard, A.D. 1981. Etched plains and braided ridges of the south polar region of Mars: Features produced by basal melting of ground ice? In *Reports of the Planetary Geology Program*. National Aeronautics and Space Administration, TM 84211, 286–288.

Howard, K.A., Head, J.W. and Swann, G.A. 1972. Geology of Hadley Rille. *Geochemica et Cosmochemica Acta*, Suppl. **3**, 1–14.

Hulme, G. 1973. Turbulent lava flow and the formation of lunar sinuous rilles. *Modern Geology*, **4**, 107–117.

Hulme, G. 1982. A review of lava flow processes related to the formation of lunar sinuous rilles. *Geophysical Surveys*, **5**, 245–279.

Huppert, H.E. and Sparks, R.S.J. 1985. Komatiites I: Eruption and flow. *Journal of Petrology*, **26**, 694–725.

Kale, V.S., Baker, V.R. and Mishra, S. 1996. Multi-channel patterns of bedrock rivers: An example from the central Narmada basin, India. *Catena*, **26**, 85–98.

Kargel, J.S. and Strom, R.G. 1992. Ancient glaciation on Mars. *Geology*, **20**, 3–7.

Kaula, W.M. 1995. Venus reconsidered. *Science*, **270**, 1460–1464.

Komar, P.D. 1979. Comparisons of the hydraulics of water flows in Martian outflow channels with flows of similar scale on Earth. *Icarus*, **37**, 156–181.

Komar, P.D. 1980. Modes of sediment transport in channelized water flows with ramifications to the erosion of the Martian outflow channels. *Icarus*, **42**, 317–329.

Komar, P.D. 1983. Shapes of streamlined islands on Earth and Mars: Experiments and analyses of the minimum-drag form. *Geology*, **11**, 651–654.

Komar, P.D. 1984. The lemniscate loop: Comparisons with the shapes of streamlined landforms. *Journal of Geology*, **92**, 133–145.

Komatsu, G. and Baker, V.R. 1994. Meander properties of Venusian channels. *Geology*, **22**, 67–70.

Komatsu, G. and Baker, V.R. 1997. Paleohydrology and flood geomorphology of Ares Vallis. *Journal of Geophysical Research*, **102**, 4151–4160.

Komatsu, G., Kargel, J.S. and Baker, V.R. 1992. Canali-type channels on Venus: Some genetic constraints. *Geophysical Research Letters*, **19**, 1415–1418.

Komatsu, G., Baker, V.R. Gulick, V.C. and Parker, T.J. 1993. Venusian channels and valleys: Distribution and volcanological implications. *Icarus*. **102**, 1–25.

Laity, J.E. and Malin, M.C. 1985. Sapping processes and the development of theater-headed valley networks in the Colorado Plateau. *Geological Society of America, Bulletin*, **96**, 203–217.

Leopold, L.B. and Wolman, M.G. 1957. River Channel Patterns: Braided, Meandering and Straight. *US Geological Survey Professional Paper* 282-B, 39–85.

Lingenfelter, R.E., Peale, S.J. and Schubert, G. 1968. Lunar rivers. *Science*, **161**, 266–269.

Lucchitta, B.K. 1982. Ice sculpture in the Martian outflow channels. *Journal of Geophysical Research*, **87**, 9951–9973.

Mars Channel Working Group 1983. Channels and valleys on Mars. *Geological Society of America, Bulletin*, **94**, 1035–1054.

Masursky, H. 1973. An overview of geologic results from Mariner 9. *Journal of Geophysical Research*, **78**, 4037–4047.

Mienert, J., Kenyon, N.H., Thiede, J. and Hollender, F.-J. 1993. Polar continental margins: Studies off East Greenland. *Eos*, **74**, 225–236.

Milton, D.J. 1973. Water and processes of degradation in the Martian landscape. *Journal of Geophysical Research*, **78**, 4037–4047.

Nanson, G.C. and Knighton, A.D. 1996. Anabranching rivers: Their cause character and classification. *Earth Surface Processes and Landforms*, **21**, 217–239.

Normark, W.R., Posamentier, H. and Mutti, E. 1993. Turbidite systems: State of the art and future directions. *Reviews of Geophysics*, **31**, 91–116.

Nummedal, D. and Prior, D.B. 1981. Generation of Martian chaos and channels by debris flows. *Icarus*, **45**, 77–86.

Oberbeck, V.R. Quaide, W.L. and Greeley, R. 1969. On the origin of lunar sinuous rilles. *Modern Geology*, **1**, 75–80.

Parker, T.J., Pieri, D.C. and Saunders, R.S. 1986. Morphology and distribution of sinuous ridges in central and southern Argyre. In *Reports of the Planetary Geology and Geophysics Program*, National Aeronautics and Space Administration, TM 88303, 468–470.

Peale, S.J., Schubert, G. and Lingenfelter, R.E. 1968. Distribution of sinuous rilles and water on the Moon. *Nature*, **220**, 1222.

Pickering, W.H. 1904. *The Moon*. Doubleday, Page, New York, 103 pp.

Pieri, D.C. 1976. Martian channels: Distribution of small channels on the Martian surface. *Icarus*, **27**, 25–50.

Pieri, D.C. 1980. Martian valleys: Morphology, distribution, age and origin. *Science*, **210**, 895–897.

Pollack, J.B., Kasting, J.F., Richardson, S.M. and Poliakoff, K. 1987. The case for a warm, wet climate on early Mars. *Icarus*, **71**, 203–224.

Potter, P.E. 1978. Significance and origin of big rivers. *Journal of Geology*, **86**, 13–33.

Quaide, W.L. 1965. Rilles, ridges and domes – clues to maria history. *Icarus*, **4**, 374–389.

Rudoy, A.N. and Baker, V.R. 1993. Sedimentary effects of cataclysmic late Pleistocene glacial outburst flooding, Altay Mountains, Siberia. *Sedimentary Geology*, **85**, 53–62.

Ruff, S.W. and Greeley, R. 1990. Sinuous ridges of the south polar region, Mars: Possible origins. In *Lunar and Planetary Science XXI*. Lunar and Planetary Institute, Houston, Texas 1047–1048.

Schubert, G., Lingenfelter, R.E. and Peale, S.J. 1970. The morphology, distribution and origin of lunar sinuous rilles. *Reviews of Geophysics and Space Physics*, **8**, 199–224.

Schultz, P.H. and Ingerson, E. 1973. Martian lineaments from Mariner 6 and 7 images. *Journal of Geophysical Research*, **78**, 8415–8427.

Schultz, P.H., Schultz, R.A. and Rogers, J.R. 1982. The structure and evolution of ancient impact basins on Mars. *Journal of Geophysical Research*, **78**, 9803–9820.

Schumm, S.A. 1970. Experimental studies of the formation of lunar surface features by fluidization. *Geological Society of America, Bulletin*, **81**, 2539–2552.

Schumm, S.A. and Khan, H.R. 1972. Experimental study of channel patterns. *Geological Society of America, Bulletin*, **83**, 1755–1770.

Schumm, S.A., Khan, H.R., Winkley, B.R. and Robbins, W.G. 1972. Variability of river patterns. *Nature*, **237**, 75–76.

Sharp, R.P. and Malin, M.C. 1975. Channels on Mars. *Geological Society of America, Bulletin*, **86**, 593–609.

Shaw, J. 1994. A qualitative view of sub-ice-sheet landscape evolution. *Progress in Physical Geography*, **18**, 159–184.

Sheehan, W. 1988. *Planets and Perception*. University of Arizona Press, Tucson, 324 pp.

Smith, D.G. and Smith, N.D. 1980. Sedimentation in anastomosed river systems: Examples from alluvial valleys near Banff, Alberta. *Journal of Sedimentary Petrology*, **50**, 157–164.

Squyres, S.W. 1989. Urey prize lecture: Water on Mars. *Icarus*, **74**, 229–288.

Squyres, S.W. and Kasting, J.F. 1994. Early Mars: How warm and how wet? *Science*, **265**, 744–748.

Swanson, D.A. 1973. Pahoehoe flows from the 1969–71 Mauna Ulu eruption, Kilauea volcano, Hawaii. *Geological Society of America, Bulletin*, **84**, 615–626.

Tanaka, K.L. and Scott, D.H. 1987. *Geologic map of the polar regions of Mars*. US Geological Survey Miscellaneous Investigations, Map I–1802C.

Taylor, S.R. 1982. *Planetary Science: A Lunar Perspective*. Lunar and Planetary Institute, Houston, Texas, 481 pp.

Taylor, S.R. 1992. *Solar System Evolution: A New Perspective*. Cambridge University Press, Cambridge, 307 pp.

Urey, H.C. 1967. Water on the Moon. *Nature*, **216**, 1094–1095.

Wilhelms, D.E. and Baldwin, R.J. 1989. The role of igneous sills in shaping the Martian uplands. *Proceedings of the Lunar and Planetary Science Conference*, **19**, 355–365.

Part 2

Bedrock and Mixed Bedrock/Alluvial Rivers

2

Bedrock Anastomosing Channel Systems: Morphology and Dynamics in the Sabie River, Mpumalanga Province, South Africa

A.W. van Niekerk,[1] G.L. Heritage,[2] L.J. Broadhurst[1] and B.P. Moon[1]

[1] Centre for Water in the Environment, University of Witwatersrand, Johannesburg, South Africa
[2] Telford Institute of Environmental Systems, Department of Geography, University of Salford, Manchester, UK

ABSTRACT

The geomorphological history of South Africa has resulted in river incision in the Lowveld zone. This has created a high level of bedrock control, which influences the geomorphological form of the Sabie River. Distributaries within the bedrock anastomosing channel type (which are common on this river system) have been shown to exploit fracture patterns and joint structures in specific bedrock lithologies. These bedrock distributaries are generally sediment-free, owing to their steep gradients and consequent high sediment transport capacities, and consist of a series of pools, rapids, cataracts and waterfalls. Localized alluvial deposition of fine sediment, carried in suspension, occurs in backwaters and on interfluve bedrock outcrops as the transport capacity declines. This is due to a dramatic increase in the flow width and reduced flow depth over topographic highs as floods fill the incised macro-channel, coupled with a reduction in the water surface slope as the flood wanes. These deposits create alluvial backwaters and bedrock core bars. Colonization by vegetation stabilizes the bar deposits and acts to encourage further deposition. The flow regime in the Sabie River is highly variable, both seasonally and annually, resulting in only a limited number of active distributaries flowing through the dry season. Much larger, infrequent events are required to inundate the entire macro-channel. This results in a highly variable pattern of inundation for the different geomorphic units across the channel. Therefore, the morphological evolution of the section is controlled by the whole flow regime, rendering conventional alluvial single thread channel concepts, such as bankfull flow, meaningless. Flow resistance investigation has generated very large Manning's n flow resistance values, which reflect the severely non-uniform flow profile, large-scale roughness elements within the bedrock distributaries and the vegetational resistance of the seasonally

Varieties of Fluvial Form. Edited by A.J. Miller and A. Gupta
© 1999 John Wiley & Sons Ltd

flooded bedrock core bars. Quantification of potential sediment transport demonstrates that bedrock anastomosing channels have a greater capacity for transporting sediment sizes supplied from the catchment than other channel types found on the river. This suggests that bedrock anastomosing channels are more resistant to alluviation. However, abstraction and modification of the flow regime, combined with increased sediment input from catchment land-use changes, is impacting severely on the frequency of inundation of individual channels and is causing progressive alluviation on bedrock core bars and in backwaters. Therefore, the long-term trend for bedrock anastomosing channels is a change to a more mixed bedrock–alluvial system.

INTRODUCTION

The Sabie River rises on the eastern slopes of the highland areas of the Drakensberg Mountain Range and flows eastwards through the flatter Lowveld before cutting through the Lebombo Mountain Range on the northeastern border of South Africa and flowing onto the coastal lowlands of Mozambique (Figure 2.1). Bedrock lithologics comprise sedimentary, and intrusive and extrusive igneous and metamorphic rocks.

The upper reaches of the rivers, from the escarpment to the Lebombo Mountain Range, are largely bedrock-controlled with varying degrees of alluviation in the Lowveld. The Lowveld represents a relatively young erosion surface which resulted from the westerly retreat of the Great Escarpment after the break-up of Gondwana through rift faulting about 100 million years ago (King, 1978). Two major cycles of rejuvenation (mid-Miocene and late Pliocene) resulted in the development of the Post-African I and II surfaces (Partridge and Maud, 1987) into which the Lowveld rivers are incised (Figure 2.2, Table 2.1). Incision of the planation surface resulted in the rivers cutting gorges in resistant rock as is seen at the Sabie River gorge through the Lebombo Mountain Range (Venter, 1991).

Thus, the river is confined to a valley cut into the bedrock, and active channel evolution and sedimentation are restricted within this zone, which is termed the macro-channel (Figure 2.3) (van Niekerk et al., 1995). The morphology of the Sabie River, in common with other bedrock-controlled rivers, is therefore controlled by the underlying geological structure and bedrock lithologies, although alluvial characteristics may occur within the macro-channel (see, for example, the Colorado River, USA (Howard and Dolan, 1981), the Burdekin Gorge, Australia (Wohl, 1992), the Nahal Paran, Israel (Wohl et al., 1994) and the Narmada River, India (Kale et al., 1996)).

The geomorphological features present in the bedrock anastomosing areas are described and a model for their development is presented based on observation and hydraulic investigation. As a result of the highly variable flow regime, a limited number of distributaries are active during the dry winter months (April–October) with many more becoming active during the wet season months (November–March). The frequency of activation strongly influences local habitat availability and controls the development of the geomorphological units.

The incision of the Sabie River through the weathered overburden on the African and Post-African surfaces, of up to 12 m, has also resulted in a large degree of bedrock control on the bed and banks of the active and macro-channels. The Sabie River has an irregular longitudinal profile with several knickpoints (Figure 2.2) reflecting underlying lithological and structural variability which is particularly pronounced at the local scale

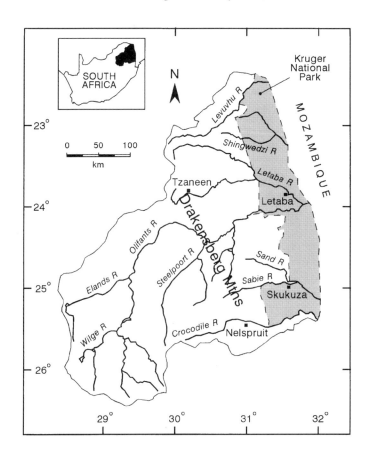

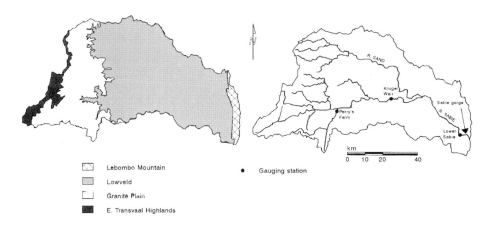

Lebombo Mountain

Lowveld

Granite Plain

E. Transvaal Highlands

● Gauging station

FIGURE 2.1 The Sabie River catchment, Mpumalanga Province, South Africa

TABLE 2.1 Summary of major geomorphological events in the region of the present Lowveld (modified from Partridge and Maud (1987) and Venter (1991))

Event	Geomorphic manifestation	Age
Climatic oscillations and glacio-eustatic sea-level changes.	Slight to moderate incision of river channels resulting in localized erosion.	Recent–0.011
Post-African II cycle of major valley incision.	Incision of coastal gorges, formation of Post-African II erosion surface.	*c.* 2.5–0.011 Ma
Major uplift (up to 900 m in eastern marginal areas).	Asymmetrical uplift of the subcontinent and major westward tilting of surfaces on interior with monoclinal warping along the eastern coastal margins.	*c.* 2.5 Ma
Post-African I cycle of erosion.	Development of imperfectly planed Post-African I erosion surface.	*c.* 18–2.5 Ma
Moderate uplift of 150–300 m	Slight westward tilting of African surface with limited monoclinal warping.	*c.* 18 Ma
African cycle of erosion (polycyclic).	Advanced planation throughout the subcontinent. Development of African surface with deep-weathered laterite and silcrete profiles.	*c.* 100–18 Ma
Break up of Gondwana through rift faulting.	Initiation of Great Escarpment owing to high absolute elevation of southern African portion of Gondwana.	*c.* 100 Ma

(Figure 2.2, inset). Outcrops of resistant rocks in the bed and banks have led to local steepening of the river and an upstream decrease in water surface gradient resulting in localized sediment accumulation.

Generally, flow in sections within the macro-channel is confined to active channels within the macro-channel (Figure 2.3). A flood of approximately 2500–3000 m^3 s^{-1} will be required to overtop the macro-channel and a flow of this magnitude has not been recorded in the last 50 years of gauging records on the Sabie River, although a 2100 m^3 s^{-1} flood passed through Lower Sabie in February 1996. The active channels carry water for a large portion of the year, although a few may run dry during the winter months.

The Sabie River displays characteristics of both bedrock and alluvial channels. Whereas alluvial channels are relatively well understood, there have been far fewer

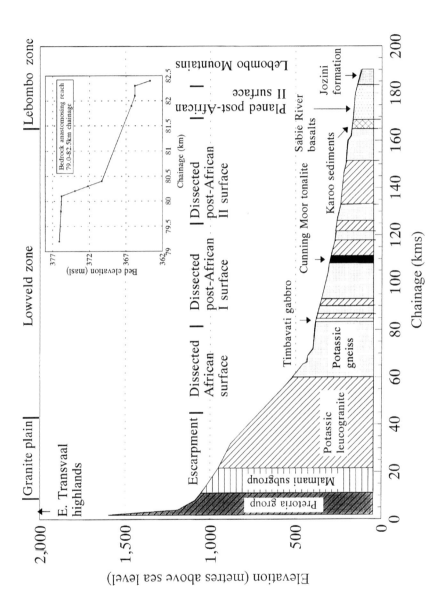

FIGURE 2.2 Longitudinal profile of the Sabie River in the Lowveld, showing underlying bedrock lithologies (strata do not dip vertically) and geomorphic zones. Inset illustrates local slope through an anastomosing reach between chainage 79 km and 82.5 km

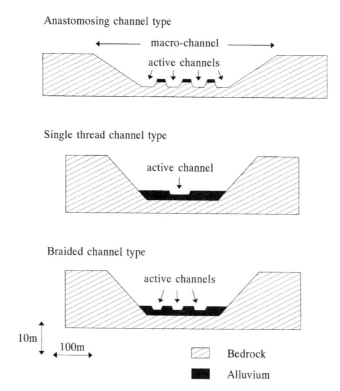

FIGURE 2.3   Diagrammatic representation of macro- and active channels in different channel types (anastomosing channels may be mixed or bedrock) along the Sabie River in the Lowveld. The terrestrial surface represents the African I and II planation levels. Scales are approximate

studies conducted on bedrock channels and it is not known how variable most bedrock channel forms can be and on what scale variations occur (Baker, 1984; Selby, 1985; Fielding, 1993; Ashley et al., 1988). For these reasons, great care must be exercised in analysing bedrock systems when using the terminology of previously published experience on alluvial rivers.

Different channel types have been identified on the Sabie River in the Lowveld, namely single thread, braided, pool–rapid and anastomosing (van Niekerk et al., 1995; see also Chapter 3). The single-thread and braided channels are alluvial, pool–rapid channels are bedrock or mixed alluvial bedrock, and the anastomosing channel types range from bedrock, through mixed to alluvial. This chapter focuses on the morphology and process characterizing the bedrock anastomosing channel types.

GEOMORPHOLOGY OF BEDROCK ANASTOMOSING CHANNELS

Along the Sabie River, the bedrock anastomosing channel type is found predominantly in Timbavati Gabbro and Potassic Leucogranite and to a lesser extent in Potassic Gneiss bedrock (Cheshire in press; see also Chapter 3). In these lithologies, outcrops of more resistant rock within the macro-channel generate tortuous laterally extensive systems of

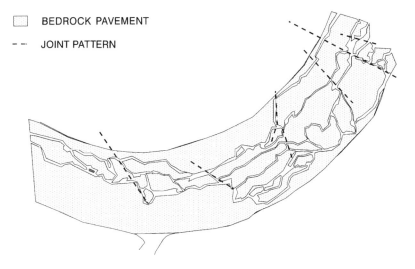

☐   BEDROCK PAVEMENT

– –   JOINT PATTERN

FIGURE 2.4   Relationship between bedrock distributary planform and the fracture pattern and joint structure in a bedrock anastomosing section downstream of Lower Sabie

distributary channels which have cut into bedrock. Their planform flowpath has been shown to be related to joint and fracture patterns in the bedrock (Figure 2.4) (Cheshire, in press), resulting in several laterally spread low-flow distributaries across a wide macro-channel, as opposed to flow being concentrated in one larger channel.

The bedrock distributary network is characteristically tortuous (Figure 2.5) with a complex interconnectivity between active channels (Figures 2.5 and 2.6). The individual channels are isolated from each other during low winter flows and summer base flows ($1 \, m^3 s^{-1}$ discharge on Figure 2.6) and occur at different elevations (Figure 2.7), whilst active distributaries join and flow as fewer channels at flood discharges ($1000 \, m^3 \, s^{-1}$ on Figure 2.6).

Individual distributary channels have distinct gradients and hydrodynamic characteristics between branch points, depending on the character of the bedrock joint and fracture patterns utilized. A variety of geomorphological units have been observed in the active channels including waterfalls, cataracts, pools and rapids (Figure 2.5, Table 2.2). Within the active channels, weathering and erosion along joints and fractures have led to the development of isolated bedrock. Sand, gravel and cobbles accumulate locally as alluvial bars, riffles and armoured areas in the distributary channels in low-energy areas upstream of bedrock controls and as lee bars downstream of bedrock outcrops (Figure 2.5, Table 2.2). More extensive bedrock outcrops between the distributary channels often form bedrock pavements (Figure 2.5, Table 2.2). Distributary channels and topographic lows in the interfluves, which become inundated when the flow level rises, form bedrock backwaters when the flow recedes. Sediment accumulates in areas within and adjacent to these bedrock backwaters leading to the development of alluvial backwaters.

Observation following elevated flows and consideration of channel hydraulics suggests that deposition on exposed bedrock occurs during the falling limb of flood hydrographs,

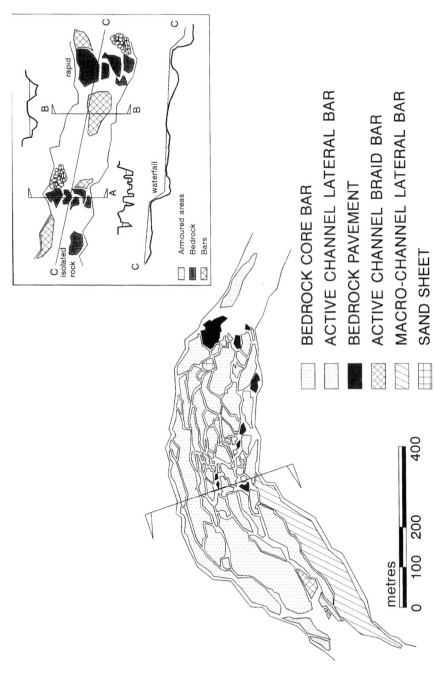

FIGURE 2.5   A bedrock anastomosing channel type on the Sabie River, 2 km upstream of Kruger Weir, showing predominant geomorphological units. Flow is in the direction of the arrows. The inset is a diagrammatic representation of the smaller geomorphologic units found in the distributary channels (distance C–C is approximately 150m)

BEDROCK CORE BAR

ACTIVE CHANNEL LATERAL BAR

BEDROCK PAVEMENT

ACTIVE CHANNEL BRAID BAR

MACRO-CHANNEL LATERAL BAR

SAND SHEET

metres

0   100   200   400

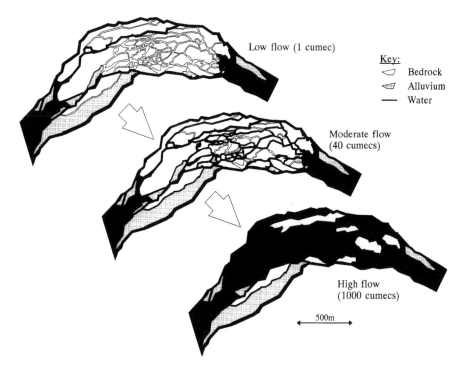

FIGURE 2.6 Interconnectivity between active channels within the bedrock anastomosing channel type for a range of discharges $(1–1000\,\mathrm{m}^3\,\mathrm{s}^{-1})$ $(1\ \mathrm{cumec} = 1\,\mathrm{m}^3\,\mathrm{s}^{-1})$

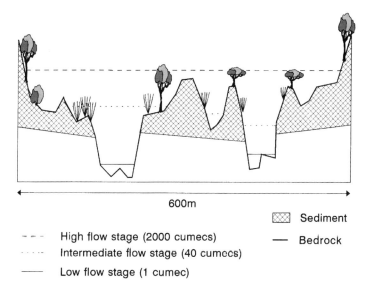

FIGURE 2.7 Schematic of a typical bedrock anastomosing channel type showing the inundation levels of a low, intermediate and high flow

TABLE 2.2   Geomorphological units associated with bedrock anastomosing channel types on the Sabie River

Morphological unit	Description
Distributary channels	
Waterfall	Abrupt vertical discontinuity in channel slope.
Cataract	Step-like successions of small waterfalls, seldom drowned out except at high discharges.
Rapid	Steep bedrock sections with concentrated flow, little or no alluvium in the channel.
Bedrock outcrop	Bedrock outcrop, single or grouped, resulting in local modification to the flow.
Pothole	Vertical cylinder drilled into bedrock by clast abrasion.
Bedrock pool	Deeper area upstream of a bedrock control with bedrock floor.
Boulder bed	Accumulation of locally derived material exceeding 0.25 m nominal diameter.
Armoured area	Accumulation of coarser sediments due to winnowing of finer material.
Bedrock backwater	Stationary bodies of water in bedrock, isolated from the active channel.
Macro-channel/interfluves	
Bedrock pavement	Extensive horizontal or subhorizontal area of exposed rock, with no alluvial cover.
Bedrock topo-graphic high	Bedrock outcrop forming topographic high point in the macro-channel between distributary channels.
Pothole	Vertical cylinder drilled into bedrock by clast abrasion.
Bedrock core bar	Accumulation of fine alluvium over sand on bedrock outcrops between distributary channels.
Macro-channel lateral bar	Large-scale, rarely inundated sedimentary features attached to the macro-channel bank.
Macro-channel island	Large-scale, rarely inundated sedimentary features detached from the macro-channel bank.

Morphological units also found include alluvial bars, riffles, alluvial pools, alluvial backwaters and sand sheets, which are commonly reported in geomorphological literature.

leading to the development of bedrock core bars (Figure 2.8). These features are influenced during seasonal high flows and lower frequency floods, which deposit fine material as the flow wanes causing the energy available for transporting suspended sediment to decrease over the topographic high points. Owing to the steep channel gradient through bedrock distributaries, these morphological units have a high potential for sediment transport at all flows and so are generally sediment-free. Rapid establishment of vegetation leads to a positive feedback whereby the bedrock core bars are stabilized and deposition is enhanced through the reduction of flow energy and trapping by vegetation. Bedrock anastomosing channel types have been observed to form one extreme of a continuum of channel types ranging from bedrock anastomosing through mixed anastomosing with varying degrees of alluviation (see Chapter 3). Additional sedimentation in the long term could potentially lead to the development of fully alluvial anastomosing channel types. Features such as macro-channel lateral bars and islands form as the anastomosing channel types become more alluvial. The probability of occurrence of the geomorphological units described in Table 2.2 varies

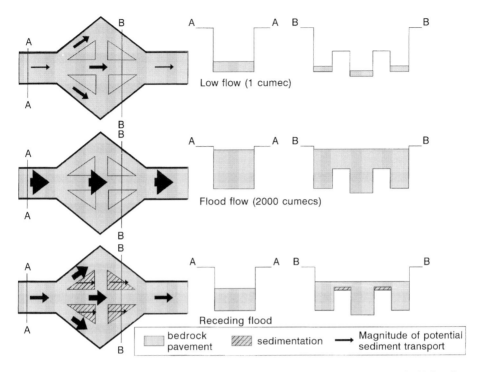

FIGURE 2.8   Inferred pattern of sediment transport and deposition on the topographic high points (bedrock core bars) as a flood flow recedes

depending on where the channel type occurs within the bedrock–alluvial continuum, which is a function of the total volume of accumulated sediment. This continuum is discussed in more detail in Chapter 3.

HYDROLOGY AND HYDRODYNAMICS

The flow regime in the Sabie River is highly variable both seasonally and annually. Wet season flows in wet years are characterized by a series of floods followed by a gradual reduction in the base flow through the dry season (Figure 2.9). During dry years, discharge is low and more uniform throughout the year (Figure 2.9). As a result of the highly variable flow regime, a limited number of primary distributaries are active during the dry winter months and in dry years. Many more become active during the wet season months of November through to March, with occasional inundation of other higher morphological units during flooding. Whether a distributary channel becomes active or not depends on conditions at the upstream branch node. A bedrock barrier, acting as an upstream control, will allow the channel to become active only once the upstream water surface elevation is high enough to overtop the barrier.

The number of active distributaries is therefore a function of the discharge (Figure 2.10) with the number of distinct individual channels initially increasing with discharge until all potential channels are active. As flow continues to increase,

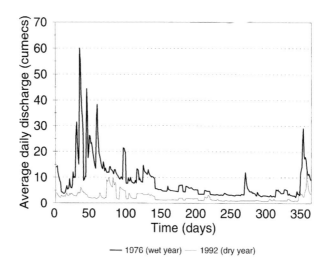

FIGURE 2.9   Flow regime for the Sabie River in the Mpumalanga Province Lowveld

channels amalgamate and the number of distributaries declines until the anastomosing system becomes one single flood channel at very high discharges (Figure 2.10).
The morphological evolution of the anastomosing channel types is therefore
controlled by the whole flow regime and application of conventional single-thread
alluvial system concepts such as bankfull flow are rendered meaningless. Individual
morphological units have widely differing inundation histories (Figure 2.11), with
active channel features commonly inundated by low discharges and macro-channel
features infrequently inundated by high discharges. Bedrock core bar elevations

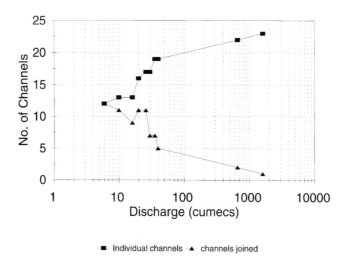

FIGURE 2.10   Number of active channels versus discharge for a bedrock anastomosing channel
type based on the mean at three cross-sections

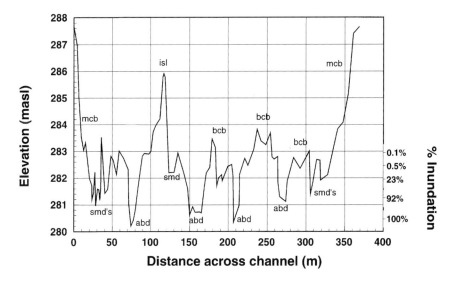

FIGURE 2.11   Inundation frequencies and elevation for different geomorphological units on a typical cross-section within a bedrock anastomosing channel (abd = active bedrock distributary, bcb = bedrock core bar; isl = island; mcb = macro-channel bank; smd = seasonal mixed distributary)

appear similar and this may be a result of their being influenced by high discharges when the individual distributary channels have combined to form a single flood flow level.

Since, at any particular cross-section, flow in the different distributaries may be at different levels (Figure 2.7), stage–discharge relationships cannot be developed for the whole cross-section and relationships need to be established for individual distributaries. A preliminary "channel overspill" model (Figure 2.12) has revealed potentially stepped stage–discharge relationships for individual channels (Broadhurst et al., 1995). Once a particular channel is filled and spills over into adjacent ones, there is no increase in stage with discharge until all adjacent distributaries reach that same level.

The Barnes (1967) methodology, formulated to calculate a reach average Manning's *n*, was applied to the different channel types, using data on stage–discharge relationships and channel geometry, along multiple cross-sectional reaches (Broadhurst et al., 1995). Use of this model revealed distinctive Manning's flow resistance values for the different channel types (Figure 2.13). The severely non-uniform long profile, large-scale roughness features and the effects of the changing numbers of active channels for different discharges in the bedrock anastomosing channel types, result in extremely variable but consistently high Manning's flow resistance. The high values at low discharges are attributed to the effects of large roughness elements and irregular channel geometry (waterfalls, rapids, cataracts, bedrock outcrops and boulders), many of which protrude above the water surface at low flows and create a stepped water surface profile. The high roughness at elevated discharges is probably the result of the influence of the dense woody vegetation outside of the active channel, which becomes inundated as stage rises. Previous field studies in steep irregular channels with a high incidence of bedrock

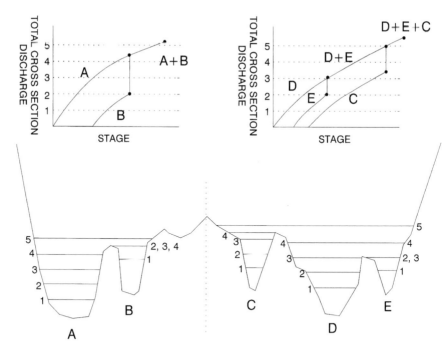

FIGURE 2.12   Conceptual diagram of the "channel overspill" model, showing stepped rating curve relationships for each individual distributary

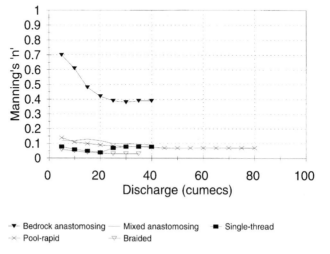

FIGURE 2.13   Manning's flow resistance coefficients for different channel types over a range of discharges

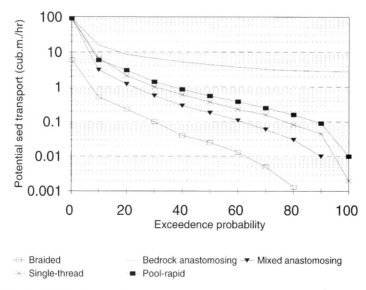

FIGURE 2.14   Probability of exceedence for sediment transport rates for 0.5 mm grain size based on a 34 year record of average daily discharges for different channel types

and boulders (Bevan et al., 1979; Bathurst, 1985; Thorne and Zevenbergen, 1985), and flume studies on densely vegetated channels (Hall and Freeman, 1994) have revealed Manning's flow resistance values within the range reported here (Broadhurst et al., 1995).

The Ackers and White (1973) total sediment load equation was used to predict potential sediment transport capacity for 0.5 mm sand (the predominant median grain size of the in-channel sediment), based on measured stage–discharge relationships and calculated flow resistance values (van Niekerk et al., 1995). Curves representing the probability of exceedence of sediment transport rate have been drawn up for representative cross-sections in the different channel types based on average daily flow records for the last 34 years (Figure 2.14). Although these calculations are based on the assumption of a horizontal water surface across all of the active distributaries, differences in potential transport rates are clearly demonstrated. The bedrock anastomosing sections are characteristically wider and steeper than the other channel types and typically have a greater capacity for transporting sediment sizes supplied from the catchment for all flows.

BEDROCK ANASTOMOSING CHANNEL PROCESS AND MORPHOLOGY

The underlying bedrock geology has a major influence on the geomorphology of the Sabie River over a wide range of scales (morphological unit, channel type and macro-channel) (Cheshire, in press). The bedrock influence on the geomorphology is greater towards the western portion of the Lowveld where dissection is more pronounced and gradients are steeper (Cheshire, in press). Downstream reduction in channel gradient, accompanied by an increased sediment input from the catchment via ephemeral tributaries, alters the character of the anastomosing channels with many displaying more alluvial features in the lower reaches of the Sabie River to the east. Field observation of

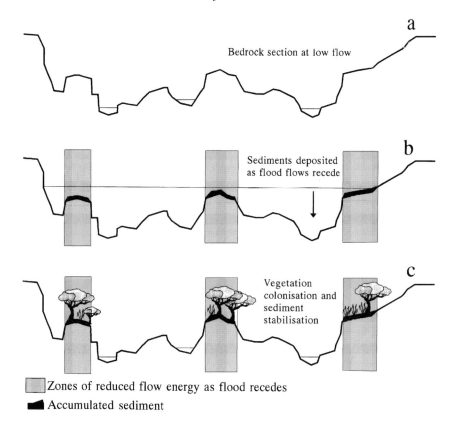

Zones of reduced flow energy as flood recedes
Accumulated sediment

FIGURE 2.15 Summary of the zones of preferential sediment deposition across a bedrock anastomosing cross-section

the range of occurrences of anastomosing channels on the Sabie River has led to the development of a model of response to increased sedimentation for bedrock anastomosing channels.

Incision into the Post-African I and II surfaces and headward erosion during river evolution resulted in a strongly bedrock-controlled river. At locally more resistant sections of the channel, vertical erosion occurs by exploiting joint and fracture patterns in the bedrock, resulting in the development of bedrock anastomosing channel sections with sediment accumulating upstream of these major bedrock controls. Since the flow regime is highly variable, distributary channels and bedrock topographic highs are periodically covered and sand and silt are deposited during the falling limb of flood hydrographs (Figure 2.15). Incipient bedrock core bars are rapidly colonized by vegetation which results in a positive feedback through bar stabilization, enhanced trapping and increased flow resistance (Figure 2.15). As sediments accumulate and the bars develop, the elevation of the feature increases and, given a stable flow regime, inundation becomes less frequent.

Bedrock high points in rapids in the distributary channels accumulate sediment leading to the development of bedrock core bars. Deposition of sediment in lower energy portions of distributary channels (upstream of rapids) and subsequent sorting results in the development of riffles and armoured areas. Locally derived boulders accumulate preferentially and boulder bed units result. Sediment transported through the anastomosing sections is deposited in the lower energy areas downstream of these major controls.

IMPLICATIONS OF CHANGE IN BEDROCK ANASTOMOSING AREAS

It is well established that different riverine fauna and flora utilize different morphological features at different stages in their life cycles (see Malanson, 1993). On the Sabie River, for example, a good relationship has been established between channel type and the associated morphological units and the presence of different plant species (van Coller et al., 1995). As a result of the wide variety of morphological units present in different distributary channels in a bedrock anastomosing channel, frequency of inundation of individual channels strongly influences local habitat availability for the riparian biota.

Since flow is confined to bedrock, there is limited subsurface interconnectivity and vegetation which is at a distance from the active low-flow channels is dependent on regular inundation. Long periods of desiccation have a far greater effect on riparian vegetation in the anastomosing channel types than in the others. A study of tree deaths along the Sabie River (van Coller, pers. comm.) as a result of the extended drought of the late 1980s and early 1990s revealed that most deaths occurred in the anastomosing channels, with a single species (*Breonadia salicina*) the most severely affected.

In the Sabie River in the Lowveld, bedrock anastomosing channels are most resistant to geomorphological change as a result of their high competence. The distribution of water through a number of distributary channels and the wide diversity of geomorphological units provide a large variety of riparian habitats and thus a high potential for supporting a wide range of riverine biota. However, the long-term trend in the Sabie River is for the bedrock anastomosing channel to become more alluvial as a result of increased water abstraction and by potentially increased sediment yield resulting from changing land-use within the catchment (see van Niekerk and Heritage (1994) and examples in Rodda (1976), Meade (1976), Graf (1979) and Blodgett and Stanley (1980)). It is suggested that the number of active channels will decrease although frequently inundated ones will probably be maintained. In particular, active bedrock-dominated distributaries will persist as a result of relatively (to other channel types) high potential sediment transport rates over the whole flow regime (Figure 2.14) despite increasing alluviation in the other channel types present on the Sabie River.

CONCLUSIONS

The geomorphological history of Southern Africa has resulted in river incision in the Lowveld resulting in a high incidence of bedrock control. Bedrock anastomosing channel types are distinguished from others within the bedrock–alluvial continuum on the Sabie River by their steep gradients, high flow resistance, high potential sediment transport rates and irregular cross-sections displaying a wide variety of geomorphological units. They are controlled by the underlying bedrock structure and lithology with individual, independent channels following joints and fractures in the bedrock. Activation of different channels is dependent on the flow regime and upstream topographic controls. Since the active channels have high sediment transport capacities relative to other channel types, they are generally sediment-free. However, sedimentation on the interfluves occurs as a function of the annual and long-term flooding regime. Establishment of vegetation plays a major role in stabilizing sediment deposits and increasing sediment trapping, resulting in a positive feedback which is driving these channel types towards a more alluvial state. The high diversity of potential biotic habitats is likely to play an important part in the ecological functioning of the Sabie River in the Lowveld. Although these channel types are probably more resistant to geomorphological change than others on the river, abstraction and modification of the flow regime, in the short term, will impact severely on the frequency of inundation of individual channels and hence diversity of habitat. Progressive alluviation is manifested by preferential deposition on bedrock core bars and backwaters, while the distributaries remain relatively sediment-free at present. Change to a mixed bedrock–alluvial system is a long-term inevitability if land-use change remains unchecked.

ACKNOWLEDGEMENTS

The funding of this project by the Water Research Commission is gratefully acknowledged. The assistance of the National Parks Board research personnel in the Kruger National Park has been invaluable in ensuring the smooth running of the project. We thank the Mazda Wildlife Fund for financial and logistical support.

During recent fieldwork, a motor vehicle accident resulted in fatal injuries to Andre van Niekerk who was the primary motivation behind this paper. Andre contributed significantly to state-of-the-art rivers research in South Africa, and his absence will be sorely felt by the research community and his colleagues. We extend our sincere sympathies to his family.

REFERENCES

Ackers, P. and White, R. 1973. Sediment transport: New approach and analysis. *Journal of the Hydraulics Division, American Society of Civil Engineers*, **99** (HY11), 2041–2060.

Ashley, G.M., Renwick, W.H. and Haag, G.H. 1988. Channel form and process in bedrock and alluvial reaches of the Raritan River, New Jersey. *Geology*, **16**, 436–439.

Baker, V.R. 1984. Flood sedimentation in bedrock fluvial systems. In E.H. Koster and R.J. Steel (Eds) *Sedimentology of Gravels and Conglomerates*. Canadian Society of Petroleum Geologists, Memoir 10, 87–98.

Barnes, H.H. 1967. Roughness characteristics of natural channels. *U.S. Geological Survey Water-Supply Paper* 1849, Washington, DC, 1–9.

Bathurst, J.C. 1985. Flow resistance estimation in mountain rivers. *Journal of Hydraulic Engineering, American Society of Civil Engineers*, **111**(4), 625–643.

Beven, K., Gilman, K. and Newson, M. 1979. Flow and flow routing in upland channel networks. *International Association of Hydrological Sciences Bulletin*, **24**(3), 303–325.

Blodgett, R.H. and Stanley, K.O. 1980. Stratification, bedforms, and discharge relations of the Platte Braided River System, Nebraska, *Journal of Sedimentary Petrology*, **50**(1), 139–148.

Broadhurst, L.J., Heritage, G.L., van Niekerk, A.W. and James, C.S. 1995. Translating discharge into local hydraulic conditions on the Sabie River: An assessment of channel roughness. *Proceedings, Seventh South African National Hydrology Symposium*, Grahamstown.

Cheshire, P.E. (in press). *Geology and geomorphology of the Sabie River in the Kruger National Park Area and its catchment area*. Water Research Commission Report No. 474/1/96, Pretoria, South Africa.

Fielding, C.R. 1993. A review of research in fluvial sedimentology. *Sedimentary Geology, Special Volume, Current Research in Fluvial Sedimentology*, **85**(1), 3–14.

Graf, W.L. 1979. Mining and channel response. *Annals of the Association of American Geographers*, **69**, 262–275.

Hall, B.R. and Freeman, G.E. 1994. Study of hydraulic roughness in wetland vegetation takes new look at Manning's n. *Bulletin of Wetlands Research Progress*, **4**(1), 1–4.

Howard, A. and Dolan, R. 1981. Geomorphology of the Colorado River in the Grand Canyon. *Journal of Geology*, **89**, 269–298.

Kale, V.S., Baker, V R. and Mishra, S. 1996. Multi-channel patterns of bedrock rivers: An example from the central Narmada basin, India. *Catena*, **26**, 85–98.

King, L.C. 1978. The geomorphology of central and southern Africa. In M.J.A. Werger and van Bruggen (Eds), *Biogeography and Ecology of Southern Africa*. Dr W. Junk, The Hague.

Malanson, G.P. 1993. *Riparian Landscapes*. Cambridge University Press, Cambridge, 296 pp.

Meade, R.H. 1976. Changes in sediment loads in rivers of the Atlantic drainage of the United States since 1900. In *Effects of Man on the interface of the hydrological cycle with the physical environment*. IASH Publication No. 113, 99–104.

Partridge T.C. and Maud, R.R. 1987. Geomorphic evolution of southern Africa since the Mesozoic. *South African Journal of Geology*, **90**(2), 179–208.

Rodda, J.C. 1976. Basin Studies, In J.C. Rodda (Ed.), *Facets of Hydrology*. Wiley, Chichester, 257–297.

Selby, M.J. 1985. *Earth's Changing Surface*. Clarendon Press, Oxford.

Thorne, C.R. and Zevenbergen, L.W. 1985. Estimating mean velocity in mountain rivers. *Journal of Hydraulic Engineering, American Society of Civil Engineers*, **111**(4), 612–624.

van Coller, A.L., Heritage, G.L. and Rogers, K.H. 1995. Linking the riparian vegetation distribution and flow regime of the Sabie River through fluvial geomorphology. In *Proceedings, Seventh South African National Hydrology Symposium*, Grahamstown.

van Niekerk, A.W. and Heritage, G.L. 1994. The use of GIS techniques to evaluate channel sedimentation patterns for a bedrock controlled channel in a semi-arid region, In C. Kirby and W. R. White (Eds), *Proceedings of International Conference on Basin Development*. Wiley, Chichester, 257–271.

van Niekerk, A.W., Heritage, G.L. and Broadhurst, L.J. 1995. The influence of changing flow regime on hydraulic parameters and sediment dynamics on the Sabie River. *Proceedings, Seventh South African National Hydrology Symposium*, Grahamstown.

Venter, F.J. 1991. Fisiese kenmerke van bereike van standhoudende riviere in die Nasionale Kruger Wildtuin. Paper presented at the First Research Meeting of the Kruger National Park Rivers Research Programme.

Wohl, E.E. 1992. Bedrock benches and boulder bars: Floods in the Burdekin Gorge of Australia. *Geological Society of America, Bulletin*, **104**, 770–778.

Wohl, E.E., Greenbaum, N., Schick, A.P. and Baker, V.R. 1994. Controls on bedrock channel incision along the Nahal Paran, Israel. *Earth Surface Processes and Landforms*, **19**, 1–13.

3

Geomorphology of the Sabie River, South Africa: an Incised Bedrock-influenced Channel

G.L. HERITAGE,[1] A.W. VAN NIEKERK[2] AND B.P. MOON[2]

[1] Telford Institute of Environmental Systems, Department of Geography, University of Salford, Manchester, UK
[2] Centre for Water in the Environment, University of Witwatersrand, Johannesburg, South Africa

ABSTRACT

The Sabie River drains a catchment of 7096 km^2 in the Mpumalanga Province, South Africa and Mozambique. It flows through the Kruger National Park for 110 km before entering Mozambique, where it has remained unaffected by direct human interference since the area was declared a National Park in 1898. The river has incised into the Post-African I and II surfaces as a result of climatic oscillations and glacio-eustatic sea-level changes. As a result there are significant areas of bedrock exposed in the river forming bedrock channels and grade controls. Geomorphological diversity is high and five principal channel types have been identified through aerial photographic analysis and field survey: bedrock anastomosing, mixed anastomosing, pool–rapid, braided and alluvial single-thread. Morphological mapping of 25 km of the river has produced an inventory of the occurrence and areal extent of the various geomorphological features. Characteristic assemblages for each channel type are detailed from these measurements. A sequence of channel types has been observed, from bedrock anastomosing and pool–rapid channels where bedrock features are common, through mixed anastomosing, braided and alluvial single-thread where bedrock features become progressively less important as alluvium builds up in the macro-channel. Aerial photographs have been utilized to study the geomorphological units present across the range of channel types. Models are proposed, based on these observations, suggesting the likely pathways of change involving progressive sedimentation from bedrock to alluvial channel types.

INTRODUCTION

Research on rivers has concentrated largely on temperate perennial alluvial systems and it is these which dominate the literature. Bedrock-influenced systems have, in contrast, received comparatively little attention (Ashley et al., 1988). Increasingly, different river types are being investigated highlighting the range of channel forms that exist. The

Varieties of Fluvial Form. Edited by A.J. Miller and A. Gupta
© 1999 John Wiley & Sons Ltd

classification of bedrock-influenced channels has been usefully addressed by Ashley et al. (1988). Bedrock influence on river channels has been noted by Sternberg and Russell (1952) and Tinkler (1993). Bedrock channel geomorphology has been described for the Colorado River by Howard and Dolan (1981), in Katherine Gorge by Baker and Pickup (1987), along Burdekin Gorge and Piccaninny Creek by Wohl (1992, 1993), and in the Sabie River by van Niekerk et al. (1995) and Heritage et al. (1998). Bedrock channel dynamics and channel change have also been studied experimentally by Shepherd and Schumm (1974) and in the field by other authors including Baker (1978), Graf (1979), Wohl et al. (1994) and Zen and Prestegaard (1994).

In order to increase the understanding of a more diverse range of rivers, other channel types need to be classified and their dynamics investigated. This paper documents the preliminary results of a five year study of the geomorphology of the Sabie River in the Kruger National Park, South Africa, where direct human impact on the channel is rare. Catchment influences, which extend beyond the Kruger National Park, have combined to create a geomorphologically diverse river system displaying both alluvial and bedrock features. Channel types on the Sabie River, South Africa, are identified and described based on quantification of their characteristic morphological components. The results indicate that the river can be categorized broadly into five distinct channel types differing according to geological influence and the degree of accumulated sediment within the incised channel.

This study utilizes the continuum concept as proposed by Brakenridge (1985) and Ashley et al. (1988) who recognized that bedrock and alluvial systems exist as opposite ends of a larger range of channel types controlled by the balance between sediment supply and channel transport capacity. The scientific basis for predicting channel change is at present poor, particularly concerning rates and magnitude of change (Kellerhals and Miles, 1996). Pathways of possible channel change as a result of progressive sedimentation are suggested based on the analysis of a single set of aerial photographs. It is assumed that a sequence of morphological change, as a result of the progressive build-up of sediment on a bedrock template, can be inferred by quantifying the geomorphological assemblage of the same channel type exhibiting differing degrees of accumulated sediment. In this way a sequence of change may be inferred that suggests a continuum of channel types from bedrock through to fully alluvial (see Figures 3.11–3.13).

CHARACTERISTICS OF THE SABIE RIVER

Topography

The Sabie River drains a catchment of 7096 km^2 in the Mpumalanga Province, South Africa and coastal Mozambique (Figure 3.1a). It rises in the Drakensberg Mountains to the west (1600 metres above sea level (m a.s.l.), descending rapidly onto the flat Lowveld (400 m a.s.l.) and Lebombo zones (200 m a.s.l.) to the east.

Geological setting and geomorphology

The geology of the catchment is diverse and this has played an important role in determining the geomorphological complexity of the Sabie River (Cheshire, in press).

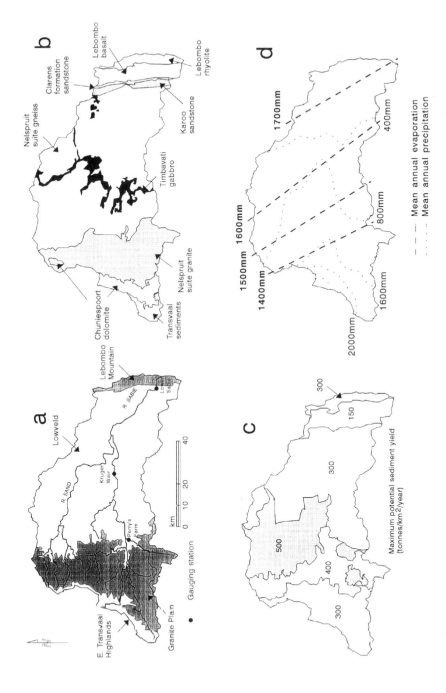

FIGURE 3.1 Physical characteristics of the Sabie River catchment, Mpumalanga Province, South Africa

56

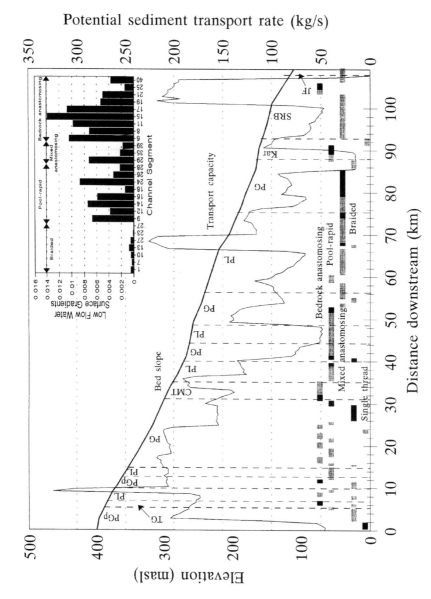

FIGURE 3.2 Geology, channel slope, sediment transport capacity and associated channel type for the Sabie River in the Kruger National Park (PGp = Potassic Gneiss with plagioclase; TG = Timbavati Gabbro; PL = Potassic Leucogranite; CMT = Cunning Moot Tonalite; PG = Potassic Gneiss; Kar = Karoo sediments; SRB = Sabie River Basalts; JF = Jozini Formation Rhyolites and Dacites). Dark bands indicate channel types that have been mapped at the morphological unit level. Inset indicates the range of channel slopes along the river grouped by channel type

Uplift in the recent geological past (mid-Miocene and late Pliocene) and subsequent incision into bedrock during the last 10 000 years, as a result of climatic oscillations and glacio-eustatic sea-level changes (Partridge and Maude, 1987; Venter, 1991), has generated a river that has steep valley sides and a "floodplain" that is restricted by the width of the incised channel (beyond which flood flows have no recorded influence). This incised feature has been termed the macro-channel (van Niekerk et al., 1995) to differentiate it from the much smaller perennial channel or channels that flow within its confines. On a regional scale (Figure 3.1b and 3.2) the Sabie River cuts through sequences of potassic leucogranites and gneisses west to east followed by sedimentary Karoo sandstone and igneous basalts and rhyolites.

At this regional scale there is no clear correlation between changing geology and channel type. However, investigation of the geology of the macro-channel has revealed relationships between lithological change, jointing and the occurrence of bedrock anastomosing and pool–rapid areas (Figure 3.3). The outcropping of potassic gneiss in reach 1 correlates with the incidence of bedrock pool–rapid and anastomosing areas. Sedimentation in the lower energy area upstream of the bedrock anastomosing channel types has led to mid-channel and lateral bar formation creating limited, confined braiding. In reach 2 the bedrock channel types are associated with potassic leucogranites and in reach 3 there is a switching from a mixed anastomosing channel to a bedrock anastomosing type as a result of a change in the geology from the Sabie River basalts to heterogeneous Jozini Formation rhyolites intersected by numerous dyke swarms. Anastomosing thus appears to occur over lithologically more resistant bedrock where the river is forced to exploit local weaknesses such as joints or dykes. Similar bedrock influences were reported for Burdekin Gorge, Australia, where lithological variation was found to be responsible for small-scale erosional features (Wohl, 1992). Kale et al. (1996) showed that bedrock anastomosing systems were strongly influenced by geology on the Narmada River, India; and Baker and Pickup (1987) noted morphological variation attributable to lithological control in the Katherine Gorge, Australia.

Global Positioning System (GPS) survey of the long section of the river (Figure 3.2) has demonstrated that the bedrock channel types are characterized by steeper gradients and act as control points on the river. These control points create backwater zones upstream which are characterized by greater amounts of sediment (braided–bedrock anastomosing sequence in reach 1 of Figure 3.3 and the single-thread bedrock anastomosing sequence in reach 2 of Figure 3.3). The sequence bedrock anastomosing, pool–rapid, braided, bedrock anastomosing was found to occur repeatedly in the upstream section of river in the Kruger National Park. The bedrock-influenced channel types also show a higher predicted potential sediment transport rate derived from the Ackers and White (1973) sediment transport formula (Figure 3.2). This can be attributed to the combined effect of steep slopes and increased channel width.

Catchment hydrology

Rainfall is concentrated in the highland areas towards the west (2000 $mm\,a^{-1}$) and declines rapidly to 450 $mm\,a^{-1}$ over a distance of 160 km, towards the border between South Africa and Mozambique. Evaporation trends are the opposite with low values to the west (1400 $mm\,a^{-1}$), increasing eastward to 1700 $mm\,a^{-1}$ (Figure 3.1d).

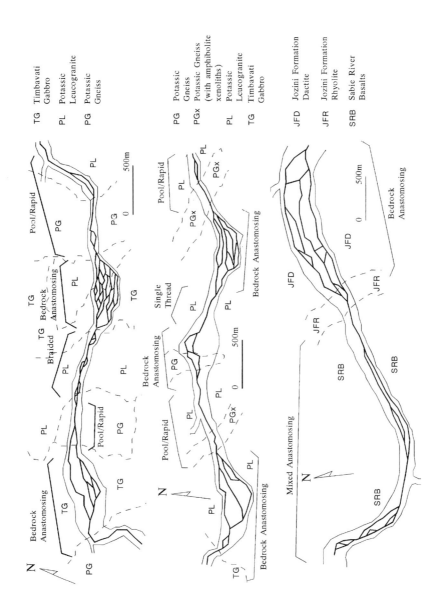

FIGURE 3.3 The influence of local geology on channel type for the Sabie River in the Kruger National Park. Reaches 1–3 (top to bottom) are at 8–12 km, 20–25 km and 104–108 km downstream, respectively; flow is from left to right

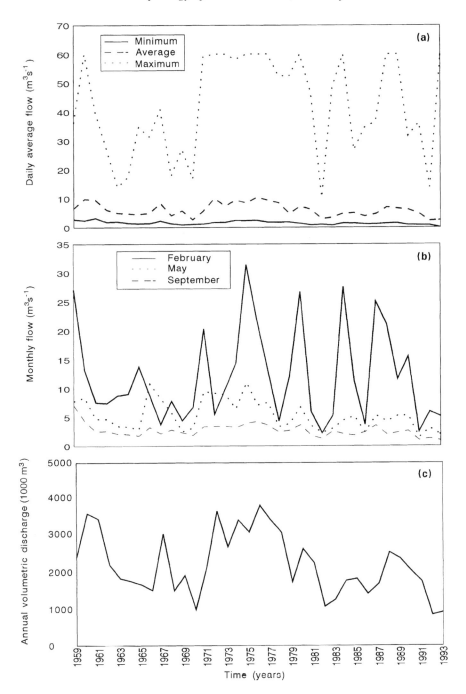

FIGURE 3.4   Summary of the hydrological characteristics of the Sabie River

Daily flow fluctuation is high; Figure 3.4a graphs the daily flow recorded at Perry's Farm gauging station (Figure 3.1a) as a daily average of the total annual flow. The mean clusters around $6\,m^3\,s^{-1}$, fluctuating between 3 and $10\,m^3\,s^{-1}$ for the period 1959–1993. Maximum annual daily flows average $41\,m^3\,s^{-1}$; however, Figure 3.4b shows that this discharge may fall to $12\,m^3\,s^{-1}$. Extreme floods in excess of $1000\,m^3\,s^{-1}$ have also been manually gauged but the weir record at Perry's Farm is only calibrated to $60\,m^3\,s^{-1}$. Minimum annual daily flows display a more constant pattern averaging $1.6\,m^3\,s^{-1}$ and fluctuating between 0.5 and $2.5\,m^3\,s^{-1}$.

Monthly averaged flows are also highly variable (Figure 3.4b). This is most marked during the wet season (February, illustrated on Figure 3.4b) where average flows may total $30\,m^3\,s^{-1}$ or fall as low as $4\,m^3\,s^{-1}$. May is an intermediate month that still displays some annual variability, and the dry season month of September shows a more constant average monthly flow of around $2\,m^3\,s^{-1}$.

Average annual flows for the gauging station also exhibit a high degree of variability (Figure 3.4c). Annual volumetric discharge varies between 1 and 4 Mm^3. The pattern correlates well with annual rainfall totals which have been shown to vary on a quasi-20-year wet–dry cycle, identified for the region based on records extending back to 1910 (Tyson, 1987).

Discharge patterns for the Sabie River primarily reflect the variability of the rainfall. The river flows perennially but variability is high, particularly during wetter climatic cycles. In addition, high magnitude events have been recorded during extended periods of low flow as a result of localized rainstorm events.

Catchment land use and sediment production

Sediment production (Figure 3.1c) in the catchments has been coarsely quantified (van Niekerk and Heritage, 1994). This study demonstrates that land degradation outside the National Park has been severe. This area corresponds to the former homelands and represents a concentration of impoverished rural communities whose impact on the landscape has been severe, with loss of vegetation cover for fuel and firewood, overgrazing of cattle and goats, and poor farming practices (van Niekerk and Heritage, 1994). Introduction of sediment into the river system is episodic, associated with summer rainfall events. This sediment enters the Kruger National Park and is distributed along the river, accumulating upstream of bedrock channel types and at locations exhibiting a low potential sediment transport capacity (Figure 3.2).

As a result of the catchment factors described above, the Sabie River displays a geomorphologically diverse character exhibiting rapid and abrupt changes in channel type as the distribution of sediment over bedrock alters (Figures 3.2 and 3.3). The geomorphological characteristics of the principal channel types are described below.

CHANNEL TYPE CHARACTERISTICS

Visual interpretation of aerial photographic records (1:10 000 scale), video footage taken during winter low flow conditions and extensive fieldwork have led to identification of five principal channel types present on the Sabie River in the Kruger National Park. Detailed mapping was conducted on 25 km (5 km bedrock anastomosing, 6.5 km pool–rapid, 8 km mixed anastomosing, 1 km single-thread and 4.5 km braided) of the river in the Kruger National Park (Figure 3.2), quantifying the area covered by each of the major morphological units associated with each channel type. Both active, perennially flowing, channel geomorphic units and those associated with the larger macro-channel were identified; some smaller features were not discernible from the aerial photographs and their presence is included based on field observation (Table 3.1).

TABLE 3.1   Morphological unit association by channel type for the Sabie River in the Kruger National Park. Figures represent in-channel and interfluve percentage areas, smaller features are described qualitatively based on field observation

Morphological unit	Braided	Bedrock anastomosing	Pool–rapid	Single-thread	Mixed anastomosing
In-channel features					
Braid bar	13.2	2.4	5.0	1.4	1.8
Lee bar	0.2	-	-	-	0.3
Lateral bar	23.6	1.1	12.2	30.5	2.1
Rapid	1.7	5.0	5.1	-	2.1
Isolated rock	0.3	1.4	1.2	0.1	2.1
Alluvial pool	49.6	14.8	31.9	68.0	11.3
Mixed pool	2.8	20.1	19.9	-	17.7
Bedrock pool	-	6.2	1.4	-	1.2
Other water	3.0	11.0	10.0	-	12.4
Rip channel	rare	absent	rare	rare	rare
Cataract	rare	absent	rare	absent	absent
Waterfall	rare	absent	rare	absent	absent
Armour area	absent	rare	rare	absent	rare
Boulder bed	absent	rare	rare	absent	rare
Dead zone	rare	rare	rare	rare	rare
Interfluve features					
Alluvial backwater	0.02	0.2	0.1	-	0.2
Bedrock backwater	-	0.6	0.1	-	0.2
Bedrock core bar	5.7	38.0	13.3	-	48.8
Terrestrial rock area	-	15.5	2.5	-	3.8
Bedrock pavement	-	1.4	0.2	-	0.2
Island	-	7.8	0.4	-	18.3
Macro-channel bank	69.5	24.9	44.7	57.9	22.8
Macro-channel lateral bar	30.3	49.7	52.0	42.1	54.6
Levee	rare	absent	rare	rare	rare
Terrace	rare	absent	absent	common	absent
Riffle	rare	absent	absent	rare	absent

Key: absent = not encountered; rare = seen at less than one in five cross-sections; common = seen at most cross-sections

Bedrock anastomosing

Bedrock anastomosing channels (Figure 3.5, Table 3.1) were first identified on the Sabie River by van Niekerk and Heritage (1993). Kale et al. (1996) have recognized similar multichannel bedrock distributary reaches on the Narmada River, India. Chemical differences in the host rock have generated resistant areas where the river is less able to erode vertically (Cheshire, in press). Typically, the incised macro-channel is widened to extend across three to four times the average width and this effect extends for several kilometres downstream, but is variable as the size of the feature is a function of the local geology. Geomorphological diversity is high, with many features occurring infrequently (Table 3.1). Numerous steep gradient, active channel bedrock distributaries exist within the incised channel, describing a tortuous route over the resistant rock (Figure 3.5). Such channels have a fixed planform as defined by the weaker pathways through the resistant outcrop (Cheshire, in press). The bedrock distributaries are steep, high energy systems (see Chapter 2); as such they display very few alluvial features, with sediment accumulation being restricted to lateral deposits and alluvium in pools in the form of armoured clastic lags and finer deposits in dead zones. Bedrock features include pools, rapids, cataracts and small waterfalls (Table 3.1).

The macro-channel is characterized by bedrock core bar deposits (see Chapter 2) and occasional larger islands that cover the areas between distributary channels. These features are strongly associated with deposition of sediment from suspension as the flow

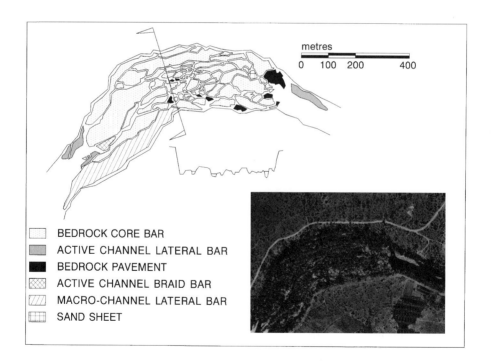

BEDROCK CORE BAR
ACTIVE CHANNEL LATERAL BAR
BEDROCK PAVEMENT
ACTIVE CHANNEL BRAID BAR
MACRO-CHANNEL LATERAL BAR
SAND SHEET

FIGURE 3.5   Detail of the morphological components of the bedrock anastomosing channel type for the Sabie River. Flow direction is indicated by flags on the line of cross-section

width expands to inundate areas outside of the active channel distributaries (see Chapter 2). Many of the distributary channels are dry for most of the year, becoming active only during elevated flows. Similarly, observation has shown that individual distributaries may be active at different elevations during the same flow as there is only a very limited lateral water table and flow is a function of upstream conditions (Broadhurst et al., 1998). Elevated bedrock areas are common and may exist as exposed bedrock pavements (Figure 3.5).

Mixed anastomosing

Mixed anastomosing channel types consist of multiple bedrock, mixed and alluvial distributary channels that divide and rejoin over a distance much greater than the distributary width (Figure 3.6). The distributary channels exhibit a high geomorphic diversity displaying many smaller alluvial bars and bedrock features that are difficult to observe from the aerial photographs. A small percentage of the active distributary channels are filled with alluvial material in the form of lateral bars attached to the banks, mid-channel bars in the centre of the channel, lee bars behind obstructions to the flow and mixed and alluvial pools (Table 3.1). The macro-channel also exhibited extensive lateral alluvial deposits (macro-channel lateral bars), islands and bedrock core bars (Figure 3.6, Table 3.1).

The multichannel planform appears to be relatively stable with the river reverting largely to its old course following floods greater than the capacity of the active channels. Extensive reed growth (*Phragmites mauritanus*) between the active distributaries

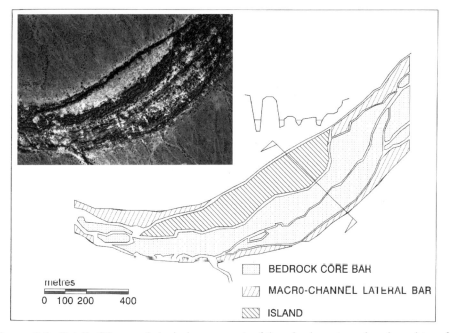

metres

0 100 200   400

BEDROCK CORE BAR

MACRO-CHANNEL LATERAL BAR

ISLAND

FIGURE 3.6   Detail of the morphological components of the mixed anastomosing channel type for the Sabie River. Flow direction is indicated by flags on the line of cross-section

increases channel resistance during these elevated flows and promotes bar growth by the vertical accretion of sediment.

Pool–rapid

The pool–rapid channel type is also geomorphologically diverse and displays many bedrock features (Figure 3.7, Table 3.1). Detailed field investigation of the geological controls revealed a number of reasons for this, including localized chemical differences similar to the bedrock anastomosing situation and differing lithologies (Cheshire, in press). These factors create active channel pool–rapid sequences, the scale of which is dependent on local geological variability and channel gradient. Typically, the rapids are free of sediment apart from occasional boulders and bedrock core bars (Table 3.1). The pool areas are more variable, ranging from sediment-free bedrock areas to bedrock-lined pools incorporating a variety of bar types, particularly mid-channel bars and lateral deposits.

The active pool–rapid geomorphological units typically occupy only a portion of the macro-channel. Large-scale consolidated sedimentary features associated with infrequent high magnitude flows have covered much of the bedrock across the rest of the incised channel (Figure 3.6), resulting in a marked increase in lateral deposits (macro-channel lateral bars) and a consequent reduction in terrestrial rock outcrop, bedrock pavement and bedrock backwaters (Table 3.1).

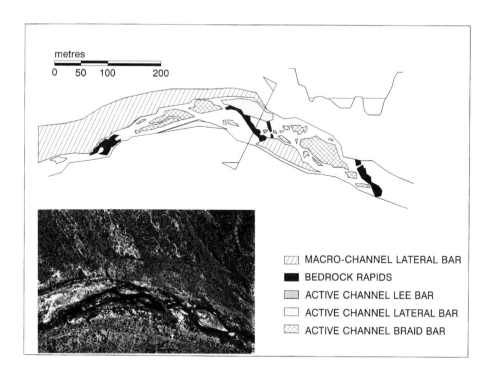

FIGURE 3.7   Detail of the morphological components of the pool–rapid channel type for the Sabie River. Flow direction is indicated by flags on the line of cross-section

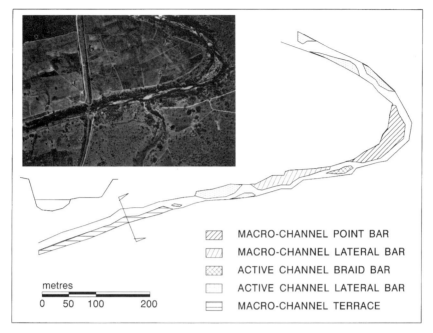

MACRO-CHANNEL POINT BAR
MACRO-CHANNEL LATERAL BAR
ACTIVE CHANNEL BRAID BAR
ACTIVE CHANNEL LATERAL BAR
MACRO-CHANNEL TERRACE

metres
0   50   100   200

FIGURE 3.8   Detail of the morphological components of the alluvial single-thread channel type for the Sabie River. Flow direction is indicated by flags on the line of cross-section

Single-thread

The alluvial single-thread channel type has developed in sections of the Sabie River where alluvium has accumulated to cover all bedrock influence in the macro-channel. The active channel may be straight or sinuous; however, the freedom to make planform adjustments is restricted to the width of the incised macro-channel (Figure 3.8). This channel type is distinct but rare on the Sabie River representing only 1 km of the 25 km of mapped river. Few geomorphological features were recorded in the active channel which was composed largely of deep alluvial pools with rare mid-channel and lateral bars. The macro-channel consisted wholly of lateral bar and bank morphologies with a complete absence of bedrock features (Table 3.1).

Braided

Braided channels are defined here as alluvial systems that exhibit channel splitting and rejoining over a distance of a few channel widths. The degree of braiding in the Sabie River, as defined by the number of braid distributaries, is low and appears restricted to the deposition of mid-channel bars (accumulations of unconsolidated sediment in the centre of the channel) and lateral bars (accumulations of unconsolidated sediment at the sides of the channel) within an active channel whose banks are well protected by vegetative cover similar to the confined braided systems classified by Mollard (1973).

   Geomorphological diversity is lower than for those channel types directly influenced by bedrock. Quantification of the features present along 4.5 km of braided channel

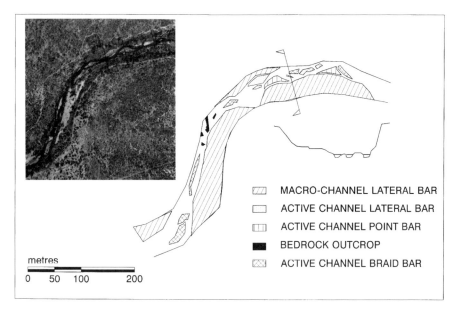

FIGURE 3.9   Detail of the morphological components of the braided channel type for the Sabie River. Flow direction is indicated by flags on the line of cross-section

revealed a significantly reduced frequency of bedrock features, these being restricted to a small area of reduced rapids. Alluvium was present over bedrock in the form of bedrock core bars and in the active channel as mid-channel and lateral deposits; all pools showed some degree of alluviation (Figure 3.9, Table 3.1). The macro-channel areas were dominated by lateral alluvial features with only very rare outcrops of bedrock.

DYNAMICS OF CHANNEL CHANGE

Differing degrees of sedimentation within the macro-channel of the Sabie River create reaches that range from almost fully bedrock through to fully alluvial, similar to the continuum noted by Ashley et al. (1988). By using aerial photographs of different examples of the same channel type that display differing amounts of accumulated sediment, a sequence of channel types can be speculated upon that defines the progression from a bedrock to an alluvial system as a result of alluviation of the river.

The geomorphological composition of 25 km of the river was assessed from aerial photographs. The study reaches were first grouped into the five channel types. These were then divided into 250 m subreaches, enabling individual assessment of the geomorphological components on a presence/absence basis. The subreach results were then averaged for each channel type to express the probability of occurrence of a particular geomorphological unit as a function of its subreach distribution.

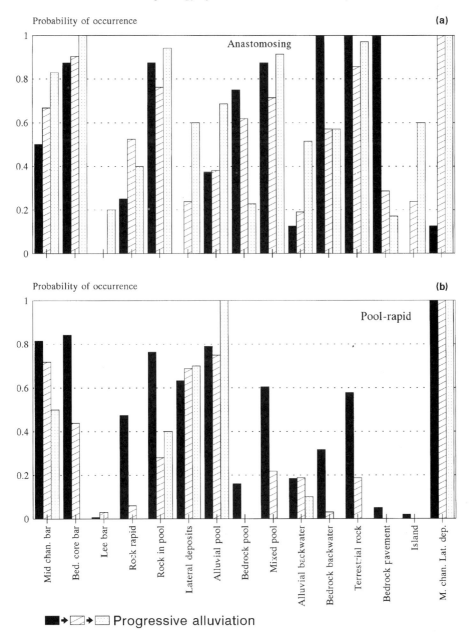

FIGURE 3.10   Inferred sedimentary evolution of the Sabie river based on an overall grouping of channel type characteristics: (a) anastomosing sequence; (b) pool–rapid sequence

Possible pathways of channel alluviation

Two pathways are proposed depending upon whether the initial bedrock channels were single- or multithread (Figure 3.10). Each pathway illustrates similar areas of river with respect to channel type, and each successive stage displays an increasing area covered by sediment (Figures 3.11–3.13). No time scale can be attached to these change pathways. However, it is argued that geomorphic change will be towards the more alluvial examples in each sequence as sediment inputs to the channel increase and flow magnitude and frequency decline in response to natural cycles in the climate (Tyson, 1987) exacerbated by anthropogenic impacts on the catchment.

Pathway 1: Multithread channel alluviation

The sequence here ranges between bedrock anastomosing, through mixed anastomosing channel types. Interpretation of the trends in geomorphological change revealed in the analysis of progressive alluviation from bedrock to mixed anastomosing (Figure 3.10a) indicates that bedrock core bars remain common as alluviation proceeds. Other alluvial features become progressively more common, especially braid bars, lateral deposits and lee bars. Rock areas in pools remain common as deposition within the active channel distributaries is not sufficient to bury these features (Figure 3.10a). Away from the active channels alluvial backwaters increase at the expense of bedrock backwaters and macro-channel lateral deposits extend to cover areas of bedrock pavement, but isolated terrestrial rock outcrops remain common (Figure 3.10a). Islands are recorded in mixed anastomosing channel types, possibly as a result of the amalgamation of smaller alluvial areas.

Pathway 2: Single-thread channel alluviation

Progressive sedimentation in pool–rapid channel types gives rise to mixed pool–rapid, braided and alluvial single-thread channels. The dominant features in these channel types are macro-channel lateral deposits, with very little bedrock pavement recorded. Occasional isolated bedrock outcrops have been observed in pool–rapid and braided channel types but these disappear in fully alluvial single-thread areas (Figure 3.10b). Backwater areas are uncommon owing to the single-thread nature of these channel types. Bedrock pools are rare even in pool–rapid channels as most pools contain some alluvium, placing them in the mixed pool category. As sedimentation proceeds, mixed pools are replaced by alluvial pools in alluvial single-thread channels (Figure 3.10b). Rock rapids and bedrock core bars are lost as sediment cover increases, but isolated rocks in pools may persist in braided areas. Braid bars remain common but may decline slightly in single-thread channels as they amalgamate with other deposits to form more extensive lateral features (Figure 3.10b).

Examples of alluviation pathways

The validity of these two general pathways inferred from the sedimentation sequence (bedrock anastomosing–mixed anastomosing; bedrock pool–rapid – mixed pool–rapid – braided – alluvial single-thread) was tested through a more detailed investigation of geomorphological changes, undertaken using specific photographic examples of progressive alluviation in anastomosing (Figure 3.11a, b), pool–rapid (Figure 3.12a, b) and braided/single-thread channel types (Figure 3.13a, b). Each photograph was arranged in a sequence displaying progressively greater areas of deposited alluvial material. Evidence of change in geomorphological composition was again determined by subsampling each photograph.

The alluviation sequence suggested for anastomosing channels (Figure 3.11a, b) indicates a gradual increase in lateral, point and braid bars in the active distributaries, with lee bars appearing only after considerable alluviation. Bedrock pools decline, being replaced by mixed pools; however, rock rapids and isolated rock outcrops in pools remain common. Outside of the main channels bedrock core bars and macro-channel lateral deposits dominate throughout and both alluvial and bedrock backwaters remain common (Figure 3.11a, b).

Geomorphological analysis of the photographic sequence of inferred pool–rapid alluviation (Figure 3.12a, b) illustrates the dominance of macro-channel lateral deposits away from the active channel, with only occasional alluvial and bedrock backwaters, bedrock pavements and terrestrial bedrock observed in the less alluviated photographs. In the active channel, bedrock pools are replaced by mixed pools and finally, in a few cases, by alluvial pools as sediments continue to accumulate (Figure 3.12a, b). Lee bars form but these are later amalgamated into increasing areas of lateral and point bars. Bedrock core bars develop to cover rapids and these eventually become amalgamated with other deposits to form larger lateral features (Figure 3.12a, b). Braid bars are seen to form during later stages of alluviation.

Reaches of the Sabie River in the Kruger National Park where alluvium has masked almost all bedrock influence were also investigated for geomorphological differences (Figure 3.13a, b). Changes due to alluviation in this sequence occurred in the active channel as the macro-channel was dominated throughout by well developed lateral alluvial deposits. Bedrock core bars, rock rapids, isolated rock outcrops in pools and mixed pools can be seen to decline to zero as braid and lateral deposits increase (Figure 3.13a, b). Initial lee bar development amalgamates into larger lateral deposits early in the sequence. Dissection of larger lateral deposits will alter the braid bar/lateral deposits ratio leading to a switching between braided and alluvial single-thread channel types (Figure 3.13a, b).

CONCLUSIONS

The Sabie River displays many characteristics that reflect the influence of bedrock control and alluviation:

- incision has created a situation where bedrock influences channel form in many places creating a diverse fluvial geomorphology;
- sedimentation within the incised macro-channel has created a range of channel types from bedrock to fully alluvial;
- five channel types have been identified based on associations of geomorphological units (bedrock anastomosing, mixed anastomosing, pool–rapid, braided and alluvial single-thread), each of which displays differing degrees of sediment accumulation within the macro-channel.

(a)

FIGURE 3.11   (a) Photographic examples of sedimentation in anastomosing channels. Flow is from right to left. (b) Interpretation of sedimentation in anastomosing channels shown in (a), (mcb = macro-channel bank; bcb = bedrock core bar; lee = lee bar; rapid = bedrock rapid; alat = active channel lateral bar; isor = isolated rock; allp = alluvial pool; bedp = bedrock pool; mixp = mixed pool; allb = alluvial backwater; bedb = bedrock backwater; terr = terrace; pave = bedrock pavement; isle = island; mlat = macro-channel lateral bar)

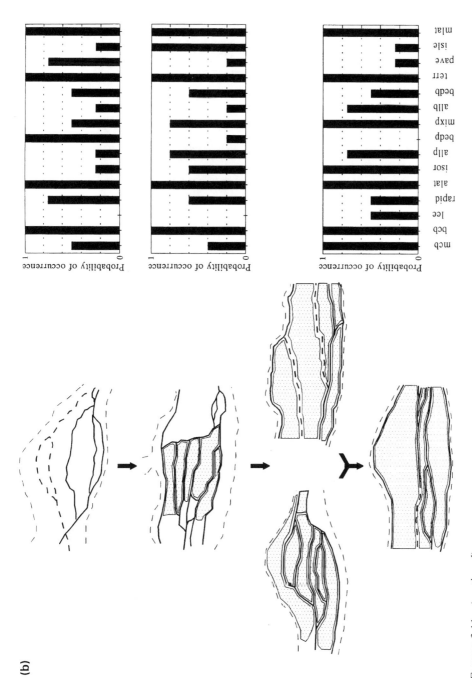

FIGURE 3.11 (continued)

(a)

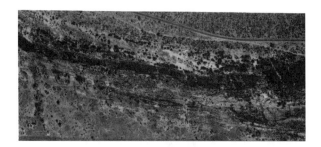

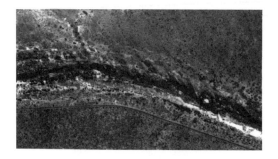

FIGURE 3.12　(a) Photographic examples of sedimentation in pool–rapid channels. Flow is from right to left. (b) Interpretation of sedimentation in pool–rapid channels shown in (a). See caption to Figure 3.11 for key

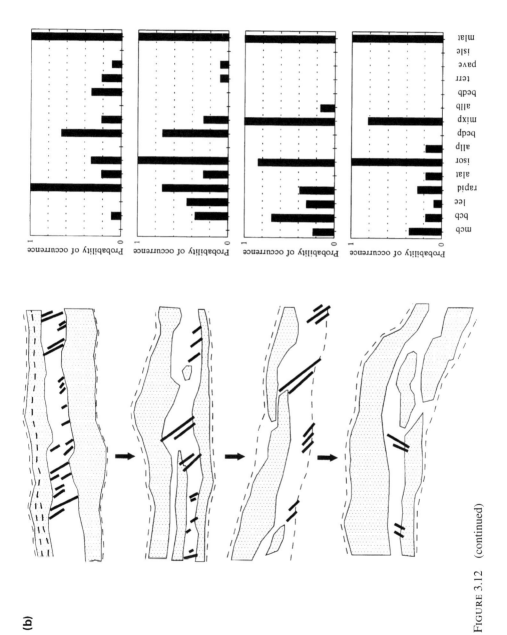

FIGURE 3.12 (continued)

(a)

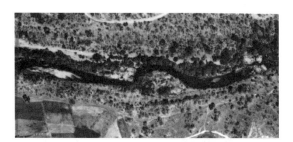

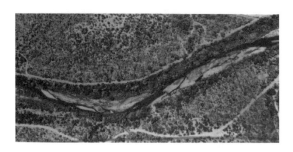

FIGURE 3.13   (a) Photographic examples of sedimentation in braided/single-thread channels. Flow is from right to left. (b) Interpretation of sedimentation in braided/single-thread channels shown in (a). See caption to Figure 3.11 for key

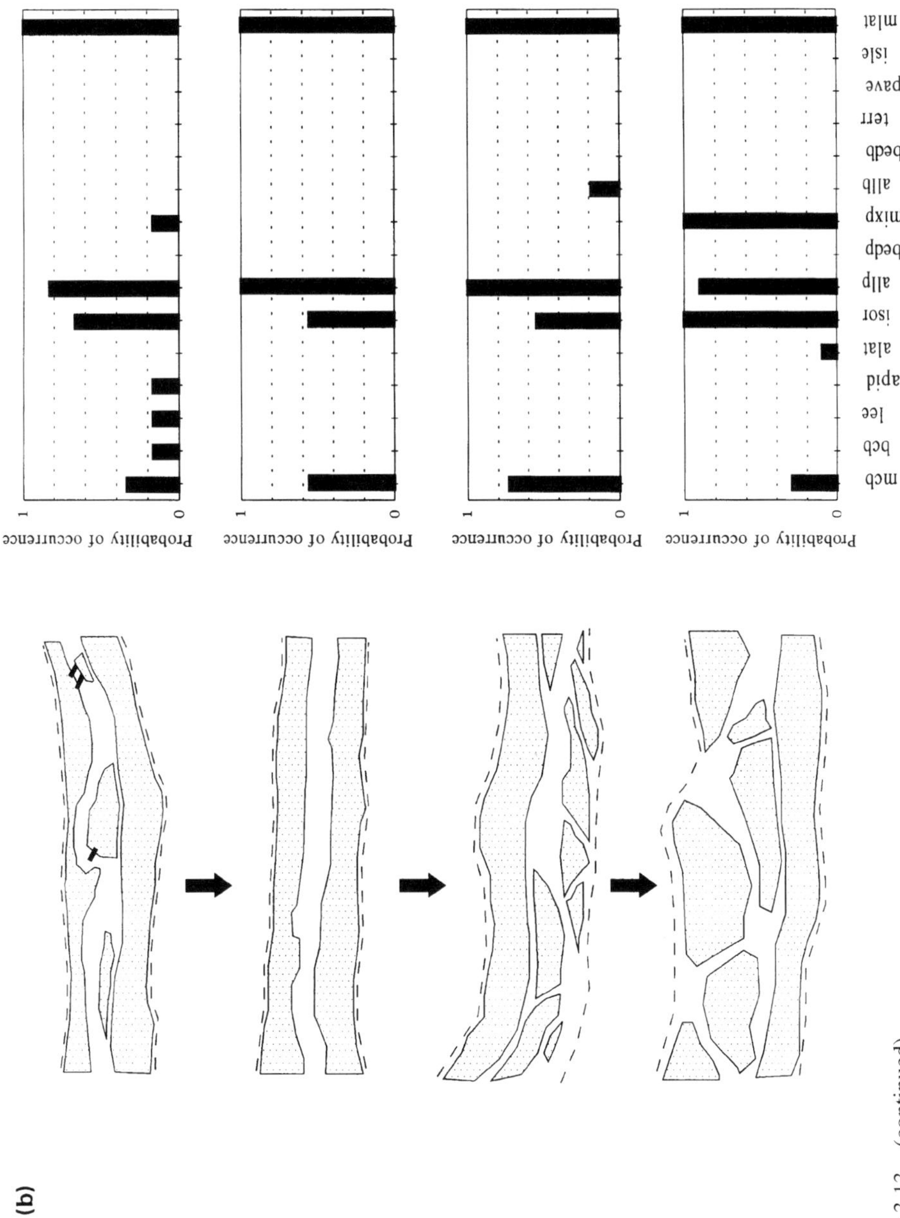

(b)

FIGURE 3.13 (continued)

TABLE 3.2 Changes to the geomorphological assemblage in anastomosing and pool–rapid channel types as a result of sediment accumulation

Increasing morphological features	Decreasing morphological features
Anastomosing channels	
Alluvial backwaters	Bedrock backwaters
Macro-channel lateral deposits	Bedrock pavement
Braid bars, point bars, lateral bars	Bedrock pools
Pool–rapid channels	
Macro-channel lateral deposits remained dominant and stable	
Alluvial pools	Bedrock pools, mixed pools
Point bars, lateral bars	Bedrock core bars, braid bars, rock rapids

An evaluation of channel dynamics and sedimentation pathways is presented that infers system change from a static picture of the river. The approach assumes that spatially isolated examples of channel areas displaying increasing alluviation can be used to generate a sequence of inferred geomorphological changes within the macro-channel as sediment accumulates.

Progressive sedimentation resulted in several changes to the geomorphological assemblages in anastomosing and pool–rapid channel types (Table 3.2).

These changes are summarized in Figure 3.14, where multichannel alluviation is seen to progress through the formation of bedrock core bars, extensive lateral deposits and islands. Single-thread channel deposition is illustrated by the build-up of lee, lateral and bedrock core bars in pool–rapid systems obscuring the influence of rock rapids. Finally, alluvial single-thread and braided systems are shown to alternate as braid and lateral bars build up and are dissected.

ACKNOWLEDGEMENTS

Ms Lucy Broadhurst and Mr Andrew Birkhead are thanked for their invaluable assistance and constructive comments on this paper, as are Dr Andrew Miller and the anonymous reviewers who helped to improve the paper's style and content. The funding by the Water Research Commission is gratefully acknowledged. The project is supported financially and materially by the Mazda Wildlife Fund and logistically by the National Parks Board. Fieldwork support was provided by G. Strydom, P. Mldlovo, G. Mauleke and J. Maboso of the Kruger National Park, and P. Frost, K. Kapur and L. Peel of King's College (London University), and is much appreciated.

During recent fieldwork, a motor vehicle accident resulted in fatal injuries to Andre van Niekerk (co-author). Andre made significant contributions to state-of-the-art rivers research in South Africa, and his absence is sorely felt by the research community and his colleagues. We extend our sincere sympathies to his family.

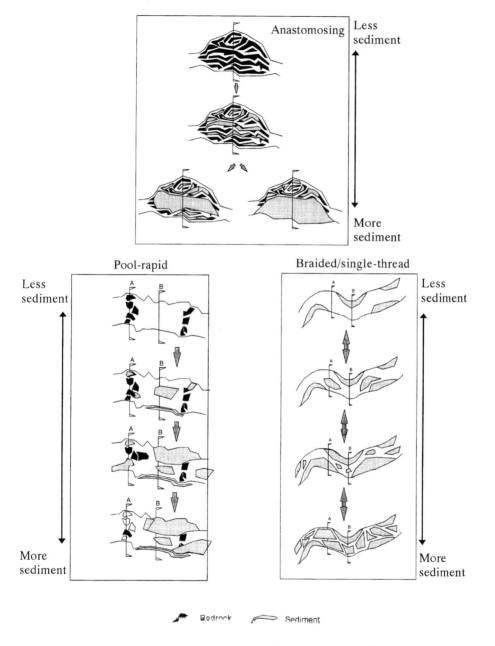

FIGURE 3.14   Generalized models of channel geomorphological evolution-based space for time substitution analysis of photographic evidence on the Sabie River in the Kruger National Park. Flow is from left to right

REFERENCES

Ackers, P. and White R. 1973. Sediment transport: new approach and analysis. *Journal of the Hydraulics Division ASCE*, **99**(11), 2041–2060.

Ashley G.M., Renwick W.H. and Haag G.H. 1988. Channel form and process in bedrock and alluvial reaches of the Raritan River, New Jersey. *Geology*, **16**, 436–439.

Baker, V.R. 1978. Palaeohydraulics and hydrodynamics of scabland floods. In V.R. Baker and D. Nummedal (Eds), *The Channeled Scabland*. NASA, Washington DC, 59–80.

Baker, V.R. and Pickup, G. 1987. Flood geomorphology of the Katherine Gorge, Northern Territory, Australia, *Geological Society of America, Bulletin*, **98**, 635–646.

Brakenridge. C.R. 1985. Rate estimates for lateral bedrock erosion based on radio-carbon ages, Duck River, Tennessee. *Geology*, **13**, 111–114.

Broadhurst, L.J., Heritage, G.L., van Niekerk, A.W. and James, C.S. (1998). *Translating discharge into local hydraulic conditions on the Sabie River: an assessment of channel roughness.* Water Research Commission Report No. 474/2/96, Pretoria, South Africa, 190 pp.

Cheshire, P.E. (in press). *Geology and geomorphology of the Sabie River in the Kruger National Park and its catchment area.* Water Research Commission Report No. 474/3/96, Pretoria, South Africa.

Graf, W.L. 1979. Rapids in Canyon Rivers. *Journal of Geology*, **87**, 533–551.

Heritage. G.L., van Niekerk, A.W., Moon, B.P., Broadhurst, L.J., Rogers, K.H. and James, C.S. (1998). *The geomorphological response to changing flow regimes of the Sabie and Letaba river systems.* Water Research Commission Report No. 474/1/96, Pretoria, South Africa, 164 pp.

Howard, A. and Dolan, R. 1981. Geomorphology of the Colorado River in the Grand Canyon. *Journal of Geology.* **89**, 269–298.

Kale, V.S., Baker, V.R. and Mishra, S. 1996. Multi-channel patterns of bedrock rivers: An example from the central Narmada basin, India. *Catena*, **26**, 85–98.

Kellerhals, R. and Miles, M. 1996. Fluvial geomorphology and fish habitat: implications for river restoration. In M. Leclerc, H. Capra, S. Valentin, A. Boudreault and Y. Côte (Eds), *Proceedings of the 2nd International Symposium on Habitat Hydraulics*. Quebec, Canada, A261–A279.

Mollard, J.D. 1973. Airphoto interpretation of fluvial features. In *Fluvial Processes and Sedimentation*, Proceedings of the Hydrology Symposium, University of Alberta, 342–380.

Partridge T.C. and Maud, R.R. 1987. Geomorphic evolution of Southern Africa since the Mesozoic. *South African Journal of Geology*, **90**(2), 179–208.

Sheperd, R.G. and Schumm, S.A. 1974. Experimental study of river incision. *Geological Society of America, Bulletin*, **85**, 257–268.

Sternberg, H. and Russell, R. 1952. Fracture patterns in the Amazon and Mississippi valleys. *Proceedings of the 8th General Assembly and 27th Congress of the International Geographical Union*, Washington DC, 380–385.

Tinkler, K.J. 1993. Fluvially sculpted rock bedforms in Twenty Mile Creek, Niagara Peninsula, Ontario. *Canadian Journal of Earth Sciences*, **30**, 945–953.

Tyson, P.D. 1987. *Climatic Change and Variability in Southern Africa.* Oxford University Press, Cape Town.

van Niekerk, A.W. and Heritage, G.L. 1993. *Geomorphology of the Sabie River: overview and classification.* Centre for Water in the Environment Report No 2/93, University of the Witwatersrand, Johannesburg, 100 pp.

van Niekerk, A.W. and Heritage, G.L. 1994. The use of GIS techniques to evaluate channel sedimentation patterns for a bedrock controlled channel in a semi-arid region. In C. Kirkby and W.R. White (Eds), *Proceedings of International Conference on Basin Development*. Wiley, Chichester, 257–271.

van Niekerk, A.W., Heritage, G.L. and Moon, B.P. 1995. River classification for management: the geomorphology of the Sabie River in the Eastern Transvaal. *South African Geographical Journal*, **77**(2), 68–76.

Venter, F.J. 1991. Fisiese kenmerke van bereike van standhoudende riviere in die Nasionale Kruger Wildtuin. Paper presented at the First Research Meeting of the Kruger National Park Rivers Research Programme.

Wohl, E.E. 1992. Bedrock benches and boulder bars: Floods in the Burdekin Gorge of Australia. *Geological Society of America, Bulletin*, **104**, 770–778.

Wohl, E.E. 1993. Bedrock channel incision along Piccaninny Creek, Australia. *Journal of Geology.* **101**, 749–761.

Wohl, E.E., Greenbaum, N., Schick, A.P. and Baker, V.R. 1994. Controls on bedrock channel incision along Nahal Paran, Israel, *Earth Surface Processes and Landforms*, **19**(1), 1–13.

Zen, E-an and Prestegaard, K.L. 1994. Possible hydraulic significance of two kinds of potholes: Examples from the palaeo-Potomac River. *Geology*, **22**, 47–50.

4

Geomorphology of the Green River in the Eastern Uinta Mountains, Dinosaur National Monument, Colorado and Utah

Paul E. Grams[1] and John C. Schmidt[2]

[1] *Department of Geology, Utah State University, Logan, Utah, USA*
[2] *Department of Geography and Earth Resources, Utah State University, Logan, Utah, USA*

ABSTRACT

The longitudinal profile, channel cross-section geometry, and locations and style of alluvial deposition along the Green River in its course through the eastern Uinta Mountains are each strongly influenced by tributary sediment delivery processes and by the bedrock lithology that is exposed at river level and in tributary basins. Lithologic resistance controls channel form directly where the channel banks are bedrock, and indirectly at the mouths of steep tributary canyons where debris fans composed of resistant boulders have formed. We surveyed channel cross-sections at 1 km intervals, mapped surficial geology, and measured size and characteristics of bed material in order to evaluate the geomorphic organization of the 70 km study reach. Steep and narrow canyon subreaches occur where resistant lithologies are exposed at river level and debris fans are abundant. Meandering reaches that are characterized by low gradient and wide channel geometry occur when river-level lithology is of moderate to low resistance and there are few debris fans. The channel is in contact with bedrock or talus along only 42% of the bank length in canyon reaches, and there is an alluvial fill of at least 12 m that separates the channel bed from bedrock at three borehole sites. Thus, the channel has been dominated by aggradation during recent geological time. Shear stress estimates indicate that bed material size and channel form and steepness are in approximate adjustment with discharges that have an approximate recurrence interval of 25 years for the unregulated streamflow regime. Downstream transport of gravel is limited; gravel bars are primarily composed of the same lithology that occurs on the adjacent upstream debris fan.

Sediment storage locations are very different in canyon and meandering reaches. Meandering reaches contain an order of magnitude more alluvium by area than canyon reaches, and most alluvium is stored in mid-channel bars and expansive flood plains. In the canyon reaches, however, about 70% by area of all fine- and coarse-grained alluvium above the low-water stage is stored in fan–eddy complexes. Less than 1% of the alluvium in meandering reaches is located in fan–eddy complexes. The depositional settings created by debris fans are similar to those described in Grand Canyon and consist of: (1) channel-

Varieties of Fluvial Form. Edited by A.J. Miller and A. Gupta
© 1999 John Wiley & Sons Ltd

margin deposits in the backwater upstream from the debris fan; (2) eddy bars in the zone of recirculating flow downstream from the constriction; and (3) expansion gravel bars in the expansion downstream from the zone of recirculating flow.

INTRODUCTION

Bedrock gorges and deep canyons are among Earth's most spectacular landscapes. Canyons are found on most continents and occur where major drainages cross topographic barriers. The geomorphic organization of canyon-bound rivers is determined by: (1) the characteristics of the bedrock into which the canyon has been eroded; (2) the characteristics and frequency of hillslope and tributary sediment delivery to the trunk stream; (3) the hydrology of the trunk stream; and (4) the characteristics and volume of the sediment load of the trunk stream. The relative influence of these factors in determining channel geometry, longitudinal profile and distribution of alluvial deposits has been investigated in only a few places, especially the 400 km long Grand Canyon in northern Arizona.

The purpose of this paper is to describe the role that bedrock lithology, tributary sediment delivery and local channel hydraulics play in determining channel and floodplain geomorphology in the canyons of the Green River in the eastern Uinta Mountains of Colorado and Utah (Figure 4.1). The geomorphic character of this stream is evaluated by examination of channel geometry, longitudinal profile, and patterns of alluvial deposition and erosion. The study area is located at the northern edge of the Colorado Plateau and includes a diverse assemblage of lithologies and geological structures that influence the river's geomorphology. Data collected and analysed for this study include detailed surficial geological maps of the river corridor, surveyed channel cross-sections and bed-material measurements. The geomorphic organization of the Green River in the Uinta Mountains is compared to Grand Canyon and generalized to the Colorado Plateau.

PREVIOUS WORK

Perhaps the most extensive canyons in the United States are found in the Colorado Plateau, a large region of uplifted sedimentary rocks. The disparity between the orientation of present stream courses and the trends of dominant geological structures causes the largest rivers of the region to traverse diverse geological formations of varying erosional resistance. The streams have carved deep and narrow canyons through more resistant formations, or they meander through broad basins and smaller parks across less resistant formations (Harden, 1990; Hunt, 1969). John Wesley Powell (1875) cited Split Mountain Canyon of the Green River (Figure 4.1) as a typical example of an antecedent stream, a river that had carved deep canyons as mountains rose in the river's path. Subsequently, geologists questioned this theory and offered alternative explanations (Davis, 1897; Jefferson, 1897; Sears, 1924; Bradley, 1936; Hansen et al., 1960; Hunt, 1969). Hansen (1986) presented evidence supporting the explanation first proposed by Sears (1924) that the canyons of the Green River were the result of superposition from a course established on Tertiary sediments that had filled local basins. Entrenchment of the canyons began in late Miocene–early Pliocene time when the Green River drainage was

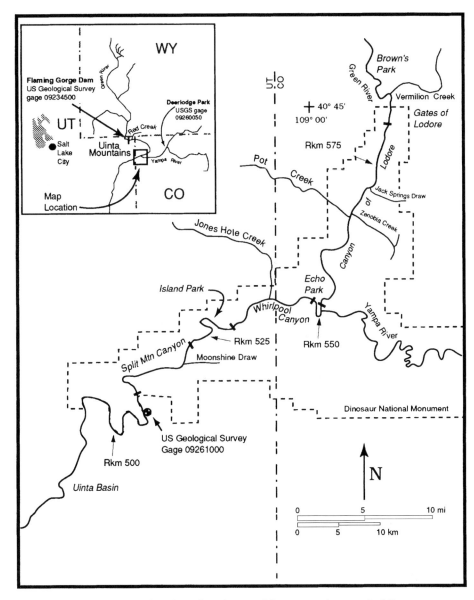

FIGURE 4.1   Map showing location of study area. Distance markers are in kilometres upstream from the Colorado River confluence (Rkm). Each geomorphic subreach is labelled, and the extent of each reach is indicated by the solid bars. The locations of major tributaries and US Geological Survey stream gauging stations are also indicated

diverted from an eastward-flowing course across the present day continental divide in Wyoming to a southward course across the eastern Uinta Mountains (Hansen, 1986). The Canyon of Lodore (hereafter informally referred to as Lodore Canyon) is about five million years old, and the incision of its gorge, 760 m deep, occurred at an average rate of at least 15 cm per thousand years (Hansen, 1986).

Geologists have also been interested in how bedrock geology controls specific fluvial morphological characteristics of modern streams within canyons. Powell (1875, p.234) anticipated dangerous reaches of river by observing the lithology of nearby rocks, noting that: "In softer strata we have a quiet river, in harder we find rapids and falls." The pool-drop pattern of the Green River was obvious to river travellers, yet the magnitude of the drops was not measured until a channel profile was surveyed during preliminary dam-site investigations made between 1917 and 1922 (US Geological Survey, 1924).

Most geomorphic studies have emphasized the importance of either (1) bedrock lithology and structure, (2) tributary processes or (3) mainstem hydraulics in determining the geomorphic organization of the large rivers of the Colorado Plateau. Leopold (1969) proposed that the Colorado River in Grand Canyon was organized by the river's own hydraulics. He argued that the average longitudinal profile of the river is nearly straight when viewed at the canyon-length scale despite the abundance of steep rapids where most of the river's elevation drop occurs. Leopold (1969) concluded that the average profile and the semi-regular spacing of rapids, gravel bars and deep pools are analogous to the riffle–pool sequence observed in small streams, and indicate that the river is in a state of quasi-equilibrium. The slope of the channel is adjusted to rework and transport the delivered sediment, however coarse, and the occurrence of rapids and pools is an inherent characteristic of transport mechanics. Later, in an investigation encompassing several large rivers of the Colorado Plateau, Graf (1979) found that the spacing of rapids was essentially random, an indication that local geomorphic conditions and Pleistocene hydrology also affected rapid and pool location. Graf (1979) found that debris fans created most of the rapids he investigated. In contrast, Dolan et al. (1978) and Howard and Dolan (1981) concluded that nearly all of the large rapids and deep pools in Grand Canyon are located at debris fans at the mouths of tributaries whose locations are determined by geological structures such as major faults, folds and fracture zones. Thus they argue that fluvial organization is structurally controlled.

Geomorphic characteristics of canyon-bound rivers are similar over long distances and do not change systematically in a downstream direction. Channel geometry in Grand Canyon has been related to bedrock lithology, although the mechanism for control has not been investigated. Schmidt and Graf (1990) examined channel cross-sections surveyed at approximately 1.6 km intervals by Wilson (1986) and identified 11 reaches of similar bedrock resistance and similar channel geometry. Smith and Wiele (1995) analysed the same data and statistically defined "typical" channel cross-sections for reaches of similar river-level geology.

Although descriptions and classifications of fluvial bedforms are common (Church and Jones, 1982; Jackson, 1975), few schemes have specifically discussed bedforms and barforms in debris fan-affected canyons. Baker (1984) described a classification of gravel bedforms in bedrock systems that emphasizes local control features, such as tributary fans, and the hydraulics of extreme floods. Howard and Dolan (1981) classified deposits in Grand Canyon based on grain size: (1) bouldery debris fans that occur at the mouths of most tributary streams and are reworked by large mainstem floods; (2) mainstem cobble bars that are transported only during floods; and (3) fine-grained alluvium deposited in eddies and along the channel margins that are transported primarily as suspended load. These aspects of the fluvial geomorphic organization of canyon-bound rivers transcend local variations in lithology.

Schmidt and Graf (1990) and Schmidt (1990) described the detailed characteristics of alluvial deposits near debris fans. Within eddies, Schmidt (1990) distinguished separation bars, formed near the point of flow separation, and reattachment bars, formed near the flow reattachment point. The sedimentology and stratigraphy of these bars were described by Rubin et al. (1990), Schmidt and Graf (1990) and Schmidt et al. (1993). Schmidt and Rubin (1995) argued that the assemblage consisting of debris fan, upstream backwater, and downstream eddy and gravel bar constitutes the fundamental geomorphic unit in canyons with debris fans and termed this unit the fan–eddy complex. For river systems where debris fans are abundant, patterns of sedimentation and erosion are largely determined by debris fan geometry, distribution and, ultimately, the frequency of fan-forming events. Sedimentology of debris fans, frequency of fan-forming debris flows, and the effects of debris fans on rapids were discussed by Webb et al. (1989) and Melis et al. (1993). The hydraulic characteristics of debris fan-created rapids and upstream pools were discussed by Kieffer (1985).

Bedrock geology, tributary delivery processes and mainstem hydrology all contribute to the geomorphic character of canyon rivers. Yet comprehensive examination of these channel-organizing elements, particularly in a system outside Grand Canyon, is lacking; and the applicability of Grand Canyon models has not been tested elsewhere. The mechanism by which bedrock geology affects channel form has not been investigated, although the association between the two has been demonstrated.

GEOLOGICAL SETTING AND DESCRIPTION OF STUDY REACH

The east–west trending Uinta Mountains span approximately 300 km in northeastern Utah, southwestern Wyoming and northwestern Colorado. Structurally, the range forms a broad arcuate anticline, separated in the middle by a structural and topographic low. The eastern Uinta Mountains range in elevation from 2960 m to 1448 m at the Green River near the mouth of Split Mountain Canyon.

Uplift and deformation of the Uinta Mountains began in latest Cretaceous time and continued into the Tertiary (Hansen, 1986). Bedrock exposures in the eastern Uinta Mountains range from highly resistant Precambrian quartzite to soft Tertiary sediments. The rocks on the south flank of the Uinta Mountains generally dip south and southwest (Hansen et al., 1983). However, folds and faults cause large variations in dip over short distances. Table 4.1 summarizes the rock formations, the reaches over which they outcrop at river level, and their bedrock resistance class. Bedrock resistance was adapted from the classification of Harden (1990). This semi-quantitative classification ranks bedrock on a scale of 1 to 9 in order of increasing resistance to erosion.

Mapped faults in the eastern Uinta Mountains generally trend NW–SE and NE–SW (Hansen et al., 1983). Faulting occurred during the Laramide uplift and post-Laramide (middle to late Tertiary) extension. In addition to the mapped faults, Hansen et al. (1983) showed an abundance of similarly trending joint sets. Nearly all of the major tributaries and many of the smaller drainages to the Green River in the north–south trending Lodore Canyon are aligned with these geological structures. The largest tributaries such as Pot Creek, Zenobia Creek and Jack Springs Draw run adjacent to mapped faults.

TABLE 4.1   Length of geometric and lithologic subreaches within study area showing the bedrock resistance classification for each formation

Reach	Rock type at river level	Upstream end of reach (Rkm)[*]	Reach length (km)	Bedrock resistance class[†]
Brown's Park	Brown's Park Formation	beyond study area		2
Lodore Canyon	Uinta Mtn Group	579.5	22.7	9
	Lodore Formation	556.8	2.4	8
	Madison Limestone	554.4	1.4	8
	Palaeozoic undif.	553.0	0.3	7.5
	Weber Sandstone	552.7	1.6	6
Echo Park	Weber Sandstone	551.0	3.2	6
Whirlpool Canyon	Palaeozoic undif.	547.8	0.5	7.5
	Uinta Mtn Group	547.3	2.4	9
	Lodore Formation	544.9	5.2	8
	Madison Limestone	539.8	4.3	8
	Palaeozoic undif.	535.4	1.8	7.5
Island Park	Mesozoic undif.	533.7	11.6	3–5
Split Mountain Canyon	Palaeozoic undif.	522.1	4.0	7.5
	Madison Limestone	518.0	3.4	8
	Palaeozoic undif.	514.7	2.1	7.5
	Weber Sandstone	512.6	2.4	6

[*] Distances are in kilometres upstream from Colorado River confluence
[†] Adapted from Harden (1990)

The Green River in the eastern Uinta Mountains is divided into six canyon-bound or "park" subreaches that were originally named by Powell (1875). These subreaches have distinct planforms, valley widths and relationships with geological structures.

The river in the canyon subreaches of Lodore Canyon, Whirlpool Canyon and Split Mountain Canyon does not meander; the river tends to flow straight for about 2 to 4 km then bends abruptly and flows straight again. Many of these bends, like tributary alignments, are coincident with geological structures. In Split Mountain Canyon, for example (Figure 4.2), the river cuts across the rock strata into the core of the anticline then turns 90° west and flows along the structural axis for about 6.5 km, and then, just as abruptly, veers back across the structural trend and flows out of the anticline into the Uinta Basin. These three subreaches each contain debris fans and are called debris fan-dominated canyons, consistent with the terminology of Schmidt and Rubin (1995). Although the river in Lodore Canyon is constantly winding around debris fans and gravel bars, the trend of the canyon is straight, punctuated by sharp bends (Figure 4.3).

Meandering reaches fall into two categories: restricted meanders and incised meanders. In the single bend through Echo Park, the channel and the valley meander at

FIGURE 4.2   After flowing through Island Park, the Green River abruptly enters the Split Mountain Anticline. While Powell's (1875) original explanation that the stream must be antecedent to the mountains was plausible, it has since been shown that the mountains are older than the course of the river (Hansen, 1986). The view of this photograph is to the southwest and streamflow direction is away from the camera. Photograph by Michael Collier

the same amplitude, characteristic of incised meanders. In the restricted meanders of Brown's Park and Island Park (Figure 4.4), the Green River valley is incised into soft Tertiary and Mesozoic sediments and the channel occasionally encounters more

FIGURE 4.3   The location where the Green River leaves the meanders of Brown's Park and enters the fan-dominated Lodore Canyon was named "The Gates of Lodore" by John Wesley Powell in 1869. The valley does not meander in the canyon; instead the channel winds around the abundant debris fans and the exposed gravel and sand bars at low water as shown in this photograph. The view of this photograph is to the south and streamflow direction is away from the camera. Photograph by Michael Collier

resistant Palaeozoic bedrock on the outsides of meander bends. The meander amplitude of the channel is less than that of the valley and only the outside margins of the meanders impinge on bedrock.

FIGURE 4.4   Photograph of Island Park reach taken during low discharge in October 1995. The view is to the northwest, and streamflow is from right to left. This photo illustrates the large mid-channel bars and wide terraces that are abundant in this reach. The bars on the right are attaching to one bank as side-channels fill. Island Park is a restricted-meander reach because channel meanders occasionally encounter bedrock

HYDROLOGICAL SETTING

The Brown's Park and Lodore Canyon subreaches are hydrologically distinct from the four downstream subreaches. Streamflow has been completely regulated since 1 November 1962 by Flaming Gorge Dam, for the 104 km distance between the dam and Echo Park (Figure 4.1). The primary functions of the dam are storage of snowmelt runoff and hydroelectric power generation. The volume of total annual runoff and the mean annual discharge have not been affected by dam operations (Andrews, 1986). Streamflow measurements at the gauging station near Greendale, Utah (station number 09234500), immediately downstream from the dam, are applicable to the upstream half of the study reach. The typical pre-dam hydrograph was dominated by spring snowmelt runoff and low autumn and winter baseflows. The pre-dam mean annual flood was 334 $m^3 s^{-1}$ and has been reduced by about 63% to 139 $m^3 s^{-1}$ since dam closure (Figure 4.5).

The unregulated Yampa River flows into the Green River at Echo Park. The mean annual flood of the Yampa River is unchanged since 1920 and is similar to that of the Green River prior to completion of Flaming Gorge Dam (Figure 4.5). The combination of regulated streamflow from the Green River and unregulated streamflow from the Yampa River results in a 26% decrease in the mean annual flood at the gauging station near Jensen, Utah (station number 09261000), located 46 km downstream from Echo Park.

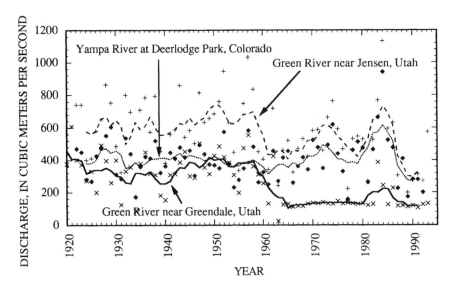

FIGURE 4.5   Annual maximum discharge of the Green River near Greendale (×) and Jensen (+), Utah, and the Yampa River at Deerlodge Park (♦), Colorado. Fitted lines are 5-year moving averages. The records for Greendale 1920–1950, Jensen 1922–1946 and Deerlodge 1922–1981 are determined by gauge-station correlation. Flaming Gorge Dam began storing water in November 1962

Sediment transport has also been affected by closure of Flaming Gorge Dam. Sediment sources downstream from the dam are limited to the bed and banks of the river and ungauged tributaries upstream from the Yampa River. The largest tributaries, Red Creek and Vermillion Creek, enter the Green River 18 km and 70 km downstream from the dam, respectively (Figure 4.1). The only site with a long-term record of sediment transport near the study area is the Jensen, Utah gauge. The mean annual load of suspended sediment has been reduced by 54%, from 6.28×10^6 to 2.91×10^6 Mg a^{-1}, since dam closure (Andrews, 1986).

METHODS

Channel cross-sections

We measured 67 channel cross-sections at approximately 1 km intervals to characterize channel geometry in each of the subreaches. These cross-sections were surveyed in 1994 during low discharge using a geodetic total station and depth-recording echo sounder. Additional water-surface elevations were surveyed at each cross-section during peak flow in June and July 1995. Characteristics such as bed and bank material, high-water marks, bankfull stage indicators and distinct geomorphic surfaces were noted and surveyed at each site. The precise location of each cross-section was recorded on aerial photographs (1:5000 scale) and topographic maps. Channel geometry characteristics at post-dam bankfull discharge were calculated from the survey data.

The bed material at the thalweg of each cross-section was determined either while wading or by interpreting fathometer traces. Sand beds produce a smooth fathometer trace, sometimes showing ripples or dunes. Coarse beds produce an uneven trace and indicate gravel. The trace over a bouldery bed is irregular and often profiles individual boulders. The bed material coarsens from bank to thalweg at most cross-sections. The bed material recorded for each cross-section is the material that occupies more than 50% of the thalweg bed.

Gravel and sand bars exposed at baseflow represent the only portion of the stream bed that is easily accessible. Bed material size was measured at 18 gravel bars; each bar was sampled at the upstream end below the high-water line. Particles were sampled at regularly spaced intervals along a tape measure for the above-water portion of the bar and by random walk for the submerged but wadable areas (Wolman, 1954). Median diameter, lithology and roundness were recorded for each particle. Roundness was visually estimated on a scale from angular to rounded.

The criteria for cross-section establishment were the desired 1 km spacing, the existence of uniform downstream flow, and feasibility of measurement. Cross-sections were generally not surveyed in rapids or in large recirculating eddies. Because these areas were avoided, few cross-sections traversed eddy-deposited sand bars, gravel bars in flow expansions, and coarse alluvium in rapids. Therefore, analyses of bed material distribution determined from cross-section data describe characteristics of uniform flow reaches and may underestimate the percentage of the bed that is composed of gravels or boulders.

Mapping of surficial geology

Surficial geology was interpreted on aerial photographs (1:5000 scale) taken in 1993 during low discharge (33.4 $m^3 s^{-1}$ and 45.3 $m^3 s^{-1}$ upstream and downstream from Echo Park, respectively). The preliminary maps were field checked and transferred to a 1:12 000 scale topographic base map using a reflecting projection table. Six geomorphic characteristics were specified for each mapped deposit or feature: (1) bedrock exposure at river level; (2) subreach in which each deposit is located (Lodore Canyon, Echo Park, Whirlpool Canyon, Island Park or Split Mountain Canyon); (3) presence of deposit in a fan–eddy complex; (4) grain size; (5) depositional environment; and (6) elevation of the deposit in relation to the low-discharge water surface. Bedrock type at river level and reach were determined from geological (Hansen et al., 1983) and topographic maps. Each discrete assemblage of debris fan, eddy bar, channel-margin deposit in backwater, and expansion bar was grouped and assigned a unique fan–eddy complex number. Deposits not in fan–eddy complexes were assigned a value of zero. Completed maps were digitized into a geographic information system (GIS) for spatial analysis of deposits.

Inventory of rapids and riffles

The association between debris fans and rapids was evaluated by an inventory of all rapids and riffles on the Green River within the study reach. Our criteria were similar to those of Graf (1979) who defined a rapid as a location where "debris particles are numerous enough or large enough to break the water surface at mean annual discharge." We defined a rapid as any discrete river segment where breaking water was visible across

most of the channel on the 1993 aerial photographs. The cause of the channel constriction
and the material of each river bank were recorded for each rapid.

RESULTS

Characteristics of alluvial and tributary fan deposits in study reach

Textural classification

Four textural classes of deposits occur within the study area and can be distinguished on
aerial photographs. Figure 4.6 is an aerial photograph showing fine- and coarse-grained
deposits in a fan–eddy complex.

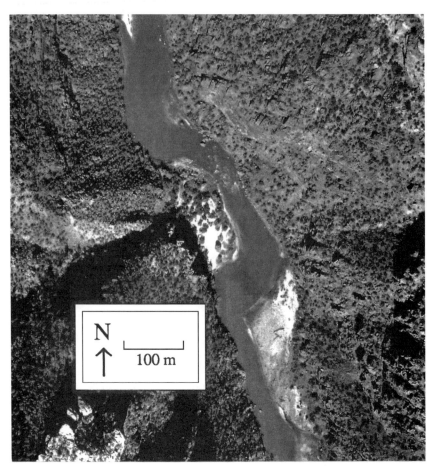

FIGURE 4.6   Aerial photograph of a fan–eddy complex in Lodore Canyon at Rkm 558.3.
Streamflow is from north to south. A channel constriction is formed where the debris fan impinges
flow against the opposite bedrock/talus bank. A second low-water constriction and rapid is formed
at the downstream gravel bar. A sand bar mantles the downstream part of the debris fan, and the
gravel bar restricts the length of the recirculation zone at high discharge. Channel-margin deposits
occur upstream from the debris fan

TABLE 4.2 Distribution of surficial deposits, rapids and fan–eddy complexes in five subreaches, measured from surficial geological maps

Geomorphic characteristic	Lodore Canyon	Echo Park	Whirlpool Canyon	Island Park	Split Mountain	Canyons	Meanders	Study area
Reach length (km)*	28.5	3.2	14.2	11.6	8.0	50.6	14.8	65.4
Area of alluvium (m²/km)	34 962	323 366	30 008	769 943	33 164	33 319	675 688	175 947
Rapids (count)	53	0	26	2	12	91	2	93
Rapid frequency (count/km)	1.9	0.0	1.8	0.2	1.0	1.8	0.1	1.4
Debris fans (count)	81	1	53	5	35	169	6	175
Fan frequency (count/km)	2.8	0.3	3.7	0.4	2.9	3.3	0.4	2.7
Total area of debris fans (m²)	739 543	5 598	259 407	13 250	281 211	1 280 161	19 118	1 299 279
Average area of fans (m²)	9 130	5 598	4 894	2 704	8 035	7 575	3 186	7 424
Percentage deposits by area								
gravel	6	23	32	32	32	16	31	29
fine-grained	62	65	55	67	27	55	67	66
mixed	32	12	13	1	41	29	2	5
Percentage deposits in fan–eddy complex†								
gravel	80	0	85	0	100	89	0	7
fine-grained	51	2	89	0	100	64	0	8
mixed	72	0	68	0	100	78	0	58
all alluvium	60	1	85	0	100	72	0	10
Percentage fine-grained material in eddy bars	34	1	63	0	44	42	0	5

* Length of mapped reach (8 km of 11.9 km long Split Mountain Canyon were mapped)
† Percentage of all deposits within each reach of indicated type that are within eddy complexes

Fine-grained alluvium includes all alluvial sands, silts and clays and typically occurs in eddy bars and channel-margin deposits. Individual deposits may contain sedimentary structures or be massive. The thickness of fine-grained deposits can range from a few centimetres to several metres. Fine-grained deposits are common in the canyon subreaches in spite of the river's high gradient, high average boundary shear stresses, and narrow alluvial valley. Between 68 and 94% of all alluvium exposed at low discharge in canyon reaches is at least partially composed of fine-grained material (Table 4.2). The wide alluvial bottoms of meandering subreaches, however, contain an order of magnitude more alluvium in areal exposure.

Coarse-grained alluvium includes all pebble- to cobble-sized deposits from 2 to 256 mm median diameter. Deposits are moderately to well sorted and contain rounded to subangular particles. Many coarse-grained alluvial deposits are completely or partially covered by a veneer of fine-grained alluvium, and are mapped as mixed fine- and coarse-grained alluvium. The proportion of the total area of alluvium in the canyon subreaches that is coarse-grained or mixed varies from 38% in Lodore Canyon to 73% in Split Mountain Canyon (Table 4.2). In the non-canyon reaches, about 35% of the area of all alluvium consists either completely or partially of gravel. Most of the exposed gravel in all subreaches occurs in mid-channel or expansion gravel bars.

Debris-flow deposits are poorly sorted, angular to subrounded and include particles ranging in size from clay to boulders. The *bedrock and talus* map unit includes rock outcrops and talus slopes. Talus is angular and dominated by large cobbles and boulders of locally derived lithology. Bedrock and talus are not distinguished from one another because overlap of talus and bedrock is common.

Depositional environment classification

The depositional environment of each deposit was classified independently of textural class. Depositional environments were identified by location, morphology and bar sedimentology. Deposit location and morphology were determined in the field and on aerial photographs. Stratigraphy was examined in the field at selected locations by excavating shallow trenches where bedform migration directions were interpreted.

Debris fans are accumulations of debris flow-deposited sediment located at the mouths of tributary streams (Figure 4.6). Debris fan size, form and composition are the products of tributary basin geology, debris-flow frequency and debris-fan reworking by mainstem floods. Cutbanks eroded into the fans by the Green River and tributary streams show that debris fans are poorly sorted, matrix-supported diamictons with angular to subrounded, pebble to boulder-sized clasts; the matrix is clay to coarse sand. Most debris fans are composed of a series of more than one debris-flow deposit.

Eddy bars form in zones of recirculation that occur downstream from channel constrictions within fan–eddy complexes. Figure 4.7 illustrates the flow patterns and deposits associated with a typical fan–eddy complex. Debris fans are the most common cause of channel constrictions, but talus slopes, bedrock outcrops and gravel bars occasionally cause flow separation and eddy deposition. The distinct separation and reattachment bars shown in Figure 4.7 are more typical of Grand Canyon fan–eddy complexes but also occur in the canyons of the eastern Uinta Mountains. However, separation and reattachment bars in the canyons of the eastern Uinta Mountains

A.

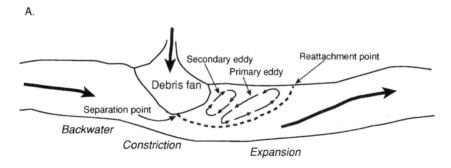

B.

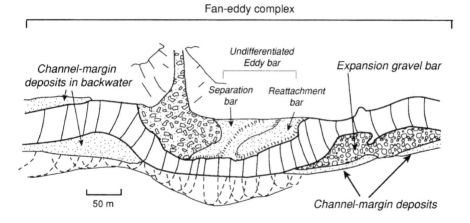

FIGURE 4.7   Illustration of a fan–eddy complex showing surface flow patterns at high discharge (A) and surficial deposits (B). The separation and reattachment bars shown are typical of Grand Canyon sand bars described by Schmidt (1990). These forms are typically merged into a single eddy bar in the eastern Uinta Mountains. The figure is adapted from Schmidt and Graf (1990)

frequently merge to form undifferentiated eddy bars. This lack of eddy bar differentiation resembles the tendency described by Schmidt and Rubin (1995) for separation and reattachment barforms to merge under conditions where sediment transport rates are high. Excavations in undifferentiated eddy bars reveal upstream and onshore migrating dunes and climbing ripples, confirming deposition in a recirculation zone. Reworking by wave action typically forms river-parallel ridge and swale topography.

Expansion bars are generally about one to three channel widths in length and form in areas of flow expansion downstream from the recirculation zone and on the opposite side of the river (Figure 4.7). The bars in this flow expansion always have a gravel core, but many also have a cap of fine-grained alluvium. These bars are mapped as fan–eddy complex deposits because they occur in most complexes and the primary source for the gravel is the upstream debris fan (discussed in the section "Adjustment of river to coarse bed material").

Channel-margin deposits are narrow strips of alluvium that are up to several channel widths long and usually occur in straight reaches of relatively uniform downstream flow. These deposits were described by Schmidt and Rubin (1995) and Schmidt and Graf (1990). Internal stratigraphy is typically horizontally bedded but may include downstream-oriented ripple-drift cross-stratification of sand and lesser amounts of silt and clay. Box elder trees (*Acer negundo*) are common on high-elevation channel-margin deposits in the canyon reaches. Some channel-margin deposits occur within fan–eddy complexes in the backwater immediately upstream from a channel constriction (Figure 4.7). *Point bars* are channel-margin deposits that are located on the inside of river bends.

Mid-channel bars are differentiated from other alluvial bars based on size, shape and location. Mid-channel bars are most common in restricted meander subreaches and are typically several channel widths long, lenticularly shaped, located in mid-channel or on channel margins (Figure 4.4). Bars that occur mid-stream but in flow expansions downstream from debris fans are mapped as expansion bars, not mid-channel bars. Former mid-stream bars that are now attached to the bank as a result of side-channel infilling are also mapped as mid-channel bars (Figure 4.4). Low-elevation bars are bare sand or covered by tamarisk (*Tamarix* sp.) or willow (*Salix* sp.). High-elevation mid-channel bars typically contain stands of mature cottonwood (*Populus fremontii*). Although vertical accretion is still an important process of bar formation, lateral accretion is indicated by scroll-bar topography not displayed in the canyon deposits.

Characteristics of fan–eddy complexes

Although there is variation in the size and characteristics of individual debris fans, eddy bars, expansion bars and channel-margin deposits in the backwater, each of these elements is present in most fan–eddy complexes. The 96 fan–eddy complexes in the mapped area occur at 79% of the debris fans. There are nearly twice as many debris fans as fan–eddy complexes because many fan–eddy complexes are formed by a debris fan on each bank; this often occurs where a structural control, such as a fault, crosses the river.

Fan–eddy complexes are the primary storage sites for alluvium in canyon subreaches. The frequency of these complexes ranges from 1.6 per kilometre in Lodore Canyon to 2.3 per kilometre in Whirlpool Canyon. In Lodore Canyon, 51% of all fine-grained alluvium and 72% of all gravel are stored in the fan–eddy complexes (Table 4.2). Recirculation zones that are part of fan–eddy complexes are the most important depositional sites for sand in canyon subreaches. Eddy-deposited sand bars contain about 42% by area of all fine-grained alluvium in the canyon reaches and less than 1% in the meandering reaches.

Geomorphic characteristics of study reach

Bedrock geology and channel form

Channel geometry is similar for subreaches where similar geological formations are exposed at river level. The ratio of channel width to channel depth at post-Flaming

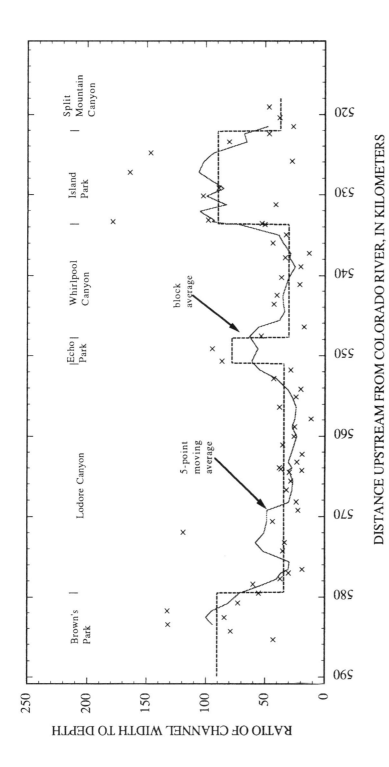

FIGURE 4.8 Downstream variation in channel width–depth ratio calculated at post-Flaming Gorge Dam mean annual flood. Solid line shows 5 km moving average and dashed line is block averages for indicated geomorphic subreaches. Channel geometry is similar within each subreach and distinct between subreaches

TABLE 4.3 Average channel geometry at bankfull discharge and average slope (US Geological Survey, 1924) for six subreaches. Channel geometry calculated from surveyed channel cross-sections

Geomorphic characteristic	Browns Park	Lodore Canyon	Echo Park	Whirlpool Canyon	Island Park	Split Mountain[†]	Canyons	Meanders	Study area
Reach length (km)[*]	3.6	28.5	3.2	14.2	11.6	11.9	54.6	14.8	69.4
Reach average slope	0.0004	0.0029	0.0006	0.0023	0.0007	0.0037	0.0021	0.0008	0.0018
Alluvial valley width (m)	819	95	246	108	721	134	101	674	285
CV[‡]	0.61	0.25	0.33	0.34	0.51	0.43	0.31	0.63	1.31
Channel width (m)	139	60	176	80	172	81	66	147	100
CV	0.38	0.24	0.20	0.18	0.40	0.11	0.25	0.48	0.61
Cross-sectional area (m^2)	215.8	115.9	406.1	223.2	365.4	183.8	147.0	316.0	209.7
CV	0.40	0.27	0.22	0.17	0.46	0.16	0.39	0.53	0.65
Mean depth[§] (m)	1.6	2.0	2.3	2.9	2.2	2.3	2.2	2.2	2.2
CV	0.15	0.24	0.19	0.27	0.34	0.21	0.30	0.28	0.32
W/D ratio	90.5	34.1	78.2	30.0	89.9	37.0	33.3	74.7	52.7
CV	0.39	0.58	0.28	0.37	0.56	0.27	0.52	0.62	0.74
Valley W/channel W	5.9	1.6	1.4	1.4	4.2	1.7	1.5	4.6	2.9

[*] Length of reach in which surveyed cross-sections are located
[†] Surveyed cross-sections include 2.5 km of Split Mountain Canyon
[‡] Coefficient of variation (standard deviation/mean) for preceding parameter
[§] Cross-sectional area divided by channel width

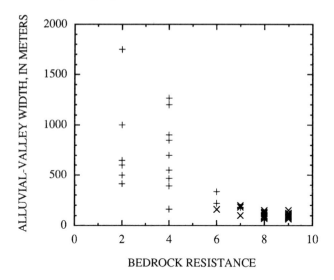

FIGURE 4.9 Illustration of the inverse relation between alluvial valley width and bedrock resistance determined at 1 km intervals through study area: +, meandering reaches; × canyon reaches

Gorge Dam bankfull discharge (122 m^3 s^{-1} upstream from the Yampa River confluence and 480 m^3 s^{-1} downstream from the confluence) is shown in Figure 4.8. Subreach-average channel-top width, cross-sectional area and width-to-depth ratio are all less in canyon than in meandering subreaches (Table 4.3). The coefficients of variation, which arise from local variability of channel characteristics, are large for most parameters. Despite this local variability, trends within subreaches occur and the coefficients of variation decrease when the sampled subset for each parameter is stratified by geologically similar subreach (Table 4.3).

Bedrock resistance and alluvial valley width are inversely related (Figure 4.9). The ratio of alluvial valley width to bankfull-channel width is higher in meandering subreaches than in canyon subreaches (Table 4.3). The coefficients of variation in the meandering subreaches (except Echo Park) are higher than in the canyon subreaches. This is a result of the large variation in proximity of bedrock to the active channel in most meandering subreaches.

The reach-length longitudinal profile of the Green River through the study reach is stair-stepped (Figure 4.10). The average slope is 0.0021 in debris fan-dominated canyon subreaches and 0.0008 in meandering subreaches (Table 4.3). The steepest subreaches occur where rocks of highest resistance are most abundant (Figure 4.10). Within subreaches, slopes are steepest where resistant rocks outcrop at river level (Figure 4.11). For example, slope is steeper in Lodore and Whirlpool Canyons where the Uinta Mountain Group quartzite is at river level than where Palaeozoic formations occur. Gradients are lowest where the Weber Sandstone, Mesozoic or Tertiary rocks occur.

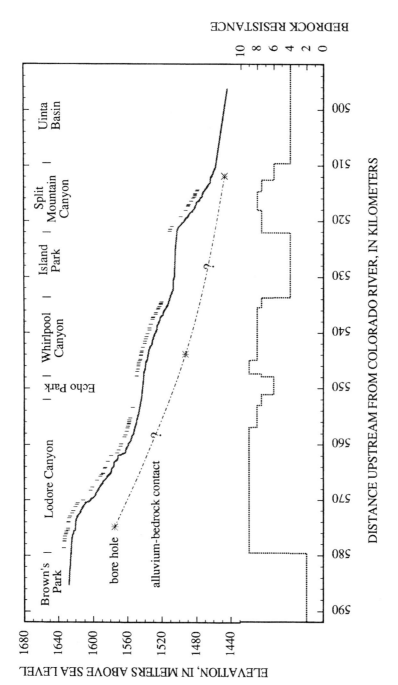

FIGURE 4.10 Longitudinal profile of the Green River through the eastern Uinta Mountains (US Geological Survey, 1924). The vertical dashes show the locations of tributary debris fans. Depth to bedrock in the river channel is shown at three borehole locations. The trend of the alluvium–bedrock contact is interpolated from the three data points. The actual profile of this contact is not known but is probably much more irregular than shown. Resistance of bedrock at river level is shown on the right axis using the bedrock resistance scale adapted from Harden (1990)

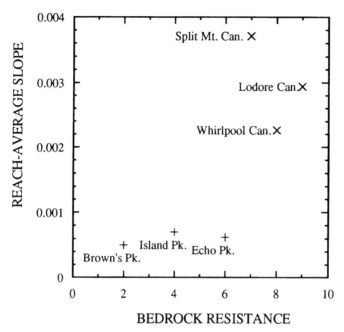

FIGURE 4.11   The relationship between resistance of bedrock exposed at river level and reach-average water-surface slope of the Green River

Bed material and channel form

The above observations indicate that bedrock geology exerts a strong influence on channel form within the study area. However, surficial geological mapping shows that in canyon reaches, bedrock and talus are the dominant bank material but occur along only about 42% of the total bank length (Table 4.4), and fine-grained alluvium occurs along about 30% of the bank length. An additional 14% of the banks are debris-fan material, 11% gravel and 4% mixed alluvium. Conversely, fine-grained alluvium is the dominant material in meandering reaches, occurring along 72% of the bank length.

Despite the abundance of bedrock and talus on the banks, rarely is the bed composed of these materials. Bed material at the surveyed cross-sections, which represent the portion of the bed exposed at low discharge, and borehole data indicate that the river does not flow directly on bedrock within the study area. Fine-grained alluvium is the dominant bed material at 39% of the 67 surveyed channel cross-sections in the study reach, and gravel is most abundant at 33% of the cross-sections (Table 4.4). A mixture of fine-grained alluvium and gravel covers 25% of the cross-sections and the remaining 3% have boulder and talus beds (Table 4.4). Fine-grained alluvium is more abundant on the channel bed in meandering subreaches and is less abundant in canyon subreaches. The distribution of bed material is similar in canyon subreaches upstream and downstream from the Yampa River confluence, indicating that bed material has not been significantly affected by flow regulation.

Borehole data show that the valley consists of river-deposited alluvium inset into bedrock. Drill holes completed for dam-site surveys at the entrance to Lodore Canyon (Wooley, 1930) and in Whirlpool Canyon (Merriman, 1941) show about 45 m of sand

TABLE 4.4 Abundance of stream bank material and bed material for each geomorphic subreach. Bank material determined by GIS analysis of surficial geological maps made for this study. Bed material determined from surveyed channel cross-sections. Fine-grained alluvium includes all sand, silt and clay

Parameter measured	Brown's Park	Lodore Canyon	Echo Park	Whirlpool Canyon	Island Park	Split Mountain Canyon	Canyons	Meanders	Study area
Length of bank (km)		41.5	10.0	16.7	32.9	10.0	68.1	43.0	111.1
Percentage of indicated material on bank*									
Gravel	na‡	5.2	6.4	18.9	6.4	19.0	11.1	6.4	9.7
Fine-grained		39.6	61.6	18.4	74.2	12.9	29.7	71.5	42.2
Mixed fine-grained and gravel		3.6	24.9	3.3	4.5	3.9	3.6	8.9	5.2
Debris		13.2	2.3	11.0	0.4	23.6	14.1	0.8	10.1
Bedrock and/or talus		38.5	4.8	48.3	14.5	40.7	41.6	12.4	32.9
Percentage of indicated material on bed†									
Gravel	0	40	0	25	54	0	33	32	33
Fine-grained	100	27	67	25	46	33	27	63	39
Mixed fine-grained and gravel	0	27	33	50	0	67	36	5	25
Boulders/talus	0	7	0	0	0	0	4	0	3

* Tabulated from surficial geological maps
† Tabulated from surveyed channel cross-sections
‡ No surficial geologic map data for Brown's Park

and gravel overlying bedrock. The dam-site survey near the mouth of Split Mountain Canyon shows 12 m of alluvium over bedrock (Merriman, 1940). The measured depths to bedrock are shown on the longitudinal profile in Figure 4.10. Individual rocks photographed in 1871 are still present in the same locations at river level (Stephens and Shoemaker, 1987), indicating that there has been no significant shift in mean bed elevation during the past century. The cross-section data and the borehole data support the observation that nowhere does the bed of the river flow in contact with bedrock, despite the occurrence of bedrock as bank material.

Debris fans, channel form and the occurrence of rapids

This characterization of the bed and bank material somewhat contradicts the notion that bedrock geology controls channel form. Stream channels must be in contact with bedrock if valleys are to be widened or deepened, or if bedrock is to control channel form directly. Bed and bank materials in the study area show that the valleys are not now deepening, but that widening may be occurring locally. Debris fans are a mechanism by which resistant bedrock may influence channel form indirectly. Debris fans are abundant in all high-gradient reaches and are uncommon in all low-gradient reaches (Figure 4.12). The canyon subreaches have an average debris-fan frequency of 3.1 fans per kilometre compared to 0.4 fans per kilometre in meandering reaches (Table 4.2). Although there are more fans per kilometre in Whirlpool Canyon, the fans are larger in Lodore and Split Mountain Canyons (Table 4.2). The subreaches that have the largest area of debris fans per kilometre of river are also the steepest (Figure 4.12).

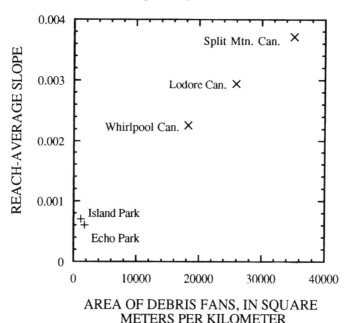

FIGURE 4.12   Relation between the area of debris fans and reach-average water-surface slope of the Green River. Debris fan area was measured from the surficial geological maps

TABLE 4.5   Channel constricting feature responsible for the formation of rapids at low discharge, by geomorphic subreach. Data collected from 1:5000 scale aerial photographs taken during low discharge

	Lodore		Whirlpool		Split Mountain		All canyons	
Cause of rapid	count	%	count	%	count	%	count	%
Debris fan only	35	66	19	73	12	75	66	70
DF–GB pair*	4	8	1	4	1	6	6	6
Gravel bar only	14	26	6	23	3	19	23	24
Totals	53	100	26	100	16	100	95	100

* Debris fan/gravel bar pair

The influence exerted by debris fans on channel form is also shown by the distribution of rapids within the study reach. We inventoried 95 rapids and riffles from aerial photographs, 45% more than Graf (1979) identified in the same reach using published river guides (Evans and Belknap, 1973; Hayes and Simmons, 1973). There are many rapids and riffles in the study area not marked on river guides. Of the 95 rapids identified, 76% (Table 4.5) are constricted by tributary fans, and 24% are constricted only by the expansion gravel bars downstream from debris fans. The percentage of rapids constricted by debris fans includes some rapids that are also impinged on the opposite bank by a gravel bar. Rockfall or a bedrock wall are frequently found on one bank at a rapid or riffle, but never on both banks; a debris fan or gravel bar is present at all constrictions. A

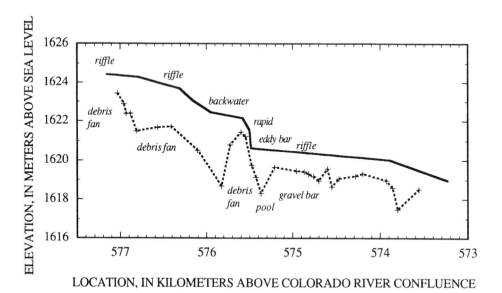

FIGURE 4.13   Water surface and bed profile through a fan–eddy complex at Rkm 575.99 (Winnie's Grotto) in Lodore Canyon. The solid line shows water surface and the dashed line shows the riverbed. The plus signs indicate measurement locations. Debris fans that occur in immediate succession sometimes lack the other features of a debris fan–eddy complex

majority of debris fans create rapids or riffles. A total of 110 debris fans were tallied on aerial photographs in the study area; 73% of these are co-located with rapids. Thus while all rapids are formed by constrictions that are caused by debris fans or downstream gravel bars, not all debris fans are associated with a rapid.

A detailed longitudinal profile of the bed and water surface through a fan–eddy complex illustrates the influence of the debris fan on local channel characteristics (Figure 4.13). These observations are similar to those made earlier by Leopold (1969) and Howard and Dolan (1981) in Grand Canyon. Pools are present upstream and downstream from the constriction, increasing depth by a factor of two to three. Depth is reduced by an equal amount at the constriction. Gravel bars are located downstream from the lower pool where thalweg depth is again reduced. The eddy occurs downstream from the constriction and adjacent to the pool.

Adjustment of river to coarse bed material

The gravel in expansion bars is derived primarily from the adjacent upstream debris fan within the same fan–eddy complex. This was investigated by a comparison of particle lithologies in debris fan–gravel bar pairs. Debris fan boulders are composed of lithologies from the contributing tributary basin, dominated by the most resistant lithologies found in that basin. For example, the debris fan at Rkm 563.6 is dominated by Uinta Mountain Group quartzite (70%) and shale (25%), Lodore Formation sandstone (5%), and rare cobbles of Madison Limestone. The expansion gravel bar immediately downstream has a similar ratio of lithologies with the more resistant formations more prevalent than the less resistant shale (Table 4.6). The outcrop area of Madison Limestone is much larger and the outcrop area of Uinta Mountain Group shale is much reduced in the tributary basin that contributes to the debris fan 1.5 km further downstream at Triplet Falls; these differences are reflected in the composition of the downstream expansion gravel bar (Table 4.6). The sandstone and limestone cobbles in these bars must be derived entirely from the immediately adjacent upstream basins, because they are the tributary basins furthest upstream in Lodore Canyon to contain outcrops of these rocks.

TABLE 4.6 Median surface-layer particle diameter and lithology of gravel bars within two fan–eddy complexes in Lodore Canyon

Site	Rkm 563.6*	Triplet Falls[†]
Location (Rkm)	563.6	562.6
D_{50} (mm)	115	130
Percentage of indicated lithology		
Uinta Quartzite	86	65
Uinta Shale	4	0
Lodore Sandstone	8	3
Madison Limestone	2	32

* Drainage basin has small outcrop of Madison Limestome and Uinta Group shale
[†] Drainage basin has large outcrop of Madison Limestone and no Uinta Group shale

The longitudinal variation of bed material sizes within the study area was compared with estimates of reach-average boundary shear stress. Shear stress was estimated from the product of the specific weight of water γ, the hydraulic radius R and the water-surface slope S:

$$\tau = \gamma RS \qquad (4.1)$$

Even in the relatively deep and narrow channel in the canyon reaches, R is well approximated by mean depth (cross-sectional area divided by channel-top width). The shear stress was calculated for the stage that inundates the most prominent geomorphic surface in the canyon and meandering reaches. This is a broad terrace with mature cottonwood trees in the meandering reaches and a narrow but distinct terrace with box elder trees in the canyons. Figure 4.14 is a typical cross-section in Lodore Canyon that shows this prominent terrace as well as the post-dam bankfull stage and corresponding floodplain. This prominent terrace was chosen because the post-dam bankfull-flow surface is much lower and not representative of the unregulated flow regime to which the gravel bars are presumably adjusted. Photographs taken during extreme events indicate that the prominent terrace is inundated by flows of about a 25-year recurrence interval in the pre-dam flow regime. Because this stage did not occur during our study, we used the US Geological Survey (1924) water-surface profile to estimate average slope for the 5 km reach centred at each cross-section. Boundary shear stress was compared with estimates of critical shear stress for median particle size of each gravel bar. A first-order approximation of flows necessary to mobilize the gravel bars was made by estimating critical shear stress using the Shields relation (Shields, 1936):

$$\tau_{c50} = \tau_{c50}^*(\gamma_s - \gamma_f)D_{50} \qquad (4.2)$$

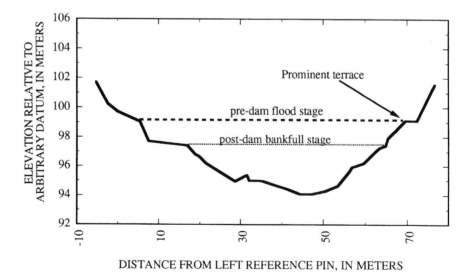

FIGURE 4.14   Example channel cross-section in Lodore Canyon at Rkm 563.0. The terrace height used to estimate the pre-regulation flood level is indicated by the heavy dashed line. This terrace is composed of fine-grained alluvium and is discontinuous through the study reach

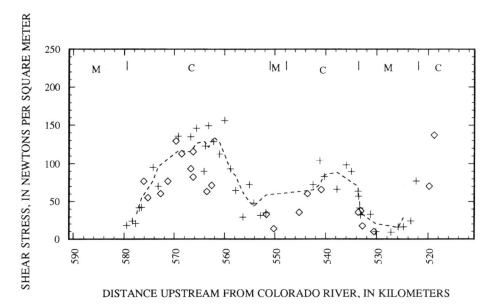

FIGURE 4.15 Comparison of average boundary shear stress at flood stage and critical shear stress of median grain size of gravel bars: +, estimated average boundary shear stress; ◊, critical shear stress. Canyon and meandering reaches are indicated by C and M, respectively. Average boundary shear stress was calculated for pre-dam bankfull flow using estimated bankfull surfaces to define channel geometry. Dashed line shows five-point moving average of average boundary shear stress. Critical shear stress was calculated from the Shields function using 0.033 for the dimensionless critical shear stress

where τ_{c50} is the critical shear stress in $(N\,m^{-2})$, τ_{c50}^* is the critical dimensionless shear stress for the median particle diameter of the bed surface, γ_s is the specific weight of the solid (in $N\,m^{-3}$), γ_f is the specific weight of water (in $N\,m^{-3}$), and D_{50} is the median particle diameter (in m) of each sampled bar. Values for τ_{c50}^* have been found to range over an order of magnitude and are affected by the bed material size distribution (Andrews, 1983). For this approximation we used 0.033, the value found by Andrews (1983) to be the most common for coarse-bedded streams.

The average boundary shear stress and the critical shear stress are larger in canyon reaches than in meandering reaches (Figure 4.15). The calculated shear stresses for the pre-dam 25 year flood approximate and sometimes exceed the estimates for critical shear stress in both canyon and meandering reaches.

DISCUSSION

Steep gradient and low channel width-to-depth ratio in the canyons of the eastern Uinta Mountains occur where river-level bedrock is most resistant and tributary debris fans are abundant. The river does not flow directly on bedrock, and less than 50% of the bank is bedrock or talus. The subreaches with the greatest area of debris fans per kilometre have the steepest gradients. Thus, lithology affects channel geometry and slope indirectly through the formation of debris fans composed of resistant boulders.

The presence of a thick alluvial fill indicates aggradation since the time when the bed of the river was bedrock. The present longitudinal profile may be either inherited from the profile established when the river did flow in contact with bedrock, or the result of differential aggradation. The profile could be inherited only if the river has aggraded the same amount everywhere. The detailed profile of the alluvium–bedrock interface is not known, but borehole data show that present depth to bedrock, and therefore aggradation since the river flowed in contact with bedrock, is not the same at each of the three measured sites. Aggradation may have occurred in response to a decrease in mainstem stream power following deglaciation of the surrounding mountains or to an increase in tributary sediment input by debris flows. Either cause is climate-related. As sediment accumulated in the canyons, local slope increased until streamflow competence matched the characteristics of tributary supply. Rapids and riffles, where gradient is steepest, all occur at constrictions formed by tributary debris fans or over gravel bars in flow expansions that consist of reworked debris-fan material. This is consistent with our observation that the gravel bars with the largest mean particle sizes occur in the reaches of highest flood-flow average boundary shear stress.

Comparisons between critical shear stress and average boundary shear stress suggest that reworking of gravel bars and downstream transport of coarse-grained alluvium probably occurs only during floods on the order of the pre-regulation 25-year recurrence interval. Observations of lithologies contained in gravel bars support the conclusion that gravel is only rarely transported through the study reach under either the current regulated streamflow regime or the historic unregulated streamflow regime. The lithological composition of gravel bars should reflect the distribution of lithologies of all upstream debris-contributing tributaries, not just the nearest upstream tributary basin, if gravel was being transported through the canyons. Our limited investigation of lithologies contained in two debris fan–gravel bar pairs indicates that gravel bars are dominated by the resistant lithology most abundant in the tributary basin immediately upstream. Thus in the debris fan-dominated canyons where deposition is determined by local fan–eddy complex depositional environments, sorting is determined by local hydraulic processes and possibly local abrasion (from debris fan to gravel bar) rather than downstream hydraulic sorting and abrasion. If abrasion were the dominant process, we would expect to find a downstream fining of lithologies, which did not occur.

Like other debris fan-dominated canyons of the Colorado Plateau (Schmidt and Rubin, 1995), fan–eddy complexes are the fundamental channel-organizing unit on the Green River in the canyons of the eastern Uinta Mountains. More than 60% of all fine- and coarse-grained alluvium is stored within these depositional units either as eddy bars, channel-margin deposits in backwaters, or expansion bars. This is in contrast to the meandering reaches, which store at least an order of magnitude more alluvium per kilometre and have no fan–eddy complexes.

The geomorphic organization of the Green River in the eastern Uinta Mountains is essentially similar to that of the Colorado River in Grand Canyon. In both systems, channel geometry and gradient vary according to the lithology exposed at river level. Debris fan frequency is higher in the eastern Uinta Mountains: there are 3.3 fans per kilometre on average in the eastern Uinta Mountains but fans are never more frequent than 2.9 per kilometre in Grand Canyon (Schmidt and Leschin, 1995). Debris fans have a strong influence on sedimentation patterns in both systems. The percentage of fine-grained alluvium contained in eddy bars is between 34 and 63% in the canyons in the eastern Uinta Mountains. Schmidt and Rubin (1995) showed that this percentage varied between 44 and 75% in Grand Canyon and between 1 and 29% elsewhere on the Green River.

The largest difference between eddy bars in the Green River and Grand Canyon systems is the lack of differentiation between separation and reattachment bars in the eastern Uinta Mountains. This may be an indication of the relative abundance of fine-grained sediment with respect to the current streamflow regime. This is consistent with the observation made by Schmidt and Rubin (1995) that, prior to the closure of Glen Canyon Dam, many eddies in Grand Canyon were filled with sand and separation and reattachment bars were merged.

The spacing of rapids is not regular, but is controlled by debris fans in both canyon systems. Other major channel elements such as gravel bars, sand bars and pools are located in association with debris fans. Tributary basin geology and hydrology, which determine debris fan size and location, strongly influence mainstem channel morphology. Deposits such as sand and gravel bars found in each specific fan–eddy complex create a quasi-adjustable, self-formed alluvial channel inset within the bedrock/talus canyon.

CONCLUSIONS

Bedrock lithology controls river form indirectly. The basic geomorphic characteristics of streams in canyons with debris fans are determined by the tributary sediment delivery processes. Longitudinal profile, channel geometry and the occurrence of rapids in the canyons of the eastern Uinta Mountains are each strongly influenced by tributary-fan frequency and average area. Bankfull channel width-to-depth ratio is lowest and gradient is steepest in the reaches that have the largest area of debris fans per kilometre of river; and all rapids are caused by debris fans or the gravel bars downstream from debris fans that are composed of reworked debris-fan material.

Expansion bars are the location of the greatest area of the coarse-grained alluvium in debris fan-dominated canyons. These gravel bars are located in the flow expansion downstream from debris fan-created channel constrictions and recirculation zones. The lithology of gravels in these bars indicates that their source is the debris fan immediately upstream and its associated tributary basin. This is an indication that the process of local sorting outweighs downstream sorting in these canyons. Estimates of average boundary shear stress for high flows and critical shear stress of gravel bars show that the channel gradient and bar-material size are in approximate adjustment with pre-dam flood conditions in both the canyon and meandering reaches of the study reach. Although the river flows alternately through reaches of extremely different geomorphic character, both are in a quasi-equilibrium condition.

Between 60 and 85% of the surface area of the alluvium in the canyon bottom exposed at low flow is contained in fan–eddy complexes. About 42% of all fine-grained alluvium in the canyons is stored in eddy bars within fan–eddy complexes. However, most of the total area of alluvium is contained in the meandering reaches where fan–eddy complexes are less important.

The fan–eddy complexes of the Green River in the eastern Uinta Mountains are similar to those described on the Colorado River in Grand Canyon by Schmidt and Rubin (1995).

Approximately the same areal proportion of fine-grained alluvium is stored as eddy bars within these depositional units in both systems.

ACKNOWLEDGEMENTS

This research was funded by the US National Park Service. Field assistance provided by E.D. Andrews, C. Bilbrough and many other volunteers is greatly appreciated. Earlier versions of this manuscript were reviewed by D.S. Kaufman, A.J. Miller, M.P. O'Neill, J. Pitlick and one anonymous reviewer. Their comments, and insightful conversations with W.E. Dietrich, significantly improved this manuscript.

REFERENCES

Andrews, E.D. 1983. Entrainment of gravel from naturally sorted riverbed material. *Geological Society of America, Bulletin*, **94**, 1225–1231.

Andrews, E.D. 1986. Downstream effects of Flaming Gorge Reservoir on the Green River, Colorado and Utah. *Geological Society of America, Bulletin*, **97**, 1012–1023.

Baker, V.R. 1984. *Flood sedimentation in bedrock fluvial systems*. Canadian Society of Petroleum Geologists, Memoir 10, 87–98.

Bradley, W.H. 1936. Geomorphology of the north flank of the Uinta Mountains. *US Geological Survey Professional Paper* 185–I, 204 pp.

Church, M. and Jones, D. 1982. Channel bars in gravel–bed rivers. In R.D. Hey, J.C. Bathurst and C.R. Thorne (Eds), *Gravel-bed Rivers: Fluvial processes, engineering, and management*. New York, Wiley, 291–338.

Davis, W.M. 1897. Current notes on physiography, Is Green River antecedent to the Uinta Mountains. *Science*, New Series, **5**(121), 647–648.

Dolan, R., Howard, A. and Trimble, D. 1978. Structural control of the rapids and pools of the Colorado River in the Grand Canyon. *Science*, **202**, 629–631.

Evans, L. and Belknap, B. 1973. *Dinosaur River Guide*. Westwater Books, Boulder City, Nevada, 64 pp.

Graf, W.L. 1979. Rapids in canyon rivers. *Journal of Geology*, **87**, 533–551.

Hansen, W.R. 1986. Neogene tectonics and geomorphology of the Eastern Uinta Mountains in Utah, Colorado, and Wyoming. *US Geological Survey Professional Paper* 1356, 78 pp.

Hansen, W.R., Kinney, D.M. and Good, J.M. 1960. Distribution and physiographic significance of the Browns Park Formation, Flaming Gorge and Red Canyon areas, Utah-Colorado. *US Geological Survey Professional Paper* 400–B, 257–259.

Hansen, W.R., Rowley, P.D. and Carrara, P.E. 1983. Geologic map of Dinosaur National Monument and vicinity, Utah and Colorado. *US Geological Survey Miscellaneous Investigations Map* I-1407, scale 1:50,000, 1 sheet.

Harden, D.R. 1990. Controlling factors in the distribution and development of incised meanders in the central Colorado Plateau. *Geological Society of America, Bulletin*, **102**, 233–242.

Hayes, P.T. and Simmons, G.C. 1973. *River runners' guide to Dinosaur National Monument and vicinity with emphasis on geologic features*. Powell Society, Denver, 78 pp.

Howard, A. and Dolan, R. 1981. Geomorphology of the Colorado River in the Grand Canyon. *Journal of Geology*, **89**(3), 269–298.

Hunt, C.B. 1969. Geologic history of the Colorado River. *US Geological Survey Professional Paper* 669–C, 59–130.

Jackson, R.G. 1975. Hierarchical attributes and a unifying model of bed forms composed of cohesionless material and produced by shearing flow. *Geological Society of America, Bulletin*, **86**, 1523–1533.

Jefferson, M.S.W. 1897. Discussion and correspondence, the antecedent Colorado. *Science*, New Series, **6**(138), 293–295.

Kieffer, S.W. 1985. The 1983 hydraulic jump in Crystal Rapid: Implications for river-running and geomorphic evolution in the Grand Canyon. *Journal of Geology*, **93**(4), 385–406.

Leopold, L.B. 1969. The rapids and the pools – Grand Canyon. *US Geological Survey Professional Paper* 669–D, 131–145.

Melis, T.S., Webb, R.H., Griffiths, P.G., McCord, V.A.S. and Wise, T.J. 1993. *Magnitude and frequency data for debris flows in Grand Canyon National Park and vicinity, Arizona.* Water Resources Investigative Report 94–4214, US Geological Survey, 144 pp.

Merriman, M. 1940. *Preliminary geological report, Split Mountain Canyon dam sites.* US Department of the Interior, Bureau of Reclamation, Split Mountain Power Investigations, Salt Lake City, 14 pp.

Merriman, M. 1941. *Geological Report, Echo Park dam site.* US Department of the Interior, Bureau of Reclamation, Middle Green River Investigations, Salt Lake City, 12 pp.

Powell, J.W. 1875. *Exploration of the Colorado River of the west and its tributaries.* US Govt. Printing Office, Washington, DC, 291 pp.

Rubin, D.M., Schmidt, J.C. and Moore, J.N. 1990. Origin, structure, and evolution of a reattachment bar, Colorado River, Grand Canyon, Arizona. *Journal of Sedimentary Petrology*, **60**, 982–991.

Schmidt, J.C. 1990. Recirculating flow and sedimentation in the Colorado River in Grand Canyon, Arizona. *Journal of Geology*, **98**, 709–724.

Schmidt, J.C. and Graf, J.B. 1990. Aggradation and degradation of alluvial sand deposits, 1965 to 1986, Colorado River, Grand Canyon National Park, Arizona. *US Geological Survey Professional Paper* 1493, 74 pp.

Schmidt, J.C. and Leschin, M.F. 1995. *Geomorphology of post-Glen Canyon Dam fine-grained alluvial deposits of the Colorado River in the Point Hansbrough and Little Colorado River confluence study reaches in Grand Canyon National Park, Arizona.* Investigative Report, US Bureau of Reclamation, Glen Canyon Environmental Studies, Flagstaff, Arizona, 70 pp.

Schmidt, J.C. and Rubin, D.M. 1995. Regulated streamflow, fine-grained deposits, and effective discharge in canyons with abundant debris fans. In J.E. Costa, A.J. Miller, K.W. Potter and P.R. Wilcock (Eds), *Natural and anthropogenic influences in fluvial geomorphology.* AGU Geophysical Monograph 89, 177–195.

Schmidt, J.C., Rubin, D.M. and Ikeda, H. 1993. Flume simulation of recirculating flow and sedimentation. *Water Resources Research*, **29**(8), 2925–2939.

Sears, J.D. 1924. Relations of the Browns Park formation and the Bishop conglomerate, and their role in the origin of Green and Yampa Rivers. *Geological Society of America, Bulletin*, **35**, 279–304.

Shields, A. 1936. Application of similarity principles and turbulence research to bed-load movement: Mitt. Preuss. Verschsanst., Berlin, Wasserbau Schiffbau. In W.P. Ott and J.C. Uchelen (translators), Report 167, California Institute of Technology, Pasadena, California, 43 pp.

Smith, J.D. and Wiele, S. 1995. Flow and sediment transport in the Colorado River between Lake Powell and Lake Mead (unpublished manuscript). US Geological Survey, Boulder, Colorado.

Stephens, H.G. and Shoemaker, E.M. 1987. *In the footsteps of John Wesley Powell.* The Powell Society, Denver, Colorado, 286 pp.

US Geological Survey 1924. *Plan and profile of the Green River from Green River, Utah to Green River, Wyoming.* US Government Printing Office, Washington, DC.

Webb, R.H., Pringle, P.T. and Rink, G.R. 1989. Debris flows from tributaries of the Colorado River, Grand Canyon National Park, Arizona. *US Geological Survey Professional Paper* 1492, 39 pp.

Wilson, R.P. 1986. Sonar patterns of Colorado riverbed, Grand Canyon. In *Proceedings, Fourth Federal Interagency Sedimentation Conference*, Las Vegas, Nevada, March, 1986, Vol. 2, 5-133–5-144.

Wolman, M.G. 1954. A method of sampling coarse riverbed material. *American Geophysical Union Transactions*, **35**(6), 951–956.

Wooley, R.R. 1930. The Green River and its utilization. *US Geological Survey Water-Supply Paper* 618, 456 pp.

5

The Narmada River, India, Through Space and Time

Avijit Gupta[1], Vishwas S. Kale[2] and S.N. Rajaguru[3]

[1] School of Geography, University of Leeds, UK
[2] Department of Geography, University of Poona, Pune, India
[3] Department of Archaeology, Deccan College, Pune, India

ABSTRACT

The Narmada River of central India flows through bedrock reaches alternating with alluvial meandering sections. This paper is based on parts of the upper and middle sections of the river. In the bedrock reaches, the Narmada usually flows through an inner channel with scablands at a higher level. In the alluvial sections, the river has a channel-in-channel physiography with high alluvial banks enclosing a large channel inside which occur point bars, a smaller channel and patches of floodplain.

The Narmada is influenced by three geomorphic processes: high-magnitude floods, high flows of the wet monsoon, and regional tectonism. High-magnitude floods, which fill the entire channel, have occurred several times in this century. These floods constitute the dominant channel process for the river, and their effects are only partially modified by the high flows of the wet monsoon. Although the high-magnitude floods control both the rocky and alluvial reaches, bed shear stress and unit stream power are much higher in the bedrock sections which are eroded only during such large floods. The derived sediment is transported downstream to be deposited in the alluvial reaches, mostly as sand. In such alluvial sections, the large floods maintain the huge channel and reorganize sediment transfer and storage. The sediment in storage ranges in size from boulders larger than 2 m in the rocky gorges to sand in the alluvial sections. Fines only occur as slackwater deposits in sheltered places. Tectonism, at least partially, determined the location of the course of the Narmada.

The Late Pleistocene alluvium of the Narmada indicates that the present river is distinctly different from its Pleistocene ancestor. The palaeoflood data suggest that the large floods started only in the Holocene. It is possible that an earlier, freely meandering river flowing on its alluvial fill was modified to form the present Narmada confined by high alluvial banks and rock gorges. It is further suggested that regional tectonism might have helped to establish the channel of the Narmada in the gorge sections.

This study of the Narmada, besides providing an account of a fascinating river, also reviews the role of high-magnitude floods in forming and maintaining river channels. Furthermore, it suggests that at least some of the present-day Indian rivers could have

Varieties of Fluvial Form. Edited by A.J. Miller and A. Gupta

acquired their forms and behaviour subsequent to climatic changes in the Holocene. This conclusion of course requires more case studies before its general acceptance.

INTRODUCTION

The Narmada River alluvium and the neotectonics of its basin have been studied for nearly 150 years by geologists and archaeologists, but publications on the river itself began to appear only in the last few years (Bedi and Vaidyanadhan, 1982; Kale et al., 1993, 1994; Rajaguru et al., 1995; Ely et al., 1996). The Narmada is primarily a flood-controlled river, although tectonic movements and seasonality of discharge also modify its channel forms and processes. We discuss (1) the channel pattern, morphology, and the operating processes of the upper and middle Narmada from Jabalpur to Rajghat (Figure 5.1); (2) the sources, transfer and storage of the sediment in the channel; and (3) the Quaternary history of the river. We also explore (1) whether the form and behaviour of the Narmada are common characteristics of Indian monsoon rivers; and (2) whether the present characteristics of the rivers can be traced back through the Quaternary.

THE DRAINAGE BASIN

Geology

The long and narrow 99 000 km^2 basin and the 1300 km course of the Narmada are located in an east–west structural low which is part of a larger lineation (the Son–Narmada–Tapi trough). Tectonic activity in the lineation has resulted in a number of cross-faults and faults running near-parallel to the basin axis. The lineament continues to be seismically active (Kale, 1989; Brahman, 1990; Verma, 1991).

   Table 5.1 summarizes the basin geology which determines the nature of Narmada's sediment. The Vindhyan Mountains and the Satpura Range (which mark the northern and southern boundaries of the basin, respectively) are formed of sandstones, conglomerates and shales of the Cambrian and Triassic periods. The rest of the basin is primarily on Cretaceous–Eocene Deccan Trap lavas. The Upper Narmada flows over an alluvial deposit of variable thickness (up to a maximum of about 300 m) which occurs over an undulating basement surface (Rajaguru et al., 1995). Rocks of various types are locally exposed in the channel of the Narmada, forming gorges, rapids and scablands. The thickly bedded and well jointed Archaean metamorphics and dolomites contribute boulder-sized material in the gorge sections, as do the Deccan Lava beds. Most of the boulders disintegrate into sand which is added to the sand derived from the Vindhyan and Gondwana rocks of the Vindhyan and Satpura Mountains. The Deccan Lava boulders ultimately break down into clay-sized particles.

Relief

The narrow, elongated basin of the Narmada consists of an alluvial valley flat (20–70 km in width) between two mountain ranges: the Vindhyans to the north and the Satpuras to the south. The Satpura Ranges are higher (800–1350 m) than the Vindhyans (<900 m), but the northern slopes are steeper, as the river is located closer to the Vindhyan

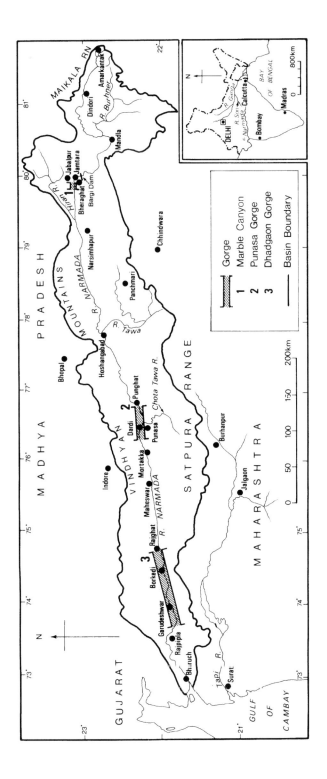

FIGURE 5.1 Location map

TABLE 5.1   The Narmada Basin: generalized stratigraphic succession

Age	Formation, etc.	Description
Quaternary	Alluvium	River valley fill, alluvial fans, Pleistocene valley fills
Eocene–Cretaceous	Deccan Traps	Horizontal basalt flow with dikes, sills and intertrap beds
Cretaceous	Bagh and Lameta beds	Fossiliferous marine and continental sediment (conglomerate, sandstone, limestone and shale); limited areal extent
Cretaceous–Triassic	Gondwana	Conglomerate, sandstone, and shale
Cambrian	Vindhyan	Deep red or purplish sandstone, conglomerate and slate
Archaean	Mahakoshal	Dolomitic marble, metamorphic rocks, dykes and breccias
Archaean	Archaean Group	Metamorphic rocks, granitoid and metasediment

Mountains. The alluvial valley flat appears to be built mostly of riverine sediment, although we do not know about deposition at depth, and near the two mountains mass movement deposits occur.

The headwaters of the Narmada (Figure 5.1) are located in the Maikala Range at an elevation of 1057 m. This is the only part of the basin where a well established concentration of drainage network occurs. Here the channel of the Narmada is mostly in bedrock, relatively narrow (230–320 m wide), about 20 m deep, and winding with several hairpin turns. The average channel gradient above Jabalpur is 0.003. Downstream from Jabalpur, three bedrock sections with gorges alternate with alluvial reaches. Near Jabalpur, the river enters the narrow 30 m deep Marble Canyon which is cut into dolomitic limestones of Archaean age. Upon exiting the gorge the river flows through a 450 m wide meandering channel with 10–15 m high banks in the first of the wide alluvial reaches. The average gradient of the Narmada in this alluvial reach ranges between 0.00002 and 0.00006. The channel widens irregularly to 755 m near Hoshangabad at the end of the alluvial reach.

Downstream from Hoshangabad, the Narmada turns abruptly south and enters the second rocky section which includes Punasa Gorge in Vindhyan quartzites and Deccan Trap basalts. Low hills approach the Narmada and two waterfalls (Punghat and Dardi) are located in the gorge. Punasa Gorge is followed downstream by the second alluvial reach extending up to Rajghat. The channel width in the alluvial reach varies between 740 and 800 m. The Narmada enters the final gorge (Dhadgaon Gorge) in Deccan Trap basalts. This gorge has a uniform width of nearly 700 m. Here the valley is at its narrowest and the Narmada flows through a long bedrock gorge which directly separates the hills of the north and south divides. Near Rajpipla, the Narmada emerges from this gorge into the coastal plains of Gujarat and enters the sea near Bharuch in a series of wide meanders.

We have studied the Narmada from Jabalpur to Rajghat which includes two bedrock and two alluvial reaches. The gradient of the Narmada changes between such sections. Over alluvium, the average gradient could be very low (0.00006), but over the bedrock, where locally the Narmada drops between 0.5 and 10 m in rapids and falls, the slope is steeper than 0.02, quite a high figure for a river of this size.

River hydrology

The Narmada is a rainfed river. The annual rainfall decreases westward from more than 1700 mm east of Jabalpur to about 800 mm near Badwani (Rajghat). The rainfall then increases to 1250 mm towards the coast near Bharuch. Nearly 90% of this rain arrives between June and September with the wet monsoon, 80% of the annual rain falling in 81 days. In an average year about 2% of the annual total comes from the one-day extreme rainfall, but this percentage increases considerably in some years (Indian Meteorological Department, 1981; Singh et al., 1988). One-day of rainfall between 300 and 500 mm has been recorded at several stations (Abbi et al., 1970).

Occasionally, very high rainfall occurs in the middle of the wet season. Tropical storms and low-pressure systems arrive from the Bay of Bengal commonly in August and September and travel downbasin (Figure 5.2), contributing to the floodprone nature of the Narmada. Although on average one monsoon storm or depression arrives over the basin each year, only a few of these produce high-magnitude floods. Again, these storms may traverse only a part of the basin, and rainfall and flood sizes from the same storm may vary over different reaches of the river. The best example of this is the storm of 1926 (Figure 5.2) which provided the highest flood markers of the upper Narmada. This storm crossed the upper basin, south to north, and did not particularly affect the lower basin. The effect of a single large flood therefore may not always be comparable along the Narmada. However, when a tropical cyclone moves directly westward down the basin axis, the flood effects are continuous and enhanced, as the peaks in storm rainfall and river discharge may move in conjunction down the channel. Such low-pressure systems may take several days to traverse the basin. The August 1968 and the September 1970 storms (Figure 5.2) did just that. The storm of 5–7 September 1970 produced three-day totals of more than 400 mm of rainfall over parts of the lower basin (Ramaswamy, 1987).

Three gauging stations on the Narmada hold discharge records for the last four decades, but the flood record has been extended further by field observation, searching of administration records, and oral history (Kale et al., 1994; Rajaguru et al., 1995). Three floods in this century (1923, 1926 and 1991) either filled the high bank channel of the upper Narmada or overtopped the river cliffs. Below Hoshangabad such channel-filling floods were recorded in 1961, 1968, 1970 and 1984. Historical records indicate that very large and devastating floods, affecting at least part of the Narmada channel, occurred earlier in the following years: 1818, 1823, 1832, 1854, 1855, 1868, 1878, 1891, 1893, 1894, 1895, 1898, 1905, 1907, 1923, 1926, 1937, 1944 and 1945.

Of the three gauging stations on the Narmada, the records at Mortakka (catchment area 67 000 km^2) and Garudeswar (89 000 km^2) go back 45 years. Monsoon peak flows at Mortakka (Figure 5.3) commonly range between 10 000 and 30 000 m^{3}s^{-1}, but peak discharges ranging between 40 000 and 55 000 m^{3}s^{-1} have been recorded in extreme floods. At Garudeshwar, the largest recorded flood (but not the highest one in historic

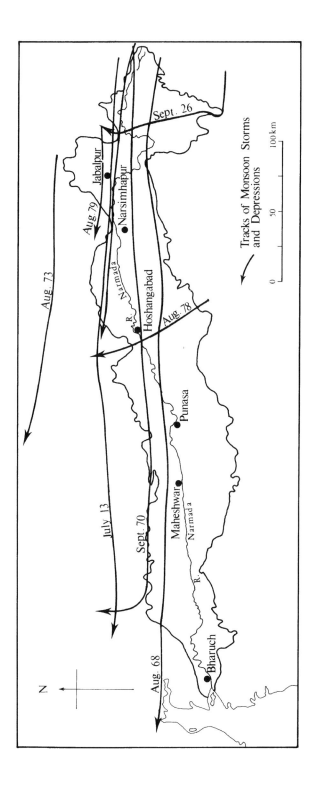

FIGURE 5.2 Tracks of tropical storms from the Bay of Bengal across the Narmada Basin. Source: Publications of the Indian Meteorology Department

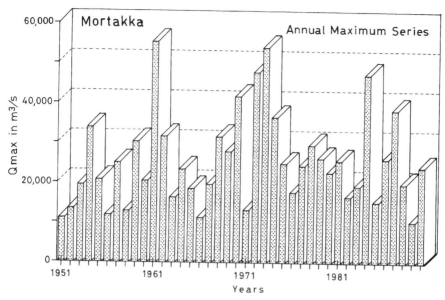

FIGURE 5.3   Peak annual discharge at Mortakka gauge. Data from Central Water Commission, New Delhi

times) occurred in September 1970 with a peak discharge of 59 400 m^3 s^{-1}. This gives a unit discharge of 0.66 m^3 s^{-1} km^{-2}. Unit discharges of large floods ranging between 0.49 and 1.25 m^3 s^{-1} km^{-2} (Kale et al., 1994) fall very close (Figure 5.4) to the envelope curve for global maximum floods as shown by Baker (1995). Maximum depths, estimated from hydraulic modelling and available gauge data, range between 13 and 60 m (depending on location) and maximum flood velocities could be locally as high as 4 to more than 11 m s^{-1} (Kale et al., 1994; Rajaguru et al., 1995). Palaeoflood studies from the central Narmada Basin show that floods larger than this have not occurred in the last 3000 years at least (Ely et al., 1996).

CHANNEL FORMS AND PROCESSES

The bedrock and alluvial channels of the Narmada differ strikingly in their morphology and scale of operating processes. Although both types of channel are primarily flood-controlled, the alluvial reaches show evidence of modification by flows of the wet monsoon. It is therefore necessary to describe the reaches separately; however, water and sediment flow from one type of reach to another, and in both, a local subreach may be partly in bedrock and partly in alluvium. The channel morphology along this long river varies strikingly, both locally and regionally, although the processes remain the same.

Bedrock channel

The channel of the Narmada in bedrock passes either through gorges of varying width and with waterfalls and rapids or through very wide sections with multiple low-flow channels separated by rock exposures.

120

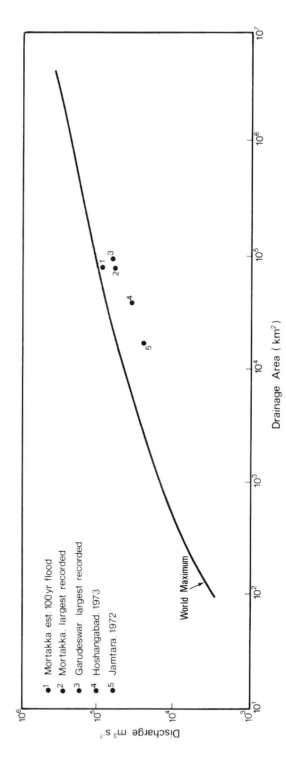

FIGURE 5.4 World flood envelope curve after Baker (1995) with values for the Narmada River plotted for comparison

Large floods in bedrock channels can lead to intense erosion arising out of macro-turbulent flow phenomena such as cavitation (Baker and Costa, 1987). Channel velocities for the Narmada computed by hydraulic modelling (Kale et al., 1994) indicate that such phenomena are possible during large floods. In the gorge sections, steep channel gradients and irregular rough boundaries capable of generating flow separation (Matthes, 1947) favour the development of kolks or macro-turbulent flows. Such flows are capable of producing a variety of erosional features seen in the bedrock reaches of the Narmada.

The slope of the river, as mentioned earlier, is steeper in bedrock sections and the highest flood depths vary between 13 m at a wider section at Guwarighat near Jabalpur and 60 m at the confined gorge of Punasa. As a result, bed shear stresses range between 300 and 3000 N m^{-2} and power per unit area between 2600 and 12 800 W m^{-2} (Rajaguru et al., 1995). The numbers towards the higher ends of these ranges are capable of bedrock erosion, and transportation of boulders as bedload and cobbles as suspended load (Williams, 1983; Baker and Pickup, 1987; Baker and Kochel, 1988). Theoretically, the high-magnitude floods which filled the channels several times in this century had the power to cause large-scale modification of the rock-cut sections.

The physical manifestations of such erosive and transporting capabilities are seen in the field at both Marble Canyon and Punasa Gorge. At Bheraghat near Jabalpur, the river flows across an exposure of Archaean Dolomite in a 15 m waterfall with a rock pavement on either side. The pavement is deeply scoured at various places. Downstream, the Narmada enters a 3 km gorge in dolomitic marble (Figure 5.5). Boulders in dolomite lie scattered over the surface of the bars about 1 km downstream of the gorge where the channel is wider. The second gorge near Punasa (Figure 5.6), where the river crosses both Vindhyan metaquartzites and Deccan Lava flows, exhibits an inner gorge flanked by pavements. The eroded nature of the pavement has caused it to be described as scabland.

FIGURE 5.5   Gorge in dolomite at Bheraghat; high-magnitude floods have been known to fill the entire gorge

FIGURE 5.6   The Narmada River at Punasa, showing bedrock erosion

Slabs of basalt and quartzites are detached along their bedding planes and joints (Figure 5.7), and cubic metre-sized blocks are disintegrated into sand by transportation downstream.

The best examples of high-magnitude flood markers, scablands, an inner gorge, flood overflow channels and boulder berms of imbricated 2 m boulders on top of the scabland, occur immediately downstream of the 10 m Dardi Falls towards the end of Punasa Gorge (Figure 5.8). Immediately upstream from the Dardi Falls, the Narmada is flanked by 10–15 m high cliffs in Pleistocene alluvium with both point bars and mid-channel bars in sand and gravel in the channel. Downstream from the waterfall, a scabland in bare metaquartzite, nearly 1 km wide, emerges from below the Pleistocene alluvium with an approximately 165 m wide inner gorge in its middle. The inner gorge appears to be the

FIGURE 5.7   The Narmada River, showing eroded blocks

creation of a retreating knickpoint as described by Wohl (1993) for bedrock streams in Australia.

Here, the Narmada flows through a shear zone, and the Archaean metaquartzites are not only thickly bedded (about 1 m), but also display a high degree of jointing and shearing. As a result, up to 2 m cuboid blocks are detached (Figure 5.9) by the river in high-magnitude floods and stacked like cushions imbricated upstream (Figure 5.10).

The forms on top of the scabland at Dardi are all flood features (Figure 5.11): outflow channels cut into metaquartzites; potholes with diameters up to 1.5 m containing pebble and cobbles; one 50 × 25 m boulder berm; scattered piles of half a dozen 2 m imbricated boulders; polished rock surfaces; grooves; and flute marks. Tributary channels which cut through the Pleistocene alluvium are graded to the top of the rock scabland and display slackwater deposits immediately inside the tributary channels. The surface of the scabland does not show any depositional feature except the boulder berm and piles of several imbricated boulders. Given such a smooth surface and depths in tens of metres (the depth in the gorge has been estimated to be 60 m), the flood velocity is certainly capable of very intense erosive phenomena and quick energy dissipation in non-uniform flow conditions. Calculations based on empirical equations derived by Williams (1983) show that the minimum bed shear stress, unit stream power and mean velocity required to transport the biggest boulder measured are 408 $\mathrm{N\,m^{-2}}$, 1550 $\mathrm{W\,m^{-2}}$ and 3.18 $\mathrm{m\,s^{-1}}$, respectively. Such figures are exceeded in the gorge in high-magnitude floods. The transported boulders are deposited downstream of a sudden change in slope or channel cross-section. Any material smaller than boulders could be carried in such floods beyond the gorge into the downstream alluvial section with lesser gradient. Cobbles, pebbles and sand are not seen in storage in either of the two gorges, except in sheltered places such as inside potholes, on rock ledges or in the lee of boulder berms. Small boulders, cobbles and pebbles can be seen to move in the gorges even during the high flows of the wet monsoon.

124

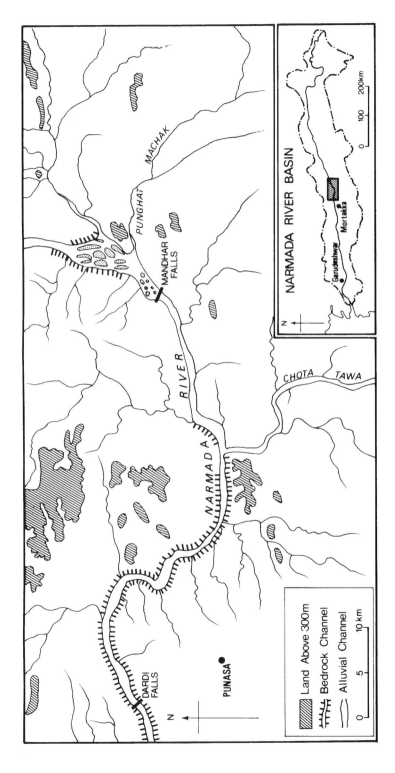

FIGURE 5.8 Map of Punasa Gorge showing the Dardi Falls and other geomorphological features. From Kale, V.S., Baker, V.R. and Mishra, S. 1996. Multi-channel patterns of bedrock rivers: an example from the central Narmada basin, India. *Catena*, **26**, 85–98. With kind permission of Elsevier Science – NL, Sara Burgerhartstraat 25, 1055 KV Amsterdam, The Netherlands

FIGURE 5.9   Jointed rocks of a shear zone are detached into metre-sized blocks by flood flows on the scabland

Different channel characteristics emerge where the river is flowing through bedrock lithologies other than dolomite or metaquartzite. The Narmada crosses lava flows in channels cut into rock scablands at several places. Near Barmanghat this scabland is extremely wide (about 900 m) and the Narmada splits into several channels separated by deposition of river sediment on high points of lava scabland which form mid-channel bars. Signs of erosion and scattered boulders and cobbles are common across the

FIGURE 5.10   Dardi, showing imbricated 2 m boulders in the inner channel

FIGURE 5.11   Dardi, showing rocky scabland and inner gorge below the waterfall; banks in alluvium

scabland. Similar intense bedrock erosion has occurred at other places: Mandla, Punghat (upstream of Punasa) and Borkedi (Figure 5.1).

Near Punghat the Narmada crosses, in an anomalous multiple-channel pattern (Figure 5.12), Archaean gneissic granite and granitic gneiss with large faults, joints and numerous dykes. The 8.5 km reach is inexplicably wide (800–2750 m). The Narmada outside this stretch is about 700–800 m across. Kale et al. (1996) have divided the reach into three parts: (1) an upstream subreach characterized by deep flows and relatively fine sediment; (2) the widest subreach with thickly forested islands and boulder berms in the middle; (3) a steep downstream subreach with rapids. It has been hypothesized that the anomalous pattern was initiated by domal uplift which caused high-magnitude floods to exploit linear weaknesses in the bedrock over steep gradients. This led to anabranching and establishment of multiple bedrock channels in a very wide reach replacing the meandering pattern of the Narmada. Abrupt changes in the channel morphology and planform suggest that adjustment to uplift is continuing via high-magnitude floods (Kale et al., 1996). It should be remembered that such features also occur at Mandla and Borkedi.

Such drastic course changes are only possible in very large floods or during tectonic disturbances. The abrupt change of the course of the Narmada abandoning its former alluvial valley to flow across an inlier of Archaean dolomite at Bheraghat, the association with shear zones at Dardi and other places in Punasa Gorge, and the very wide channel at Punghat cumulatively suggest that tectonism has played a part in shaping the bedrock reaches of the Narmada.

Alluvial channel

The alluvial Narmada is a meandering river. The sinuosity index of the Narmada increases to 1.5 locally in contrast to that in the bedrock reaches, where it is close to 1.

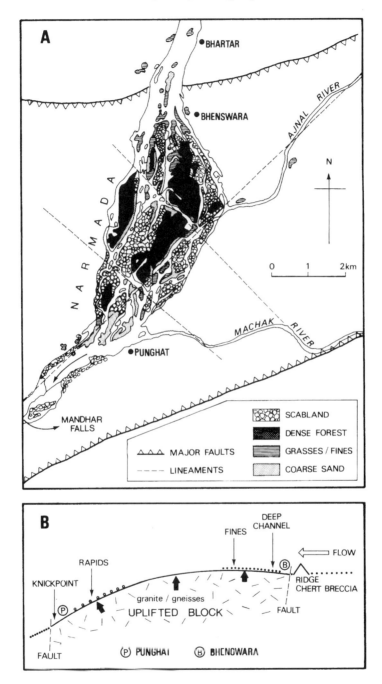

FIGURE 5.12   Characteristics of the wide reach near Punghat. From Kale, V.S., Baker, V.R. and Mishra, S. 1996. Multi-channel patterns of bedrock rivers: an example from the central Narmada basin, India. *Catena*, **26**, 85–98. With kind permission of Elsevier Science – NL, Sara Burgerhartstraat 25, 1055 KV Amsterdam, The Netherlands

The alluvial fill, over which the Narmada flows, is generally thick (at places up to 300 m) but this thickness is quite variable. Even within an alluvial reach the river had to incise its bed across several rocky bands. The entire channel of the river cuts through small outcrops of Vindhyan or Gondwana sandstones in a series of rapids. An inner channel has not been excavated. These rock-cut sections act as fixed points in the middle of the alluvial reach. The Narmada therefore cannot move freely across the valley flat until its course is changed locally by avulsion, thereby allowing the river to bypass a former anchor point. This restriction to movement is reflected in the shape of its elongated point bars. Past avulsions are shown by several palaeochannels which run near-parallel to the present river. The gradient of the stream on the alluvium is very low, on average 0.00006.

In the alluvial section, the river width increases downstream from about 350 m near Jabalpur to about twice that width. The river is bounded on both sides by steep cliffs in Quaternary sand and sandy silt, commonly about 10 m high but in places rising to 25 m (Figure 5.13). The cliffs become the banks of the river during high-magnitude floods when the channel is full. These cliffs are cut into by amphitheatre-like gully heads, the 10–15 m wide gullies ending abruptly at the base of the cliff. The common form of the gullies and the presence of slumped blocks at their bases indicate their formation by post-flood flowage and slumping towards the river (Rajaguru et al., 1995). The cliffs also hinder the meandering river from freely moving sideways. and confine the deposition process within the channel itself. The width–depth ratio of the entire channel may vary locally, but usually stays around 35:1.

The major channel features of the alluvial Narmada are depositional, and are found within its high banks. These include:

- a smaller channel bounded by a point bar on one side and a high cliff or an inset floodplain on the other

FIGURE 5.13   High cliffs of the Narmada River in the alluvial reach, Guwarighat

- sandy point bars
- fine-grained floodplains.

The river has a channel-in-channel physiography with the inner channel carrying the flow most of the time and the entire channel bounded by the steep high banks operating as a high-magnitude flood-discharge conduit on rare occasions. In comparison, during typical high flows of a wet monsoon, the inner channel remains full, most of a point bar is submerged, and the water rises several metres against the alluvial cliff on the concave side of the meander bend.

The long, narrow point bars are the distinctive depositional forms of the Narmada (Figure 5.14). The bars are stable, as the channel is largely non-migratory. These bars are very large, up to 5 km in length. They display flood features such as mega-ripples and concentration of coarse grains on top. The mega-ripples, separated by parallel low areas, have been identified as subaqueous bedforms superimposed on remnants of point bars during flood stages. The sedimentary stratigraphy of the point bars consists of planar cross-beds overlain by horizontally bedded sand, and the surface grains show a downstream fining along the bar. Immediately downstream of the low bedrock rapids which locally occur within the alluvial reaches, small boulders, cobbles and pebbles are found. Such material is rarely found either on or inside the bars of alluvial reaches. The first bar downstream from such rapids frequently has gravel exposed on the surface at its upstream end. It is possible, however, that coarse material is submerged in the present channel (Rajaguru et al., 1995).

The feature most resembling a floodplain along the Narmada occurs within the cliffs (Figure 5.15). It is identified as a floodplain on the basis of its periodic destruction and recurrence, the flat surface, and the fine-grained nature of its sediment which is almost all silt and clay. It is the floodplain formed by the inner channel during the high flows of the wet monsoon. The distribution of the floodplain is discontinuous and found only in places

FIGURE 5.14   Narmada point bar

FIGURE 5.15   Narmada floodplain in silt and clay inside the cliffs

where there is flow expansion and/or slowing of channel velocity. As such, it is found hugging the base of cliffs either at crossovers or on the concave side of meander bends but away from the axis of the bend. The dimensions of this floodplain-type feature are in hundreds of metres, very much smaller than those of the point bars.

The characteristics of the alluvial Narmada described here have been reported for other rivers, e.g. for channels from semi-arid Australia by Bourke (1994).

Geomorphic processes

Three sets of information help to conceptualize the processes currently operating in the Narmada River.

1.   There is a hierarchical pattern of channel processes which includes high-magnitude floods, the high flows of the wet monsoon, and the low flows of the dry monsoon.
2.   The efficiency of these processes varies between the bedrock and alluvial reaches; the Narmada is the cumulative result of both types of reaches and the processes operating therein.
3.   Both the location of the channel and deviations from the general pattern (waterfalls, extremely wide multibranch reaches, rock bands forming rapids across the entire channel) are related to tectonic disturbances.

The Narmada is primarily a flood-controlled river. High-magnitude floods (which fill the entire channel and very rarely overtop the cliffs) maintain the large channel, erode the rocky outcrops and sandy point bars, and transport huge amount of sediment downstream. No reliable measurement of the sediment transport by the Narmada in these floods exists but calculations of bed shear stress and unit power indicate that the Narmada in channel-

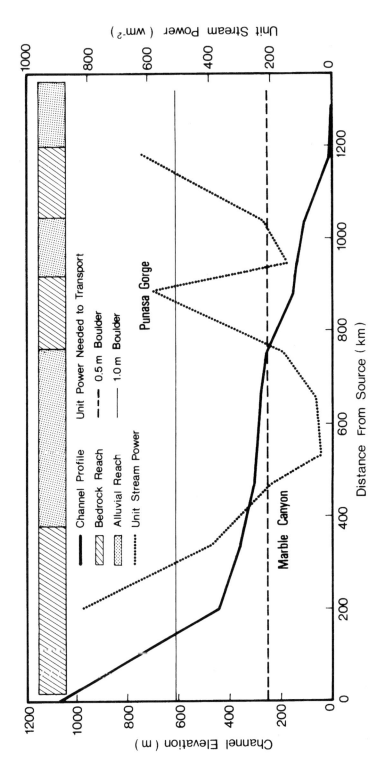

FIGURE 5.16 Longitudinal variation in unit stream power along the Narmada River (after Kale and Mishra, 1995)

filling flood is capable of immense erosion and large-scale sediment transport. The unit stream power exhibits a wide range. The variation in the unit stream power along the river between bedrock and alluvial reaches is shown in Figure 5.16. In the gorges the river is capable of eroding bedrock and transporting boulder-sized material as bedload and everything else as suspended load. This conclusion is supported by the boulders in Marble Canyon and also from Punasa Gorge as described for the Dardi Falls area. After such a flood, therefore, the gorges are swept of everything else but large boulders. In the alluvial reaches the river is capable of removing cobbles, pebbles and sand (Kale and Mishra, 1995).

During these floods the depositional forms in the Narmada channel should be severely depleted, although the alluvial reaches would receive a high amount of sediment derived from the erosion of the upstream bedrock reaches. At least the coarser material should be deposited in the channel due to the lessening of the gradient. Such material probably forms the cores of future point bars.

During the wet monsoon the Narmada fills the inner channel and overflows the floodplain but the water surface only rises up to the lower levels of the cliffs and also the point bars. The unit stream power, however, drops to barely 10^2 W m^{-2} in the alluvial reaches. The gorges can still move gravel but not the huge 2 m boulders. In the alluvial sections modification of sandy point bars and building of the in-channel floodplain are possible. A significant part of the work is probably carried out during several days of high discharge when a depression moves down the basin (Figure 5.17). These are not high-magnitude floods that fill the entire channel, but flows are much greater than the modal wet season discharges.

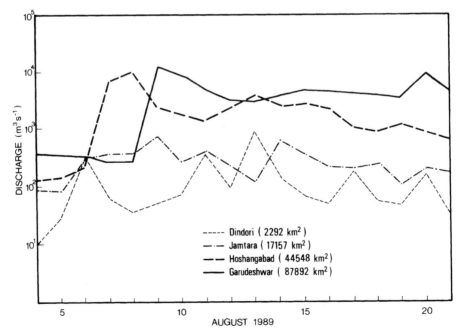

FIGURE 5.17   Wet season discharge peaks moving downstream. Note that the vertical axis is in log-scale. Data from Central Water Commission, *Water Year Book 1989–90, Narmada Basin*

Very little work is done during the low flows of the dry monsoon when most of the channel emerges. Sand is moved as bedload. Wohl (1992) has reported similar variations in sediment transport in the gorge of the Burdekin River in Australia. The Burdekin moves boulders in extreme floods (10^5 m^3 s^{-1}), large volumes of sand and silt in the wet season (mean discharge 1260 m^3 s^{-1}), and only very fine silt and clay in suspension in the dry season (mean discharge 60 m^3 s^{-1}).

The channel adjusts to the variations in the size of high-magnitude floods, the wet monsoon high flows, and the dry monsoon low flows by changing its width–depth ratio. The width–depth ratios for the same section but at different stages show an inverse relationship with discharge. The channel becomes more efficient by getting relatively deeper and therefore capable of carrying extraordinary flood and sediment discharges. This explains the high banks and channel-in-channel physiography of the rivers in regions affected by high-magnitude floods (Gupta, 1995). Hydraulic geometry equations for the Narmada at Rajghat are as follows:

$$w = 496.3 Q_p^{0.037} \qquad (5.1)$$

$$y = 0.03 Q_p^{0.46} \qquad (5.2)$$

$$v = 0.07 Q_p^{0.50} \qquad (5.3)$$

where Q_p is the annual peak discharge, w is width, y is mean depth and v is mean velocity (Kale et al., 1994).

The Narmada has a box-shaped channel. The river gets deeper and faster to carry the high-magnitude floods.

The major sources of the Narmada's sediment are:

- the rocky gorges and bedrock outcrops in the channel
- the tributaries
- the high banks on both sides
- the point bars.

The first source contributes the coarse fraction mainly as bedload, which is gradually diminuted to sand size. Almost all the depositional features in the gorges are boulder berms or small piles of imbricated boulders. The bedrock surface is otherwise bare. Evidently a considerable amount of material has been swept out of the gorges into the alluvial reaches. Only where the gorges widen locally, some sand and gravel bars occur. The slackwater deposits in the gorges indicate that the Narmada carries sand and fines suspended in flood but it is swept out of the gorges and deposited only in sheltered places. The accumulation of gravel immediately downstream of the gorges and the prevalence of sand in the alluvial reaches suggest that those come from the upstream gorge. Sorting of coarse material into sand and small pebbles occurs within a short distance. The load of the tributaries varies texturally, but again sand is the commonest fraction. The high banks release sand and silty sand with a small proportion of pebbles and silt. This, primarily the sediment of an earlier river, is eroded only during the higher wet season flows and the high-magnitude floods. The reworked material of the point bars, which are extensively eroded during high-magnitude floods, is again predominantly sand.

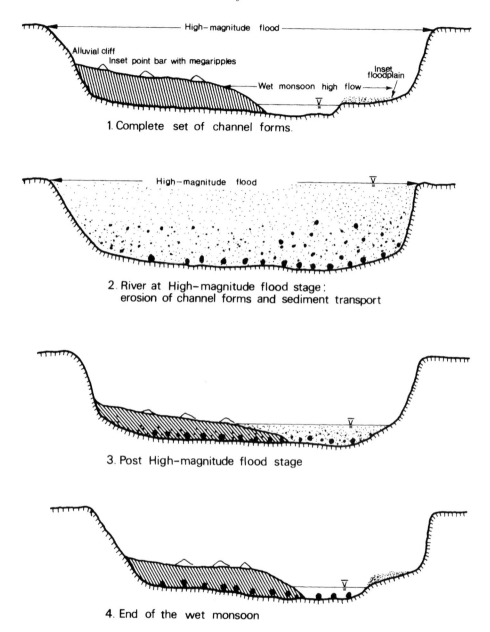

FIGURE 5.18   Sketch showing the proposed sequence of channel form construction in the Narmada River

   Some of silty sand and silt, however, is deposited in sheltered places by the falling stages of the wet monsoon to build a floodplain inside the high banks of the Narmada. Basically the bedrock reaches contribute the sediment which is deposited in the alluvial reaches. This sand moves down the channel in cycles of storage and transfer (Figure 5.18).

Finally, tectonism is suspected to have acted and to be continuing to act as an important control in determining channel location and formation of certain channel features. Punasa Gorge either follows lines of major faults or runs at a right angle to these (Kale, 1989). The Narmada near Jabalpur cuts across a hard Archaean dolomite in a 15 m gorge (Rajaguru et al., 1995). Near Punghat (Figure 5.12), the meandering Narmada changes into a multibranched multi-island section 2.5 times its modal width (Kale et al., 1996). Channel establishment across hard rock exposures may be due to a combination of tectonic movements and high-magnitude floods. For example, the waterfall and the gorge at Bheraghat near Jabalpur lie across a small outlier of Archaean dolomite. This outlier of hard rock is located in the middle of an alluvium-filled valley indicating that the river must have had a different course at other times. A wide palaeochannel skirts the dolomite with its 10–15 m high banks in Pleistocene material. Dolomitic clasts appear in the Narmada alluvium only in the Holocene sediment (Rajaguru et al., 1995). Other evidences from the field also suggest that the Narmada could have been different during earlier times.

THE NARMADA IN THE QUATERNARY

The high cliffs of the Narmada are in Pleistocene alluvium. The alluvium exposed on these cliffs is distinctly different from the sediment which is currently being deposited by the river. Furthermore, the evolution of the present river is intricately related to the excavation of the bedrock gorges of the Narmada. In this section, we attempt to reconstruct the Quaternary history of the Narmada River. We suggest that the Narmada changed dramatically some time during the Holocene to acquire its present form and behaviour.

The Pleistocene alluvium of the Narmada has been studied intensively over the last 150 years because of the abundant presence of Pleistocene mammal fossils and archaeological artefacts (Princep, 1832; De Terra and Patterson, 1939; Joshi et al., 1981; Badam et al., 1986). The recent discovery of a hominid skull (Sonakia, 1985; Kennedy et al., 1991) has renewed interest in the Narmada alluvium.

Table 5.2 shows the generalized alluvial stratigraphy commonly seen on the banks of the upper and middle Narmada. The alluvium (except for the boulder bed) does not indicate deposition by a river of high energy. Very little evidence of flood deposition is seen in the field while examining the alluvial banks of the river. We measured in detail six stratigraphic sections in the field: three at Pawla, two at Bheraghat and one at Mortakka. Two sections from Pawla and the one from Mortakka are shown in Figure 5.19. At Pawla the alluvium is that of a major river, but 85% of the material consists of point bars, channel alluvium or sand wedge. Trough cross-bedded channel sand occurs over eroded bases, with several channels displaying root casts on top, indicating shallowness and rapid channel alteration. The planar cross-bedded (between 22° and 30°) sand of the section is a remnant of the point bars of the ancient Narmada. This cross-bedded sand in Figure 5.19 is identified as the upper Pleistocene sediment of the Narmada on the basis of ^{14}C dating and the presence of mammalian fossils and common artifacts (Table 5.2). At Pawla only two thick beds could be identified as flood deposits, on the basis of the coarseness of the clasts and the presence of horizontal laminations. The biggest clasts seen were 3–4 cm in diameter and included transported calcrete nodules.

TABLE 5.2   The Narmada bank alluvium: generalized stratigraphic succession

Stratigraphy	Thickness (m)	Description
Black cotton soil	8–10	Holocene
Yellow brown silt with carbonate concretions	15–20	Terminal Pleistocene (?); Upper Palaeolithic artifacts, molluscan and ostrich egg shells, animal fossils; carbonate cement
Pebbly sand with cross-bedding	5–8	Sedimentary structures of fluvial origin; Late Pleistocene fossils and Middle Palaeolithic artifacts. Freshwater mollusc shells ^{14}C dated to 32 750 (+1770–1580) years BP (Agarwal and Kusumgar, 1974)
Boulder bed	2–3	Conglomerate bed with pebbles, cobbles and boulders and carbonate cement; broken and rolled fossils and Acheulian artifacts
Red silt and clay	4–5	Clay and silt with tubular concrete nodules, a few sand beds.

Usually not exposed below this at most sections. At places hard rocks appear (Deccan lavas or Gondwana sedimentary rocks) below one of these units. The sections at Pawla show the details of "Yellow Brown silt with carbonate concretions" and "Pebbly sand with cross-bedding". The flood deposits of Bheraghat post-date the "Yellow brown silt with carbonate concretion".

The Narmada of the Late Pleistocene appears to have been a mobile meandering river which carried large quantities of sand. This is also the main fossiliferous part of the Narmada alluvium (Badam and Grigson, 1990), and the fauna is represented by terrestrial animals (*Bos namadicus, Cervus* sp., *Equus namadicus, Antelope cervicapra*), semi-aquatic animals (*Bubalus palaeoindicus, Hexaprotodon* spp., *Elephas* spp.) and aquatic animals (*Crocodylus* spp., *Trionynx* sp., *Lessesnya* spp.). It has therefore been suggested (Badam et al., 1986) that the Narmada valley was a mosaic of dense forests, park savannas and waterbodies.

The section measured at Mortakka is the upper part of the alluvial cliff approximately 10 m above the low-flow level of the present river and 200 m from the low waterline. Most of it is identifiable as a flood channel cut across a silty sand bar. The flood channel is filled in by about 50 thinly laminated silt units without coarse material at their bases, indicating ponded drainage on top of a bar. This is overlain by a massive clay channel-fill. Rootlets, however, are present and vertical structures, possibly from dewatering of the original mud, have developed. This indicates a river carrying a large amount of silt and clay in suspension and small overflow and bartop channels.

In contrast, the alluvium at Bheraghat indicates the arrival of a very large flood on top of the cross-bedded pebbly sand similar to the sand at Pawla (Figure 5.20). The site of the section is shown in Figure 5.21. A boulder bed rich in damaged mammalian fossils overlies the sand, indicating the occurrence of a single high-energy event: an unusually large flood. This flood, which added dolomitic clasts to the Narmada alluvium for the first time, is also evidenced in an abandoned palaeochannel at Bheraghat (Figure 5.21).

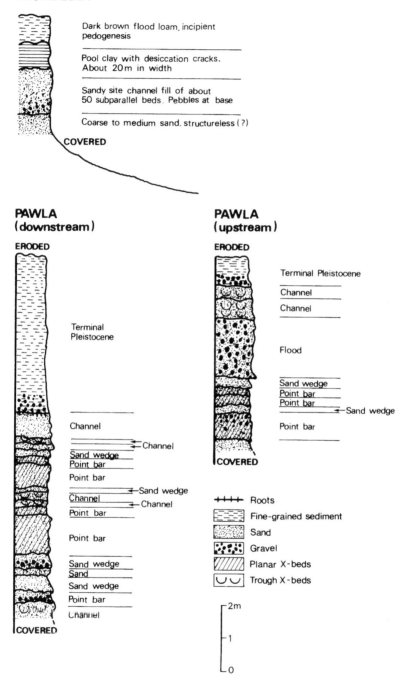

MORTAKKA

Dark brown flood loam, incipient pedogenesis

Pool clay with desiccation cracks. About 20 m in width

Sandy site channel fill of about 50 subparallel beds. Pebbles at base

Coarse to medium sand, structureless (?)

COVERED

PAWLA (downstream)

ERODED

Terminal Pleistocene

Channel

Channel

Sand wedge
Point bar

Point bar

Sand wedge

Channel
Channel

Point bar

Point bar

Sand wedge
Sand

Sand wedge

Point bar

Channel

COVERED

PAWLA (upstream)

ERODED

Terminal Pleistocene

Channel

Channel

Flood

Sand wedge
Point bar
Point bar
Sand wedge

Point bar

COVERED

++++ Roots

Fine-grained sediment

Sand

Gravel

Planar X-beds

Trough X-beds

2m
1
0

FIGURE 5.19   Stratigraphic sections: Narmada alluvium at Pawla and Mortakka

BHERAGHAT

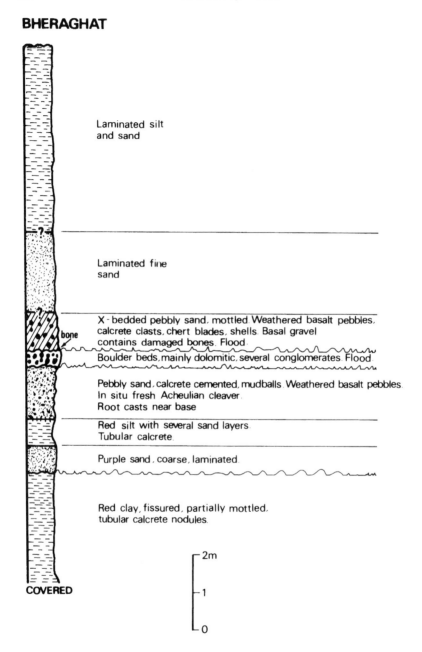

Laminated silt
and sand

Laminated fine
sand

X - bedded pebbly sand, mottled. Weathered basalt pebbles,
calcrete clasts, chert blades, shells. Basal gravel
contains damaged bones. Flood.

bone

Boulder beds, mainly dolomitic, several conglomerates. Flood.

Pebbly sand, calcrete cemented, mudballs. Weathered basalt pebbles.
In situ fresh Acheulian cleaver.
Root casts near base

Red silt with several sand layers.
Tubular calcrete.

Purple sand, coarse, laminated.

Red clay, fissured, partially mottled,
tubular calcrete nodules.

2m

1

0

COVERED

FIGURE 5.20   Stratigraphic section at Bheraghat showing flood sedimentation

The palaeochannel is cut into alluvium resembling that at Pawla and with fine-grained
terminal Pleistocene material on top of the banks. This channel was abandoned after the
occurrence of the above-mentioned large flood. All deposits within the channel are flood
deposits, starting with dolomitic cobbles and boulders derived from the dolomitic inlier,

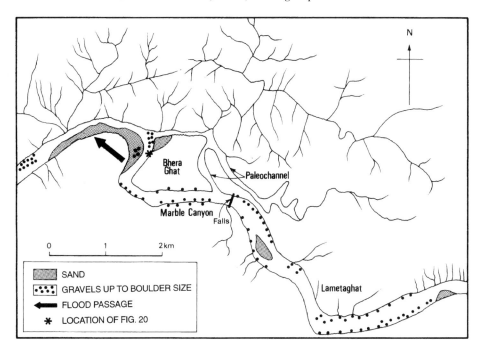

FIGURE 5.21   Map of the palaeochannel at Bheraghat (from Rajaguru et al., 1995); flow from east to west. Copyright John Wiley & Sons Limited. Reproduced with permission

mentioned earlier, which the Narmada cut through in a 3 km gorge (Rajaguru et al., 1995). The Narmada by avulsion cut its new channel through the dolomitic inlier and continued to stay inside a deep gorge (Marble Canyon, Figure 5.5). Old boulder berms and gravel bars are found in the palaeochannel. This palaeochannel is currently used by overflows at its upstream end during high-magnitude floods. The water is ponded back at the downstream end of the channel by the main flow coming out of the Narmada gorge in dolomitic marble. After the 1991 flood, the banks of the channel (near the downstream end) were coated with the fine slackwater deposits.

It is likely that during the last glacial advance (isotope stages 2 and beyond), when parts of western India were arid, the Narmada was a meandering river carrying sand and silt. The monsoon would have been weak and the drop in sea temperature would have resulted in fewer or no tropical storms, which are the source of the high-magnitude floods of present-day Narmada. Tropical storms are normally formed when the sea-surface temperature is above 27°C. The estimated drop in sea-surface temperature for the Bay of Bengal is about 2°C for the month of August (Ruddiman, 1984) which puts it below this limit. The alluvium of the high banks of the Narmada reflects this hydrological environment.

The large floods that maintain the channel of the Narmada at present probably started to appear in the Holocene (Kale et al., 1993, 1994; Baker et al., 1995), and have continued to be the dominant channel control since then. Slackwater deposits related to large palaeofloods indicate the occurrence of high-magnitude floods for the last 3000 years on the Narmada (Ely et al., 1996; Kale et al., 1994) and for 5000 years on one of her tributaries (Kale et al, 1993). At the moment we do not have much information regarding the hydrology of the Narmada in the early Holocene, although it is tempting to conclude that with the warming of the surface waters of the Bay of Bengal, tropical cyclones started to bring heavy rainfall over India. The monsoon system should also become stronger with higher temperature of the landmass in summer. High-magnitude floods would enlarge the channel cutting into the Pleistocene sediment of the Narmada. The present-day channel is therefore likely to have appeared in the Holocene, with high banks cut into Pleistocene river alluvium and building channel forms which relate both to very large floods and seasonality in discharge.

Changes in channel form and behaviour following hydrological changes in the late Quaternary have been suggested for other rivers of India. For example, Williams and Clarke (1984), studying the Late Quaternary environments in the Son and Belan valleys immediately to the northeast of the Narmada, described a change from sparse vegetation and high sediment yield in the glacial period of 25 000–17 000 years BP to dense vegetation and low sediment yield in the post-glacial period. In the Son channel this is indicated by overbank clays deposited above cross-bedded sand. In the Early Holocene, channel incision occurred into the Pleistocene floodplain, and a channel with different dimensions was formed. The sequence of events in the Narmada Basin is approximately the same, and as these changes are related to variation in regional climate, this could be true for a number of Indian rivers.

The Narmada at present flows through three bedrock gorges whose excavation requires explanation. We have not worked on the lower Narmada and therefore are not in a position to discuss the longest and most downstream of the gorges (Dhadgaon Gorge). The most upstream of the gorges, Marble Canyon near Jabalpur, was formed in the Holocene. This is evidenced by the occurrence of a palaeochannel with comparable dimensions which curves round the gorge (Figure 5.21), the presence of thick terminal Pleistocene sediment on top of the banks of the palaeochannel, the relatively unweathered dolomite gravel in the channel, the use of the palaeochannel in current high-magnitude floods, the short (3 km) length of the gorge below the waterfall, and the fresh dolomitic walls of Marble Canyon with inconspicuous soil formation on top. Downstream of the gorge, where the palaeochannel rejoins the present Narmada, the bed of the palaeochannel is approximately at the level of the top of the Narmada banks.

The middle gorge (Punasa Gorge) is longer and more difficult to explain. It includes two large waterfalls. The Narmada has developed an inner channel with scablands on both sides, and the tributaries are graded to the top of the scabland surface, not to the inner gorge. The bedrock emerges from below the Pleistocene alluvium on both sides (Figure 5.11). Assuming a continuity of this alluvium, the gorge must have been excavated in the Holocene. The well developed flood-related forms of Punasa Gorge, as described earlier in this paper, suggest that this gorge is older than Marble Canyon. The location of both gorges and the occurrence of shear zones allowing the break-up of the bedrock into blocks at Punasa suggest tectonic activities. It is possible that a combination of tectonism and large floods excavated the gorges.

The size and inventory of flood forms within Punasa Gorge are impressive. Given a shear zone and a number of high-magnitude floods in the Holocene it is possible that this gorge was excavated in several thousand years. On the other hand, as the proposed history of the river suggests, aridity-related aggradation occurred in the channel during the last glacial advance followed by erosion due to higher rainfall and appearance of large tropical storms. Punasa Gorge could therefore be a more complicated landform which was filled in at least partially during arid glacial periods and re-excavated during the wetter interglacials. Given the present state of knowledge, we are not in a position to eliminate either of the two hypotheses. Marble Canyon, on the other hand, is definitely a Holocene product as is the channel form and behaviour of the present-day Narmada.

CONCLUSIONS

The channel form and processes of the Narmada are controlled by (1) high-magnitude floods, (2) the seasonal pattern of Indian monsoon, and (3) the tectonics of the basin.

The Narmada, being extremely flood-dominated, has certain characteristic features which are quite distinctive, for example the scablands and the large size of the entire channel. On the other hand, a number of its features are shared by many Indian rivers, such as the channel-in-channel physiography controlled by two different sets of discharges (Gupta, 1995). The number of case studies on Indian rivers is limited, but it seems likely that channel size in many rivers is controlled by high-magnitude floods, whereas depositional channel forms are built as insets against the river banks by the high flows of the wet monsoon. In the case of the flood-dominated Narmada, these insets are also formed during the falling stage of the flood hydrograph. The texture of the sediment transported by the stream determines the adherence to and variations from this pattern to a significant extent. As most Indian rivers carry sand as common bedload, the post-high-magnitude flood adjustments are not difficult to achieve, especially as the high-magnitude floods arrive in the wet season when the river is high. The post-flood flows transport sand to build insets against the banks of a flood-widened channel. Variations from this model are caused by rivers which carry a high amount of cobbles and boulders or clay, and also by rivers that are tectonically controlled. The Narmada carries the signature of a seasonal river in its depositional forms, but the huge magnitude of its floods is responsible for its large dimensions, very high banks and channel immobility. Tectonic movements contributed to local variations.

The valley alluvium exposed on the high banks of the Narmada suggests that the high-magnitude floods do not go back beyond the Holocene. Similar interpretations have been forwarded for at least one more large Indian river. The fascination of the Narmada therefore lies firstly in understanding how its three channel processes work in combination, and secondly in determining its flood history, which could prove to be a signature of the climatic change on the Indian subcontinent some time early in the Holocene.

ACKNOWLEDGEMENTS

Avijit Gupta would like to thank the National University of Singapore for granting him sabbatical leave which made part of the fieldwork possible. The alluvial sections at Pawla were measured with the help of R.K. Ganjoo. Vishwas Kale thanks the officials of Central Water Commission (India) for supplying discharge data and the Department of Science and Technology, New Delhi, for fieldwork grants. Many discussions on the Narmada with Vic Baker and Sheila Mishra, at many localities including on the banks of the Narmada, have been especially beneficial.

REFERENCES

Abbi, S.D.S., Gupta, D.K. and Subramanian, S.K. 1970. On some hydrometeorological aspects of Narmada Basin. *Indian Journal of Meteorology and Geophysics*, **21**, 539–553.

Agarwal, D.P. and Kusumgar, S. 1974. *Prehistoric Chronology and Radiocarbon Dating in India*. Munshiram Manoharlal, New Delhi.

Badam, G.L. and Grigson, C. 1990. A cranium of Gaur, *Bibos gaurus* (Bovidae, Mammalia. from the Pleistocene of India. *Modern Geology*, **15**, 49–58.

Badam, G.L., Ganjoo, R.K. and Salahuddin 1986. Preliminary taphonomical studies of some Pleistocene fauna from the central Narmada valley, Madhya Pradesh, India. *Palaeogeography, Palaeoclimatology, Palaeoecology*, **53**, 335–348.

Baker, V.R. 1995. Global paleohydrological change. *Quaestiones Geographicae*, Special Issue **4**, 27–35.

Baker, V.R. and Costa, J.E. 1987. Flood power. In L. Mayer and D. Nash (Eds), *Catastrophic Flooding*. Allen and Unwin, Boston, 1–22.

Baker, V.R. and Kochel, R.C. 1988. Flood sedimentation in bedrock alluvial systems. In V.R. Baker, R.C. Kochel and P.C. Patton (Eds), *Flood Geomorphology*. Wiley, Chichester, 123–137.

Baker, V.R. and Pickup, G. 1987. Flood geomorphology of the Katherine Gorge, Northern Territory, Australia. *Geological Society of America, Bulletin*, **98**, 635–646.

Baker, V.R., Ely, L.L., Enzel, Y. and Kale, V.S. 1995. Understanding India's rivers: late Quaternary paleofloods, hazard assessment and global change. In S. Wadia, R. Korisettar and V.S. Kale (Eds), *Quaternary Environment and Geoarchaeology of India*. Geological Society of India, Memoir 32, 61–77.

Bedi, N. and Vaidyanadhan, R. 1982. Effect of neotectonics on the morphology of the Narmada river in Gujarat, Western India. *Zeitschrift für Geomorphologie, NS*, **26**, 87–102.

Bourke, M.C. 1994. Cyclical construction and destruction of flood dominated flood plains in semiarid Australia. In L.J. Olive, R.J. Loughran and J.A. Kesby (Eds), *Variability in Stream Erosion and Sediment Transport*. IAHS Publication No. 224, Wallingford, UK, 113–123.

Brahman, N.K. 1990. Seismic hazard at Narmada Sagar Dam. *Current Science*, **59**, 1209–1211.

De Terra, H. and Patterson, T.T. 1939. *Studies on the Ice Age in India and associated Human Cultures*. Carnegie Institute of Washington, Publication No. 493.

Ely, L.L., Enzel, Y., Baker, V.R., Kale, V.S. and Mishra, S. 1996. Changes in the magnitude and frequency of late Holocene monsoon floods on the Narmada River, Central India. *Geological Society of America, Bulletin*, **108**, 1134–1148.

Gupta, A. 1995. Magnitude, frequency, and special factors affecting channel form and processes in the seasonal tropics. In J.E. Costa, A.J. Miller, K.W. Potter and P. Wilcock (Eds), *Nature and Anthropogenic Influences in Fluvial Geomorphology*. American Geophysical Union Monograph 89, Washington, DC, 125–136.

Indian Meteorological Department. 1981. *Climate of Madhya Pradesh*. Office of the Additional Director General of Meteorology (Res.), Pune.

Joshi, R.V., Badam, G.L. and Pandey, R.P. 1981. Fresh data on the Quaternary animal fossils and stone age cultures from the central Narmada valley, India. *Asian Perspectives*, **21**, 164–181.

Kale, V.S. 1989. Significance of Riphean stromatolites from the Kishengad (Bijawar) group, Dhar Forest inlier, Central Narmada Valley. *Himalayan Geology*, **13**, 63–74.

Kale, V.S. and Mishra, S. 1995. Modern and palaeoflood geomorphology of central Narmada basin. *SERC Research Highlights*, Department of Science and Technology Publications, 85–88.

Kale, V.S., Mishra, S., Baker, V.R., Rajaguru, S.N., Enzel, Y. and Ely, L. 1993. Prehistoric flood deposits on the Choral River, Central Narmada Basin, India. *Current Science*, **65**, 877–878.

Kale, V.S., Ely, L.L., Enzel, Y. and Baker, V.R. 1994. Geomorphic and hydrologic aspects of monsoon floods on the Narmada and Tapi rivers in Central India. *Geomorphology*, **10**, 157–168.

Kale, V.S., Baker, V.R. and Mishra, S. 1996. Multi-channel patterns of bedrock rivers: an example from the central Narmada basin, India. *Catena*, **26**, 85–98.

Kennedy, K.A.R., Sonakia, A., Chiment, J. and Verma, K.K. 1991. Is the Narmada hominid an Indian homo erectus? *American Journal of Physical Anthropology*, **86**, 475–496.

Matthes, G.H. 1947. Macroturbulence in natural stream flow. *Transactions, American Geophysical*

Union, **28**, 255–262.

Princep, J. 1832. Note on the Jabalpur fossil bones. *Journal of the Asiatic Society of Bengal*, **1**, 456–458.

Rajaguru, S.N., Gupta, A., Kale, V.S., Mishra, S., Ganjoo, R.K., Ely, L.L., Enzel, Y. and Baker, V.R. 1995. Channel form and processes of the flood-dominated Narmada River, India. *Earth Surface Processes and Landforms*, **20**, 407–421.

Ramaswamy, C. 1987. *Meteorological aspects of severe floods in India 1923–1979.* Indian Meteorological Department Meteorological Monograph, Hydrology No. 10, 358 pp.

Ruddiman, W.F. (Co-ordinator and compiler) 1984. The last interglacial ocean. CLIMAP project members. *Quaternary Research*, **21**, 123–224.

Singh, N., Soman, M.K. and Kumar, K.K. 1988. Hydroclimatic fluctuations of the Upper Narmada catchment and its association with break–monsoon days over India. *Proceedings, Indian Academy of Sciences (Earth and Planetary Sciences)*, **97**, 87–105.

Sonakia. A. 1985. Skull cap of an early man from the Narmada Valley alluvium (Pleistocene) of Central India. *American Anthropologist*, **87**, 612–616.

Verma, R.K. 1991. *Geodynamics of the Indian Peninsula and the Indian Plate Margin*. Oxford University Press.

Williams, G.P. 1983. Paleohydrological methods and some examples from Swedish fluvial environments, I. cobble and boulder deposits. *Geografiska Annaler*, **65A**, 227–243.

Williams, M.A.J. and Clarke, M.F. 1984. Late Quaternary environments in North–central India. *Nature*, **308**(5960), 633–635.

Wohl, E.E. 1992. Bedrock benches and boulder bars: floods in the Burdekin Gorge of Australia. *Geological Society of America, Bulletin*, **104**, 770–778.

Wohl, E.E. 1993. Bedrock channel incision along Piccaninny Creek, Australia. *Journal of Geology*, **101**, 749–761.

6

Drainage Evolution and Morphological Development of the Late Cenozoic Sundays River, South Africa

JOHAN HATTINGH[1] AND IZAK C. RUST[2]

[1] Council for Geoscience, Geological Survey, Port Elizabeth, South Africa
Present address: Trans Hex Mining Ltd, Private Bag X3, Alexander Bay 8290, South Africa
[2] Department of Geology, University of Port Elizabeth, Port Elizabeth, South Africa

ABSTRACT

The Jurassic break-up of Gondwana led to complete rejuvenation of drainage systems along the southern margin of Africa, causing northward-draining river systems to reverse their flow southward towards the newly formed Indian Ocean. After the fragmentation of the supercontinent, the Sundays River was probably one of the first major rivers to form in what is now the Eastern Cape, South Africa. During the prolonged period of the African erosion cycle, from the break-up of Gondwana to the Miocene, sedimentation rates on the continental shelf showed an overall decline from Early Cretaceous to Early Tertiary times. Late Cenozoic Post-African uplift events, however, resulted in renewed erosion cycles leading to increased sedimentation in alluvial basins.

An almost continuous erosional and depositional chronology spanning the Late Miocene to Holocene records the fluvial development of the Late Cenozoic Sundays River. Valley morphology and remnants of river terraces in the Sundays River middle reaches indicate widespread river capture that significantly altered river patterns in this region. The Sundays River captured substantial portions of drainage basins of adjoining river systems and lost large sections of its original drainage basin to river piracy. As a result, the system not only experienced changes to its middle-reach drainage pattern and channel morphology, but also established a new provenance. Owing to ease of bedrock erosion in the lower Sundays River valley, this part of the valley proved to be a very sensitive palaeoindicator of changes in the dynamic regime of the drainage system. A flight of 13 fluvial terraces flanks the lower Sundays River valley. The Late Miocene to Late Pliocene Higher Terraces lie between 220 and 40 m above the present river bed and the deposits consist mainly of gravel. The Pleistocene to Holocene Lower Terraces, with silt and fine-grained sand deposits, occur between 25 m and 3 m above the present river bed. From the highest terrace levels, unpaired

Varieties of Fluvial Form. Edited by A.J. Miller and A. Gupta
© 1999 John Wiley & Sons Ltd

terraces descend, mainly in an eastward direction, to a few metres above present river level in an almost complete set of steps.

Variation in valley morphology and composition of terrace deposits clearly records the influence of sea-level change, climate variation and tectonic activity. These erosional and depositional features reveal the Sundays River's response to extrinsic and intrinsic influences. The Sundays River responded to these influences mainly by adjusting its channel form. Continuous decline in sea level, tectonic uplift in the sediment source area and high-discharge conditions during the Late Miocene to Late Pliocene produced a steeply sloping braided Sundays River with boulder-sized gravel as bed material. The Neogene mixed-load braided Sundays River transformed into a meandering, suspended-load river in the Quaternary. The abrupt change from gravel-dominated Higher Terraces to fine-grained sand- and silt-dominated Lower Terraces in the lower Sundays River valley probably resulted from sea-level rise during the Quaternary interglacial periods when the gently sloping Sundays River was competent only to transport fine-grained sediments.

INTRODUCTION

Erosional and depositional features in river valleys are effective indicators of long-term channel morphology and river pattern adjustments. Variation in channel morphology and river pattern are commonly recorded as changes in channel entrenchment or sediment accumulation trends during degradational and aggradational processes. Modification of prevailing river slope, discharge regime, sediment supply and size of bed material generally cause changes to existing river morphology. In the Sundays River, Eastern Cape, South Africa, such adjustments resulting from intrinsic as well as extrinsic influences are particularly well recorded by valley morphology and associated terrace deposits.

The present-day Sundays River originates at the Great Plateau Escarpment, in the Great Karoo, more than 200 km into the Eastern Cape hinterland. The river flows southwards over near-horizontal to northward-dipping Karoo shale and sandstone before breaching the Klein Winterhoek Mountains 80 km from the coast (Figure 6.1). The Klein Winterhoek Mountains form part of the eastern limit of the Cape Fold Belt Mountain ranges. Strata of the Cape Fold Belt consist of well indurated quartzite and sandstone. The present-day breach of the Sundays River through the erosion-resistant Klein Winterhoek Mountains lies in a steep-sided, deeply incised gorge. The upstream boundary of the lower Sundays River valley is defined by the Klein Winterhoek Mountains. Downstream of this narrow mountain range, the lower Sundays River enters the Mesozoic Algoa Basin in which the well entrenched lower Sundays River valley developed on highly erodible Cretaceous mudstone and shale. In the distal drainage basin, the lower Sundays River drains the Algoa Basin, a half graben (Tankard et al., 1982) formed with a Mesozoic fault immediately north of the lower Sundays River valley during fragmentation of Gondwana. Owing to the distinctive geology underlying the lower Sundays River valley, this part of the valley proved to be a very sensitive indicator of changes in the dynamic regime of the drainage basin.

The purpose of this paper is to report on mechanisms controlling the development of the Sundays River drainage system and the influences of these mechanisms on fluvial channel morphology and river pattern adjustments. Attention will be focused on the ability of erosional and depositional features to reveal the long-term fluvial landform development of the Sundays River system.

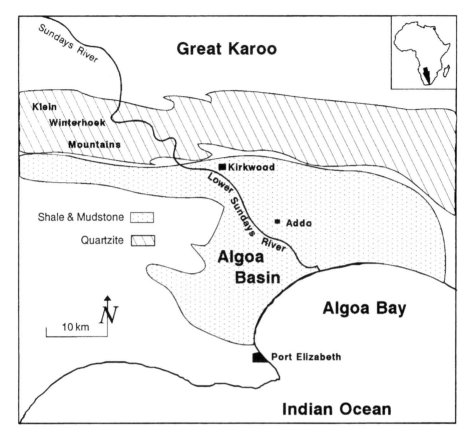

FIGURE 6.1   Location map of the lower Sundays River catchment showing prominent landscape and geological features

LANDSCAPE DEVELOPMENT AND RIVER EVOLUTION

Complete rejuvenation of drainage systems along the southern margin of Africa took place during the Jurassic break-up of Gondwana when well established northward-draining river systems (Rust, 1962) reversed their flow southward towards the newly formed Indian Ocean (Tankard et al., 1982). After the fragmentation of the supercontinent the Sundays River was probably one of the first major rivers to form in what is now the Eastern Cape (Hattingh, 1996). Southern Africa occupied a central position in the Gondwana supercontinent. As a result, the elevated inland landmass of this newly formed subcontinent extended to the coast (Dingle et al., 1983), and the proto-Sundays River, and other rivers draining this elevated plateau, had steep gradients. These steep rivers generated flows with very high stream power, indicated by the large boulder size (up to 800 mm in diameter; Haughton et al., 1937) of the fluvial conglomerate of the Cretaceous Enon Formation. This fluvial conglomerate accumulated in rift basins such as the Algoa Basin during the break-up and early rifting phase of Gondwana (Tankard et al., 1982). High erosion rates, especially during this post-break-up African erosion cycle, caused the plateau escarpment to retreat rapidly northwards, reaching its present position

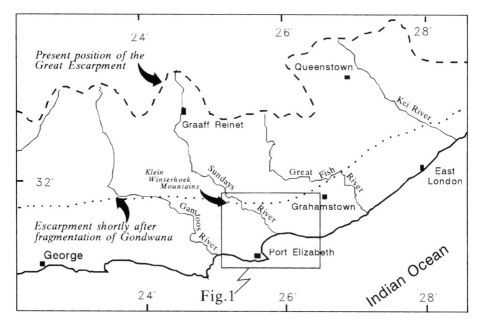

FIGURE 6.2   The positions of the Great Escarpment indicating the northward retreat of the escarpment from its position in the south shortly after the fragmentation of Gondwana to its present position. Note the indentations into the escarpment due to river erosion

as the Great Escarpment, 200 km inland (Figure 6.2), by the Miocene (Partridge and Maud, 1987). The lack of Early Tertiary erosional or depositional evidence along the southern margin of Africa (Hattingh, 1996) suggests that during this period the erosion rate decreased substantially, probably due to the more gentle gradients assumed by the rivers after the effective denudation of the land surface.

Continental uplift of major proportions (up to 900 m; Partridge and Maud, 1987) during the Post-African erosion events renewed the erosion cycle during the Late Miocene (Hattingh, 1996). This epeirogenic uplift event caused steepening of slopes, resulting in higher erosion rates and a consequent increase in sediment load of rivers draining the southern African continental margin. Remnants of the fluvial terrace deposits formed during this period are still present along the flanks of the lower valleys of major rivers in this region. These terrace deposits represent a valuable record of the fluvial and landscape evolution during the Late Cenozoic.

FLUVIAL DEVELOPMENT IN THE MIDDLE REACHES

Upon leaving the mountains and Great Escarpment, headwater streams enter the mountain foothills as middle-reach streams where their beds widen as more water arrives from the headwater tributaries. Reduced channel gradients in these middle reaches nevertheless still display erosive flow in these channels. In the middle-reach part of the Sundays River system, discharges throughout its Late Cenozoic development have been well channelized and flowed energetically towards the gently sloping alluvial basin occupied by the lower reaches. In these middle-reach parts a final attempt at growth of the drainage network occurred on a large scale, resulting in stream capture in many places. Such stream capture commonly occurred in the middle reaches of large rivers flowing through the mountains of the Cape Fold Belt.

Stream capture in the Cape Fold Belt Mountains was first identified by Haughton et al. (1937) and was later also described by Taljaard (1949), as well as Rust and Illenberger (1989). The Cape Fold Belt Mountains in the catchment area of the Sundays River have an east-southeasterly orientation. Consequently, many large subsequent rivers display similar orientation, following the mainly east-southeastward trending valleys. The middle reaches of the large rivers crossing the Cape Fold Belt in the immediate vicinity of the Sundays River drainage basin, locally deviate from this eastward flow direction. Instead, in many places, the rivers make an almost 90° turn to the north or south after flowing for some distance in an easterly direction (Figure 6.3).

Initially four major drainage systems drained this area as parallel yet separate systems before stream capture altered the drainage pattern. Strong evidence for this process of stream capture is the presence of wind gaps, underfit streams and numerous remnants of gravelly fluvial terraces in the former valleys of these systems, confirming the presence of once-large rivers in parts where only underfit streams now exist. These fluvial terrace remnants in wide palaeovalleys are totally detached from any of the modern river systems. The proto-upper Sundays–Great Fish River system flowed in the northeastern-most drainage basin. Fluvial terrace remnants manifesting as gravel-capped hills bordering the underfit stream between the present-day Sundays River and Great Fish River indicate the existence of this palaeosystem. In an adjacent valley to the southwest, similar evidence indicates the presence of a proto-Groot–Kariega–lower Sundays River system that used to flow in the underfit valley occupied by the present-day Kariega River. Prior to widespread stream capture, the proto-Baviaanskloof–Gamtoos River system flowed in the adjacent valley to the south, presently occupied by the underfit Baviaans River in the upper reaches and Gamtoos River in the lower reaches. Similarly, in the southernmost drainage basin the proto-Krom–Kouga River system existed.

High mountain ranges of the Cape Fold Belt, comprising erosion-resistant sandstone and quartzite forming apparently unbreachable ridges, separated these drainage systems. However, well developed sets of north–south oriented fault and joint planes (Hill, 1988) provided zones of weakness across these mountain ranges. Tributaries exploited these zones of weakness by concentrated erosion enabling interception of the drainage system of an adjacent valley (Ver Steeg, 1935). Headwater erosion exploiting the north–south oriented fault and joint planes in the Cape Fold Belt Mountains was accelerated during the Post-African epeirogenic uplift events. Large, steep-sided, relatively straight, north–south oriented gorges aligned at about 90° to the trend of the major valleys provide evidence for widespread stream capture in the Cape Fold Belt in the immediate vicinity of the Sundays River drainage basin (Figure 6.4).

By this process of stream capture, a tributary of the proto-Baviaanskloof–Gamtoos River system captured the proto-Krom–Kouga River system so that the Kouga River became a major contributor to the newly formed Gamtoos River system (Figure 6.3). As a result, the Baviaanskloof River became a less significant contributor to the Gamtoos River system. Totally separated from the Kouga River, the Krom River continued to exist as a separate system. The proto-Baviaanskloof–Gamtoos River system also captured the

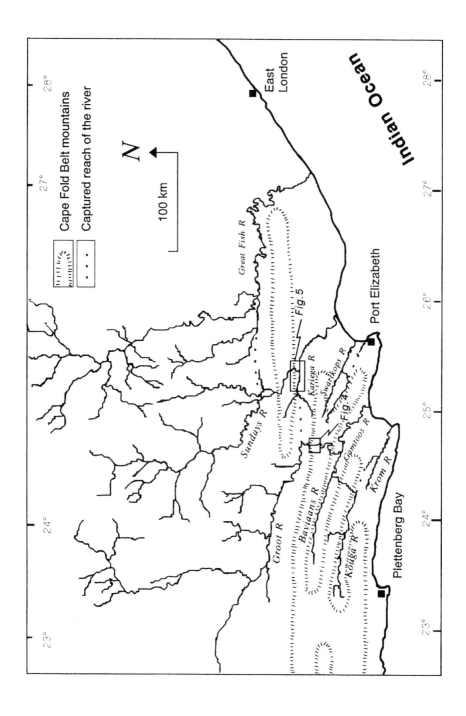

FIGURE 6.4   One of the deeply incised, steep-sided gorges through which the Sundays River has lost a part of its drainage basin to a neighbouring drainage system

Groot River from the proto-Groot–Kariega–lower Sundays River system by a northward-eroding tributary. A tributary of the proto-Groot–Kariega–lower Sundays River exploited joints across the Klein Winterhoek Mountains to the north of the lower Sundays River and in the process captured the discharge of the upper Sundays River, which before then was part of the proto-upper Sundays–Great Fish River system. This stream capture finalized the Sundays River system in its present-day configuration with the Kariega River as a relatively minor tributary. The Great Fish River basin (that lost the upper Sundays basin) now exists as a system on its own. Some unique geographical distribution patterns of the redfin *Barbus* species (*Pisces, Cyprinidae*) observed by Skelton (1980) in the Eastern Cape can only be explained by assuming that river piracy in the larger fluvial systems of this area occurred on a large scale.

Stream capture in the middle-reach streams significantly changed the drainage style in the immediate vicinity of the Sundays River, especially north of the Cape Fold Belt. During the northward retreat of the Great Escarpment, the Cape Fold Belt was exhumed and left behind as a hard, erosion-resistant barrier after the surrounding land reached an advanced state of denudation by the Miocene (Partridge and Maud, 1987). Late Cenozoic rivers, draining the area below the escarpment in the interior, flowed southward towards the seemingly impregnable Cape Fold Belt Mountain ranges. On approaching the

FIGURE 6.3   *(opposite)* The major drainage systems in the Eastern Cape originate north of the Cape Fold Belt and south of the Great Escarpment. Upon reaching the Cape Fold Belt, river courses were initially strongly influenced by the Cape Fold Belt mountains, which confined them to valleys parallel to the east-southeasterly structural strike. By the process of stream capture the rivers managed to breach the Cape Fold Belt mountain ranges through north–south oriented gorges. Dotted lines indicate the positions of the captured reaches, presently occupied by underfit streams

mountain range, the rivers were led to follow the structural strike draining eastwards and continued to flow in that direction until they found a zone of weakness along which to breach (proto-Groot–Kariega River), or flowed around the obstruction caused by the mountains (proto-Sundays–Great Fish River). Stream capture created breaches through this barrier along which the rivers from the interior were led along shorter routes to the Indian Ocean. Stream capture severely modified drainage systems, with the headwater and middle reaches of these systems being switched between drainage basins.

FLUVIAL DEVELOPMENT IN THE LOWER REACHES

Prior to stream capture, the provenance of the lower Sundays River lay to the west of the present-day catchment area in the region presently being drained by the Groot River. After stream capture the provenance shifted to the north, and since the Middle Pliocene (Hattingh, 1996) the lower Sundays River has received the bulk of its water and sediment discharge via the gorge at Korhaans Poort (Figure 6.5). Terrace deposits and valley morphology comprehensively record the fluvial development of the lower Sundays River during the Late Cenozoic as an almost continuous erosional and depositional record spanning the period from Late Miocene to Holocene. Terrace sediment properties and associations give detailed insight into former conditions within the fluvial system whereas the geographical positions of terraces and erosional surfaces provide valuable information regarding palaeoriver and palaeolandscape morphology. During the Late Miocene to Holocene, fluvial development in the Sundays River valley and landscape evolution in the surrounding areas were strongly influenced by external factors such as sea-level change, climate variation and neotectonic activity.

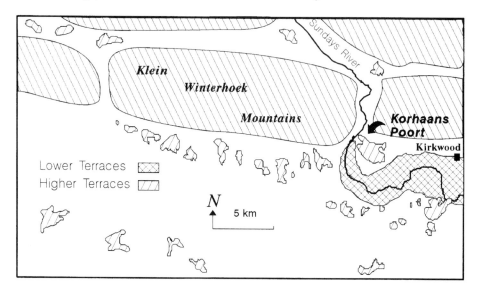

FIGURE 6.5   Abandoned terrace remnants in the palaeo-Sundays River valley, west of the present entrance of the Sundays River into the Algoa Basin at Korhaans Poort, indicate the existence of large-scale fluvial activity in this region prior to the capture of the Groot River by the Gamtoos River

Terrace development in the lower Sundays River valley

Substrate geology of the lower Sundays River valley had a controlling influence on the formation of the valley and its terraces. The upstream boundary of the lower Sundays River valley lies at the Klein Winterhoek Mountains consisting of erosion-resistant quartzite. This mountain range at the gorge acted throughout the Late Cenozoic as a fixed knickpoint in the Sundays River profile. Upstream of the knickpoint the Sundays River valley is underlain by a combination of erosion-resistant and erosion-prone Karoo rocks. Fluvial deposits in this area upstream of the fixed knickpoint display poor preservation and occur as isolated terrace features. Downstream of the Klein Winterhoek Mountains the lower Sundays River valley is underlain by highly erodible mudstone and shale into which the river can cut readily during intrinsic or extrinsic adjustments to the fluvial system. The lower Sundays River valley is therefore a much more sensitive indicator of changes in the dynamic regime of the drainage basin than the Sundays River valley upstream of the knickpoint. Despite the relatively erodible nature of the lower Sundays River valley substrate, terrace features were nevertheless well preserved owing to the fact that the river experienced continuous eastward migration throughout the Late Cenozoic as a result of tectonic tilt.

A flight of 13 fluvial terraces flanks the lower Sundays River valley, mainly on its western side (Figure 6.6). The highest terrace level lies 220 m above and 15 km west of the present river. By projecting fluvial terrace profiles onto marine terraces bordering Algoa Bay, these fluvial terraces were correlated to the marine terraces with known ages, enabling the relative dating of the lower Sundays River valley terraces. Based on relative age, sedimentological and spatial characteristics, the 13 terrace levels logically fall into two groups. The Late Miocene to Late Pliocene Higher Terraces lie between 220 and 40 m above the present river bed and occur mostly on the western side of the valley. The Pleistocene to Holocene Lower Terraces are limited to the present-day floodplain region between 25 m and 3 m above the present river bed. The oldest terraces are considerably dissected by tributaries of the Sundays River, reducing these once-vast terrace features to gravel-capped hills.

Terrace deposits

Fluvial deposits, varying in thickness between 3 m and 12 m, in places cover extensive remnants of straths in the lower Sundays River valley. Deposits on the Higher Terraces (levels Tτ1 to Tτ9) consist of gravel with subordinate very coarse-grained sand. The Lower Terraces (Qτ10 to Qτ13) consist of fine-grained sand and silt, with subordinate coarse-grained sand and gravel. Terrace gravel clasts consist mainly of Devonian quartzite and sandstone, an appreciable percentage of Karoo diamictite, hornfels, dolerite and sandstone, and a very small fraction of Cretaceous lithotypes. Sedimentary structures and fabric of the gravel deposits on the nine Higher Terraces display a distinctive downvalley variation in composition and character.

The best, most proximal exposure of the Higher Terraces is a 10 m high gravel cliff extending 340 m laterally, 3 km south of Kirkwood (Figure 6.7). At the proximal end of the lower Sundays River valley, immediately south of the Klein Winterhoek Mountains, poorly sorted Higher Terrace gravel units consist of polymodal assemblages of clast-supported

154

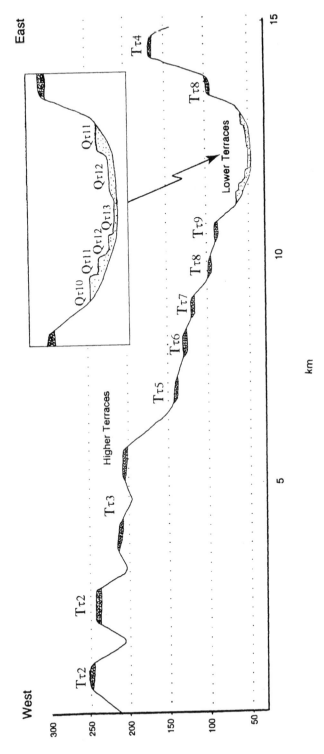

FIGURE 6.6 A cross-section of the Sundays River valley 20 km downvalley from Kirkwood, showing those terraces present along the cross-section line, flanking the valley mainly on the western side. Terraces Tτ2 to Tτ9 contain gravel whereas deposits on terrace levels Qτ10 to Qτ13 consist of silt and fine-grained sand. Terrace Tτ1 is absent at this locality. The eastward migration of the Sundays River is clearly indicated by the stepped terraces. The presence of terrace Tτ4 on the eastern side of the valley possibly displays a hiatus in the migration rate. Terrace Tτ8 gives an indication of the width of the palaeo-Sundays River braid belt

FIGURE 6.7   A 10 m high cliff face exposing gravel deposits of one of the Higher Terraces. Here a lag deposit of large boulders in a channel structure can be seen at the base of the terrace deposit overlying the Cretaceous shale

boulder and pebble units up to 0.4 m thick. The proximal gravel deposits contain subrounded clasts and a large fraction of very coarse-grained sandy matrix. Bedding is not apparent because of the poor grading in individual gravel units. Layers of very coarse-grained sand interbedded with horizontally bedded gravel units are present in a few places.

Approximately 30 km downvalley from the location where the proximal Higher Terrace deposits were described, deposits of the medial lower Sundays River valley are exposed in a road cutting. The well graded gravelly deposits of the medial part of the lower Sundays River valley display very distinct horizontal bedding and cross-stratification. The subrounded to rounded clasts in these medial gravel deposits are much better sorted and the gravel has a lower matrix content than that of the proximal deposits. The matrix consists of medium- to coarse-grained sand. Coarse-grained sand units are absent from these medial lower Sundays River valley deposits.

Distal Higher Terrace deposits are exposed approximately 27 km downvalley from the medial lower Sundays River valley outcrops in a 5 m high railway cutting. At the distal end of the Sundays River valley, terrace gravel contains rounded and well sorted clasts. Cross-stratified gravel is more common than at the more proximal localities. A marked reduction in grain size at the distal outcrops is also apparent. Horizontally bedded gravel units with clasts up to cobble size are in places overlain by large-scale cross-beds up to 4 m thick. Maximum grain sizes in these large-scale cross-bed units are large pebbles. Medium-grained sand commonly occurs as horizontal beds or as lenses interbedded in the gravel (Hattingh, 1994).

The Lower Terraces in the proximal part of the lower Sundays River valley developed on both sides of the present river course and are confined to the immediate vicinity of the

FIGURE 6.8   An abrupt change is evident from the gravel-dominated Higher Terraces to the fine-grained sand- and silt-dominated Lower Terraces. This change is interpreted as resulting from transgressions occurring during the Quaternary interglacial periods when higher base levels produced gentler river gradients and thereby limited the flow competence to transporting comparatively fine-grained sediment particles only. Gravel units in these deposits may indicate periods of glacially driven sea-level lowering, which subsequently resulted in steeper gradients due to river incision

present-day Sundays River floodplain. In the distal reaches, Lower Terraces are preserved to the west of the present-day river course. Here, the youngest terraces are immediately to the west of the river adjoining the river course and the oldest are on the western limit of the alluvial valley. Terraces $Q\tau 10$ and $Q\tau 11$ are very well developed in the proximal part of the valley whereas $Q\tau 12$ and $Q\tau 13$ formed much further downvalley. Remnants of older channels and oxbow lakes filled with sediment are clearly evident on terrace $Q\tau 13$. The maximum combined thickness of the Lower Terraces exceeds 7 m. Horizontal bedding is the primary depositional feature, but channel structures are common in vertical section. Deposits of the Lower Terraces consist predominantly of unconsolidated silt and fine-grained sand units with subordinate gravel and coarse-grained sand layers (Figure 6.8). Lower Terrace gravel layers are best developed at the proximal part of the lower Sundays River valley. However, thick gravel units present as channel fills in the distal lower Sundays River have been encountered in boreholes close to the river mouth.

Extrinsic influences on the development of the lower Sundays River

Fluvial response to sea-level change

The oldest fluvial terrace remnants in the Sundays River valley have been correlated to a Late Miocene relative sea-level stand of approximately 300 m above present sea level

(Hattingh, 1996). Sea levels at any time during the Pliocene were unlikely to have been higher than approximately 35 m above present sea level (Dowsett et al., 1994). The very high Late Miocene and Pliocene relative sea-level stands recorded in the Algoa Bay area (Figure 6.9) reflect subsequent tectonic uplift in this region. Relative sea level in Algoa Bay fluctuated from approximately 300 m above present-day sea level, to well below present-day sea level, with an overall fall in sea level during the Late Cenozoic (Le Roux, 1989). Sea-level lowering was the major driving force behind the formation of fluvial terraces in the Sundays River valley. Attempts to relate past sea levels to river terraces have usually proved less than entirely successful, owing to poor preservation of the marine to fluvial transition zone (Dawson and Gardiner, 1987). However, the relationship between marine terraces and fluvial terraces in the northern part of Algoa Bay and the immediate hinterland serves as conspicuous confirmation of their association, despite the absence of a direct link. This close correlation between marine terraces and fluvial terraces can be visualized by means of a contour map of this area (Figure 6.9). Shortly after each sea-level decline, the river established a new profile, as the initial response of a

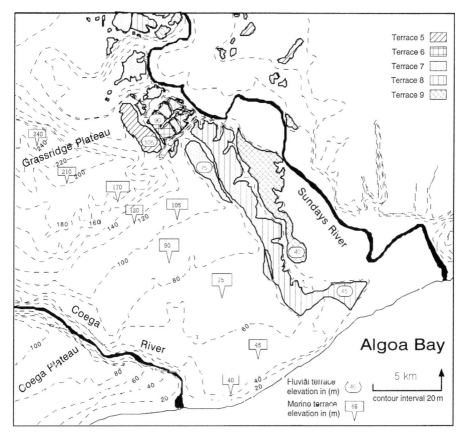

FIGURE 6.9   A contour map of the Sundays River mouth area. The close relationship between the marine and fluvial terraces is clearly shown by the positions and elevations of the marine terraces in relation to those of the fluvial terraces of the Sundays River valley

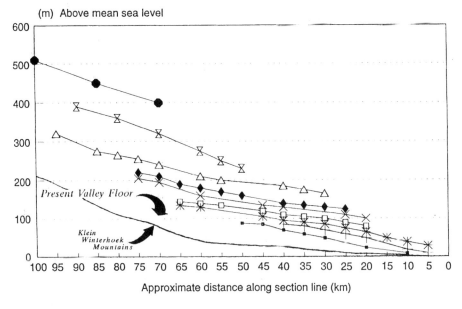

(m) Above mean sea level

Approximate distance along section line (km)

--- TERRACE 9 + TERRACE 8 ✳ TERRACE 7 ⊡ TERRACE 6 ✕ TERRACE 5
◆ TERRACE 4 △ TERRACE 3 ✕̄ TERRACE 2 ● TERRACE 1

FIGURE 6.10   Downvalley longitudinal profiles of the nine Higher Terraces. The thick solid line indicates the present-day river profile and the profiles of the Lower Terraces in the lower Sundays River valley. Stream capture altered the route of the Sundays River shortly after the establishment of terrace Tτ5 when discharge from the upper Sundays River reached the lower Sundays River via the breach through the Klein Winterhoek Mountains

fluvial system to a lowering in sea level is one of vertical incision (Wood et al., 1993). Soon after each lowering event, a new knickpoint formed near the river mouth and progressively migrated upstream towards the middle reaches of the river, establishing a new river profile at a level below the existing floodplain. Subsequent drops in sea level led to the development of successive terraces, each formed at a lower level.

The fixed knickpoint at the quartzite gorge through the Klein Winterhoek Mountains acted as a hinge in the river profile limiting the influence of sea-level change to the lower reaches of the Sundays River. The proximity of this fixed knickpoint to the coast produced very steep gradients in the lower Sundays River (Figure 6.10) that generated the high stream power values, evident from the large boulder bedload associated with the Higher Terraces. The cobble to boulder gravel sizes of the nine Higher Terraces imply steep river gradients (average 0.0023) maintained in the lower Sundays River by a continuous fall of sea level throughout the Late Miocene to Late Pliocene.

Rising sea level to above that of the present sea level characterized interglacial periods experienced mainly during the Quaternary. These transgressive events partly drowned the lower reach of the Sundays River. These high sea-level stands caused fluvial and marine sediment to choke the lower Sundays River as it accumulated in an estuary, newly established in the lower 20 to 30 km reach of the river by the elevated base level. The

Sundays River, with its gentle gradient (0.0011) and relatively low discharge during these transgressions, was competent only to transport a suspension load (Hattingh, 1996). These conditions were responsible for deposition of floodplain sediments comprising mainly fine-grained sand and silt by the Quaternary Sundays River. Miall (1991) pointed out that elevation of base level may only affect the lower reaches of a river, the effect not being felt far upstream. Fine-grained deposits formed when four interglacial transgressions (Le Roux, 1989) presumably buried the gravel units deposited during Quaternary glacial regressions. After each cycle of sea-level fall and sea-level rise, the shoreline formed at a slightly lower level during subsequent interglacial periods. Progressive lowering in sea level produced four silt and fine-grained sand terraces at levels below the nine gravel terraces before sea level stabilized at its present level shortly after the Early Holocene (Le Roux, 1989).

The effect of tectonic tilt on fluvial behaviour

By Late Pliocene, sea level had fallen from approximately 300 m above present-day sea level to 30 m above present-day sea level (Le Roux, 1989), leaving nine gravel-bearing fluvial terraces on the western flank of the Sundays River valley. An asymmetric valley with unpaired terraces formed by eastward migration of the Sundays River (Figure 6.6) through neotectonic activity. The Algoa Basin is a half-graben; an east–west striking fault, exceeding 80 km in length, forms the northern boundary of the basin (Figure 6.11). Since their initiation, some of the major strike faults on the northern borders of the Cretaceous basins, such as the Algoa Basin, have had long histories of periodic extensional movement (Hattingh and Goedhart, in press). The neotectonism affecting the Sundays River produced an eastward tilt in the Algoa Basin due to reactivation of this Mesozoic fault (Hattingh, 1996).

For a part of its course, in the proximal lower valley, the Sundays River flows parallel to this fault and for the rest of its course, in the distal lower valley, perpendicular to the structural strike. Although this half-graben was established during the Mesozoic, the fault remained active throughout the Cretaceous and Cenozoic owing to continuous subsidence as a result of basin-filling (Hattingh and Goedhart, 1997). Syndepositional tectonic tilt caused the east-northeastward migration of the Sundays River, covering a distance of at least 15 km lateral to the flow axis since the Late Miocene. The unpaired terraces of the lower Sundays River occur both parallel and perpendicular to the fault strike, confirming tilt in an east-northeasterly direction.

Analysis of the lower Sundays River valley using a technique presented by Cox (1994) satisfactorily quantifies transverse topographic basin symmetry. The technique is well suited to assess recent tectonic activity in regions where surface faults are poorly exposed or absent. Data obtained by this technique permit discrimination between random river migration and regionally preferred river migration, as well as determining the direction of maximum migration. The deflection of the active river from the drainage basin midline is used to define an asymmetry vector for each valley segment of a given length. In this calculation, transverse topographic symmetry (T) is given by $T = D_a/D_d$ where D_a = the distance from the active river to the basin midline, and D_d = the distance from the basin divide to the basin midline. Average T values were calculated for straight lines fitted to 10 km segments along the medial lower Sundays River. The bearing of the deflection of

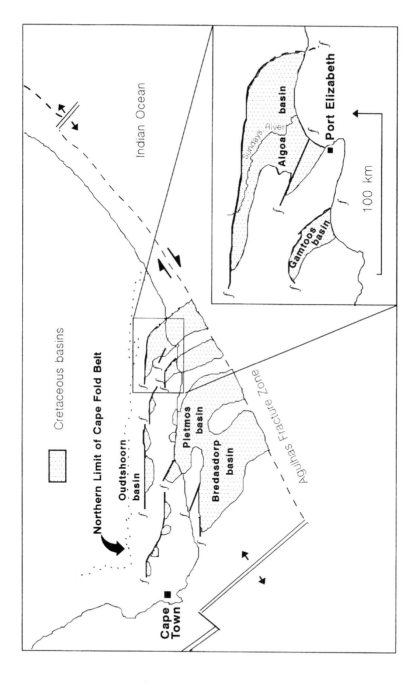

FIGURE 6.11 Mesozoic fault-controlled basins of Southern Africa in relation to the direction of regional tension (large arrows) established during the Late Jurassic–Early Cretaceous. Inset shows the location of the Sundays River in the Algoa Basin. f = FAULT

the river from the basin midline was then measured perpendicular to the fitted line. Asymmetry vectors therefore have magnitude (T value) and direction (bearing of deflection). From these data a mean vector was calculated. Statistical analysis was employed to determine the probability of whether the mean vector is the result of random processes or not. The application of this technique on the lower Sundays River valley indicated the asymmetry vector to be 0.47, 52° for the medial lower Sundays River valley, demonstrating strong evidence against random processes causing river migration.

The spatial arrangement of Sundays River terraces suggests that river migration was strongly episodic, since several hiatuses in river migration cyclically followed more active phases of movement. The response of the lower Sundays River to tilting depends strongly on the relative balance between the magnitude and frequency of the tilting event, and the available stream power and sediment supply, i.e. the potential for the system to adjust to the tilting. The rates of active tilting and resulting adjustment in the Sundays River system must have maintained an equilibrium throughout the Late Cenozoic development of the Sundays River. Even the Quaternary period represents a period of active tilting as indicated by the Lower Terraces of the distal Sundays River valley and the mouth of the present-day Sundays River lying at the easternmost limit of the valley.

The most significant influence of rejuvenation of the Algoa Basin fault on development of the Sundays River, apart from river migration, was the uplift of the source area. Episodic uplift and associated sedimentation pulses increased deposition of coarse-grained material in terrace deposits of the lower Sundays River valley in the Algoa Basin. Gravels are relatively uncommon in the sedimentary record. Thus, the appearance of gravel in the sedimentary record commonly indicates tectonic activity in the source area (Rust and Koster, 1984). Gravel deposition in alluvial basins is controlled mainly by tectonism in the source area increasing relief, thus increasing sediment supply rates, stream power and, as a result, the ratio of the gravel to fine-grained sediment fractions.

Fluvial response to climate variation

Climatic conditions in Southern Africa during the Late Cenozoic varied considerably. Warm, tropical conditions with very high rainfall prevailed throughout most of the Neogene (Tyson, 1986), presumably resulting in high discharge floods in the Sundays River between the Late Miocene and Late Pliocene. The prevailing high discharge rates, coupled with steep gradient, were responsible for mass sediment transport during formation of the Higher Terraces. The warm, humid climate also encouraged weathering and erosion in the drainage basin and increased sediment supply to the river. Alluvial fans entering the valley from the surrounding mountains fed large amounts of sediment and water into the valley. The large areal extent of the nine Higher Terraces (Hattingh, 1996) suggests that from the Late Miocene to the Late Pliocene, the Sundays River consisted of a wide braid belt occupying the entire 2 to 3 km wide floor of the alluvial valley (Figure 6.6), indicating exceedingly high discharge flows (Hattingh, 1994). Downstream of the sediment source area, the channels were not confined and vast alluvial plain piedmont gravel deposits formed on the floodplain.

Dramatic global changes in climate affected Southern Africa during the Quaternary, resulting in a steep decline in temperature and increased aridification due to major glacial periods (Tyson, 1986). The cooler, dry climate accompanied a drop in sea level to well below the present sea level. A sea-level minimum stand in Algoa Bay occurred at 13 000 BP when a shoreline formed 110 m below present-day sea level (Bremner et al., 1991). During such glacially induced lowering of sea level, the valley became deeply incised, with the Sundays River assuming a very steep gradient. These steep gradients rendered the river competent to transport a coarse-grained bedload. The change in the climate and temperature prevented tropical cyclone-induced rain storms in the catchment area. Discharge declined in the Sundays River, but the steep slope of the river profile ensured the maintenance of gravel transport, although only pebble-sized gravel clasts were transported. No outcrop of gravel deposits, formed during these regressions, occurs amongst the Lower Terraces, as the gravel units were buried by silt and fine-grained sand deposited during subsequent transgressions. Evidence for this was obtained from four boreholes, 5 km upstream from the present river mouth, showing three gravel units at 16, 22 and 32 m below surface, interbedded with silt units (Hattingh, 1996). This evidence is supported by the submerged, gravel-bearing fluvial terraces of the Sundays River discovered by Bremner et al. (1991) in Algoa Bay near the Sundays River mouth.

FLUVIAL FORM DEVELOPMENT IN THE LATE CENOZOIC SUNDAYS RIVER

Mean annual flood and bankfull discharge strongly influences channel morphology in the middle reaches (Schumm, 1985). Together with the type and volume of available sediment, the discharge parameters therefore determine the size and morphology of the river channel best suited to accommodate the discharge of middle-reach streams. The fluvial channel pattern typical of the Neogene Sundays River middle reach is a bedrock-controlled bedload channel with a classic braided stream character. Prevailing climatic conditions and high erosion rates caused high sediment and water discharge. The bars and thalweg of the Neogene Sundays River middle reach probably shifted within an unstable channel, having a large mean particle size, mostly boulder sized, and high sediment load. Coarse bedload preserved as gravel terrace remnants flanking these valleys confirms the steep gradients of these braided streams.

By the Quaternary, regional denudation had reduced river gradients and, with drier climatic conditions, produced a meander–braided transition stream pattern (Schumm, 1985) in the middle reaches. Sediment volumes, although still large, reduced drastically from the Neogene levels, and sand, gravel and cobbles formed a significant fraction of the sediment load. Channel morphology and terrace sedimentological analysis showed that channel width varied but was relatively wide compared to depth. The stream gradient was steep, but less so than during the Neogene, as indicated by the size reduction of Quaternary bedload material.

In the lower reach or floodplain reach, a typical alluvial channel characterized this part of the river throughout its Late Cenozoic development. Higher Terrace deposits in the proximal lower Sundays River valley indicate that braided streams, hyper-concentrated flood flows, stream floods and sheetflood conditions resulted in the accumulation of Higher Terraces gravel deposits in this part of the valley. The Sundays River experienced major floods when tropical cyclones managed to move in over the catchment area. Higher

Terrace gravel probably accumulated during exceptional floods triggered by brief, yet intense rainstorms, producing cyclone-induced event beds (Hattingh, 1996). The high fraction of quartzite clasts in the terrace deposits and preservation of palaeoalluvial fans suggest that alluvial fans debouching from the Klein Winterhoek Mountains probably supplied the bulk of coarse clastic sediment to the palaeo-Sundays River valley (Hattingh, 1994). In this region, steep mountain slopes, high sediment accumulation rates and high precipitation rates produced debris-flow and sheetflood conditions supplying sediment to the fans and braided streams that joined the palaeo-Sundays River (Figure 6.12). Sediment transported and deposited in this manner was partly redistributed in the aftermath of the main flood event. Higher Terrace deposits in the proximal valley very likely formed by combinations of these processes.

Sedimentological conditions were notably different in the medial part of the valley. Gravel deposits there are texturally mature and well stratified, probably the result of stronger channelized flow, more continuous runoff and effective contemporaneous reworking in a well developed braid plain. During conditions of high discharge and sediment supply, existing gravel lags probably developed into longitudinal or diagonal bars with horizontal stratification or, under conditions of lower discharge and a decreased sediment supply, into transverse bars with cross-stratification (Hattingh, 1994). These conditions are typical of a braided river.

Whereas perennial flow with continuous runoff in well defined channels characterized the palaeo-Sundays River in the medial part of the valley, the river gradient in the distal parts of the valley became more gentle, as indicated by the downvalley profiles of the Higher Terraces (Figure 6.10), where wider reaches and deeper flow would occur. A high-sinuosity (meandering) stream pattern was likely to have prevailed, suggested by planar cross-stratification and finer-grained point bar gravels in outcrops of the distal valley (Hattingh, 1994). The transition from coarse-grained gravel to the overlying finer-grained planar cross-stratified gravel was probably the result of river slope adjustment after sea level fell from the level of one terrace to the succeeding lower level.

The Neogene mixed-load, braided Sundays River transformed into the meandering, suspended-load river towards the Quaternary. The abrupt change from the gravel-dominated Higher Terraces to the fine-grained sand- and silt-dominated Lower Terraces is interpreted as resulting from transgressions during Quaternary interglacial periods following the Neogene overall regression phase. Higher base levels during the interglacial phases produced gentler river gradients and thereby limited the flow competence to transport large sediment particles. The cool and dry periods during the Quaternary coincided with the termination of a cyclone-dominated climate regime (Tyson, 1986), causing moderate flow conditions to prevail during deposition of the Lower Terraces. Reduced runoff due to increasing aridification contributed to the marked reduction in grain size of the Lower Terraces.

At present, a suspended-load channel forms the lower 25 km reach of the river on the coastal platform before entering the ocean. Here the river channel is relatively straight with a uniform width. The Sundays River carries a clay and silt suspended load and a very small bedload of sand and gravel. Gradient is low, and the channel is relatively narrow and deep. Falling sea levels, experienced during Quaternary glacial periods, caused the shoreline to retreat oceanward. During this regression the lower reaches of rivers along the Southern African coast, including the Sundays River, extended

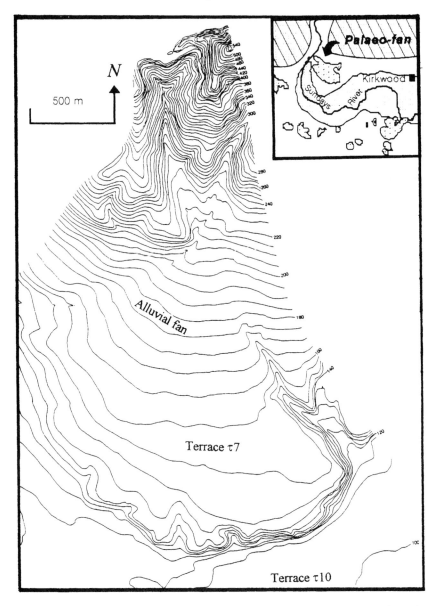

FIGURE 6.12   Palaeofan features, such as the one shown by this contour map situated on the southern slopes of the Klein Winterhoek Mountains, are still evident just above the terraces in the proximal lower Sundays River valley

oceanwards at the same rate as the shoreline, straightening the lowermost reaches to form a straight channel pattern, in contrast to the meandering channels in the hinterland. Subsequent rises in sea level during the interglacials flooded the lower river reaches. Present-day coastal rivers are still flooded by the sea, displaying the straight, drowned 5 to 30 km lower courses of these rivers.

CONCLUSIONS

Drainage basin substrate was probably the single most important factor controlling the development of fluvial systems and therefore channel morphology of the rivers draining the southern margin of Africa. This influence of bedrock lithology and structural fabric on drainage patterns is well illustrated by the variation in fluvial development styles in the middle and lower reaches of the Sundays River. River gradient, discharge regime, sediment size and sediment volume played a significant role in determining channel form during the Late Cenozoic development of the palaeo-Sundays River. These intrinsic factors were in turn modified by extrinsic influences such as base-level change, tectonic activity and the prevailing climate conditions. Fluvial terrace deposits of the Sundays River valley record the systematic development of the Sundays River drainage basin since the Late Miocene. Not only offshore but also onshore deposits, formed during the Late Cenozoic erosion cycles, serve as reliable records of the fluvial processes that prevailed in the Sundays River drainage basin. The combined influence of sea-level change, tectonic tilt and climatic variability produced complex driving forces responsible for the variation in river development. During the Late Miocene to Late Pliocene the lower Sundays River assumed a very steep gradient and experienced extremely high discharge floods due to tropical rains. Sediment production rates in the source area were high and uplift of the source area facilitated the supply of a coarse gravel-sized bedload.

Gravel-bearing, steep-gradient rivers (>0.01) differ from sandy, gentle-gradient rivers (<0.001) by generating higher stream power values and a greater tendency towards braiding. The abrupt change from gravel-dominated Higher Terraces to the fine-grained sand- and silt-dominated Lower Terraces is interpreted as the Sundays River's response to Quaternary sea-level rises that resulted in gentler gradient and reduced stream power values in the lower Sundays River. These gentle flow conditions coupled with cool, dry periods subsequently introduced fine-grained sediment to a meandering lower Sundays River during the Quaternary.

In conclusion, changes in the fluvial style, both between the depositional character of the Higher Terraces and Lower Terraces, and within the Higher Terraces themselves, are direct responses to allocyclic controls, the most important of which is sea-level change, coupled with channel form adjustment to climatic variations. Tectonism in the source area not only facilitated and sustained the supply of large volumes of sediment throughout the Late Cenozoic, but the eastward migration of the river due to tilting also ensured the preservation of the depositional record in the lower Sundays River valley.

REFERENCES

Bremner, J.M., Du Plessis, A., Glass, J.G.K. and Day, R.W. 1991. Algoa Bay – marine – geoscientific investigation. *Geological Survey of South Africa. Bulletin*, **100**, 173pp.

Cox, R.T. 1994. Analysis of drainage-basin symmetry as a rapid technique to identify areas of possible Quaternary tilt-block tectonics: An example from the Mississippi Embayment. *Geological Society of America, Bulletin*, **106**, 571–581.

Dawson, M.R. and Gardiner, V. 1987. River terraces: the general model and a palaeohydrological and sedimentological interpretation of the terraces of the lower Severn. In K.J. Gregory, J. Lewin and J.B. Thornes (Eds), *Palaeohydrology in Practice*. Wiley, London, 269–305.

Dingle, R.V., Seisser, W.G. and Newton, A.R. 1983. *Mesozoic and Tertiary Geology of Southern Africa*. A.A. Balkema, Rotterdam, 375pp.

Dowsett, H.J. and PRISM Project Members. 1994. Characterization of middle Pliocene marine climatic conditions, sea surface temperature, sea ice distribution and sea level. *Eos (Transactions of the American Geophysical Union)*, **75**(16), 206.

Hattingh, J. 1994. Depositional environment of some gravel terraces in the Sundays River Valley, Eastern Cape. *South African Journal of Geology*, **97**, 156–166.

Hattingh, J. 1996. *Late Cenozoic drainage evolution in the Algoa basin with special reference to the Sundays River valley*. PhD Thesis, University of Port Elizabeth, 181pp.

Hattingh, J. and Goedhart, M.L. (1997). Neotectonic control on drainage evolution in the Algoa basin, South-eastern Cape Province. *South African Journal of Geology*, **100**, 43–52.

Haughton, S.H., Frommurze, H.F. and Visser, D.J.L. 1937. *The geology of a portion of the coastal belt near the Gamtoos Valley, Cape Province*. Expl. Sheet 151 (Gamtoos River), Geological Survey of South Africa, 64pp.

Hill, R.S. 1988. Quaternary faulting in the south–eastern Cape Province. *South African Journal of Geology*, **91**, 399–403.

Le Roux, F.G. 1989. *The lithostratigraphy of Cenozoic deposits along the south-east Cape coast as related to sea-level changes*. MSc Thesis, University of Stellenbosch, 247pp.

Miall, A.D. 1991. Stratigraphic sequences and their chronostratigraphic correlation. *Journal of Sedimentary Petrology*, **61**, 497–505.

Partridge, R.B. and Maud, R.R. 1987. Geomorphic evolution of Southern Africa since the Mesozoic. *South African Journal of Geology*, **90**, 179–208.

Rust, B.R. and Koster, E.H. 1984. Coarse alluvial deposits. In R.G. Walker (Ed.), *Facies models* (2nd Edn.). Geoscience Canada, Ainsworth Press, Kitchener, Ontario, Reprint series 1, 53–69.

Rust, I. C. 1962. On the sedimentology of the Molteno sandstones in the vicinity of Molteno. *C. P. University of Stellenbosch Annals*, **37**, 165–234.

Rust, I.C. and Illenberger, W.K. 1989. Geology and geomorphology of the Baviaanskloof. In G.I.H. Kerley and C.M. Els (Eds), *The Kouga–Baviaanskloof complex*. RSA Directorate Nature and Environmental Conservation, 42–45.

Schumm, S.A. 1985. Patterns of alluvial rivers. *Annual Reviews Earth Planet Science*, **13**, 5–27.

Skelton, P.H. 1980. Aspects of freshwater fish biogeography in the eastern Cape. *The Eastern Cape Naturalist*, **24**, 17–22.

Taljaard, M.S. 1949. *A Glimpse of South Africa*. University Publishers, Stellenbosch, 226pp.

Tankard, A.J., Jackson, M.P.A., Eriksson, K.A., Hobday, D.K., Hunter, D.R. and Minter, W.E.L. 1982. *Crustal Evolution of South Africa*. Springer-Verlag, New York, 523 pp.

Tyson, P.P. 1986. *Climatic Change and Variability in Southern Africa*. Oxford University Press, Cape Town, 220pp.

Ver Steeg, K. 1935. Wind gaps and water gaps – their value as indicators of erosion surfaces. *American Journal of Science*, Fifth series, **30**, 98–105.

Wood, L.J., Ethridge, F.G. and Schumm, S.A. 1993. *The effects of rate of base–level fluctuation on coastal–plain, shelf and slope depositional system: an experimental approach*. International Association of Sedimentologists, Special Publication 18, 43–53.

7

Fluvial Evolution in Areas with Volcanic and Tectonic Activity: the Armería River, Mexico

DAVID PALACIOS[1] AND ARMANDO CHÁVEZ[2]

[1]Department of Physical Geography, Complutense University, Madrid, Spain
[2]Faculty of Geography, University of Guadalajara, Guadalajara, Mexico

ABSTRACT

The Armería River runs along the western side of the Colima Rift, located on the western end of the Trans-Mexican Volcanic Belt (Mexico). The Colima Volcanic Complex emerges from the floor of the rift, and one of its volcanoes, Volcán de Fuego, is active. Two factors have contributed to the development of the Armería valley: a tectonic block that borders the west side underwent intense uplifting and formed the Sierra de Manantlán, while the east received a constant flow of volcanic material that emanated from Volcán de Fuego. This chapter analyses the valley's volcanic–tectonic context and its morphology. The conclusions propose a model that explains the evolution of the valley.

INTRODUCTION

Fluvial dynamics can be altered by neotectonic processes, and their results have been relatively well studied in terms of tectonic geomorphology. Less attention, however, has been given to the effects that volcanic processes may have on stream dynamics, perhaps because the simultaneous occurrence of tectonic and volcanic activity can generate very unusual and practically unknown fluvial morphology. This is the case of the Armería River that flows into the Pacific Ocean, and is located at 19°30′ N and 103°37′ W, in the western region of the Republic of Mexico, in the states of Jalisco and Colima (Figures 7.1 and 7.2).

The river drains the Colima Rift in the western sector of a graben that was very active during the Quaternary. This chapter will examine the length of the river that flows through the central sector of the rift. The western edge of the sector is one of Mexico's most active areas at present. A group of Quaternary stratovolcanoes called the Colima Volcanic Complex emerges on the far side of the central part of the graben, and one of the group, Volcán de Fuego, is one of Mexico's most active volcanoes.

Varieties of Fluvial Form. Edited by A.J. Miller and A. Gupta
© 1999 John Wiley & Sons Ltd

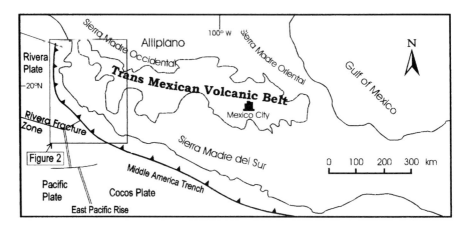

FIGURE 7.1   Location of the study area

The above factors contribute to the Armería River's evolution, which has been constantly altered by the effects of volcanism and active tectonics. The rift and the volcanoes of Colima are part of a major volcanic–tectonic feature called the Trans-Mexican Volcanic Belt. Studies conducted on the origin and activity in this area, especially the rift, provide valuable information on how the river evolved.

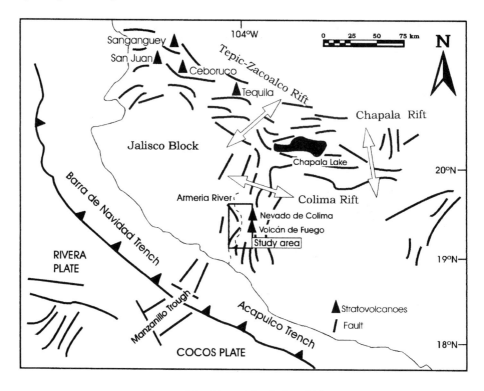

FIGURE 7.2   Map of the Colima Rift area

The results of fieldwork have been used to define a series of geomorphological units which suggest that very recent volcanic and tectonic activity has had a great influence on the dynamics of the Armería River. Based on neotectonics texts and a few articles that deal with the interrelationship between volcanic processes and fluvial dynamics, a series of phases has been identified that explains the evolution of the river. Although no precise dates are available, some absolute dating has been done and suggests that probably three phases of evolution occurred during the late Pleistocene and Holocene; these may provide a new understanding of the landforms that are typical of fluvial–volcanic–tectonic processes.

THE VOLCANIC–TECTONIC CONTEXT: THE TRANS-MEXICAN VOLCANIC BELT

The Trans-Mexican Volcanic Belt (TMVB) is a volcanic, tectonic and geomorphological feature, located between 19° and 21°N (Mooser, 1972). The belt extends across the Republic of Mexico in roughly an east–west direction from the Pacific Ocean to the Gulf of Mexico, and is approximately 1000 km long and 25–150 km wide (Figure 7.1). The great andesitic stratovolcanoes, that are characteristic of the area, are commonly organized in a linear north–south direction. Groups of monogenetic volcanoes tend to align themselves along the northeast–southwest fault lines. This fault system is normally responsible for the many tectonic grabens that form the TMVB (Demant, 1978).

The origin of the TMVB is traditionally associated with the subduction of the Cocos Plate along the Middle American Trench (MAT) (Mooser, 1972). The process generates calc-alkaline volcanism, which is predominant in the TMVB, and produces large andesitic volcanoes (Mooser, 1972). The characteristic that differentiates the TMVB from the other volcanic belts in Central and South America, however, is that it is not parallel to the MAT, but lies at an angle to it of about 15°. Although there are many theories that try to explain this anomaly (see review in Verma (1987) as an example), one aspect is certain, and that is that the fracture systems found throughout the TMVB exert a major influence on the belt's volcanic activity (Urrutía and Böhnel, 1988). Some authors believe that the TMVB can be interpreted as a transtension sector that forms the boundary between the North American Plate and a microplate that separated from it. According to this theory, the transtension is the result of a difference in velocity between the North American Plate (2.6 cm a^{-1}) and the southern microplate (2.7 cm a^{-1}) (Shubert and Cebull, 1984).

The basement of the TMVB is composed of Palaeozoic metamorphic rocks and Mesozoic intrusive granites and sedimentary rocks, mainly limestone. An initial cycle of volcanic activity dating from the Oligocene to the Miocene produced rhyolitic and ignimbritic rocks that appear today in the west of Mexico (Sierra Madre Occidental) and also form part of the TMVB basement. It would appear then that this volcanic phase is disassociated from the formation of the TMVB (Demant, 1978). Studies show that the TMVB dates to no earlier than the Pliocene–Quaternary transition (Demant and Robin, 1975), so its volcanic–tectonic formation would have begun about 2 Ma BP, and is still active today.

Alternating grabens and horsts define the geomorphology of the TMVB, and are associated mainly with normal and strike-slip faults, the majority of which register seismic activity (Johnson and Harrison, 1990). Studies of the neotectonic activity in specific areas of the TMVB have been conducted (Suárez and Singh, 1986; Martínez-Reyes and Nieto-Samaniego, 1990; Suter et al., 1992), but very little has been written about the geomorphological consequences (Mooser and Ramírez-Herrera, 1989; Ortíz-Pérez and Bocco-Verdinelli, 1989; Ramírez-Herrera et al., 1994).

The TMVB underwent a long period of volcanic–tectonic activity that constantly thwarted any attempts to reorganize the drainage system. As a result, lacustrine conditions are common in most of the TMVB and any drainage to the tributaries of major rivers that flow outside of the volcanic belt is weak.

NEOTECTONIC AND VOLCANIC ACTIVITY ON THE COLIMA GRABEN

The western sector of the TMVB is marked by a triple-junction graben whose nexus is located at 20°15′ N and 104°20′ W, 50 km southwest of the city of Guadalajara (Figure 7.2). Towards the east is the Chapala graben, 100 km long and up to 40 km wide, which is occupied by Chapala Lagoon. A series of normal faults trending east–west create a boundary between the graben and a group of blocks parallel to it, which results in the formation of a rift (Allan, 1986). The fracture system has generated a large field of monogenetic volcanoes that emerge mainly to the south and southeast of the rift and is called Michoacan Volcanic Field.

The Tepic-Zacoalco graben extends to the northeast and is more than 200 km long and up to 50 km wide. It is divided into three small basins that are confined by faults that are normally of a strike-slip nature, trending northwest–southeast, and are parallel to the transform faults of the Gulf of California. In addition to the fields of monogenetic volcanoes and volcanic calderas, large stratovolcanoes also occupy the Tepic-Zacoalco graben (San Juan, Sanganguey, Tequila, etc.), some of which are active (Ceboruco).

The Colima graben, which is 100 km long and up to 30 km wide, lies to the south and is bounded by an inward-dipping normal fault (Allan, 1986). It registers a great deal of seismic activity that had an impact in the 1985 and 1995 earthquakes. The most important volcanic complex of the western sector of the TMVB is found on this graben and consists of the Colima group of andesitic volcanoes (Cantaro, Nevado de Colima and Volcán de Fuego) which has a volume of approximately 850 km^3 (Luhr and Carmichael, 1980). Volcán de Fuego is presently one of the most active volcanoes in Mexico.

Axes of the three graben structures form angles of 100°, 115° and 145° with each other, counterclockwise from Colima graben. The feature is regarded as an example of a Y-shaped rift structure (Allan et al., 1991).

The basement of the rift is composed of Mesozoic granodiorite and limestone and Miocene ignimbrites. The volcanic activity associated with the formation of the rift began in the early Pliocene with calc-alkaline andesites, typical of subduction arcs. Subduction of the Rivera Plate formed the northern sector of the MAT termed the Barra de Navidad Trench and subduction of the Cocos Plate in the southern sector of the MAT produced the Acapulco Trench (Figure 7.2). In the latter sector, the calc-alkaline volcanism was accompanied by a significant number of alkaline volcanoes dating from the early Pliocene (Allan and Carmichael, 1984). This material and the calc-alkaline products were introduced along the flanks of the grabens and in volcanoes like the stratovolcanoes of the Colima Complex.

Alkaline volcanism is typical of the birth of continental rifts and is associated with the separation of a continental block from the mainland. In this respect the spreading of the Jalisco block from the mainland of Mexico would be like the formation of the Peninsula of California (Luhr et al., 1985; Allan, 1986; Allan et al., 1991; Delgado, 1993). Absolute dating has recently detected a period of slow spreading rate, coinciding with continuous normal faulting and alkaline volcanism in the rifts, especially in the Colima Rift (1.4–0.2 Ma BP) (Delgado, 1993).

The Colima Rift began to form in the early Pliocene, and its growth since then has been associated with alternating periods of calc-alkaline and alkaline volcanism. Formation of the great stratovolcanoes of the Colima Complex are generally related to a fast convergence rate in Barra de Navidad Trench, and phases of mainly monogenetic alkaline volcanism, with a slow convergence rate and the formation of normal faults. The Colima Rift is composed of a system of inward-facing, high-angle normal faults, trending north-northeast (Allan, 1986). The most active faults are found on the western border of the central graben (Allan et al., 1991). It is possible that the rift continues into the Manzanillo trough that separates the Barra de Navidad Trench from the Acapulco Trench (Bourgois et al., 1988; Bandy et al., 1988).

Three large stratovolcanoes which are aligned in a N–S chain in the centre of the middle sector of the Colima Rift form the Colima Volcanic Complex. Cantaro is the most northern volcano and has been inactive since the mid-Pleistocene. To the south is Nevado de Colima, dating from at least about 0.6 Ma BP and whose final eruptive episodes are associated with the late Pleistocene. The third volcano, Volcán de Fuego, erupted from the southern slope of Nevado de Colima and formed around 0.2 Ma BP (Robin et al., 1987). Unlike Nevado de Colima, Volcán de Fuego is still active.

Nevado de Colima underwent three major "Mount St Helens" type events, each of which was responsible for the formation of a caldera and the emission of great debris avalanches (Robin et al., 1987). The most recent of these occurred about 18 520 ± 260 years BP and moved in primarily an easterly and then southerly direction for at least 120 km, reaching the sea (Stoopes and Sheridan, 1992).

Volcán de Fuego emerged following the destruction of an earlier volcano (Palaeofuego) that suffered a slide of 10 km^3 of upper cone. Consequently, Volcán de Fuego generated a horseshoe-shaped caldera with a 4 km diameter. Several collapses formed massive volcanic debris avalanches that moved south to a distance of at least 70 km from the source and buried more than 1550 km^2 of surface area. The new volcano emerged from the centre of the old caldera and grew rapidly in subsequent years because of the accumulation of volcanic domes.

GEOMORPHOLOGICAL CONTEXT OF THE STRETCH OF THE ARMERÍA RIVER LOCATED IN THE CENTRAL SECTOR OF THE COLIMA GRABEN

There are several recent studies on the geomorphology of the Armería River Valley. Some focus on the geomorphology to the east of the river where the Colima Volcanic Complex is located (Lugo et al., 1993), while others analyse the general aspects of the influence of tectonics on the river (Ortíz-Pérez, 1990) or the effects of the strike-slip fault, downstream from the sector selected for this chapter (Ortíz-Pérez et al., 1993).

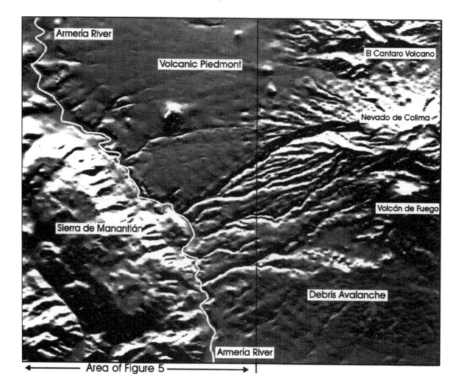

FIGURE 7.3   Geomorphological units of western Colima Rift (central sector)

The Armería River flows through the central sector of the Colima Rift at an altitude of 700–400 metres above sea level (m a.s.l.) and hugs the base of an uplifted western block called the Sierra de Manantlán (maximum altitude 2520 m a.s.l.) (Figures 7.3, 7.4 and 7.5), which basically follows the fault line. This explains why the western valley wall is very high (2300 m) and steep (average slope 65%).

The eastern edge of the valley is formed by the volcanic piedmont of Nevado de Colima and Volcán de Fuego de Colima. The gradient on this side is low (5–10%), but at the base of the stratovolcanoes it abruptly increases, climbing steeply to the summits of Nevado de Colima (4260 m) and Volcán de Fuego (3960 m).

The Sierra de Manantlán is formed by massive horizontal layers of Cretaceous limestone. The horst is bordered by sets of normal faults, trending NNW–SSE and N–S. The enormous uplift of this block was reported in tectonophysics studies (Allan et al., 1991).

The rapid uplifting and the distension exerted on the slopes caused the edges of the horst to fracture into a series of small secondary blocks that generally follow simple slip planes. Sometimes sub-blocks formed by distension dip inward, but usually they tilt towards the outside and assume the appearance of a tectonic anticline. The plateau-like upper surface of the Sierra is subject to intense karst development.

The Sierra has not been affected by stream incision, which is why the slopes are nearly vertical and why there is little talus which is characteristic of slope erosion processes. The

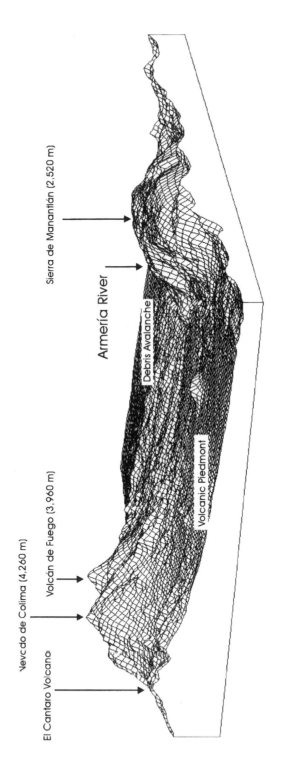

Figure 7.4   Block diagram of western Colima Rift (central sector)

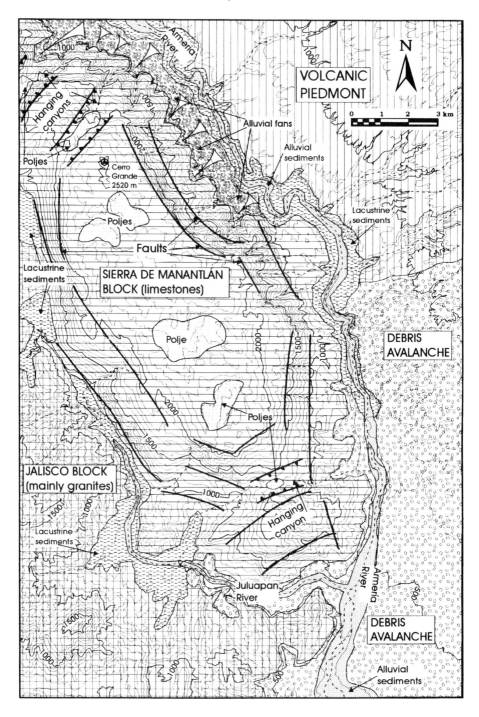

FIGURE 7.5   Geomorphological map of the study area

northeast side of the Sierra is the only place where a few alluvial fans formed and were later downcut by the Armería River. These fans are attached to the graben walls and have no apparent relationship to the present stream or torrent channels that originate in the Sierra.

The western boundary of the Sierra de Manantlán is also defined by a network of fault lines that separate it from a Mesozoic granodioritic complex called the Jalisco block. The geomorphology of this granitic sector is heavily influenced by chemical weathering and intense fluvial dissection. A series of valleys has eroded headward towards the northeast and encountered the wall of the Manantlán horst. Endoreic basins of lacustrine sediments form where these valleys and the western wall of Manantlán meet. The basins that lie furthest to the south are currently drained by a tributary of the Armería River known as the Juluapan River, which borders the Sierra de Manantlán to the south and follows the major fault lines.

The volcanic piedmont on the eastern bank of the Armería River can be divided into two sectors. The northern portion is formed mostly by pyroclastic deposits from Nevado de Colima. They have various origins including ash falls, ash flows and nuée ardente-type flows. The incision of the Armería River reveals many palaeovalleys, where a sequence of very complex stream incisions was later buried by volcanic materials. The great debris avalanches that were responsible for the most recent destruction of Nevado de Colima, and formed its upper caldera at the end of the Pleistocene, flowed towards the eastern side of the volcano and left no remains on the western slope except for pyroclastic materials, mainly ash fall, that are associated with the genesis of the caldera.

The southern sector is composed mainly of material from the great debris avalanches that destroyed the volcano called Paleofuego de Colima, and formed its caldera. The present-day cone of Volcán de Fuego emerged from this caldera. The avalanches were related to several distinctive time intervals, and early research determined the following corresponding ages: 4280 ± 110 years BP (Luhr and Prestegaard, 1988), 4360 ± 140 years BP (Luhr and Carmichael, 1990) and 4350 ± 100 years BP (Robin et al., 1984). Other studies revealed that the materials from the same avalanche could be dated from 9370 ± 400 years BP (Robin et al., 1987). These data confirmed the existence of two different avalanches (Martin del Pozzo et al., 1990). Recent analyses of the charcoal found in the avalanche materials dated it from 2690 ± 40 years BP (Siebe et al., 1992), indicating that there had been multiple collapses. Finally, the latest efforts suggest that there were at least 10 avalanche episodes associated with the following ages: 2500, 3600, 7000, 9700, 14 000, 18 000, 21 500, 28 000, 39 100 and 45 000 years BP (Komorowski et al., 1994). The relatively fresh appearance of the landforms on the upper surface of the avalanche, including numerous hummocks, closed undissected basins and little, if any, stream incision, supports the theory that the last avalanche is recent (Luhr and Prestegaard, 1988). More detailed geomorphological studies reveal, however, that there are other, more eroded surfaces that might belong to much older avalanches (Lugo et al., 1993).

GEOMORPHOLOGICAL EVIDENCE FOR UPLIFT OF THE TECTONIC BLOCK OF THE SIERRA DE MANANTLÁN

In addition to proof of seismic activity along the faults that border the Manantlán horst, there is also clear geomorphological evidence for rapid uplift of the block (Figures 7.5 and 7.6).

176

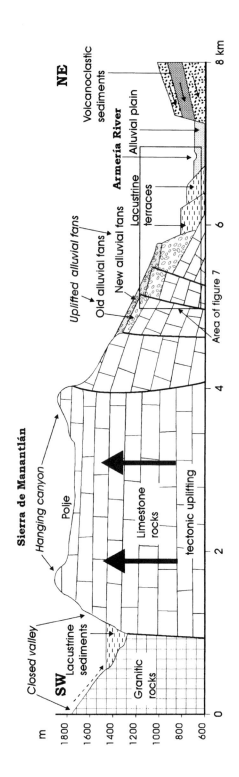

FIGURE 7.6   Cross-section of the Sierra de Manantlán-Armería valley, San Pedro de Toxín area

The heads of the valleys of the granitic mountains of the Jalisco block encounter an abrupt obstruction formed by a fault escarpment on the southwestern slope of Sierra de Manantlán, which tectonically seals off the valleys. The extensions of these valleys are the palaeocanyons that have been uplifted with the upper surface of Manantlán. The floors of these old canyons lie 500–1000 m above the former channel. Violently uplifted and stripped of their streams, the canyons were exposed to intense karst processes and transformed into poljes. The valleyheads were filled with lacustrine sediments and were partially drained by the Manantlán subterranean network.

The stream channels evolved towards the interior of the Colima graben and deposited sediments at the mouths of the canyons in the form of broad alluvial fans. The Armería River eroded the middle and distal sectors of the fans, so that today only the apices of some of them survive. The fan deposits are cemented together by a calcareous crust. The lithological composition of these deposits is characterized by a mixture of crystalline and limestone materials. The dislocation of the alluvial remnants along the fault lines is traced on the surface of the crust that binds the deposits together. In the larger canyons, small alluvial fans are superimposed on the broader ones, and they consist exclusively of limestone materials without a calcareous crust. This second generation of fans may reflect the period just prior to the termination of water flowage from the headwaters in the granitic sector.

The closed valleys, the persistence of the canyon geomorphology despite the subterranean drainage that transformed it into a polje, and the sudden separation of the apices of the alluvial fans from their channels, are all geomorphological features indicating that the Manantlán block was uplifted rapidly, which typically occurs along the edge of an active rift (Gerson et al., 1985; King and Ellis, 1990).

The rapid uplift, the compactness of the limestone block and the tendency for the Armería River to erode the base of its eastern side and to remove the material that falls from the slopes have all contributed to the fault escarpment's vertical profile, which under normal conditions would have eroded to a more gentle incline (Crozier and Pillans, 1991).

GEOMORPHOLOGICAL EVIDENCE FOR BLOCKAGE OF THE ARMERÍA RIVER VALLEY BY AVALANCHES FROM VOLCÁN DE FUEGO

The volcano Paleofuego was destroyed by at least one, and probably multiple debris avalanches, between 45 000 and 2500 years BP. Geomorphological features reveal that these avalanches completely filled the ancient Armería River valley. Downstream from the area chosen for this study, the river meanders in a braided channel pattern over avalanche material and does not downcut (Figure 7.5).

Upstream, however, the valley is incised where the avalanche spread against the Sierra de Manantlán and here a hummock projects from the valley floor. In this area the river generally downcuts to 10–15 m and the avalanche surface forms a terrace, but in the area where the avalanche piled into a higher hummock, it exhibits as much as 50 m of downcutting.

Further upstream from the point discussed above and in an area unaffected by the avalanche, the valley is filled with lacustrine sediments that can be separated into two distinct terraces, one at +12 m and the other at +20 m above the present river bed. These terraces converge upstream.

The fact that the lacustrine sediments are terraced seems to support the theory that there were two avalanches, but in order to prove this more fieldwork must be conducted and more reliable information, including absolute dating, must be obtained to distinguish between the volcanic units that generate each lacustrine terrace.

The avalanche also filled the Juliapan River, a tributary of the Armería, and formed a large barrier where the two joined. Upstream from the barrier, a great, closed basin was formed and filled with a single layer of lacustrine sediments. The present-day river cuts 50 m through the avalanche barrier thus recovering a drainage outlet for its headwaters.

EVIDENCE IN THE SEDIMENT ANALYSIS OF SAN PEDRO DE TOXÍN AREA FOR TECTONIC UPLIFT IN SIERRA DE MANANTLÁN AND BLOCKAGE OF THE ARMERÍA RIVER

The sector of the Armería River near the town of San Pedro de Toxín is an excellent spot to observe the geomorphological consequences of two merging processes: the tectonic uplift of Sierra de Manantlán and the blockage of the river (Figures 7.6 and 7.7).

An alluvial fan descends from Sierra de Manantlán through a canyon that was uplifted during the time of the formation of the Manantlán horst complex (Figure 7.7, Spots 6 to 9). The vertical movement caused the apex of the alluvial fan to disconnect from the rest of the formation. The fan is composed of several superimposed layers, the oldest of which were most affected by faults. The results of granulometry of sands for each alluvial layer reveal extremely heterometric forms, caused by poor sorting and cumulative size distributions, and display steeply inclined logarithmic curves, indicating little transport development. The size sorting of the clasts yields similar results: the classification is very heterometric, although the overall grain size tends to be large. The gorge now incises the alluvial fan (Spot 10), and the deposits found on the floor have a more homogeneous classification of clasts and sands, indicating more effective sorting as a result of transport processes.

Two terraces appear below the front of the alluvial fan that is incised by the Armería River. The granulometry of sands (Spots 1 to 4) reveals an abundance of fines (>0.06), which clearly reflects the lacustrine origin of the terraces. The graphic representation of the analytical results displays well defined hyperbolic cumulative curves. There are almost no fragments larger than 2 cm.

The properties of the existing deposits in the present alluvial plain of the Armería River suggest that there has been intensive transport of materials (Spot 5). The cumulative curve of the sand fraction has a sigmoidal aspect with a steeply sloping central part. It should also be noted that there were almost no fine or very coarse materials. These findings denote that free accumulation has taken place, which is typical of fully evolved and completely removed sediments from a very active fluvial environment.

The sedimentological analyses carried out in the San Pedro de Toxín sector confirm that the alluvial fans formed as a consequence of tectonic uplift in Sierra de Manantlán, and that the lacustrine terraces of the Armería River, very different from

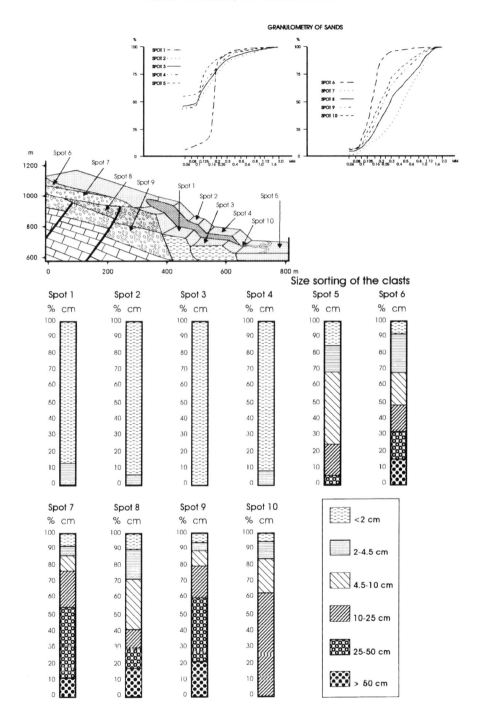

FIGURE 7.7   Granulometry of sands and size sorting of clasts in Armería Formation

the present-day river sediments, were formed by blockage of the river by volcanic avalanches.

TENDENCIES FOR LATERAL DISPLACEMENT OF RIVERS CAUSED BY VOLCANIC ACTIVITY

Lava or pyroclastic flows tend to follow the interior of valleys and ravines, partially filling them. When this happens, the drainage system must reorganize and new, major stream beds form along the edge of the volcanic unit. This phenomenon creates a feature called relief inversion, causing the mass of lava or pyroclastic flow to stand as a ridge in the location of the former stream channel.

Inverted relief is discussed in most geomorphology textbooks and is usually associated with differential erosion processes. Cotton (1952) is the only author, to date, to have published a text on volcanic geomorphology. He states that this process causes the valley walls to retreat, thus making them very steep. First, a valley forms, then it is covered by volcanic flows. Most of the water of the stream system network comes from the slopes, so as the water flows towards the inside of the valleys it creates new stream beds at the base of the slopes. Stream incision undermines the slopes, and they begin to retreat. This creates two lateral channels at the base of vertical walls on either side of the valley and effectively isolates the volcanic fill, so that it stands out from the rest of the relief. Cotton refers to a single cycle of interaction between erosion and volcanic flow.

An area, however, could be affected by many cycles of interaction. The volcanic relief is explained by alternating phases of volcanic flows and erosion. The fragility of the volcanic material causes intense and rapid erosion, so the drainage system must constantly reorganize itself (Shelley, 1989). Inverted relief and the recession of the valley walls is normal, and it could recur through many cycles. Stream incision is continually at work at the base of the walls, causing them to become unstable and to recede. The process continues until new volcanic material fills the channel and the cycle is renewed. The walls tend to retreat so rapidly that streams cannot cut down fast enough to reach base level and are left hanging (Figure 7.8) (Palacios, 1994).

The process explains why some valleys on the Canary Islands have such flat floors and are so wide (10 km), and also why their sides are vertical and their floors are covered with recent lava flows. The only two stream incisions that are visible are found at the base of the vertical walls (Figure 7.9) (Palacios, 1994).

Initially, these landforms were mistaken for a graben, because of their similarity. In the case of the Armería River, the course of the river actually does follow the edge of a graben, which is bordered by highly active normal faults. The process by which the river bed was formed, however, is similar to the one described for inverted relief. Nevado de Colima and Volcán de Fuego lie to the east of the Armería and their frequent volcanic flows have choked the river and even caused its fluvial morphology to disappear. In fact, this occurred a little more than 2500 years ago, and coincided with an avalanche and the formation of Volcán de Fuego's most recent caldera.

The drainage system reorganizes rapidly. Debris from the slopes is transported from the east and west and concentrates at the base of the block of the Sierra de Manantlán, where stream incision is renewed. Since the last avalanche, the Armería River has severely cut down through the deposits to more than 50 m in some locations.

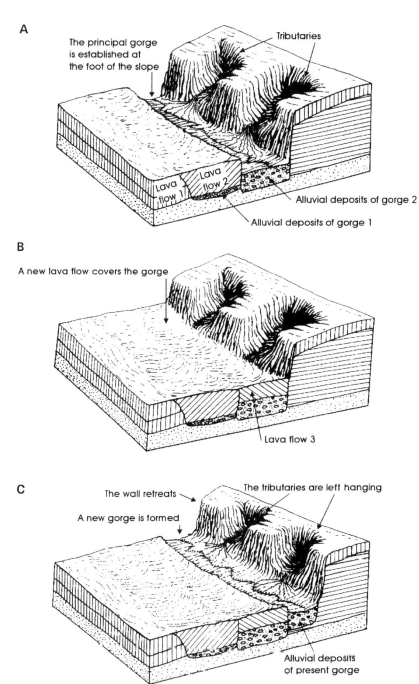

A
The principal gorge is established at the foot of the slope
Tributaries
Lava flow 1
Lava flow 2
Alluvial deposits of gorge 2
Alluvial deposits of gorge 1

B
A new lava flow covers the gorge
Lava flow 3

C
The wall retreats
The tributaries are left hanging
A new gorge is formed
Alluvial deposits of present gorge

FIGURE 7.8   Repeated alternations of stream incision at the foot of the slope and burial of the gorges by lava flows

A

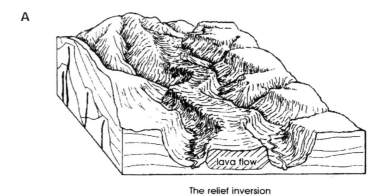

The relief inversion

B

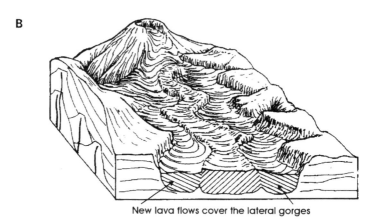

New lava flows cover the lateral gorges

C

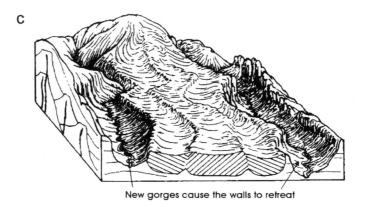

New gorges cause the walls to retreat

FIGURE 7.9   The origin of lateral expansion valleys. Reprinted from Palacios, D. 1994. The origin of certain wide valleys in the Canary Islands. *Geomorphology*, **9**, 1–18, with permission from Elsevier Science

As the Armería River valley undergoes reorganization, the eastern side of the Sierra de Manantlán tends to be undermined and forms a natural division between the volcanic material from the Colima Complex and the limestone from Manantlán. Stream incision downcuts through ancient alluvial fans and allows the sidewalls to maintain their vertical profile, despite the frequent rockslides and other erosion processes taking place on the slopes.

CONCLUSIONS: THE EVOLUTION OF THE ARMERÍA RIVER

The Armería River drains a continental rift, and is constantly affected by the tectonic and volcanic activity that is characteristic of this type of area (Frostick and Reid, 1989).

With the use of available data, it is possible to propose a model for the evolution of this river valley, based on two concurrent processes (Figure 7.10). The first is the uplift of the horst of the Sierra de Manantlán, which probably began in the early Pleistocene with the formation of the rift. Although uplifting continues even now, it accelerated, presumably in the late Pleistocene, and provoked the disappearance of most of the river's tributaries on the west side.

The second process involves the flow of volcanic materials from the Colima Complex, which probably began in the early Pleistocene when the volcanic complex was formed. The last major event was the collapse of the volcano Paleofuego in the Holocene and the formation of a debris avalanche that filled most of the Armería River Valley and blocked the rest.

Earlier (Figure 7.10A), the Armería River flowed much more towards the interior of the graben (Lugo et al., 1993). Canyons evolved in the Sierra de Manantlán and channelled material into great alluvial fans, which are commonly associated with grabens. Later (Figure 7.10B), the Manantlán horst was rapidly uplifted, which raised the canyons and disconnected them from their valleyheads. Closed basins formed on the far side of the horst from what had originally been the heads of the valleys. Also, the alluvial fans were separated from their channels and fractured. The emission of volcanic material from Nevado de Colima and Volcán de Fuego pushed the Armería River towards the side of the graben, causing the erosion of the alluvial fans and of other sediments worn from the slopes of Manantlán. Following the destruction of Paleofuego (Figure 7.10C), the valley was partially filled and two terraces obturated the upper reaches of the river. Today (Figure 7.10D), the river has downcut through the largest part of the debris that accumulated when the avalanche collided with the Sierra de Manantlán and has also incised the sediments that sealed off the course of the river.

A

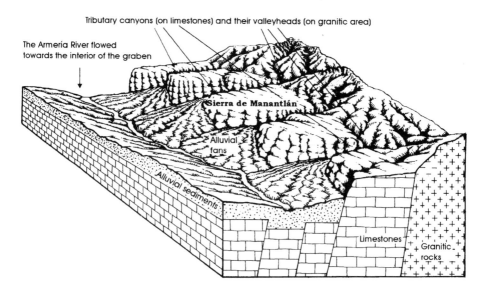

Tributary canyons (on limestones) and their valleyheads (on granitic area)

The Armería River flowed
towards the interior of the graben

Sierra de Manantlán

Alluvial
fans

Alluvial sediments

Limestones

Granitic
rocks

B

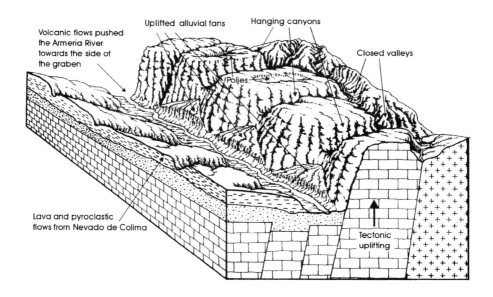

Volcanic flows pushed
the Armería River
towards the side of
the graben

Uplifted alluvial fans

Hanging canyons

Closed valleys

Poljes

Lava and pyroclastic
flows from Nevado de Colima

Tectonic
uplifting

FIGURE 7.10   Geomorphological evolution of the study area, eastern slope of the Sierra de Manantlán

C

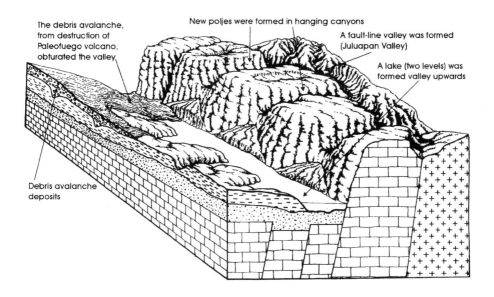

The debris avalanche, from destruction of Paleofuego volcano, obturated the valley

New poljes were formed in hanging canyons

A fault-line valley was formed (Juluapan Valley)

A lake (two levels) was formed valley upwards

Debris avalanche deposits

D

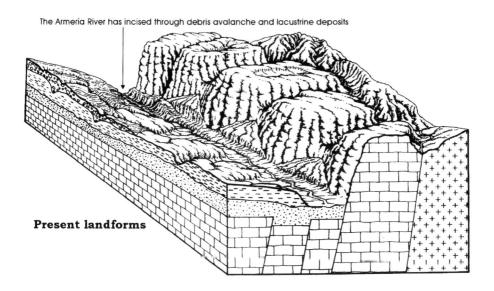

The Armería River has incised through debris avalanche and lacustrine deposits

Present landforms

FIGURE 7.10   (*continued*)

REFERENCES

Allan, J.F. 1986. Geology of the northern Colima and Zocoalco grabens, southwest Mexico: Late Cenozoic rifting in the Mexican Volcanic Belt. *Geological Society of American, Bulletin*, **97**, 473–485.

Allan, J.F. and Carmichael, I.S.E. 1984. Lamprophic lavas in the Colima Graben, SW Mexico. *Contributions to Mineralogy and Petrology*, **88**, 203–216.

Allan, J.F., Nelson, S.A., Luhr, J.F., Carmichael, I.S.E., Wopat, M. and Walace, P.J. 1991. Pliocene–Holocene rifting and associated volcanism in southwest Mexico: an exotic terrane in the making. *Memoirs, American Association of Petroleum Geologists*, **47**, 421–445.

Bandy, W.T., Hilde, W.C. and Bourgois, J. 1988. Redefinition of plate boundaries between the Pacific, Rivera, Cocos and North American plates at the north end of the Middle American trench. *EOS Transactions, American Geophysical Union*, **70**, 1342.

Bourgois, J., Renard, V., Aubouin, J., Bandy, W., Barrier, E., Calmus, T., Carfatan, J., Guerrero, J., Mammerickx, J., Mercier de Lepinay, B. Michaud, F. and Sosson, M. 1988. Fragmentation en cours du bord Ouest du continent Nord-Américain: Les frontières sous-marines du Bloc Jalisco (Mexique). *C.R. Acad. Sci. París*, Series II, **307**, 1121–1130.

Cotton, C.A. 1952. *Volcanoes as Landscape Forms*. Hafner, New York, 399 pp.

Crozier, M.J. and Pillans, B.J. 1991. Geomorphic events and landforms response in south-eastern Tarnaki, New Zealand. *Catena*, **18**, 471–487.

Delgado, H. 1993. Late Cenozoic tectonics offshore western Mexico and its relation to the structure and volcanic activity in the western Trans-Mexican Volcanic Belt. *Geofísica Internacional*, **32**(4), 543–559.

Demant, A. 1978. Características del Eje neovolcánico Transmexicano y sus problemas de interpretación. *Revista Inst. Geología (UNAM)*, **2**(2), 172–178.

Demant, A. and Robin, C. 1975. Las fases del vulcanismo en México; una síntesis en relación con la evolución geodinámica desde el Cretácico. *Revista Inst. Geología (UNAM)*, **75**(1), 70–82.

Frostick, L. and Reid, I. 1989. Is structure the main control of river drainage and sedimentation in rifts? *Journal of African Earth Sciences*, **8**, 165–182.

Gerson, R., Grossman, D. and Bowman, D. 1985. Stages in the creation of large rift valley-geomorphic evolution along the Dead Sea rift. In M. Morisawa and J.T. Hack (Eds), *Tectonic Geomorphology*. Allen and Unwin, Boston, 53–73.

Johnson, C.A. and Harrison, C.G.A. 1990. Neotectonics in Central Mexico. *Physics of the Earth and Planets International*, **64**(2–4), 187–210.

King, G. and Ellis, M. 1990. The origin of large local uplift in extensional regions. *Nature*, **348**, 689–693.

Komorowski, J.C., Navarro, C., Cortes, A. and Siebe, C. 1994. The repetitive collapsing nature of colima volcanoes (Mexico): problems related to the distinction of multiple deposits and the interpretation of 14C ages with implications for future hazards. *IV Reunión Internacional del Volcán de Colima*, Colima, Mexico, 12–18.

Lugo, J., Martín del Pozzo, A.L. and Vázquez, L. 1993. Estudio Geomorfológico del complejo volcánico de Colima. *Geofísica Internacional*, **32**(4), 633–641.

Luhr, J.F. and Carmichael, I.S.E. 1980. The Colima Volcanic Complex, Mexico I: Post-caldera andesites from Volcan Coloma. *Contributions to Mineralogy and Petrology* **71**, 343–372.

Luhr, J.F. and Carmichael, I.S.E. 1990. *Geology of Volcán de Colima*. Instituto de Geología, México, 100 pp.

Luhr, J.F. and Prestegaard, K.L. 1988. Caldera formation at Volcán Colima, Mexico, by large Holocene volcanic debris avalanche. *Journal of Volcanic Geothermal Research*, **35**, 335–348.

Luhr, J.F., Nelson, S.A., Allan, J.F. and Carmichael, I.S.E. 1985. Active rifting in southwestern Mexico: manifestations of an incipient eastward spreading-ridge jump. *Geology*, **13**, 54–57.

Martín del Pozzo, A.L., Lugo, J. and Vázquez, L. 1990. Multiple debris avalanche events in Colima, Mexico. *EOS Transactions, American Geophysical Union*, **71**(43), 1720.

Martínez-Reyes, J. and Nieto-Samaniego, A.F. 1990. Efectos geológicos de la tectónica reciente en la parte central de México. *Revista Inst. Geología (UNAM)*, **9**(1), 33–50.

Mooser, F. 1972. The Mexican Volcanic Belt: Structure and tectonics. *Geofísica Internacional*, **12**, 15–22.

Mooser, F. and Ramírez-Herrera, M.T. 1989. Faja Volcánica Transmexicana: Morfoestrutura, tectónica y vulcanotectónica. *Boletin de la Sociedad Geologica Mexicana, T. XLVIII*, **2**, 75–85.

Ortíz-Pérez, M.A. 1990. Algunos aspectos de la geomorfología del curso medio de los ríos Armería y Tuxpan-Coauayana relacionados con la tectónica. *Actas de la Segunda Reunión Nacional "Volcán de Colima"* Universidad de Colima, 41–42.

Ortíz-Pérez, M. A. and Bocco-Verdinelli, G. 1989. Análisis morfotectónico de las depresiones de Ixtlahuaca y Toluca, México. *Geofísica Internacional*, **28**(3), 507–530.

Ortíz–Pérez, M.A., Zamorano, J.J. and Bonifaz, A. 1993. Reconocimiento de una falla reciente de tipo transcurrente en Colima, México. *Geofísica Internacional*, **32**(4), 569–574.

Palacios, D. 1994. The origin of certain wide valleys in the Canary Islands. *Geomorphology*, **9**, 1–18.

Ramírez-Herrera, M.T., Summerfield, M.A. and Ortíz-Pérez, M.A. 1994. Tectonic geomorphology of the Acambay graben, Mexican Volcanic Belt. *Zeitschrift für Geomorphologie NF*, **38**(2), 151–168.

Robin, C., Camus, G., Cantagrel, J.M., Gourgand, A., Mossand, P. and Vicent, P.M. 1984. Les volcans de Colima (Mexique). *Bull. PIRPSEV –CNRS INAG*, **87**, 1–98.

Robin, C., Mossand, P., Camus, G., Cantagrel, J.M., Gourgaud, A. and Vincent, P.M. 1987. Eruptive history of the Colima Volcanic Complex (Mexico). *Journal of Volcanic and Geothermal Research*, **31**, 99–113.

Shelley, D. 1989. Anteconsequent drainage: an unusual example formed during constructive volcanism. *Geomorphology*, **2**, 363–367.

Shubert, D.H. and Cebull, S.E. 1984. Tectonic interpretation of the Trans-Mexican Volcanic Belt. *Tectonophysics*, **101**, 159–165.

Siebe, C., Rodriguiez, S. Stoopes, G., Komorowski, J.C. and Sheridan, M.F. 1992. How many debris avalanche deposits at the Colima volcanic complex? *Tercera Reunión Nacional "Volcán de Colima"* Universidad de Colima, 15.

Stoopes, G.R. and Sheridan, M.F. 1992. Giant debris avalanches from the Colima Volcanic Cmplex, Mexico. Implications for long-runout landslides (>100 km) and hazard assessment. *Geology*, **20**, 299–302.

Suárez, G. and Singh, S.K. 1986. Tectonic interpretation of the trans-Mexican volcanic belt: discusion. *Tectonophysics*, **127**, 155–160.

Suter, M., Quintero, O. and Johnson, C.A. 1992. Active faults and State of Stress in the Central Part of the Trans-Mexican Volcanic Belt, Mexico. *Journal of Geophysical Research*, **97**(8), 11983–11993

Urrutia, J. and Böhnel, H. 1988. Tectonics along the Trans-Mexican Volcanic Belt according to paleomagnetic data. *Physics of the Earth and Planets International*, **52**, 320–329.

Verma, S.P. 1987. Mexican Volcanic Belt: present state of knowledge and unsolved problems. *Geofísica Internacional*, **26**(2), 309–340.

8

Boulder Bedforms in Jointed-bedrock Channels

RAINER WENDE

School of Geosciences, University of Wollongong, NSW, Australia

ABSTRACT

Bedrock channels cut by hydraulic plucking into jointed rocks reveal characteristic erosional and depositional bedforms. Numerous positive (upstream-facing) steps are typical erosional forms for channel reaches where joints in the bedrock produce layers of platy to bladed slabs which dip gently downstream relative to the channel slope. Such erosional bedrock steps initiate the deposition of coarse particles of the bedload which slide, pivot or probably even saltate along the channel bed. Resulting depositional bedforms are of three types: deposits of single clasts leaning on the steps; clusters of imbricated boulders formed only on the stoss side of the steps; and clusters which extend upstream and downstream beyond the bedrock step. The periodic occurrence of erosional rock steps and depositional boulder clusters seems to indicate a hydraulic control of these bedforms.

Cluster bedforms associated with bedrock steps in natural channels occur at various scales, and very large examples from northwestern Australia can comprise boulders with an intermediate axis of more than 8 m. To erode and transport such large rock slabs, high-magnitude floods with high flow velocities are required. Threshold values for entrainment of imbricated boulders can, under certain conditions, exceed those for bedrock erosion along channel reaches and possibly delay further erosion.

INTRODUCTION

Studies of fluvial processes and forms have traditionally concentrated on alluvial channels. In such channels, characteristic bedforms result from fluid shear over a movable channel during flows that are competent to move all or part of the bed material. Modern alluvial bedforms can serve as analogues for interpreting the ancient sedimentary record and many fluvial studies have been conducted for this purpose. Furthermore, alluvial channels have the ability to react to changes in process rates by negative feedback mechanisms, and such fluvial process–response systems can have relaxation times well within the historical record. Often it is possible to monitor the forming processes and to measure the actual process rates in natural alluvial channels.

Varieties of Fluvial Form. Edited by A.J. Miller and A. Gupta
© 1999 John Wiley & Sons Ltd

TABLE 8.1   Classification of depositional and erosional bedforms in bedrock and boulder-bed channels (after Baker, 1978a), and correlation with other classifications of sediment storage elements

	Depositional examples	Erosional examples	Other classifications of sediment storage elements		
			Ashley (1990)	de Jong and Ergenzinger (1995)	Grant et al. (1990)
Macro-forms	Point bars, braid bars	Step–pool sequences, bedrock anabranches	Channel forms, braid bar complexes	System roughness	Channel unit
Meso-forms	Dunes, transverse ribs, boulder clusters	Rock steps, longitudinal grooves, inner channels	Unit bars, bedforms	Form roughness	Subunit and particle
Micro-forms	Sand ripples, pebble clusters	Flute marks	Ripples		Particle

In stark contrast to this are bedrock channels for which our knowledge of processes and characteristic forms remains very limited. This lack of understanding results because bedrock channels are less common, more difficult to study and are often seen as being less significant sedimentologically than are alluvial channels (cf. Baker, 1984).

Bedrock systems rarely exhibit channel boundaries formed by bare bedrock throughout their entire length, but consist of boulder-bed and bedrock channel reaches (Knighton, 1984, p.86) which grade into each other. They are generally supply-limited, for the rate of sediment input by localized bedrock erosion and from upstream cannot satisfy the rate of potential sediment transport. Such systems are therefore in a constant state of disequilibrium (Ahnert, 1994). Since bedload flux along bedrock channel reaches is usually low, most resulting bedforms are erosional rather than depositional.

To date, few attempts exist to classify erosional and depositional bedforms in bedrock channels. Baker (1978a) modified a scheme developed by Jackson (1975) and applied it to both depositional and erosional bedforms of bedrock channels (Table 8.1). The largest bedforms in the hierarchy are macro-forms. These are the large form elements of the channels that contribute to the system roughness (de Jong and Ergenzinger, 1995), such as depositional braid bar complexes, or step–pool sequences scoured along bedrock canyons. Macro-forms are related to long-term hydrological factors and their scale is controlled by channel width. An order of magnitude smaller are meso-forms which contribute to the form roughness (de Jong and Ergenzinger, 1995) of the river bed. The periodic depositional forms of this group, such as dunes, are scaled to flow depth, and solitary forms (such as unit bars) are controlled by local hydraulic conditions (Ashley, 1990). Typical erosional meso-forms are longitudinal grooves and inner channels (Baker, 1978a). Micro-forms, the smallest bedforms, are abundant in sand-bed channels (e.g.

ripples) and gravel-bed rivers (e.g. pebble clusters), but depositional micro-forms are not present in many bedrock and boulder-bed channels owing to the large particle size of the bedload (Baker, 1984). However, erosional micro-forms, such as flute marks (Allen, 1982, Vol.2, pp. 253–291) are common in many bedrock channels.

Studies on bedforms in modern bedrock and boulder-bed systems have concentrated on apparent macro-forms, in particular on step–pool sequences (e.g. Wohl, 1992) or various types of gravel bars (e.g. Baker, 1984). Several other studies are detailed investigations of erosional and depositional macro- and meso-forms associated with the sudden release of floodwaters from Pleistocene lakes (e.g. Baker, 1973; O'Connor, 1993).

This paper is concerned with special types of meso-forms rarely mentioned in the literature, in particular clusters of imbricated boulders which are deposited along positive (upstream-facing) steps in the bedrock channel. These erosional steps occur along channels cut into jointed rocks and are closely linked to geological structure. At several field sites, such erosional steps formed by a process known as hydraulic plucking, and the slabs of rock eroded by this process form the source material for depositional boulder clusters.

THE STUDY SITES

The examples of boulder bedforms presented here are from the Kimberley region in monsoonal northwestern Australia. The semi-arid climate of this region leads to highly seasonal flow in the rivers, a convenient condition for the study of the channel bed morphology. Annual rainfall (median) is highest along the coastal areas in the northwest of the Kimberley and decreases inland, reaching some 600–800 mm near the study sites. However, during the summer wet season, tropical storms and cyclones can cause heavy rain resulting in violent floods.

The study sites (Figure 8.1) are from geologically different regions, but have in common that the channels are actively incising into little-disturbed and well jointed sedimentary rocks. The examples of boulder bedforms from the Durack River and an unnamed creek (Figure 8.1) are from channels cut into a landscape of cuestas and structural plateaux developed on Proterozoic sedimentary rocks with a dominance of sandstones. The soils are generally only skeletal soils, and rock outcrops are abundant. Near the site on the Ord River, the landscape is characterized by gently undulating plains formed on little-disturbed Palaeozoic shales and limestones which are covered by deep calcareous and loamy soils. The vegetation at all study sites consists only of scattered trees or shrubs and grass communities with a variable degree of soil cover. During the Quaternary, the region was not affected by any glaciation or periglacial conditions, but, from evidence from elsewhere in northern Australia, was probably affected by substantial variations of the flow regime (Nott and Price, 1994; Nott et al., 1996).

EROSIONAL PROCESSES

Bedrock erosion results from a combination of weathering and erosional processes that cause the decomposition, disintegration and removal of rock. The wear of bedrock by transported sediment (abrasion), the removal of rock fragments by hydraulic forces (hydraulic plucking), and possibly the physical damage of the rock by imploding

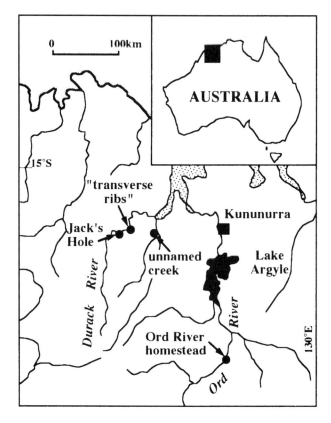

FIGURE 8.1   Study sites

cavitation bubbles (Allen, 1982, Vol.1, pp.69–74; Baker, 1988), can cause mechanical destruction and removal of bedrock mass. The chemical action of water (corrosion) results in the direct erosion of rock by dissolution. Furthermore, chemical and biochemical weathering of bedrock works to reduce the resistance of the rock against the physical erosional processes mentioned above. The contribution of individual erosional processes to the total rate of bedrock erosion along a channel reach depends largely on the material properties of the channel boundaries which control the thresholds of erosion, and on the hydraulic conditions that control the exerted forces of the erosional physical processes. In some channels, this relationship can be quantified by correlating the rate of energy dissipation of the flow to an erodibility classification of the rock mass (Annandale, 1995).

Under turbulent conditions, the hydrodynamic forces acting on a coarse particle are proportional to the dynamic pressure ($1/2\ \rho v^2$) and the area over which the pressure acts. They are commonly expressed as drag (F_D) and lift (F_L) force acting parallel and perpendicular to flow, respectively (e.g. Komar and Li, 1988; Ergenzinger and Jüpner, 1992; Denny, 1993):

$$F_D = \frac{1}{2} C_D A_f \rho v^2 \tag{8.1}$$

$$F_L = \frac{1}{2} C_L A_p \rho v^2 \qquad (8.2)$$

with C_D, C_L the drag and lift coefficients, A_f and A_P the area perpendicular and parallel to flow, respectively, ρ the fluid density, and v the flow velocity.

In channels cut into well jointed lithologies, hydraulic plucking can be an important if not the dominant process of channel erosion. Individual joint blocks forming part of the channel bed and bordered on their downstream side by a neighbouring block, can only be removed if the lift force exceeds all counteracting forces, especially the submerged weight of the block. The lift force acting on the block is the result of pressure differences at the top and the base of the block. In deep fast flows, a very low water pressure on the surface of the block and resulting intense suction can be produced by a large-scale upward vortex (Baker, 1973), a type of macro-turbulence phenomenon observed in natural channels (Matthes, 1947). Extensive Pleistocene flood erosion has been attributed to the action of hydraulic plucking under macro-turbulent conditions (Baker, 1973; Baker, 1978b). Intense suction can also occur in shallow flows under the influence of impinging water jets, but is limited in its erosional significance to blocks that are short in respect to flow direction (Otto, 1990). However, besides upward-acting forces due to suction on the block, high dynamic uplift pressure can result from high dynamic pressure in the horizontal joints, separating the block from the underlying rock mass. Such joint pressures can be created by high-velocity flow which impacts on a protrusion of the rock block and is transmitted into the open cracks (Renius, 1986). The resulting mechanism of erosion is the ejection of blocks from the rock mass that forms the channel bed. This occurs by high pressures in the subhorizontal joint "like a piston of a hydraulic jack moved by high fluid pressure" (Otto, 1990, p.900). Annandale (1995) used the mechanism of hydraulic plucking as a conceptual model to link the erosive power of flow to the rate of energy dissipation.

To quantify uplift forces acting on particles, experimental studies have frequently been conducted to determine numerical values for the uplift coefficient C_L in Equation (8.2), which itself is a function of various parameters. Renius (1986) modelled flow parallel to a jointed rock surface and found that the uplift coefficient varied with the inclination of the subsurface joints relative to the channel slope. Coefficients for the total uplift force acting

TABLE 8.2   Coefficient C_N of the uplift force for various dip angles (α) of the bedding joints

	Inclination α (degrees)						
	-18	-2.9	0	2.6	9	17.5	33.5
Lift coefficient C_N	-0.22	-0.07	0.16	0.31	0.46	0.45	0.21
Surface irregularity D_b/t	3	20	75	22	6	3	?
Froude number	2.7	5.5	3.4	6.5	2.7	3.1	3.6

The coefficient C_N refers to the force acting normal to the top surface of a rectangular joint block subject to fully turbulent flow parallel to the channel bed. The lift coefficient (C_L) is related to the normal force coefficient (C_N) by $C_L = \cos\alpha C_N$. Each column represents a different experiment and only the maximum value for the lift coefficient for each experiment is given. The parameter t denotes the vertical protrusion of the test block into flow and $D_b = 15$ cm is the length of the test blocks. All data from Renius (1986)

perpendicular to the top surface of the rock block (C_N) increased from about 0.15 for subsurface joints oriented parallel to the channel to a maximum value of 0.48 for joints with dip angles of 12° in a downstream direction (Table 8.2). Negative pressure coefficients, and therefore negative lift forces, acted to hold the particle on the bed if the joint-bounded blocks dipped upstream.

EROSIONAL BEDFORMS

The following erosional and depositional bedforms have been identified along the Durack River and other bedrock channels of the Kimberley region, but are believed to be representative of a variety of bedrock channels with horizontal or gently dipping rock structure.

In channels cut into sedimentary or igneous rocks that have intersecting sets of horizontal and vertical joints, joint block erosion can result in characteristic erosional bedforms. Along channel reaches where the dip angle of the planes of the horizontal joints is low, and the dip direction varies from upstream to downstream relative to channel slope, channel bed morphology is largely controlled by the strike and dip of the planes along which erosion occurs (Miller, 1991; Wende, in revision). Along channels where strata strike is across the channel and dip is upstream relative to the channel slope (*anti-dip channel reaches*), downstream-facing steps or knickpoints are characteristic erosional bedforms produced by the removal of joint blocks from the channel bed (Figure 8.2). Along channels where strata dip is nearly parallel with the channel slope (*dip-parallel channel reaches*), the stream beds are frequently developed along a continuous bedding plane or a single horizontal joint over a considerable distance (Figure 8.2). Of particular interest in this paper are channels where strata dip is downstream relative to channel slope (*dip channel reaches*). Numerous successive upstream-facing steps which extend more or less transverse to the channel are typical for these dip channel reaches (Figure 8.2). Individual steps are produced by hydraulic plucking of blocks from upstream of these steps, while the succession of steps is the result of the dip of the strata. The height of individual steps and the spacing of the steps along the channel in a downstream direction is strongly influenced by the spacing of the truncated horizontal joints.

The planform of bedrock steps is controlled by the vertical joint pattern of the rock mass forming the channel bed. If only a single roughly parallel set of vertical joints

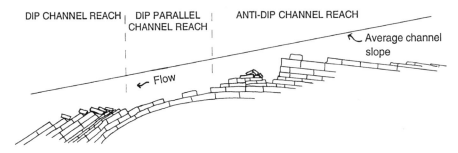

FIGURE 8.2   Channel bed morphology and structural control. The dominant process of bedrock erosion along the channel is hydraulic plucking. Characteristic channel bed morphologies are the result of variations of strata dip relative to channel slope. Modified after Miller (1991)

traverses the channel, straight or linear steps are the result, while intersecting sets of vertical joints result in non-linear steps, frequently following a zig-zag pattern.

Steps in natural bedrock channels cut into sedimentary rocks are rarely oriented perpendicular to the channel, but most often are oblique (e.g. Pohn, 1983; Miller, 1991). They frequently exhibit a rather complex planform pattern which can be due to many factors, like an irregular fracture pattern or differential erosion across the channel. For example, blocks, not yet detached from the bedrock mass by an open horizontal joint, or groups of adjoining blocks that are too thick to be removed, can remain little affected by hydraulic plucking, while adjacent areas of the channel bed with a lower threshold of erosion are eroded at a higher rate.

IMBRICATED BOULDER BEDFORMS: TERMINOLOGY AND GENERAL DESCRIPTION

Locally plucked boulders can contribute a very coarse fraction to the bedload of streams and accumulations of coarse clasts can form distinct bedforms along bedrock channel reaches. The depositional bedforms studied here are closely linked to positive joint steps in dip channel reaches predominantly eroded by hydraulic plucking (Figures 8.3–8.6). In contrast to bedforms superimposed on larger depositional forms, such as large dunes on gravel bars, these features rest on bedrock. They consist of deposits of imbricated platy particles associated with bedrock steps along dip channel reaches, and they have been identified in channels ranging in size from small creeks to large rivers. There appears to be a scale continuum of these structurally controlled bedforms in bedrock streams, with the maximum size of clast ranging from pebbles to extremely large boulders measuring several metres along their intermediate axis.

Descriptions of groups or sequences of gravel-sized particles aligned parallel to flow are numerous (e.g. Laronne and Carson, 1976; Billi, 1988; de Jong, 1991; de Jong and Ergenzinger, 1995), and they have been called a pebble cluster (Dal Cin, 1968) or more generally a particle cluster (e.g. Brayshaw et al., 1983). The bedforms associated with such clusters may be referred to as cluster bedforms, or more specifically as imbricate-type cluster bedforms (Brayshaw, 1984). Following this terminology, the large-scale examples of clusters of imbricated boulders presented in this paper can be referred to as imbricate-type boulder clusters.

Generalizing, a succession of three principal kinds of deposits associated with bedrock steps have been identified in channels of various sizes. (1) Single clasts are imbricated against positive steps in the bedrock channel, although several such clasts may be aligned along the step (Figures 8.3A and 8.4). (2) Subsequent accumulation of clasts on the stoss side of the step leads to the development of contact-type imbrication (Johansson, 1976) in the form of an imbricated *stoss-side boulder cluster* (Figures 8.3B and 8.5). The downstream end of these clusters is well defined by the bedrock step and no deposition occurs on the downstream or lee side of the cluster. However, if the first clast located upstream of the bedrock step projects downstream beyond the step, minor deposits of finer grained material are often found in the lee of this clast. (3) The third principal kind of accumulation is a large *step-covering boulder cluster* which extends upstream and downstream of this step, obscuring its position (Figures 8.3C and 8.6). Along a succession of bedrock steps, stoss-side or step-covering clusters can extend far upstream

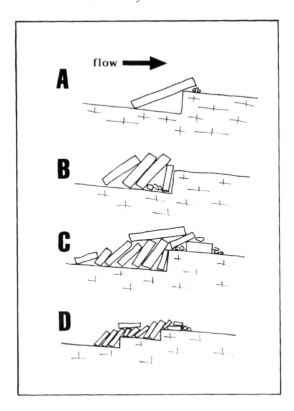

FIGURE 8.3   Boulder deposits associated with bedrock steps: (A) single boulder imbricated against bedrock step; (B) stoss-side boulder cluster; (C) step-covering boulder cluster; (D) combined boulder cluster

such that they link up with other boulder clusters to form *combined boulder clusters* (Figure 8.3D).

To describe the geometry of cluster bedforms, Brayshaw (1984) distinguished between stoss side, obstacle clasts, and the wake or lee side of the cluster. He found that the relative length of stoss to lee side is largely dependent on the shape characteristics of the sediment. Where imbrication occurs, long trains of clasts can form on the stoss side of an initial obstacle, which is not necessarily the largest particle in the cluster. For structurally influenced clusters, the positive step in the channel represents the obstacle that initiated deposition. In general with gravel clusters, stoss-side accumulations grow by deposition on the upstream side while lee-side deposits grow in a downstream direction. Many boulder clusters consist only of stoss-side accumulations. However, minor deposits of much smaller calibre are generally found in the lee of individual large clasts forming a train of imbricated boulders. Step-covering clusters exhibit a lee side, but the lee deposits frequently form only a minor part of the whole cluster. Generally they are just short tail deposits of small boulders, gravel and sand downstream of the last large boulder. However, in combined boulder clusters, it is difficult to define a lee side since a potential initial obstacle can often not be clearly identified. In fact, they can form elongated

FIGURE 8.4   Examples of single boulders imbricated against positive bedrock steps. (A) Jack's Hole, Durack River. Note the likely origin of this boulder in the foreground. (B) Unnamed creek with a catchment of 3 km², a local channel slope of 0.058, a bankfull width of 25 m and a maximum depth of 1.8 m. Note the linear steps across the channel. The two downstream successive steps in the centre of the photo have a height of 0.4 m and 0.6 m. The large boulders on the left and right side of the channel measure $1.7 \times 1.5 \times 0.20$ m and $1.4 \times 1.3 \times 0.15$ m for the long, intermediate and short axis, respectively

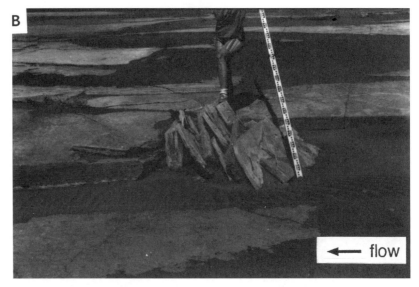

FIGURE 8.5   Examples of stoss-side clusters: (A) Jack's Hole, Durack River. Accumulation of
rock slabs on the stoss side of a positive step in the bedrock channel. Note the absence of lee
deposits. (B) Ord River Homestead, Ord River. Stoss-side cluster. Note the flow marks around the
cluster

deposits (longitudinal boulder bars) that resemble in shape central bars typically found in
braided rivers carrying a coarse bedload.

The planform of stoss-side and step-covering boulder clusters is often closely linked to
the planform of bedrock steps along which they are deposited. A regular pattern of
bedrock steps can result in an even pattern of cluster bedforms. However, as mentioned

A

B

FIGURE 8.6   Examples of step-covering clusters (all Jack's Hole, Durack River). (A) The cluster has a total length of 18 m and a height of 2.5 m. The bedrock step with a height of about 1 m is completely covered. The accumulation of boulders on the stoss side of the step is 6.5 m long. Downstream of the step, the cluster can be subdivided into a stoss and wake deposit separated by an obstacle clast (3.1 × 1.5 × 0.7 m). The wake side measures 4.5 m, while the stoss side is 6 m long and links up with the accumulations upstream of the step. The inclination of the 10 largest clasts in the cluster ranges from 5 to 50°. The largest boulder in the cluster measures 5.8 × 3.2 × 0.4 m. The photo shows the boulder cluster on the stoss side of the step marked with an arrow. (B) Edge of a step-covering cluster

before, bedrock steps in natural channels often have a complex pattern, and the planforms of boulder clusters that mimic those steps are therefore also highly variable.

THRESHOLD CONDITIONS FOR ENTRAINMENT OF LARGE ROCK SLABS

Two principal approaches to the problem of threshold conditions for movement of coarse particles are apparent from the literature (for reviews see Novak, 1973; Baker and Ritter, 1975; Bradley and Mears, 1980; Costa, 1983; Komar, 1988, 1989). First, empirical studies on channel stability and flow competence are based on observations or experiments (e.g. Carling, 1983; Costa, 1983; Williams, 1983). Second, the formulation of threshold criteria for initiation of particle movement is based on consideration of the forces acting on a particle (e.g. Rusnak, 1957; Helley, 1969; Naden, 1987; Komar and Li, 1988; James, 1990).

While most of these latter theoretical studies have focused on the particular problem of particle entrainment in sand- or gravel-bed streams, the emphasis in the following discussion is placed on the entrainment and stability of a rectangular block on a rock bed. Estimation of threshold velocities for the entrainment of a single large rectangular block on a rock-bed channel is considered here for three principal situations: a flat channel bed; upstream of a positive step; and imbricated against a positive step. The longest, intermediate and shortest side of the block are denoted respectively by D_a, D_b, and D_c. To illustrate the implications of particle shape, size and orientation for entrainment and stability, numerical values of the critical velocities were calculated for several rock blocks with dimensions similar to those found at the study site along the Durack River (Figure 8.1, Tables 8.3 and 8.4).

The principal forces involved in the entrainment of a block are its submerged weight F_G and the friction force F_F resisting motion, and the drag F_D and lift F_L as the driving forces (Figure 8.7A). Drag and lift are given by Equations 8.1 and 8.2 and F_G and F_F by:

$$F_G = (\rho_s - \rho)gD_aD_bD_c \tag{8.3}$$

$$F_F = \mu_f F_N \tag{8.4}$$

where ρ_s is the density of the block, g is the acceleration due to gravity, μ_f is the coefficient of friction, F_N is the resultant force of F_G and F_L acting normal to the surface, and all other parameters are as defined above. For reasons of simplicity, the water surface slope was neglected. In the following, α is assumed to be small so that $\cos \alpha \approx 1$, which allows that all forces can be considered as acting perpendicular to the respective surface of the block (Figure 8.7A).

For the coefficient of static friction μ_f in Equation 8.4 of a platy rock on a flat rock bed, Carling (1995) and Carling and Grodek (1994) used $\mu_f = 0.9$, while Bradley and Mears (1980) as well as Allen (1942) suggested a value of 0.6. The higher value appears to be preferable, especially considering the likelihood that even on a smooth bed some minor interlocking can occur due to the surface roughness of the block and the channel bed.

The drag coefficient C_D (Equation 8.1) for a rectangular block subject to flows of high Reynolds numbers ($Re > 10^6$) varies with the shape of the block and its orientation. Platy to compact rock blocks characteristically have values between 1.0 and 1.2 (Hoerner,

TABLE 8.3a   Comparison of critical velocities v_c needed to entrain rock blocks of differing shape by sliding, pivoting or hydraulic plucking, with $\alpha = 0°$ or $3°$ and $C_N = 0.15$ or 0.3

D_b (m)	D_c (m)	Sliding v_c (m s^{-1})		Pivoting v_c (m s^{-1})		Plucking v_c (m s^{-1})	
		$\alpha = 0°$ $C_L= 0.15$	$3°$ 0.3	$0°$ 0.15	$3°$ 0.3	$0°$ 0.15	$3°$ 0.3
3	0.2	5.2	4.1	6.5	4.6	6.6	4.6
3	0.5	6.6	5.6	9.4	7.0	10.4	7.3
3	3	8.1	7.7	8.5	8.0	25	18
1	0.5	4.4	4.1	6.0	5.2	10.4	7.3
1	1	4.7	4.4	4.9	4.6	14.7	10.4

Lift coefficients (C_L) were taken from Table 8.2. The density of the block was assumed to be $\rho_s = 2.65$ g cm^{-3}, the density of water $\rho = 1$ g cm^{-3}, the acceleration due to gravity $g = 9.8$ m s^{-1}, the coefficient of friction $\mu_f = 0.9$, and the drag coefficient $C_D = 1.2$. The block is assumed to rest flat on the channel bed with its long axis transverse to flow

TABLE 8.3b   Critical velocities v_c needed to entrain a block of differing shape by pivoting from a position ahead of a low rock-step with height H

D_b (m)	D_c (m)	H (m)	v_c (m s^{-1})
3	0.2	0.05	4.6
3	0.5	0.2	7.3
3	1	0.2	9.2
3	3	0.2	8.5
1	0.5	0.2	6.7
1	1	0.2	5.6

The lift coefficient was assumed to be $C_L=0.3$ which corresponds to a dip angle of about $\alpha = 3°$. All other constants as noted in Table 8.3a. The block is assumed to rest flat on the channel bed with its long axis transverse to flow

TABLE 8.4   Summary of size and shape characteristics of 50 large boulders forming boulder clusters at Jack's Hole, Durack River

	D_a(m)	D_b(m)	D_c(m)	D_c/D_a	$(D_a-D_b)/$ (D_a-D_c)	$(D_c^2/D_aD_b)^{1/3}$
Mean, $n = 50$	4.4	2.8	0.5	0.14	0.42	0.31
Minimum	2.1	1.0	0.2	0.05	0.03	0.14
Maximum	12.9	8.8	1.0	0.30	0.78	0.54

Boulder size is given as long (D_a), intermediate (D_b) and short (D_c) axis, respectively. Shape indices are those of Sneed and Folk (1958). Boulder shape ranges from very platy (minimum values) to very elongated (maximum) with the average shape being very bladed (mean)

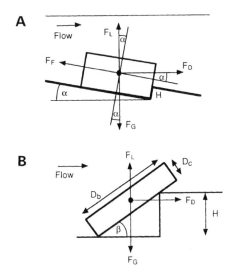

FIGURE 8.7   Schematic diagrams showing principal forces acting on a rectangular rock block resting flat on a rock bed (A), and imbricated against a positive rock step (B)

1965; Engineering Sciences Data, 1976), the latter value being used for platy rock blocks in this study and elsewhere (e.g. Carling and Grodek, 1994).

Somewhat more problematic is the question for the appropriate empirically determined lift coefficient C_L (Equation 8.2). As with drag, lift depends, under given flow conditions, on shape and orientation of the rock block. Table 8.2 gives some information on likely lift coefficients for rectangular blocks which are part of the channel bed. Little is known about the appropriate lift coefficients for the situation of a platy block resting on a flat rock bed. However, in view of the absence of experimental data for this case, a reasonable assumption may be to use the lift coefficients determined by Renius (1986) for the similar case of flow parallel to the surface of a jointed rock mass (Table 8.2) (cf. Carling and Grodek, 1994; Carling, 1995).

Of importance, especially for the mode of motion of the rock blocks (sliding, pivoting or saltation), is the location of the centre of pressure, or point of attack, of drag and lift relative to the centre of gravity of the block. However, for reasons of simplicity, in this study lift and drag are assumed to act at the centre of the block (cf. Komar and Li, 1988).

(1) *Rectangular block on flat channel bed.* Movement of the block can either be by sliding along the channel bed or by pivoting about its downstream lower edge, depending upon which threshold velocity is lower. Threshold velocities can be determined from a simple balance of forces in the case of sliding ($F_D = F_F$), and a balance of the moments about the pivoting point in the case of overturning ($F_G m_G = F_D m_D + F_L m_G$). The critical velocity for initial sliding of a rectangular block that lies with its long axis transverse and its intermediate axis parallel to flow can be expressed as (e.g. Bradley and Mears, 1980; Carling and Grodek, 1994):

$$v_c^2 = \frac{2(\rho_s - \rho)gD_b}{\rho} \frac{\mu_f}{C_D + [C_L \mu_f (D_b/D_c)]} \tag{8.5}$$

The critical velocity for overturning is given by:

$$v_c^2 = \frac{2(\rho_s - \rho)gD_b}{\rho} \frac{1}{C_D(m_D/m_G) + C_L(D_b/D_c)} \tag{8.6a}$$

where m_D and m_G are the moment arms according to:

$$m_D = 0.5D_c \tag{8.6b}$$

$$m_G = 0.5D_b \tag{8.6c}$$

For overturning to occur before sliding, the critical velocity of Equation 8.6a must be less than that of Equation 8.5. This condition simplifies to the inequality $(m_G/m_D) < \mu_f$, or, inserting Equations 8.6b and 8.6c, to $\mu_f > (D_b/D_c)$. For a given friction factor close to unity, this indicates that for blocks with an intermediate and short axis of about equal length the critical velocity required to initiate sliding is about the same as that to initiate pivoting (Table 8.3a). Indeed, flume experiments conducted by Allen (1942), who also made a theoretical analysis of the criteria for overturning versus sliding, showed that cubic concrete blocks slid as well as pivoted along the concrete floor of the flume.

In comparison, platy blocks are much more likely to slide along a smooth channel bed than to pivot (Table 8.3a). If, however, the blocks are supported on their downstream side by an obstacle, such as a bedrock step, sliding becomes impossible.

(2) *Rectangular block upstream of a positive bedrock step.* A platy block located upstream of a bedrock step can either pivot about its downstream support (Figure 8.8A), or be uplifted and carried beyond the step without overturning (Figure 8.8B). The critical velocity for pivoting about the edge of the step downstream of the block can be calculated from Equation 8.6a with the moment arms of lift and weight remaining the same as given in Equation 8.6c, and that for the drag modified to include the step height H to become:

$$m_D = 0.5D_c - H \tag{8.6d}$$

If the step is higher than the level of the line of attack of the drag force ($H > 0.5D_c$), the block cannot pivot under the sole influence of the drag force. In this situation, the block can only be entrained if the lift force exceeds the submerged weight of the particle. The

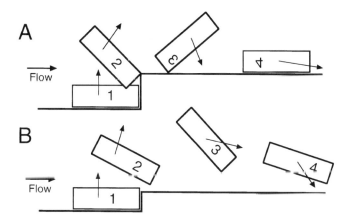

FIGURE 8.8 Schematic diagrams showing alternative modes of motion for a very platy rock block located ahead of a positive rock step: by pivoting about its downstream point of support (A), and by saltation without overturning (B)

critical entrainment velocity then becomes that of hydraulic plucking which can be obtained from a balance of the vertical forces ($F_L=F_G$, with friction along the vertical joints considered negligible). This is equivalent to ignoring drag ($C_D=0$) in Equation 8.6a, which then reduces to:

$$v_c^2 = \frac{2(\rho_s - \rho)g}{\rho} \frac{D_c}{C_L} \tag{8.7}$$

The critical velocity of plucking is thus proportional to the square-root of the thickness of the block, while the long and intermediate axis of the block only affect the shape-dependent lift coefficient.

For small step heights H, the critical velocity of pivoting is only a little higher than that for pivoting on a plane bed, and for compact blocks it remains well below the critical velocities of plucking (Tables 8.3a and 8.3b). However, for very platy blocks, plucking may not require much higher velocities than are needed for pivoting (Tables 8.3a and 8.3b), indicating the possibility of entrainment without overturning of these blocks.

In fact, such a mode of entrainment of a rock slab located ahead of a rock step was observed in a flume simulation which forms part of ongoing research. The tests were conducted in a recirculating glass-walled flume with a length of 8 m, a width of 0.4 m and maximum depth of 0.3 m. In the simulations, a block of glass with the dimensions $10 \times 9 \times 0.4$ cm was used to model a very platy rock block. The surface of the block was coated with masking tape to add surface roughness. The positive rock step transverse to the channel was simulated by a long plastic panel as wide as the flume, which was placed on the bottom of the flume. The thickness of the panel was varied to model steps of differing height. Flow depth was about 15 cm for all runs, and flow generally remained subcritical. It is stressed that the observations presented here are only meant to complement the results obtained from the theoretical considerations.

It was observed that in the flume the platy block upstream of a positive step higher than the slab thickness was generally uplifted at its upstream edge first, increasing the dip of the slab. Approaching the threshold velocity, the slab started to vibrate, a common and well documented motion of particles just prior to entrainment (Tipper, 1989). When the instantaneous velocity exceeded the threshold, the slab rapidly pivoted about its upper downstream edge, overturned, and was pushed back onto the channel bed. During the process the slab was displaced only a short distance downstream, generally only one or two slab lengths beyond the step (Figure 8.8A). In other flume studies, such pivoting has been observed to be a common mode of initial motion and transport of particles of various shapes (e.g. Carling et al., 1992).

However, if the slab was located upstream of a low step with a height in the order of the thickness of the slab or less, the slab was frequently entrained and displaced without being flipped over. As in the situation of the high step described above, the slab started to pivot about its downstream support, but instead of being overturned, the slab was released from the step. While the slab was then even further uplifted and also displaced significantly downstream, its inclination against the flow increased to a maximum just before it started to sink back to the channel bed, gradually returning to a flow-parallel orientation (Figure 8.8B). It has been argued that, under the influence of a strong lift force, platy particles can "jump" away from the channel bed (Johansson, 1976), or that they can be transported in a state of quasi-suspension (Bradley et al.,

1972). However, initial motion and transport by saltation appears to be the exception, even for platy particles (Carling et al., 1992), and the occurrence of saltation of platy particles may be restricted to high-magnitude flow events, when the effects of lift are significant.

(3) *Rectangular block imbricated against a positive bedrock step.* For a very platy rock slab, this situation can be subdivided into two principal cases. In the first case, the level of attack of the drag force is below the edge of the step (Figure 8.7B). In this situation, the slab can only be entrained if the lift force is positive and the resultant force acts away from the channel bed. Alternatively, the drag force could become negative, i.e. directed upstream, such as in regions of strong backflow. In the second principal case, the level of attack of the drag force is above the edge of the step, and the slab can be entrained by a positive drag force. However, if the line along which the lift and the gravitational forces act is located behind or downstream of the step, the slab is unstable. In such situations, the slab is likely to pivot under the influence of its own weight, or during flows of low velocity.

For a thin rock slab, the criterion for the level of the drag force being below the edge of the step can be approximated by:

$$\sin\beta < H/0.5D_b \qquad (8.8)$$

For imbrication angles as defined by Equation 8.8, the slab is relatively stable as it can only be entrained under the influence of a large positive lift force or a negative drag force. Appropriate lift coefficients for the particular case of a rectangular block imbricated against a positive transverse step are not available. However, the force coefficients listed in Table 8.2 indicate that the lift force is likely to be negative, i.e. to act towards the channel bed. This suggests that a platy rock slab with an imbrication angle smaller than that given by Equation 8.8 could only be entrained if the drag force becomes transiently negative.

The actual flow past the block and over the positive step is likely to be complex, with flow separation occurring in three dimensions. An additional important factor is certainly the large velocity fluctuations created by the high turbulence intensity at the step and the slab. Allen (1982, Vol.2, pp.101–131) outlined the likely effects of a positive step across the channel bed on the turbulent flow. A zone of flow separation and backflow is characteristically found ahead of the step, with high bed pressures occurring near the point of separation. Such high bed pressure (low velocity) is also typically found ahead of isolated large particles on the channel bed together with high velocities at the flanks of the particle (Brayshaw et al., 1983). According to Brayshaw et al. (1983), a spherical particle located upstream of such a stationary particle can experience higher positive lift forces than a single unobstructed particle, but the effects of particle shape were not included in their study. Field evidence from gravel streams elsewhere suggests that localized flow patterns can even result in the upstream migration of gravel-sized particles from clusters (de Jong, 1991), indicating the importance of localized flow pattern for particle entrainment. Nevertheless, it is well documented that imbricated particles are generally more stable than the same particles not in imbricated positions (e.g. Laronne and Carson, 1976; Brayshaw, 1984; Carling et al., 1992). Similarly, rock slabs imbricated in stable positions against positive bedrock steps (Equation 8.8) are likely to be more stable than the same slabs in other orientations. For an imbricated very platy block, the critical velocity for entrainment may even be higher than that required to hydraulically pluck a slab of equal size from the channel bed. The calculated low velocities needed for plucking of such a rock slab support this view (Table 8.3a).

Summarizing the discussion above, the critical entrainment velocity and the mode of motion of a rectangular block on a rock bed are strongly influenced by the shape and orientation of the block. On a smooth bed, a platy block will slide while a compact block could pivot at the same critical velocity (Table 8.3a). If the bed dips weakly downstream, the platy block will be entrained at a lower velocity than a more compact block. In the presence of a positive step transverse to the channel, most blocks will tend to pivot about their downstream support. However, on downstream-dipping beds, very platy blocks located ahead of low positive steps can also move by saltation. Platy blocks imbricated against positive steps can be very stable with stable imbrication angles approximated by Equation 8.8. For a given step, or obstacle height, short platy blocks can have steeper imbrication angles than long blocks before they become unstable and eventually pivot under the influence of their own weight.

From these theoretical considerations, it may be concluded that once a platy block has been dislodged from the channel bed by hydraulic plucking, it probably overturns, and then slides along the channel bed (Figure 8.8A). At the next downstream obstacle, which is likely to be a boulder cluster, it will pivot again, this time probably to a stable imbricated position (Figure 8.3A–D). Alternatively, very platy blocks may even saltate after being plucked (Figure 8.8B), and they could be deposited at locations significantly above the average level of the rock bed, such as on top of bars or large clusters (Figure 8.3C).

FIELD CHARACTERISTICS OF GIANT BOULDER BEDFORMS ALONG THE DURACK RIVER

A dip channel reach downstream of Jack's Hole (Figure 8.1), a prominent waterhole along the Durack River during the dry season, reveals an abundance of upstream-facing steps or rises in the channel bed produced by hydraulic plucking (Figures 8.9 and 8.10). The channel reach forms a local convexity, or bedrock high, in the longitudinal profile of the river and flow derived from a catchment of about 12 000 km^2 is funnelled through a gorge, 500 m wide and about 1.5 km long, cut into very resistant quartz sandstone. Along the reach, large hydraulically plucked rock slabs showing contact-type imbrication are frequently deposited on the stoss side of the upstream-facing steps (Figures 8.5A and 8.6). They represent an extremely coarse fraction of the bedload along the channel reach. The shape of most of these boulders is very platy to very bladed (cf. Sneed and Folk, 1958) which is largely the result of the joint spacing in the local source rock (Table 8.4). The large boulders have on average a long axis of more than 4 m, and exceptional clasts can have a long axis of nearly 13 m (Table 8.4). Alignment of individual clasts forming the imbricate-type boulder clusters is predominantly with the long axis transverse and the intermediate axis parallel to flow, and only about 10% of the large clasts are aligned with the long axis parallel to flow. This preferential orientation of clasts has been observed in other natural channels (e.g. Gustavson, 1974) and in experiments (e.g. Johansson, 1976). However, published evidence on preferential orientation of particles appears to be somewhat inconsistent (cf. Allen, 1982, Vol.1, pp.228–229).

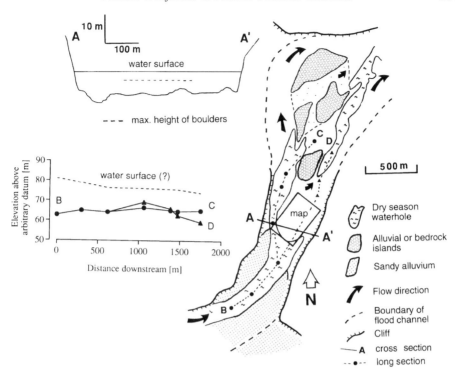

FIGURE 8.9   Geomorphology, cross- and longitudinal profiles of the study reach at Jack's Hole on the Durack River. Water-surface profiles represent probable maximum water-surface elevations as inferred from geomorphic evidence such as trim-lines and slackwater deposits. The box around "map" indicates the area enlarged in Figure 8.10

The inclination of the D_a–D_b plane of individual clasts forming the imbricate-type boulder cluster is very variable and ranges from horizontal to near-vertical. From the previous discussion about entrainment and stability, it is clear that the maximum stable angle of imbrication of a single slab is largely controlled by its shape and size relative to the size of the obstacle. Surely, for smaller clasts in a cluster, the inclination of the supporting downstream obstacle becomes important. Other factors influencing the final inclination of individual clasts include the initial position of deposition, interlocking effects, and post-depositional modification by impacting clasts. At Jack's Hole, however, very large clasts generally have low angles of imbrication, while clasts in near-vertical positions are always relatively small, rarely larger than the height of the next downstream obstacle clast or bedrock step in the channel bed. This confirms the importance of the position of the downstream point of support for a clast relative to the line of attack of the lift, drag and gravitational forces (Equation 8.8).

The height of boulder clusters varies with the size of the incorporated boulders and ranges from less than one metre to several metres, with the highest cluster found to reach a height of 4 m above the channel bed immediately upstream. The length of individual boulder clusters is variable, but stoss-side clusters are generally shorter than 20 m, while combined clusters can be tens of metres long. On top of some combined clusters, very

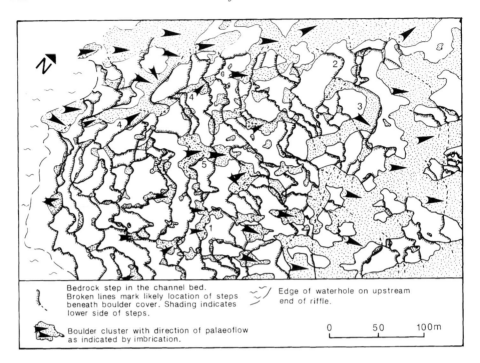

FIGURE 8.10   Map of principal pattern of erosional steps and associated boulder clusters at Jack's Hole, Durack River. The map was prepared from aerial photos and field mapping. The probable extension of bedrock steps underneath boulder deposits is indicated where possible. Numbers on the map refer to examples mentioned in the text

large and platy boulders with very low angles of imbrication are found. As outlined before, these slabs possibly moved to their present position by saltation under the influence of a large positive lift force.

In planform, imbricate-type boulder clusters display a complex pattern largely influenced by the planform of the bedrock steps that run in an irregular course across the channel (Figure 8.10). Stoss-side clusters or step-covering clusters are frequently more extensive transverse to flow than parallel to flow. In other words, they are usually wider than they are long. Along roughly straight segments of bedrock steps, such boulder clusters can have a roughly linear planform (Figure 8.10, no. 1). Steps that are broadly concave to the flow direction (Figure 8.10, no. 2) mark scour channels which are deeper than the surrounding channel bed and, therefore, have the tendency to concentrate flow. Deposits of imbricated boulders along such concave step segments frequently form stoss-side clusters which display the same concave planform as the bedrock step (Figure 8.10, no. 3). Combined boulder clusters formed along a succession of bedrock steps are generally elongated, and a link between their shape and the planform of bedrock steps is not immediately apparent (Figure 8.10, no. 4). However, as they consist of a combination of individual clusters along single bedrock steps, individual parts of the combined clusters can still display the planform of the steps along which they were deposited (Figure 8.10, no. 5).

FIGURE 8.11   Oblique aerial photo showing train of transverse boulder deposits (dark stripes, highlighted by dashed lines) along the Durack River. On the eastern side of the river, at least the first four rows of these depositional features are step-covering boulder clusters formed along oblique linear rock steps. The individual rows of imbricated boulders are separated by smooth bedrock surfaces free of any sediment (bright stripes). The very dark areas are bodies of standing water present at the time of the photograph during the dry season

Remarkable examples of boulder bedforms formed along regular linear bedrock steps are found about 5 km downstream of the Jack's Hole site along the Durack River (Figures 8.1 and 8.11). Here the strata dip less than 5° downstream and oblique to the course of the river (strike, dip: N40E, 4N). The channel is about 500 m wide on the upstream side of the 2 km long reach and has approximately twice that width at the downstream end of the reach. A train of oblique transverse depositional features can be identified along the reach, some with a peculiar, roughly V-shaped planform, with the base of the V pointing upstream (Figure 8.11). On the eastern side of the river, a train of at least eight more-or-less linear depositional features which extend obliquely across about two-thirds of the channel can be identified. These deposits consist of tightly packed imbricated clustered boulders forming more-or-less straight rows or ridges.

The boulders forming these deposits are generally larger on the stoss side of the ridges compared to the downstream side. Furthermore, the size of boulders of successive ridges appears to decrease in a downstream direction. The largest boulder found has dimensions of 5 × 1.5 × 0.6 m and is located on the stoss side of the second ridge near the centre of the channel. Other large boulders ($n = 10$) along the stoss side of this ridge measured on average 2 × 1.5 × 0.5 m while large boulders on the downstream side measured only 1 × 0.5 × 0.2 m. The height of the ridges is 2–3 m and the width 20–50 m. Between the individual rows of imbricated boulders are smooth, downstream-dipping bedrock

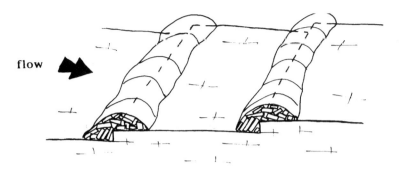

flow

FIGURE 8.12   Schematic diagram of rows of step-covering boulder clusters formed along linear steps

surfaces, often tens of metres long and with no clasts, which clearly separate the ridges from each other. The maximum spacing between ridges measured perpendicular to their alignment ranges from 70 to 90 m with an average of 75 m, but streamwise spacing of the ridges is somewhat difficult to determine since the ridges converge towards the eastern channel bank (Figure 8.11).

At least the first four of these ridges on the *eastern* side of the channel are clearly related to laterally straight positive steps in the bedrock channel and can be interpreted as step-covering boulder clusters, as illustrated schematically in Figure 8.12. The ridges on the *western* side of the channel, however, appear not to be related to any bedrock control. As a whole, the features appear to have several aspects in common with the regularly spaced transverse ribs of clustered pebbles reported from gravel streams (e.g. McDonald and Banerjee, 1971; Gustavson, 1974; Allen, 1982; de Jong and Ergenzinger, 1995). A review by Allen (1982, Vol.1, pp.383–394) of transverse ribs and their possible origins indicates that they are probably related to phenomena that accompany critical to supercritical flow. Furthermore, Allen argues that the development of transverse ribs is associated with hydraulic jumps, and that a train of transverse ribs could form downstream of any transverse row or pile of clasts of sufficient size to cause a hydraulic jump. Besides such a spreading of transverse ribs *downstream* of an initial rib (obstacle), flume experiments conducted by McDonald and Day (1978) indicate that transverse ribs can also be formed by the *upstream* migration of a hydraulic jump. The explanations above are restricted to situations where supercritical flow occurs, but transverse ribs may also form in association with standing waves or antidunes where flow is close to critical (e.g. Koster, 1978). If it is assumed that the transverse features along the Durack River developed under the influence of standing waves and flow conditions close to critical, the palaeoflow velocity (v) can be estimated from their wavelength (L) according to $v^2 = gL/2\pi$ (e.g. Koster, 1978). Using a spacing of $L = 75$ m, this estimate yields a flow velocity of about 10 m s^{-1} for the observed train of ribs. Interestingly, such high-velocity floods are also required to cause the hydraulic plucking of very large boulders observed at the Jack's Hole site 2 km upstream where the channel has roughly the same width as at the entrance to the rib section (Wende, in revision).

It is not clear whether the features described here are in fact giant transverse ribs, possibly related to critical or even supercritical flow conditions, or just exceptional examples of step-covering boulder clusters associated with regularly spaced bedrock steps. The problem with the latter interpretation is that the oblique ridges on the western side of the valley have not been shown to be associated with similarly oblique bedrock steps that could have initiated deposition on that side. While similarities with features caused by oblique waves during supercritical flows in sand beds (cf. Allen, 1982, pp.395–405) are remarkable, at this stage an interpretation of the processes causing these features would be entirely speculative.

Hydraulic considerations

It is evident that the extremely coarse fraction of the bedload along parts of the Durack River can only be moved during exceptional flood events (Wende, in revision). Such events are also necessary to activate erosion of thick rock slabs by hydraulic plucking, the process that produces the typical stepped channel morphologies along dip channel reaches. The uplift forces acting on the blocks during those floods have to be greater than the submerged weight of the blocks to cause their removal and can be estimated from Equation 8.7 using the uplift coefficient given in Table 8.2. These calculations indicate that hydraulic plucking along the channel reach downstream of Jack's Hole on the Durack River was probably caused by rare floods with instantaneous flow velocities up to 10 m/s (Tables 8.3a and 8.4). Inferring from geomorphic evidence along the reach, maximum flow depths at the site reached about 10 m (Figure 8.9), and regional estimates of extreme floods suggest that such high-velocity flows can occur along the reach under the present flow regime (Wende, in revision). However, it cannot be ruled out that some boulder bedforms are the product of a prior Quaternary flow regime characterized by flood events of much greater magnitude than have been experienced during the Holocene (cf. Nott and Price, 1994; Nott et al., 1996).

Nevertheless, as the field evidence shows, exceptional floods are capable of lifting those large joint blocks out of the channel bed. Based on the previous discussion about entrainment thresholds, the probable mode of motion of the large rock slabs at Jack's Hole may be outlined. Once a platy block is hydraulically plucked from the rock bed, it probably overturns about its downstream contact with the rock mass. Overturned blocks, recognized by erosion features such as potholes on their underside, are relatively common in the cluster bedforms at the site. The overturned blocks are then either imbricated against an obstacle, such as a step, or come to rest parallel to the channel bed. However, isolated platy to bladed boulders deposited flat-lying on the channel bed are very rare at the study sites, supporting the finding that the critical velocity for sliding is smaller than that needed for their plucking (Equations 8.5 and 8.7, Table 8.3a).

Platy boulders imbricated against a positive step with an imbrication angle in the stable range as defined by Equation 8.8, can only be entrained by flows with velocities considerably higher than those needed to entrain the particles resting parallel on the channel bed. If their threshold of entrainment is even higher than that for the hydraulic plucking of blocks forming the bedrock step itself, such boulder clusters formed along upstream-facing steps may delay further erosion of those steps. The size and likely longevity of boulder clusters, and also their impact on the flow hydraulics of floods below threshold conditions, make these combined erosional and depositional features important fluvial forms along certain reaches of the Durack River.

FIGURE 8.13   Step-covering cluster and cluster associated with obstacle clasts on the Ord River near the Ord River Homestead. Note the planform of the bedrock steps controlled by the vertical joint pattern

Structurally influenced boulder clusters not only form in channels subject to extreme flood conditions, but are also present in channels where the threshold of erosion and entrainment of sediment is low and frequently exceeded by floods. Such examples of boulder clusters are found on the Ord River just downstream of a gauging station near the Ord River Homestead (Figure 8.1). Along this reach, shales and jointed marine sedimentary rocks are exposed, striking slightly oblique to the channel and dipping less than 2° downstream. Hydraulic plucking of rock slabs along a bench in the channel bed has produced positive steps with a maximum height of 0.3 m. In planform, these steps reflect the pattern of the vertical joints which run roughly perpendicular and transverse to the channel (Figure 8.13). Boulders found along the reach are very platy to very bladed, and exceptional clasts can be as large as 4.5 × 1.9 × 0.3 m. Also present are several stoss-side and step-covering clusters, as well as clusters initiated by single obstacle clasts, with the latter two generally consisting of the larger boulders (Figure 8.13).

The channel along this reach of the Ord River has a rectangular cross-section with a bankfull width of some 255 m and a bankfull maximum depth of 14 m (Figure 8.14). The two largest floods since records began in 1969 measured 24 600 and 15 700 $m^3 s^{-1}$ (Department of Public Works Western Australia, 1984; Peter Clews, Water Authority of Western Australia, pers. com.), with the first flood exceeding bankfull flow along the reach, and the second staying well below. The exact water-surface elevation along the reach during these floods is not known, nor is the hydraulic gradient between the reach and the gauging site some 200 m further upstream. However, a minimum estimate of flow velocities can be obtained by neglecting the hydraulic gradient along the reach (i.e. assuming a horizontal water surface between the gauging site and the study cross-

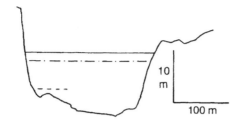

FIGURE 8.14 Cross-section at Ord River Homestead on the Ord River. Water-surface elevations represent maximum (solid line) and minimum (dash–dot line) estimates for a discharge of 15 700 $m^3 s^{-1}$ recorded at the site. Dashed line indicates maximum height of cluster bedforms in section

section), and a maximum estimate obtained by assuming critical flow conditions along this channel reach where flood flows actually are likely to remain subcritical. The largest flood below bankfull would then have had an average flow velocity between a minimum of 7.2 $m s^{-1}$ and a maximum of 8.7 $m s^{-1}$, with maximum depths of 12 m and 10.4 m respectively (Figure 8.14). To balance the uplift force produced by such flow velocities (Equation 8.7), an *in situ* rock slab which is bounded by horizontal and vertical joints, must be between 0.25 and 0.35 m thick, if the density of the rock is assumed to be 2.65 $g cm^{-3}$ and that of the water to be 1 $g cm^{-3}$, and C_L is assumed to be 0.15 (Table 8.2). This compares well with the observed size of hydraulically plucked boulders along the reach. Recent moderate floods, well below the level of the floods mentioned above, left several boulders imbricated against trees with the largest example measuring 2 × 1.4 × 0.12 m. This clearly indicates that this river, in contrast to the Durack River, is competent to move much of its coarse bedload even during moderate floods.

SUMMARY AND CONCLUSIONS

Along channel reaches where the joint system of a rock mass produces layers of platy to bladed rocks which dip gently downstream relative to the channel slope, hydraulic plucking can be a very effective erosional process if threshold values of erosion are exceeded during flood events. Threshold values are largely controlled by the degree of inclination of the rock mass relative to the channel slope, and by the platiness of the rock slabs. Generally, a thin slab of rock is more easily eroded than a thicker slab with the same surface area because the thin plate has the smaller submerged weight but the same effective area for the uplift forces to act on. If threshold values for hydraulic plucking are high and exceeded only by rare floods, erosional and depositional bedforms formed during such events are likely to persist unaltered for long periods of time.

Hydraulic plucking along dip channel reaches produces characteristic positive steps in the channel bed which are, in height and planform, largely controlled by the joint pattern of the exposed rock mass. These typical erosional bedforms present obstacles for coarse bedload particles which slide, pivot or, in exceptional cases, even saltate along the channel bed. Locally plucked platy boulders, or clasts supplied from further upstream, are imbricated against the steps, and subsequent accumulation of further rock slabs leads to the development of stoss-side clusters. Boulder clusters can extend laterally reflecting the general pattern of the steps. They can also increase in length by lee-side deposition, and such step-covering clusters can link up with other clusters further upstream or downstream to form longitudinal boulder deposits.

Threshold values for incipient motion of boulders deposited in imbricate-type cluster bedforms are most likely to be higher than those for boulders resting parallel on the channel bed. Importantly, it seems possible that threshold values for imbricated clasts can exceed those required for bedrock erosion by hydraulic plucking. Stable cluster bedforms can then delay or even prevent further erosion along the protected bedrock steps.

Owing to the conspicuous periodical occurrence of many stoss-side and step-covering clusters, it seems likely that the spacing of the erosional rock steps and the subsequent development of depositional boulder clusters are both controlled by local hydraulic conditions similar to those causing the development of dunes in alluvial channels. Clearly, more field studies and experimental investigations are needed to explain, quantify and classify the processes and forms of bedrock channels.

ACKNOWLEDGEMENTS

This research was undertaken with support from an Australian Research Council grant to Gerald Nanson, who also generously reviewed and contributed to an earlier draft of this paper. Many thanks to Christine Obermeier-Wende for her assistance in the field, and to Victor Baker, Paul Carling and Andrew Miller for reviewing the manuscript and providing helpful comments for its improvement. David Martin helped with some of the diagrams.

REFERENCES

Ahnert, F. 1994. Equilibrium, scale and inheritance in geomorphology. *Geomorphology*, **11**, 125–140.

Allen, J. 1942. An investigation of the stability of bed materials in a stream of water. *Journal of the Institute of Civil Engineers*, **5**, 1–34.

Allen, J.R.L. 1982. *Sedimentary Structures: their Character and Physical Basis* (2 vols). Elsevier, Amsterdam.

Annandale, G.W. 1995. Erodibility. *Journal of Hydraulic Research*, **33**, 471–495.

Ashley, G.M. 1990. Classification of large-scale subaqueous bedforms: a new look at an old problem. *Journal of Sedimentary Petrology*, **60**, 160–172.

Baker, V.R. 1973. Erosional forms and processes for the catastrophic Pleistocene Missoula floods in eastern Washington. In M. Morisawa (Ed.), *Fluvial Geomorphology*. State University of New York, Binghamton, 123–148.

Baker, V.R. 1978a. Large-scale erosional and depositional features of the scabland floods. In V.R. Baker and D. Nummedal (Eds), *The Channeled Scabland*. National Aeronautics and Space Administration, Washington, DC, 81–115.

Baker, V.R. 1978b. Paleohydraulics and hydrodynamics of Scabland floods. In V.R. Baker (Ed.), *Catastrophic flooding: The origin of the Channeled Scabland*. Dowden, Hutchinson and Ross, Stroudsburg, Pennsylvania, 255–275.

Baker, V.R. 1984. Flood sedimentation in bedrock fluvial systems. In E.H. Koster and R.J. Steel (Eds), *The Sedimentology of Gravels and Conglomerates*. Canadian Society Petroleum Geologists, Memoir 10, 87–98.

Baker, V.R. 1988. Flood erosion. In V.R. Baker, R.C. Kochel and P.C. Patton (Eds), *Flood Geomorphology*. Wiley, London, 81–95.

Baker, V.R. and Ritter, D.F. 1975. Competence of rivers to transport coarse bedload material. *Geological Society of America, Bulletin*, **86**, 975–978.

Billi, P. 1988. A note on cluster bedform behaviour in a gravel-bed river. *Catena*, **15**, 473–481.

Bradley, W.C. and Mears, A.I. 1980. Calculations of flows needed to transport coarse fraction of Boulder Creek alluvium. *Geological Society of America, Bulletin*, **91**, 1057–1090.

Bradley, W.C., Fahnenstock, R.K. and Rowekamp, E.T. 1972. Coarse sediment transport by flood flows on Knik River, Alaska. *Geological Society of America, Bulletin*, **83**, 1261–1284.

Brayshaw, A.C. 1984. Characteristics and origin of cluster bedforms in coarse-grained alluvial channels. In E.H. Koster and R.J. Steel (Eds), *Sedimentology of Gravels and Conglomerates*. Canadian Society of Petroleum Geologists, Memoir 10. 77–85.

Brayshaw, A.C., Frostick, L.E. and Reid, I. 1983. The hydrodynamics of particle clusters and sediment entrainment in coarse alluvial channels. *Sedimentology*, **30**, 137–143.

Carling, P.A. 1983. Threshold of coarse sediment transport in broad and narrow natural streams. *Earth Surfaces Processes and Landforms*, **8**, 1–18.

Carling, P.A. 1995. Flow-separation berms downstream of a hydraulic jump in a bedrock channel. *Geomorphology*, **11**, 245–253.

Carling, P.A. and Grodek, T. 1994. Indirect estimation of ungauged peak discharges in a bedrock channel with reference to design discharge selection. *Hydrological Processes*, **8**, 497–511.

Carling, P.A., Kelsey, A. and Glaister, M.S. 1992. Effects of bed roughness, particle shape and orientation on initial motion criteria. In P. Billi, R.D. Hey, C.R. Thorne and P. Tacconi (Eds), *Dynamics of gravel-bed rivers*. Wiley, New York, 23–39.

Costa, J.E. 1983. Paleohydraulic reconstruction of flash-flood peaks from boulder deposits in the Colorado Front Range. *Geological Society of America, Bulletin*, **94**, 986–1004.

Dal Cin, R. 1968. Pebble clusters: their origin and utilization in the study of palaeo-currents. *Sedimentary Geology*, **2**, 233–241.

de Jong, C. 1991. A reappraisal of the significance of obstacle clasts in cluster bedform dispersal. *Earth Surfaces Processes and Landforms*, **16**, 737–744.

de Jong, C. and Ergenzinger, P. 1995. The interrelations between mountain valley form and river-bed arrangement. In E.J. Hickin (Ed.), *River Geomorphology*. Wiley, New York, 55–91.

Denny, M.W. 1993. *Air and Water: the biology and physics of life's media*. Princeton University Press, Princeton.

Department of Publics Works Western Australia. 1984. *Streamflow Records of Western Australia to 1982*, Volume 3.

Engineering Sciences Data. 1976. *Fluid Mechanics, External Flow*, Volume 2. Engineering Sciences Data Unit, London.

Ergenzinger, P. and Jüpner, R. 1992. Using COSSY (CObble Satellite SYstem) for measuring the effects of lift and drag forces. *Erosion and Sediment Transport Monitoring Programmes in River Basins*, Proceedings of the Oslo Symposium, August 1992, IAHS Publication, 210, 41–49.

Grant, G.E., Swanson, F.J. and Wolman, M.G. 1990. Pattern and origin of stepped-bed morphology in high-gradient streams, Western Cascades, Oregon. *Geological Society of America, Bulletin*, **102**, 340–352.

Gustavson, T.C. 1974. Sedimentation on gravel outwash fans, Malaspina glacier foreland, Alaska. *Journal of Sedimentary Petrology*, **44**, 374–389.

Helley, E.J. 1969. Field measurement of the initiation of large bed particle motion in Blue Creek near Klamath, California. *US Geological Survey Professional Paper*, 562-G, 1–19.

Hoerner, S.F. 1965. *Fluid Dynamic Drag*. Hoerner Fluid Dynamics, Bricktown, NJ.

Jackson, R.G.I. 1975. Hierarchical attributes and a unifying model of bed forms composed of cohesionless material and produced by shearing flow. *Geological Society of America, Bulletin*, **86**, 1523–1533.

James, C.S. 1990. Prediction of entrainment conditions for nonuniform noncohesive sediments. *Journal of Hydraulic Research*, **28**, 25–41.

Johansson, C.E. 1976. Structural studies of frictional sediments. *Geografiska Annaler*, **58A**, 201–301.

Knighton, D. 1984. *Fluvial Forms and Processes*. Edward Arnold, London.

Komar, P.D. 1988. Sediment transport by floods. In V.R. Baker, R.C. Kochel and P.C. Patton (Eds), *Flood Geomorphology*. Wiley, New York, 97–111.

Komar, P.D. 1989. Flow-competence evaluations of the hydraulic parameters of floods: an assessment of the technique. In K. Bevan and P.A. Carling (Eds), *Floods: hydrological, sedimentological and geomorphological implications.* Wiley, New York, 107–134.

Komar, P.D. and Li, Z. 1988. Applications of grain-pivoting and sliding analyses to selective entrainment of gravel and to flow-competence evaluations. *Sedimentology*, **35**, 681–695.

Koster, E.H. 1978. Transverse ribs: their characteristics, origin, and paleohydraulic significance. In A.D. Miall, (Ed.), *Fluvial Sedimentology.* Canadian Society of Petroleum Geology, Memoir 5, 161–186.

Laronne, J.B. and Carson, M.A. 1976. Interrelationships between bed morphology and bed-material transport for a small, gravel-bed channel. *Sedimentology*, **23**, 67–85.

Matthes, G.H. 1947. Macroturbulence in natural stream flow. *Transactions of the American Geophysical Union*, **28**, 255–262.

McDonald, B.C. and Banerjee, I. 1971. Sediments and bedforms on a braided outwash plain. *Canadian Journal of Earth Sciences*, **8**, 1282–1301.

McDonald, B.C. and Day, T.J. 1978. *An experimental flume study on the formation of transverse ribs.* Geological Survey of Canada, Paper 78-1A, 441–451.

Miller, J.R. 1991. The influence of bedrock geology on knickpoint development and channel-bed degradation along downcutting streams in south-central Indiana. *Journal of Geology*, **99**, 591–605.

Naden, P. 1987. An erosion criterion for gravel-bed rivers. *Earth Surfaces Processes and Landforms*, **12**, 83–93.

Nott, J. and Price, D.M. 1994. Plunge pools and paleoprecipitation. *Geology*, **22**, 1047–1050.

Nott, J.F., Price, D.M. and Bryant, E.A. 1996. A 30 000 year record of extreme floods in tropical Australia from relict plunge-pool deposits: implications for future climate change. *Geophysical Research Letters*, **23**, 379–382.

Novak, I.D. 1973. Predicting coarse sediment transport: the Hjulström curve revisited. In M. Morisawa (Ed.), *Fluvial Geomorphology.* State University of New York, Binghamton, 13–25.

O'Connor, J.E. 1993. *Hydrology, Hydraulics and Geomorphology of the Bonneville Flood.* Geological Society of America, Special Paper 274, 83pp.

Otto, B. 1990. The effect of high in-situ rock stresses on the scour potential of sheet-jointed rocks below two dams in Australia. In H.P. Rossmanith (Ed.), *Mechanics of Jointed and Faulted Rock.* Balkema, Rotterdam, 899–904.

Pohn, H.A. 1983. The relationship of joints and stream drainage in flat-lying rocks of south-central New York and northern Pennsylvania. *Zeitschrift für Geomorphologie* NF, **27**, 375–384.

Renius, E. 1986. Rock erosion. *Water Power and Dam Construction*, **38**, 43–48.

Rusnak, G.E. 1957. The orientation of sand grains under conditions of 'unidirectional' fluid flow. *Journal of Geology*, **65**, 384–409.

Sneed, E.D. and Folk, R.L. 1958. Pebbles in the lower Colorado River, Texas, a study in particle morphogenesis. *Journal of Geology*, **66**, 114–150.

Tipper, J.C. 1989. The equilibrium and entrainment of a sediment grain. *Sedimentary Geology*, **64**, 167–174.

Wende, R. (in revision). Bedrock channel erosion: hydraulic plucking in monsoonal northwestern Australia. *Earth Surfaces Processes and Landforms*.

Williams, G.P. 1983. Paleohydrological methods and some examples from Swedish fluvial environments. *Geografiska Annaler*, **65A**, 227–243.

Wohl, E.E. 1992. Gradient irregularity in the Herbert Gorge of northeastern Australia. *Earth Surfaces Processes and Landforms*, **17**, 69–84.

Part 3

Arid-Region Rivers

9

Floodouts in Central Australia

Stephen Tooth

School of Geosciences, University of Wollongong, Wollongong, Australia
Present address: Department of Geology, University of the Witwatersrand,
Johannesburg, South Africa

ABSTRACT

Floodouts are a widespread but little-studied feature of the arid and semi-arid areas of Australia. This paper assesses floodouts in arid central Australia and focuses on the Sandover, Sandover–Bundey and Woodforde Rivers, three drainage systems on the Northern Plains in the Alice Springs region. In the *floodout zone* there is a marked reduction in channel capacity compared with reaches upstream and an increasing proportion of flood flows are diverted overbank. The floodout zone consists of this upstream *channel tract* and the *floodout*, which occurs at the terminus of the channel where floodwaters spill across adjacent alluvial surfaces. Floodouts occur across a wide range of scales in central Australia (c. 1–1000 km^2) but it is the larger examples at the downstream end of ephemeral rivers that are of prime concern in this paper. Floodouts in central Australia result from four main causes: downstream reductions in discharge; aeolian barriers to flow; hydrologic/alluvial barriers to flow; and structural (bedrock) barriers to flow. At the local scale, however, downvalley burial of older terraces by younger alluvial deposits is a key factor in the location of the floodout zone. The *emergence point* is where the channel leaves the confines of the upstream terraces. Downstream of here, splays and distributary channels develop and there is a rapid downstream decline in the capacity of the trunk channel which eventually terminates in the floodout. A distinction can be made between *intermediate* floodouts, downstream of which channels reform, and *terminal* floodouts, where floodwaters ultimately dissipate. Features characteristic of both include fluvial–aeolian interactions, waterholes and transverse bedforms. Floodout zones are difficult to classify, possessing features common to both floodplains and fans. However, the low gradients and the fine-grained deposits typical of floodout zones distinguish them from alluvial fans. Furthermore, their common location on medial or distal reaches of a channel system and a general absence of extensive distributary patterns of channel breakdown make application of the term "terminal fan" inappropriate. Floodouts represent a highly distinctive variety of fluvial form and are best regarded as part of the continuum of floodplains.

Varieties of Fluvial Form. Edited by A.J. Miller and A. Gupta
© 1999 John Wiley & Sons Ltd

INTRODUCTION

In many arid and semi-arid regions of the world, headwater channels debouch onto lowland plains where few tributary contributions are received. As a result of a combination of diminishing downvalley flows, an over-supply of sediment relative to the capacity for onward transport, declining gradients and sometimes aeolian, structural or hydrologic obstructions to flow, many channels fail to reach the lowest point in the drainage basin and channelized flow largely, or completely, disappears. This phenomenon of channel breakdown is a common occurrence in the arid and semi-arid areas of Australia, with many early explorers (e.g. Sturt, 1849) referring to the "gradual exhaustion", "termination" or "failure" of many of the desert rivers.

Although the phenomenon whereby channels increasingly lose definition downstream has been described in various geomorphic contexts from a number of locations around the world (e.g. Dubief, 1953; Glennie, 1970; Karcz, 1972; Mukerji, 1976; Sneh, 1983; Rodier, 1985; Ori, 1989), it is only in Australia that the term "floodout" or "flood out" has been used. This term is in widespread use in Australia and has been used both in the context of arid-zone fluvial systems (e.g. Slatyer and Mabbutt, 1964; Sullivan, 1976; Mabbutt, 1977, 1986; Pickup, 1991; Thompson, 1991) and discontinuous gullies (e.g. Erskine and Melville, 1983, 1984; Pickup, 1986; Crouch, 1990). However, there is no rigorous definition of "floodout", and the term has been used to refer both to a process (e.g. channels "flooding out") and to a fluvial form (e.g. the location of a "floodout"). As a result, the term "floodout" has acquired various process connotations and a number of qualifying adjectives (such as multiple, lateral, intermediate, terminal and compounded) but, surprisingly, there have been few detailed studies of these processes and forms.

This paper is an assessment of floodouts in arid central Australia and has four main aims: first, to establish a terminology for floodouts in order to enable comparative studies; second, to outline the broad factors controlling the establishment and development of floodouts; third, to illustrate some of the characteristic surficial features of floodouts; and finally, to consider the depositional style of floodouts. Although specific examples are drawn from floodouts in central Australia, it is likely that much of the discussion will pertain to floodouts elsewhere in arid and semi-arid Australia.

FLOODOUTS OF CENTRAL AUSTRALIA: REGIONAL SETTING

This study is concerned with the Alice Springs area of central Australia (Stewart and Perry, 1962; Figure 9.1), an arid region with ephemeral streams which remain dry for much of the year and which only occasionally flow throughout their length. The mean annual rainfall at Alice Springs is around 276 $mm\,a^{-1}$ with a summer maximum and compares with a pan evaporation of about 3000 $mm\,a^{-1}$ (Commonwealth of Australia, Bureau of Meteorology, 1996). A substantial rainfall gradient occurs across the region, from a minimum of around 125 $mm\,a^{-1}$ at the southeastern extremity to around 300 $mm\,a^{-1}$ along the northern margin, and there is great variability in rainfall from year to year. Heavy rainfalls (>100 $mm\,24\,h^{-1}$) can result when the predominating high-pressure system is destabilized by an influx of moist air, such as by the incursion of monsoonal depressions from the north and northwest of the continent or by the easterly passage of frontal weather systems. Such conditions can lead to widespread flooding with

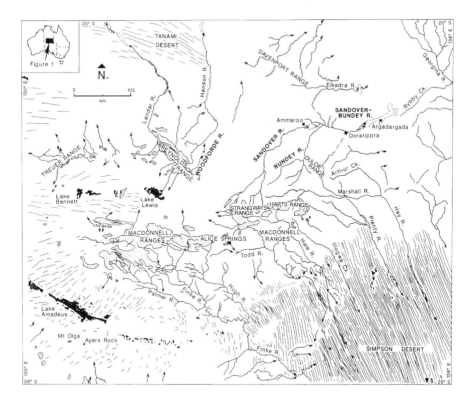

FIGURE 9.1   Ephemeral drainage systems in the Alice Springs region of central Australia. Stippled areas represent land over 750 m in elevation

floods in the last 30 years including some of the largest events since European settlement in the region (Williams, 1970a; Baker et al., 1983).

Topographically, the Alice Springs region of central Australia is dominated by a belt of sedimentary and crystalline ranges which were folded and faulted in the Devonian–Carboniferous Alice Springs Orogeny (Thompson, 1991). These ranges trend west–east across the middle of the region (Figure 9.1). The ranges are broken by intermontane lowlands and are flanked by the Northern Plains and Southern Desert Basins (Mabbutt, 1962): extensive low-lying plains of erosional or depositional origin overlying Precambrian shield or younger cover rocks with uplands more restricted in extent (Figure 9.1).

Drainage in the region is largely internal, with floodwaters from the larger rivers only occasionally persisting beyond the Alice Springs area (Figure 9.1). Many of the larger drainage systems disappear either in the longitudinal dunes of the western Simpson Desert (Illogwa Creek and the Todd, Hale, Plenty and Hay Rivers) or in aeolian sandplain of the Tanami Desert (the Hanson and Landar Rivers). Previous studies of the geomorphology of the drainage systems in the region (e.g. Williams, 1969, 1970a, 1971; Baker et al., 1983; Pickup et al., 1988; Pickup, 1991; Patton et al., 1993; Bourke, 1994; Nanson et al., 1995) have concentrated either on systems close to the ranges or on short reaches in the middle of larger river systems.

The findings presented in this paper are drawn largely from study of the Sandover, Sandover–Bundey and Woodforde Rivers, three drainage systems on the Northern Plains (Figure 9.1). The catchments of these drainage systems are approximately 10 600, 11 000 and 550 km² respectively, and the headwaters of the rivers are largely in Proterozoic crystalline ranges. In all three catchments, networks of low-order channels carrying bedloads of poorly sorted sand and gravel converge on the lowland plains to give rise to well defined channel trains flanked by higher alluvial surfaces and extensive areas of aeolian sandplain. Beyond the limit of further tributary contributions, bankfull capacities progressively decline in the downvalley direction due to the reduction in discharge that results from transmission losses and the attenuation of floodwaves. The channels remain well defined for much of their length but capacities eventually decline to the point at which an increasing proportion of the larger floods reaching the distal reaches is diverted overbank. Eventually, channelized flow and bedload transport of sand and gravel largely cease but large floods continue across typically broad, low-gradient, alluvial surfaces which are lightly vegetated with a mixture of trees, shrubs and grasses, and which are surrounded by aeolian sandplain, older alluvial surfaces or low-relief hills.

On the Sandover River, the channel initially disappears on Ammaroo station, some 250 km from the headwater ranges (Figure 9.1). Large floods disperse across an area of approximately 200 km² before a definable network of channels is again formed which carries floodwaters to the confluence with the Bundey River. Although it is the Bundey that provides the vast majority of the flow and sediment downstream of the confluence, the channel continues as the Sandover River. It is here referred to as the Sandover–Bundey River in order to avoid confusion with the Sandover River upstream of Ammaroo (Figure 9.1). The Sandover–Bundey continues some 55 km further to Ooratippra station where it breaks into several distributary channels which decrease in capacity and eventually disappear (Figure 9.1). Large floods disperse across an area of approximately 800 km² with flows occasionally continuing to Bybby Creek and thence to the Georgina River (Figure 9.1), part of the centripetal drainage of Lake Eyre.

On the far smaller Woodforde River, the channel initially disappears around 45 km from the ranges (Figure 9.1). Large floods disperse over an area of approximately 40 km², before a network of channels is again formed. These channels carry flows a further 22 km before channelized flow again disappears near to the confluence with the larger Hanson River (Figure 9.1). It is the forms and processes in the distal reaches of these and similar systems that are the principal concerns of this paper.

DEFINITION OF "FLOODOUT"

Consideration of the ephemeral channel system as a whole reveals that it is useful to think in terms of a *floodout zone*, taken as that part of the system where there is a marked reduction in channel capacity compared with reaches upstream and where overbank flows become increasingly important. The floodout zone consists of the upstream *channel tract* and the *floodout*, which occurs where channels finally terminate. Hence, a *floodout* can be defined as: "*a site where channelized flow ceases and floodwaters spill across adjacent alluvial surfaces*". Floodouts occur on a wide range of scales throughout the Australian arid zone (*c.* 1–1000 km²) and such a definition is inclusive both of floodouts described for discontinuous gullies (e.g. Erskine and Melville, 1983, 1984; Pickup, 1986;

Crouch, 1990) and of the larger examples at the downstream end of ephemeral river systems that are of prime concern here. As for many definitions of floodplain (e.g. Bates and Jackson, 1987; Nanson and Croke, 1992), some process is implicit in the definition of a floodout but only in so far as it describes the changing nature of flood flows from a channelized to a largely unchannelized form.

The definition of a floodout provided above is a general one and finer distinctions can be drawn between different types of floodouts based on their location in the drainage network, such as between *intermediate* and *terminal* floodouts. Such terms have been used in the literature before (e.g. Mabbutt, 1962, 1977, 1986; Adamson et al., 1987; Pickup, 1991) but have not been clearly defined. Other loosely used adjectives such as "multiple", "lateral" and "compounded" are unnecessary and it is suggested that their use be discontinued in the description of floodouts.

Thus, an *intermediate floodout* is defined as *"a site where channelized flow ceases and floodwaters spill across adjacent alluvial surfaces but downvalley of which flow channelizes again"*. These floodouts occur on the medial or distal reaches of channel systems and consist both of a *distributary* part (where floodwaters spill across alluvial surfaces) and a *contributory* part (where flow channelizes again). Good examples of intermediate floodouts are found on the Sandover River at Ammaroo station and on the Woodforde River (Figure 9.1), with floodwaters initially dispersing across the plains before rechannelizing further downvalley.

A *terminal floodout* is defined as: *"a site where channelized flow ceases and floodwaters spill across adjacent alluvial surfaces and ultimately dissipate"*. These floodouts are found at the downstream end of channel systems and consist of a *distributary* part only. The second floodout of the Woodforde River and that of the Sandover–Bundey at Ooratippra are examples of terminal floodouts, for while very large floods occasionally reach the Hanson and Georgina Rivers (Figure 9.1), in neither case are there continuous channels linking the drainage systems and, for most events, the floodouts represent terminal sites for floodwaters. Good examples of terminal floodouts also occur in the dunefields of the western Simpson Desert (Figure 9.1) where several large drainage systems terminate.

Clearly, floodouts are just one possible end result of the broader processes of channel breakdown and termination that have been reported elsewhere in the literature. As such, there is a need to distinguish floodouts from the instances whereby channels terminate in a playa or a semi-permanent body of standing water such as a swamp (e.g. Williams, 1970b; Ori, 1989; Nanson et al., 1993; O'Brien and Burne, 1994; Linton and McCarthy, 1995). Floodouts are predominantly alluvial features which are normally dry except after flood events or heavy local rains, and which differ from the saline environments of playas or the organic-rich, waterlogged nature of swamps. In the Alice Springs region of central Australia, channel breakdown is usually associated with floodouts although some drainage systems terminate in playas such as Lake Lewis (Figure 9.1).

CONTRIBUTORY FACTORS TO THE ESTABLISHMENT AND DEVELOPMENT OF FLOODOUTS

The common occurrence of floodouts throughout the Australian arid zone can be largely related to Cenozoic climate changes. Increasing desiccation of the Australian continent during late Tertiary and Quaternary times has resulted in the retraction and disintegration of formerly better integrated drainage networks throughout much of the inland (Mabbutt, 1962, 1967; Sullivan, 1976; Arakel and McConchie, 1982) leading to the widespread establishment of floodouts. Within this climatic context, however, more regional and local-scale factors must be responsible for the subsequent development of floodouts.

Channel breakdown in the distal reaches of ephemeral stream systems has generally been considered to result from factors such as diminishing flows, increasing sediment loads relative to the capacity for onward transport, and an overall downstream decline in channel slope (Dubief, 1953; Glennie, 1970; Karcz, 1972; Mukerji, 1976; Sneh, 1983; Ori, 1989). In combination, these act to decrease flow velocities and hence to reduce the transporting capacity of the stream which eventually leads to channel termination. In detail, however, channel breakdown has sometimes been noted to begin across a zone of slightly steepened slopes (Sullivan, 1976; Rodier, 1985). In other instances, aeolian barriers to flow have been cited as the principal reason for channel breakdown (e.g. Lancaster and Teller, 1988; Ward, 1988; Langford, 1989). Hydrologic/alluvial barriers to flow (such as where a trunk channel in flood results in local ponding and sediment deposition on an adjacent tributary channel) and structural (bedrock) barriers are additional reasons for channel breakdown (Sullivan, 1976; Mabbutt, 1977). All of these factors contribute to the development of floodouts in the Alice Springs region of central Australia (Figure 9.2). Hence, floodouts establish as the result of four main causes: downstream reductions in discharge, aeolian barriers to flow, hydrologic/alluvial barriers to flow, and structural barriers to flow.

These factors are not mutually exclusive and some floodouts may result from a combination of one or more influences. For instance, downstream reduction in discharge is common to all floodouts but the *geographic location* of the floodout zone is often determined by other factors such as aeolian, hydrologic/alluvial and/or structural barriers. Thus, floodouts result from a number of causes and occur in different physiographic settings, making generalizations difficult. In the case of the Sandover, Sandover–Bundey and Woodforde drainage systems, floodout development is largely the result of downstream reductions in discharge and hence this type of floodout is the prime concern of this paper.

In the absence of any obstructions to flow, a key issue is the point at which the factors of downstream reductions in discharge, sediment transport and changes in slope interact to result in channel breakdown and in the associated establishment of a floodout. A major limitation in this regard is the paucity of data regarding flow and sediment transport for the channels in central Australia. Where flow gauging records do exist they are generally of limited duration and gauging stations often have not been reliably rated. Hence, the anecdotal accounts of local pastoralists are sometimes the only available evidence of the nature of flows reaching the floodouts. For the Sandover River, which receives few tributary contributions for some 135 km upstream of the terminus of the channel (Figure 9.1), floods appear to reach the floodout every three to four years. For the Sandover–Bundey River, which is joined by several small tributaries at various points along its length (Figure 9.1), flows appear to reach the floodout at a more frequent interval of one to two years, as is the case for the smaller Woodforde River where the floodout is located closer to the upland sources of runoff. Assessment of the frequency of flows reaching other floodouts in arid Australia (Sullivan, 1976; Mabbutt, 1977) reveals a similarly varied picture.

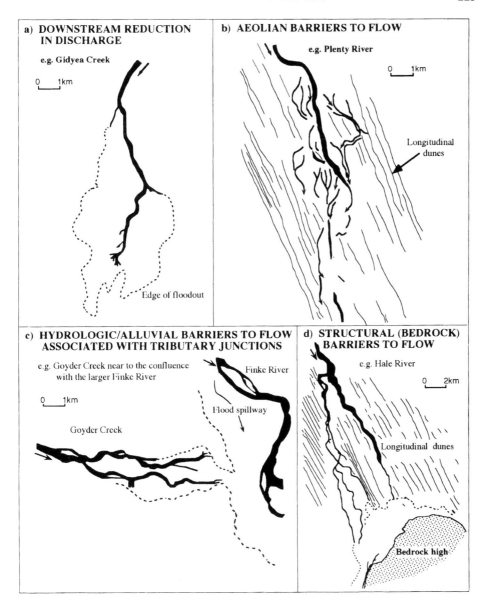

FIGURE 9.2 Contributory factors to floodout development in central Australia. Scales are approximate only. The size of channels approaching floodouts is variable and ranges from a few metres to several tens of metres in width. Floodouts range in area from a few square kilometres at the downstream end of smaller rivers up to around 1000 km^2 for larger rivers

Indirect estimates of discharge for the channels in the floodout zone are made difficult by the uncertain roughness coefficients resulting from the variable distribution of in-channel trees such as river red gums (*Eucalyptus camaldulensis*) (Graeme and Dunkerley, 1993). Furthermore, although accounts by local pastoralists indicate that sediment transport occurs during flood events, there are no data from which to evaluate the relative importance of suspended and bedload transport.

There are few signs of substantial declines in channel-bed slopes as the floodouts are approached. Bed slopes approaching the floodout of the Woodforde River are around $0.0014 \, \mathrm{m \, m^{-1}}$ but for the longer Sandover and Sandover–Bundey Rivers they are lower at around $0.0006 \, \mathrm{m \, m^{-1}}$. These relatively low slopes are typical of many rivers in arid and semi-arid Australia. Slopes across the floodouts appear to be in the same order of magnitude as the upstream channel-bed slopes, ranging from around 0.0005 to $0.001 \, \mathrm{m \, m^{-1}}$. Descriptions of flows across floodouts suggest that they take the form of a shallow (typically <30 cm deep), steadily creeping flow of water (D. Winstanley, T. Mahney, pers. comm.). Hence, owing to the relatively low slopes and the limited depths and velocities of the unchannelled flows across floodouts, they are best regarded as "sheetflows" (*sensu* Hogg, 1982) rather than as "sheetwash" (Pickup, 1991) or "sheetfloods". Following Hogg (1982), the term "sheetwash" is now considered redundant and has been superseded by the term "rainwash" (defined as the washing action of rain on slopes), whereas the term "sheetflood" describes deeper (up to *c.* 1 m depth), faster moving (up to $10 \, \mathrm{m \, s^{-1}}$), unconfined floodwaters which commonly originate in steeper areas (slopes $> 0.2 \, \mathrm{m \, m^{-1}}$).

Quantitative explanations for the development of floodouts will not be possible until detailed flow and sediment transport data are collected. Nevertheless, qualitative explanations of the location of floodouts are possible by considering their broader fluvial geomorphic context.

"EMERGENCE POINTS" OF THE WOODFORDE, SANDOVER AND SANDOVER–BUNDEY RIVERS

Figure 9.3 is a schematic diagram to illustrate the downstream pattern of the various sedimentary units along the Sandover, Sandover–Bundey and Woodforde Rivers. For all three drainage systems, the same broad patterns are evident. Once clear of the uplands, the channels flow through low-lying depositional plains of varying age and characteristics. In the piedmont, mottled red (2.5 YR 4/8 – 10R 4/8) terraces of gravel, sand and silt typically flank the contemporary channels and floodplains (Site A, Figure 9.3). These terraces are highly indurated, often contain abundant pedogenic carbonate and, in conjunction with numerous bedrock outcrops and remnants of Tertiary age lateritic weathering profiles (Mabbutt, 1965, 1967), laterally confine the channels.

Further downstream, less indurated red (2.5YR 4/8 – 10R 4/8) terraces of silt and sand envelope the contemporary channels and floodplains (Sites B and C, Figure 9.3). Although sometimes inundated during large flood events, these surfaces are here referred to as terraces rather than floodplains as they are a legacy of a former flow regime. In many locations, the terraces appear in a highly degraded and gullied form or they are obscured by aeolian sandplain and source-bordering dunes. Nevertheless, they commonly provide a resistant boundary for the channels which are typically characterized by a series of low-amplitude meanders.

This pattern of channels and floodplains inset within older terraces is a key factor in maintaining the integrity of the drainage systems as they traverse the low-gradient plains. Processes of erosion, transportation and sedimentation generally occur within the

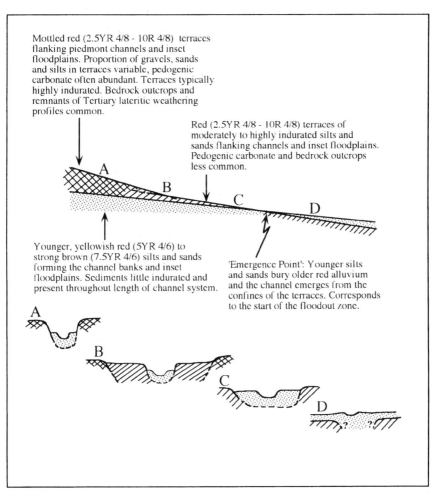

FIGURE 9.3 Schematic illustration of the downstream pattern of alluvial sedimentary units along the Sandover, Sandover–Bundey and Woodforde Rivers. The channels and adjoining inset floodplains commonly bear an open woodland of large phreatophytes, principally river red gums (*Eucalyptus camaldulensis*) and coolibah (*Eucalyptus microtheca*). Older terraces typically support a sparser vegetation of trees and shrubs, especially mulga (*Acacia aneura*), teatrees (*Melaleuca* spp.) and ironwood (*Acacia estrophiolata*). Vegetation variations between the different sedimentary units are often gradational, however, and they do not permit easy distinction between the different sedimentary units

confines of the terraces and consequently there is little potential for channel migration or for the development of splay and distributary channels. Further downstream, however, the terraces and floodplains gradually converge. Inset floodplains are less well developed and shallow burial of the terraces by younger, yellowish red (5YR 4/6) to strong brown (7.5YR 4/6) silt and sand eventually occurs. The younger deposits, now forming the channel perimeter (Site D, Figure 9.3), are little indurated and are relatively erodible. On the Woodforde River, where the burial of the terraces first occurs, a large splay has breached the channel banks (Figure 9.4) exposing the red alluvium beneath 1–1.5 m of

FIGURE 9.4   (a) The channel tract in the floodout zone of the Woodforde River showing the approximate location of the emergence point. The direction of flow is towards the north. The low-amplitude meanders characteristic of the river upstream of the floodout zone are just visible in the lower portion of the photograph. A short distance downstream of the emergence point a prominent splay channel has breached the right-hand bankline, overbank flows spread for greater distances on either side of the channel margins, and channel capacity progressively declines towards the terminus of the channel (located to the north of the area shown). The floodout zone is bordered by low-relief, red earth plains and aeolian sandplain (Napperby SF53-9, 1989, NTc 1114, Run 4, 007. Reproduced by permission of Northern Territory Department of Lands, Planning and Environment). (b) Typical bank exposure in the vicinity of the emergence point of the Woodforde River. Up to 2 m of indurated, red alluvium underlies 1.5 m of younger, browner alluvium

younger silty sand. Further exposures, which typically show an abrupt contact between the two units (Figure 9.4b), are provided on the cut banks of several pronounced meanders or where there is marked widening or deepening of the main channel. Downstream of this point, few further exposures of the red alluvium are found, the banks are formed entirely of the younger silty sands and the channel is flanked by low-relief alluvial plains (Figure 9.4a).

The term *emergence point* is used here to refer to the emergence of the channel from the confines of the previously enveloping terraces as a result of burial by more recent alluvial deposits (Figure 9.3). Sullivan (1976) considered that a zone of aggradation where bedload is deposited represents the upper limit of the floodout zone. For the Woodforde River, the upper limit of the floodout zone is where the burial of the terraces first occurs (Figure 9.4), even though bedload transport continues beyond this point. Where the channel emerges from the confines of the terraces, overbank flows spread for greater distances on either side of the channel margins and there is greater potential for channel migration. Hence, this sedimentological change is a key factor in the location of the floodout zone.

A similar emergence point can also be identified for the larger Sandover River. For the Sandover–Bundey River, however, unequivocal evidence of a solely alluvial emergence point is less forthcoming due to the influence of a low-relief dolostone plain abutting the southern margin of the channel. Both the bedrock plain and older alluvial surfaces of red (2.5YR 4/8) silty sand would appear to influence the upper limit of the floodout zone for, where the channel temporarily meanders away from the bedrock, a complex pattern of channel breakdown ensues which ultimately results in the adoption of a distributary channel pattern. Beyond this point, the older alluvial surfaces disappear and the channels of the Sandover–Bundey River traverse a broad plain consisting predominantly of yellowish red (5YR 4/6) silty sand.

Patterns of channel breakdown leading into floodouts

As implied in much of the preceding discussion, the disappearance of confining alluvial (or bedrock) surfaces in the distal reaches of the Woodforde, Sandover and Sandover–Bundey Rivers has major implications for fluvial forms and processes in the channel tract. For all three drainage systems, downstream of the emergence point channel breakdown rapidly ensues before floodwaters spill across the floodout. Nevertheless, there are differences in the pattern of channel breakdown for each drainage system.

On the relatively short Woodforde River, the channel undergoes a general decrease in size downstream of the emergence point. Cross-sections surveyed at various points show the decline in channel capacity to be roughly progressive (Figure 9.5) and to take place by both narrowing and shallowing of the channel. Width–depth ratios of the channel are generally in the order of 25–40 and remain relatively constant downstream. Only in the last 500 m of the channel is there any appreciable evidence of infilling by sands and gravels. Eventually, the channel disappears, bedload transport ceases and floodwaters spill across the floodout.

In contrast, on the larger Sandover and Sandover–Bundey Rivers, the patterns of channel change in the floodout zone are far less regular. Downstream of the emergence point, the channels undergo a series of marked fluctuations in width and depth superimposed on an overall decline in channel capacity (Figures 9.6 and 9.7). Width–depth ratios range between 30 and 250 over distances of 10 km or less, with significant steepening of channel-bed slope occurring in several of the short reaches where the width–depth ratio decreases to a minimum. In the case of the Sandover River, the decline in bankfull capacity is especially rapid over the last few kilometres of the channel (Figure 9.6). In contrast, the Sandover–Bundey adopts a distributary channel pattern with two

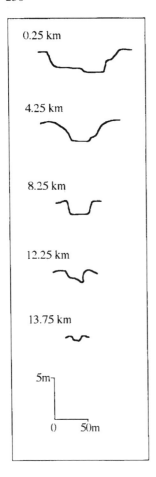

0.25 km

4.25 km

8.25 km

12.25 km

13.75 km

5m⌐

0   50m

FIGURE 9.5   Channel changes through the floodout zone of the Woodforde River. For each section, the approximate distance downstream of the emergence point is indicated

channels that terminate some 4–5 km apart. The width–depth ratio of these distributary channels is generally between 20 and 40 and the channels show a roughly progressive downstream decline in capacity (Figure 9.7). As for the Woodforde River, channel capacities of both the Sandover and Sandover–Bundey Rivers decrease by narrowing and shallowing and there are few signs of infilling by the sand and gravel that are transported to the termini of the channels.

In summary, therefore, the patterns of channel breakdown leading into the floodouts of the three study systems are remarkably varied given that they represent adjacent drainage systems. Nevertheless, some similarities are evident which allow general conclusions to be drawn. First, in all three instances, the disappearance of channelized flow does not occur due to channel aggradation (e.g. Schumm, 1961) but results instead from downstream decreases in the cross-sectional area of the channels that are cut into the flanking alluvial surfaces. Second, bedloads of poorly sorted, slightly gravelly sand are transported to the termini of the channels and thus contrast with the descriptions of the highly sinuous, suspended-load distributary channels in the floodout zone of the Finke River (Mabbutt, 1977). Statements that *all* floodouts result from the deposition of bedload followed by the splitting of the main channel into a number of smaller

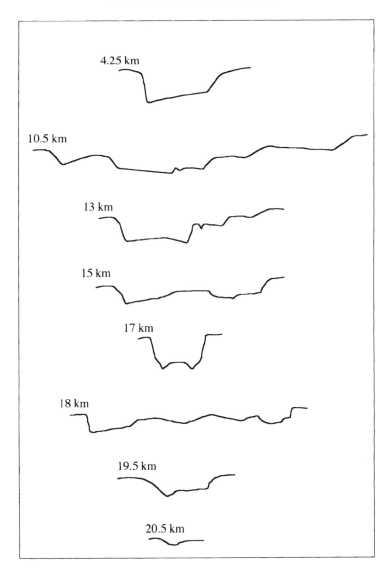

FIGURE 9.6   Channel changes through the floodout zone of the Sandover River. For each section, the approximate distance downstream of the emergence point is indicated. See Figure 9.5 for scale

distributary channels of suspended-load form (Sullivan, 1976; Mabbutt, 1977) are clearly incorrect. Third, the rapidity of channel breakdown downstream of the emergence point is apparent on all three systems. A major contributory factor in this regard is the dispersal of water and sediment from the channel via splay or distributary channels but, as with the patterns of channel breakdown, these take a different form on each drainage system. For instance, the Woodforde River displays a series of splay and distributary channels (Figure 9.4) that branch from the trunk channel, the Sandover River displays a series of poorly defined swales that breach the channel banks, and the Sandover–Bundey has also

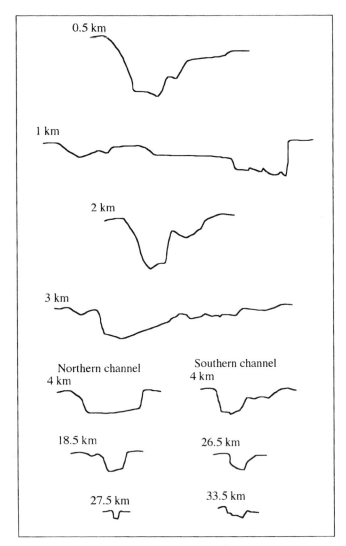

FIGURE 9.7   Channel changes through the floodout zone of the Sandover–Bundey River. For each section, the approximate distance downstream of the emergence point is indicated. See Figure 9.5 for scale

developed a number of small splay channels that branch from the distributary channels in the floodout zone. As splay and distributary channels are virtually absent upstream of the emergence point, their development clearly accelerates the downstream decline in the capacity of the trunk channel as the floodout is approached.

Emergence points as identified elsewhere in central Australia

Similar relationships between older terraces and contemporary channel–floodplain systems to those described here have been noted on a number of other drainage systems in central Australia. For instance, the tendency for younger drainage systems to be incised into older alluvium in the proximal and medial reaches has often been referred to (Mabbutt, 1962, 1967; Shaw and Warren, 1975; Arakel and McConchie, 1982; Pickup, 1991; Bourke, 1994). Furthermore, although the implications for floodout development have not been explicitly recognized, the pattern of older units dipping beneath younger alluvial deposits has been alluded to in the distal reaches of other drainage systems in central Australia (Mabbutt, 1962, 1966; Hays, 1967; Litchfield, 1969).

In addition, several systems draining towards Lake Lewis (Figure 9.1) have been observed by the present author to show the pattern of a terrace and inset channel–floodplain in their proximal and medial reaches together with the tendency for the distinction to become more blurred downvalley. Where the distinction between the terraces and floodplain is at its weakest, splay and distributary channels develop, the capacity of the trunk channels rapidly declines and the channels finally terminate in floodouts. Even on these relatively short systems, however, the burial of the terraces is a very gradual process taking place over some tens of kilometres, and stratigraphic exposures to rival those of the Woodforde and Sandover Rivers have yet to be found.

It remains to be seen, however, if the concept of the emergence point can be advanced for other floodouts in the Alice Springs region (such as where they form as a result of aeolian, hydrologic/alluvial or structural barriers) and for floodouts elsewhere in the inland. In other areas of arid and semi-arid Australia, floodouts have been noted to occur where streams emerge from confining bedrock valleys onto low-gradient plains. Nevertheless, the overall control on floodout development is the same: namely, the transition from a relatively confined channel–floodplain system to a relatively unconfined system.

MORPHOSEDIMENTARY FEATURES OF THE FLOODOUT ZONE

As with the patterns of channel breakdown and the characteristics of the splay and distributary channels, the floodout zones of the Sandover, Sandover–Bundey and Woodforde Rivers each have a somewhat unique array of surficial morphological and sedimentary features, resulting from the interaction of the floodwaters and alluvial sediments with local topography. Once again, however, some common patterns can be recognized. Although many features occur widely, both throughout the broader floodout zone and on the floodout *sensu stricto*, for descriptive purposes it is nonetheless useful to consider the distinction between the distributary and contributory parts of intermediate and terminal floodouts.

Features characteristic of the distributary parts of intermediate and terminal floodouts

Fluvial–aeolian interactions

A number of researchers have referred to fluvial–aeolian interactions in the distal reaches of arid environment drainage systems, both in Australia (e.g. Mabbutt, 1963, 1967; Nanson et al., 1995) and in the deserts of Africa and North America (e.g. Glennie, 1970; Langford, 1989; Lancaster and Teller, 1988; Ward, 1988), but the details of these interactions and the resulting deposits have received relatively little attention. In the floodout zones of arid central Australia, where fluvial processes are competing against the aeolian domain, there are a number of different fluvial–aeolian interactions, particularly where aeolian obstruction is the principal cause of floodout development (Figure 9.2). In the case of the Sandover, Sandover–Bundey and Woodforde Rivers, however, where the floodouts result primarily from downstream reductions in discharge and are located in low-relief settings far from the ranges, the interactions are very subtle in their characterization.

The principal aeolian deposits on the Northern Plains of the Alice Springs region are extensive areas of sandplain (Mabbutt, 1967) – low-relief areas of sand that have few definable dune forms – and source-bordering dunes. For all three drainage systems, sandplain and source-bordering dunes interact with the deposition of overbank fines along the margins of the floodout zone. Thin (<0.01 m) layers of alluvial silt and clay are deposited over the aeolian sand following flood events but owing to the infrequent and shallow flows in the floodout zones the rates of aggradation are extremely slow and sedimentary structures rarely preserved. Furthermore, in the floodout zone of the Sandover–Bundey River, narrow swales carry floodwaters from the channels to a number of pans (Goudie and Wells, 1995) isolated in the sandplain. Such pans can be found up to 4 km from the channels, are generally elliptical to rounded in shape and are usually less than 0.25 km^2 in area. Trees such as coolibah (*Eucalyptus microtheca*) can be found around the margins of the pans with gilgai soils and cracking clays in the centre of the depression.

In addition to interactions at the margins of the floodout zones, large patches of aeolian sandplain or source-bordering dunes up to 5 m high and 1.5 km^2 in area have been isolated in the middle of the floodout zones where they have been surrounded by more recent alluvial deposits. Hence, local aeolian relief is slowly being incorporated into the alluvial sedimentary sequences by the vertical accretion of sand and mud.

The floodout zones of the Sandover, Sandover–Bundey and Woodforde Rivers thus demonstrate a number of fluvial–aeolian interactions but it must be noted that other floodouts, particularly those in the extensive dunefields of the Simpson Desert, are likely to provide a vast range of other examples. In such situations, where declining downvalley flows are competing against the aeolian domain, a continuum of subtle patterns and forms dependent on local circumstances is likely to be the result. As such, the interactions and the resulting forms are likely to defy easy classification.

Waterholes

Waterholes are natural holes, hollows or small depressions that contain water (Bates and Jackson, 1987) and can be found in various geomorphic settings in arid and semi-arid Australia (Argue and Salter, 1977; Knighton and Nanson, 1994). Although there have been no previous references to waterholes on floodouts, prominent waterholes up to 1.5 km in length and 2–3 m deep are found at several different locations on the extensive terminal floodout of the Sandover–Bundey River and on a number of other floodouts in the Alice Springs region. Waterholes are not present on the floodouts of the Woodforde or Sandover Rivers but, in the case of the Sandover, numerous narrow depressions lined with silt and clay occur between patches of aeolian sandplain (Figure 9.8) and hold water for short periods after floods. Thus, waterholes and related features also appear to be a common though localized feature of floodouts in the Alice Springs region.

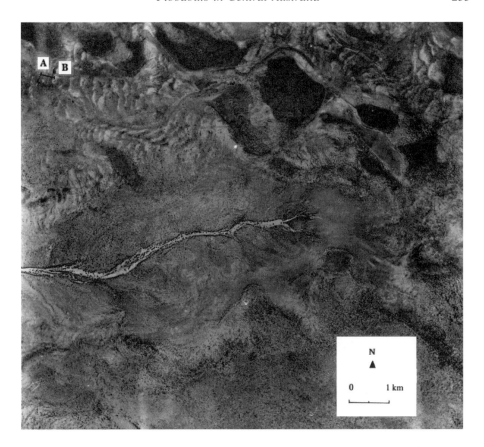

FIGURE 9.8   Part of the transverse bedform field in the floodout zone of the Sandover River as evident in 1950. The ripple-like features are visible to the north and northeast of the Sandover River, which flows from west to east and terminates in the middle centre of the photograph. The succession of stoss and lee slopes on each ripple is characterized by alternately darker and lighter tones, respectively. Where flow constriction has occurred between areas of sandplain, narrow depressions have scoured into the underlying sands (visible at top centre of photograph). See Figure 9.9 for section A–B (Elkedra SF53-7, 1950, SVY 873, Run 12, 5202. Reproduced by permission of the General Manager, Australian Surveying and Land Information Group, Department of Administrative Services, Canberra, ACT)

Transverse bedforms

One prominent feature of the floodout zone of the Sandover River is the existence of a "transverse bedform field" (Patton et al., 1993, p.207). These features are most visible on aerial photographs taken in 1950 (Figure 9.8) before the expansion of large-scale pastoralism in the area obscured many of the subtle surficial forms in the floodout zone.

Figure 9.8 illustrates that the features assume the form of a series of wavy, sinuous crested "ripples". They start just downstream of the emergence point and are widely distributed throughout the floodout zone, particularly to the north of the Sandover River. Where they are well defined, wavelengths are typically in the order of 100–250 m with amplitudes of 0.3–0.4 m. Up to 20 ripples can occur within each train and crestline

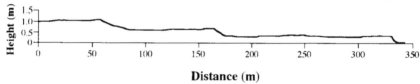

FIGURE 9.9   Topographic survey of transverse bedforms in the floodout zone of the Sandover River. See Figure 9.8 for location of surveyed transect

lengths of individual ripples are generally between 350 and 700 m. Although hard to find on the ground, aerial photographs and topographic surveys reveal gentle stoss slopes and steeper lee slopes (Figure 9.9).

Similar features have been identified in a number of arid and semi-arid locations in Australia where they have been interpreted either as essentially erosional features (scalds) resulting from occasional sheetfloods (Mabbutt, 1977, 1986) or as the deposits of rare, very large sheetfloods (Rust and Gostin, 1981; Pickup, 1991; Patton et al., 1993). Pickup (1991, p.466) refers to such features as "megaripple systems" but the term "megaripple" would seem to have a more specific meaning in the sedimentological literature (e.g. Reineck and Singh, 1975). Nevertheless, a fluvial origin for these features is also the favoured explanation here, for two main reasons. First, the ripple trains are generally oriented in the direction of downvalley flows, and second, the paths taken around, and between, patches of aeolian sandplain (Figure 9.8) would tend to rule out an aeolian origin. The distribution and pattern of the ripples throughout the northern part of the floodout zone far from the present-day channel, however, suggests that they are either the product of very rare flood events or that they are unrelated to contemporary flood flows.

Shallow (0.5 m deep) trenches excavated on the lee slopes reveal an upper 4–8 cm of poorly sorted sand and granule gravels in a silt matrix overlying structureless silty sand. In two of three examples (transverse bedforms 1 and 2, Figure 9.10a), a considerable fraction of very coarse sand and granules in the -0.5 to -2 phi (1.4–4.0 mm) range is noticeable. The average size distribution, compared to sands from the bed of the Sandover River in the floodout zone, further demonstrates the poor sorting and coarse nature of the material in the ripples (Figure 9.10b).

In general, the stoss slopes of the ripples are fairly well vegetated with a mixture of taller trees over shrubs and grasses. The sinuous crests and lee slopes of the ripples are less vegetated and the coarse sand is often scattered over the surface, in some instances being reworked by the activity of ants. A sharp junction between this sand and a bare surface of silty sand at the base of the lee slope is evident in many places, with the swale between the vegetated stoss slope of the next ripple often taking the appearance of a large scald.

On the basis of the coarse sand and granules, the ripples at Ammaroo are interpreted as depositional bedforms constructed by one or more very large floods, although erosion of the swales between the ripples by subsequent smaller floods and/or local rainfall events is not discounted. Unfortunately, as the deposits are very thin and subject to bioturbation they cannot be dated using thermoluminescence and, furthermore, they contain no original organic material for radiocarbon dating. As a consequence, little can be said

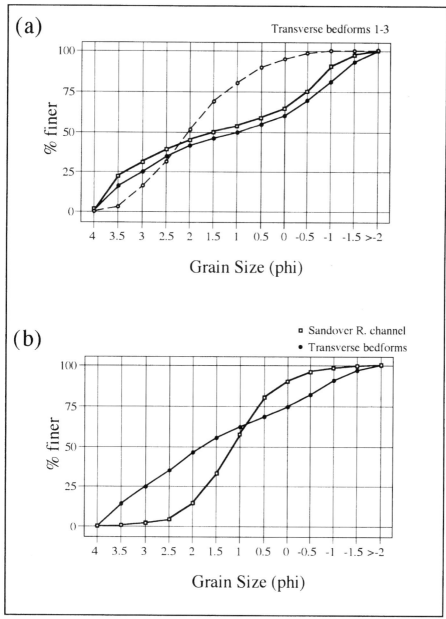

FIGURE 9.10 (a) Grain size curves for material in upper 4–8 cm of transverse bedforms in the floodout zone of the Sandover River. Key: □, —, transverse bedform 1; ●, —, transverse bedform 2; ○, – – –, transverse bedform 3. (b) Grain size curves for material in upper 4–8 cm of transverse bedforms compared to material from the bed of the Sandover River. In each case, the curves represent the average of three samples. The curve for the transverse bedforms is the average of the three lines shown in (a). Grain size samples for the transverse bedforms were collected from the surveyed transect (Figure 9.9) and the samples for the Sandover River were collected from the channel bed at different points throughout the floodout zone, including near to its present-day terminus

about the age or nature of the floods responsible for the formation of these features, other than to echo Pickup (1991) by commenting that the floods must have been considerably larger than those experienced in the 100–110 years of European settlement in the region. Previous descriptions of such features in the Alice Springs region of central Australia have been from piedmont settings close to the ranges (e.g. Pickup, 1991; Patton et al., 1993) but the features at Ammaroo are unusual in that they are found some 240 km from the headwater ranges.

Features characteristic of the contributory parts of intermediate floodouts

The distinguishing feature of intermediate floodouts is the persistence of largely unchannelled flows through the distributary part of the floodout which concentrate into more defined channels downvalley. Rechannelizing flow occurs on a number of floodouts in central Australia but only Pickup (1991) has briefly referred to the phenomenon. Such *reforming channels* tend to occur as a result of two main causes: tributary inflows and flow constriction (Figure 9.11). Where floodouts occur close to uplands, the extra volume of floodwater supplied by tributaries may be a significant factor in promoting rechannelization (Figure 9.11a) by enhancing the potential for erosion of the floodout deposits and by enabling the newly incised channels to persist downvalley. Where floodouts occur far from the uplands, however, constriction of the unchannelled flows across floodouts by aeolian or alluvial deposits or by bedrock highs appears to be the key factor in their formation (Figure 9.11b).

Tributary inflows and flow constriction are not mutually exclusive, however, and in many instances channels can reform as a result of both factors. For example, the reforming channels of the Sandover and Woodforde Rivers occur largely as a result of flow constriction but nevertheless receive runoff from higher surfaces surrounding the floodouts. Given the absence of flow data for the reforming channels of the Sandover and Woodforde Rivers, quantitative statements regarding the magnitude and frequency of flows cannot be made. The accounts of local pastoralists, however, suggest that flows only reach the reforming channels at infrequent intervals: about every four to five years for the Sandover River and every three to four years for the Woodforde River.

A number of general conclusions can be drawn regarding the typical characteristics of reforming channels downstream of floodouts. First, one or more channels can reform downstream of intermediate floodouts which have different channel morphologies, perimeter sediments and vegetation relations compared to the upstream channels. Surveyed cross-sections from the reforming channels of the Sandover and Woodforde Rivers illustrate the characteristic channel morphologies (Figure 9.12a, b). Where channels first start to reform, the limited channel capacities, reduced bedloads, typically indurated nature of the perimeter sediments and the greater proportion of tree species such as coolibah (*Eucalyptus microtheca*) and mulga (*Acacia aneura*) among the riparian vegetation are all in stark contrast to the channels approaching the floodout. Reforming channels can either join a larger drainage system, disappear in an intermediate floodout and reform again, or dissipate in a terminal floodout.

Second, in many instances, reforming channels are associated with noticeably steepened bed slopes. For example, bed slopes approaching the floodout of the Sandover River are around $0.0006 \, \mathrm{m \, m^{-1}}$ but increase to 0.0008 to $0.001 \, \mathrm{m \, m^{-1}}$ in reaches of the

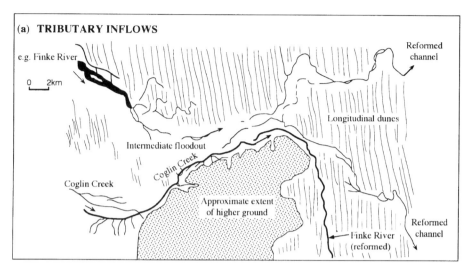

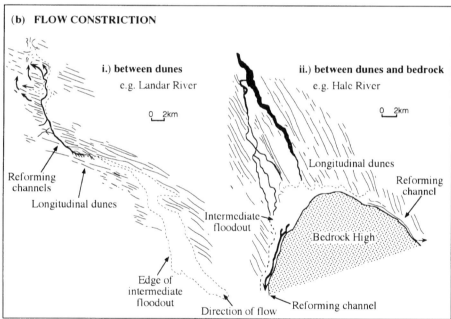

FIGURE 9.11   Contributory factors to the development of reforming channels in central Australia. Scales are approximate only. The size of reforming channels is variable, ranging from a few metres to several tens of metres in width, but they are typically smaller than the channels upstream of the floodouts. In some instances, several channels can reform downstream of the floodout, with the allocation of floodwaters between these channels controlled by the size of the flood events and by local physiography, such as bedrock outcrops or aeolian dunefields. For instance, while the main reforming channel of the Finke River (a) probably carries the majority of floodwaters during relatively small flood events, during larger events floodwaters are also diverted into two other reforming channels that follow courses amongst extensive longitudinal dunes (a)

reforming channels. Similar increases in bed slope occur in the reforming channels of the Woodforde River. In both instances, the increases in channel-bed slopes seem to correspond with slight steepening of the valley gradient which may provide the energy necessary for the largely unchannelled alluvial tracts on the downvalley side of the floodout to rechannelize (Figure 9.12). On the reformed Sandover River, the steepened bed slopes persist downstream and the channel continues to narrow and deepen as far as its junction with the Bundey River (Figure 9.12a). In contrast, on the reformed channels of the Woodforde River, bed slopes become more gentle towards their terminal floodout and the channels broaden and shallow once more (Figure 9.12b).

Finally, analysis of aerial photographs for the Alice Springs region taken in the 1950s indicates that reforming channels are often relatively stable features of floodouts. Although many rivers in central Australia have undergone significant change since 1950 due to a recent series of large floods (Pickup, 1991), on the Sandover and Woodforde Rivers very little change has occurred to the network of reforming channels.

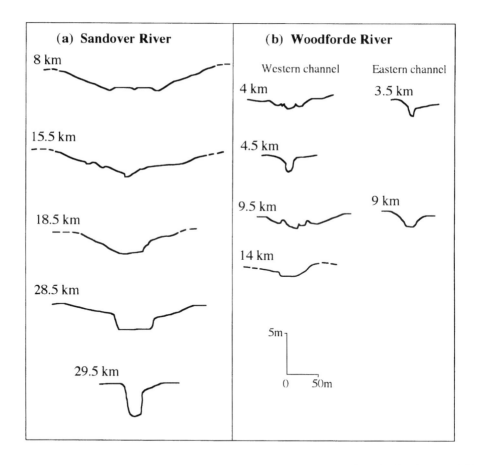

FIGURE 9.12   Cross-sections surveyed along the reforming channels of (a) the Sandover and (b) the Woodforde Rivers. For each section, the approximate distance downstream of the intermediate floodout is indicated

FLOODOUT ZONES OF CENTRAL AUSTRALIA: FLOODPLAINS OR FANS?

Alluvial fans and floodout zones of central Australia

In the Alice Springs region, evidence of prior channel avulsions downstream of upland ranges (Litchfield, 1969; Freeman, 1986) and the aggradational tendencies of many floodplain systems (Pickup, 1991) both have similarities with the processes on alluvial fans. Furthermore, there are similarities between the emergence points of floodout zones and the intersection points on alluvial fans, which describe the point where an incised channel emerges onto the fan surface (e.g. Hooke, 1967; Wasson, 1974; Bowman, 1978). The tendency for flow and sediment processes to "fan out" downstream of the emergence point (Figure 9.4), the eventual disappearance of channelized flow and features such as transverse bedform fields and rechannelizing flow, also have parallels with the forms and processes on alluvial fans. As a result, there is much confusion over the depositional style of the drainage systems in the region. Indeed, many authors (Mabbutt, 1962, 1967; Litchfield, 1969; Freeman, 1986; Stidolph et al., 1988) seem unsure whether to define the drainage systems as floodplains or fans, and the two terms have often been used interchangeably. To date, the most comprehensive statement on this issue has been provided by Pickup (1991) who considers that the central Australian drainage systems have the characteristics of both floodplains and alluvial fans.

To a large extent, the confusion over the definition of the central Australian drainage systems may be a reflection of the vigorous debate in the geomorphic literature as to the definition of alluvial fans and whether they can be distinguished from other fluvial systems. The term "alluvial fan" has generally been used to refer to steeply sloping, sedimentary deposits which radiate downslope from the point where channels emerge from uplands (e.g. Bull, 1977; Rachocki, 1981). However, the term "alluvial fan" (or "subaerial fan") is now commonly applied to a broad spectrum of fluvial geomorphic features that deposit sediment in a fan-shaped form, including types of relatively low-gradient, braided, distributary and meandering fluvial systems (e.g. Schumm, 1977; Boothroyd and Nummedal, 1978; Miall, 1978; Riley and Taylor, 1978; Adamson et al., 1987; Wells and Dorr, 1987a,b; Gohain and Parkash, 1990; Stanistreet and McCarthy, 1993).

Recently, there has been an attempt to restrict the definition of alluvial fans and to re-establish the distinctiveness from rivers (McPherson and Blair, 1993; Blair and McPherson, 1994a,b). Although these suggestions are not beyond contention (e.g. Nemec and Steel, 1988; McCarthy and Cadle, 1995), on the basis of the morphologic, hydraulic and sedimentary criteria outlined by Blair and McPherson (1994a), the drainage systems of the Alice Springs region are clearly best described as *riverine* systems rather than as alluvial fans. By virtue of the axial lengths, hydrology, sedimentary processes and the predominantly fine-grained deposits, the depositional tracts in the proximal and medial reaches of the drainage systems are undoubtedly characterized by *floodplain* formation. The floodout zones in the distal reaches are more difficult to classify, possessing features common to both floodplains and alluvial fans. However, the geomorphic setting in low-gradient plains, the often large areal extent and the relatively fine-grained deposits typical of floodout zones differentiate them from alluvial fans. Even where floodout zones occur closer to the uplands, the low gradients and fine-grained deposits still distinguish them from alluvial fans.

Terminal fans and floodout zones of central Australia

A limitation to a more restricted definition of alluvial fan, however, is that past use is likely to ensure that the term "fan" will continue to be applied to a wide range of fluvial features that deposit sediment in a fan-shaped form yet may depart considerably from the criteria of Blair and McPherson (1994a). As a case in point, Mukerji (1976) introduced the term "terminal fan" (or "inland fan") to characterize the distal reaches of inland streams that occur widely in the Sutlej–Yamuna Plains of the Indian subcontinent. Following Mukerji's (1975, 1976) and Parkash et al.'s (1983) work on the Markanda terminal fan, similar modern depositional systems have been recognized (Abdullatif, 1989) and many ancient sedimentary sequences have also been interpreted as the deposits of terminal fans and related drainage networks (e.g. Friend, 1978; Tunbridge, 1984; Olsen, 1987; Kelly and Olsen, 1993a,b; Sadler and Kelly, 1993).

In terms of geomorphic setting, as well as in aspects of morphology and sedimentology, terminal fans appear to have similarities with the floodout zones in central Australia. Despite some similarities, however, there are also a number of differences between floodout zones and terminal fans. First, floodout zones generally lack the extensive distributary channel networks defining terminal fan systems (Olsen, 1987; Kelly and Olsen, 1993b), although there are exceptions, such as the Landar River (Figure 9.1) which breaks down in an extensive distributary/ anabranching network approaching its intermediate floodout. Both the Woodforde and Sandover–Bundey Rivers show varying degrees of distributary development in the floodout zone but even in these instances the fan-shaped form of deposition is only weakly developed (Figure 9.4), largely due to lateral confinement by aeolian sandplain or red earth plains. Second, terminal fans form largely as a result of downstream reductions in discharge (Kelly and Olsen, 1993b) but floodout zones also result from factors such as aeolian, hydrologic/alluvial and structural barriers to flow (Figure 9.2). Finally, and most importantly, the common location on medial or distal reaches of drainage systems, where channels reform downstream of the floodout, is a key characteristic differentiating many floodout zones from terminal fans. In short, floodout zones are not necessarily terminal nor does deposition necessarily adopt a fan-shaped form.

In view of the differences from alluvial and terminal fans, and to avoid the problematic use of the term "fan", it is suggested here that floodouts are best regarded as part of the *continuum of floodplains*. In the floodout zone, the disappearance of channelized flow means that floodplains (*sensu* Nanson and Croke, 1992) grade into floodouts in the downvalley direction. Hence, in many respects floodouts can be regarded simply as channel-less floodplains which vertically accrete fine-grained alluvial sediments. Graf (1988, pp.235–236) has described the broad "flow zones" without clearly marked banks found in areas of southwest Arizona but, as with floodouts, the absence of channels means that it is difficult to reconcile these distinctive varieties of fluvial form with traditional definitions of floodplains (e.g. Bates and Jackson, 1987) or with floodplain classifications (e.g. Nanson and Croke, 1992).

CONCLUSION

This paper has illustrated a range of features characteristic of the distal reaches of ephemeral drainage systems in the Alice Springs region of central Australia, including many that have not previously been described and for which no adequate terminology presently exists. Floodouts can establish as a result of a number of factors (Figure 9.2), but in the absence of any obstructions to channelized flow, the burial of upstream terraces by younger alluvial deposits is a key factor in the location of the floodout zone. The emergence point describes the location where the channel leaves the confines of the terraces. This may occur over a short distance, such as on the Woodforde River, or it may involve a gradual transition over tens of kilometres, such as on the larger Sandover and Sandover–Bundey systems. Downstream of the emergence point, overbank flows and sedimentation processes spread for greater distances on either side of the channel margins, a situation which has major implications for channel–floodplain forms and processes.

Declining channel capacity through the floodout zone eventually results in a floodout, defined as the point where channelized flow disappears and floodwaters spill across adjacent alluvial surfaces. Olsen (1987) considers that a rapid downstream decrease of streamflow occurs within a limited area in three different types of depositional systems: deltas, alluvial fans and terminal fans. Hence, the floodout zones of central Australia can be seen as an important addition to this spectrum. Floodouts can establish both on intermediate and terminal parts of channel systems and represent a variety of fluvial form distinctive from alluvial fans and terminal fans. Hence, floodouts are best considered as part of the continuum of floodplains.

The Sandover, Sandover–Bundey and Woodforde Rivers illustrate the typical characteristics of floodouts on the Northern Plains of the Alice Springs region, but the study of a greater range of floodouts in different geomorphic settings, such as in the western Simpson Desert, will undoubtedly reveal other patterns of channel breakdown and associated morphosedimentary features. In such low-gradient settings far from the ranges, local topographic and random environmental influences tend to predominate over hydraulic factors leading to a multiplicity of fluvial features that are often very subtle in their characterization. Nevertheless, it is suggested that the concept of the floodout zone, the broad factors influencing the development of floodouts and the distinction between intermediate and terminal floodouts are likely to be of more general applicability.

ACKNOWLEDGEMENTS

Fieldwork was financially supported by the Quaternary Environments Research Centre, School of Geosciences, University of Wollongong, and by the Australian Research Council via grants to G.C. Nanson and D.M. Price. The author wishes to thank G.C. Nanson for many suggestions and comments which improved earlier versions of this paper. The comments of G. Pickup and an anonymous reviewer are also gratefully appreciated. Figures were drafted by D. Martin, School of Geosciences, University of Wollongong.

REFERENCES

Abdullatif, O.M. 1989. Channel–fill and sheet–flood facies sequences in the ephemeral River Gash, Kassala, Sudan. *Sedimentary Geology*, **63**, 171–184.

Adamson, D., Williams, M.A.J. and Baxter, J.T. 1987. Complex Late Quaternary alluvial history in the Nile, Murray-Darling, and Ganges basins: three river systems presently linked to the Southern Oscillation. In V. Gardiner (Ed.), *International Geomorphology 1986: Part II*, Wiley, Chichester, 875–887.

Arakel, A.V. and McConchie, D. 1982. Classification and genesis of calcrete and gypsite lithofacies in paleodrainage systems of inland Australia and their relationship to carnotite mineralization. *Journal of Sedimentary Petrology*, **52**, 1149–1170.

Argue, J.R. and Salter, L.E.M. 1977. Waterhole development: a viable water resource option in the arid zone? *Hydrology Symposium 1977: The Hydrology of Northern Australia*. Institution of Engineers, Australia, National Conference Publication No. 77/5, 35–39.

Baker, V.R., Pickup, G. and Polach, H.A. 1983. Desert palaeofloods in central Australia. *Nature*, **301**, 502–504.

Bates, R.L. and Jackson, J.A. 1987. *Glossary of Geology (3rd. edn.)*. American Geological Institute, Alexandria.

Blair, T.C. and McPherson, J.G. 1994a. Alluvial fans and their natural distinction from rivers based on morphology, hydraulic processes, sedimentary processes and facies assemblages. *Journal of Sedimentary Research*, **A64**, 450–489.

Blair, T.C. and McPherson, J.G. 1994b. Alluvial fan processes and forms. In A.D. Abrahams and A.J. Parsons (Eds), *Geomorphology of Desert Environments*. Chapman and Hall, London, 354–402.

Boothroyd, J.C. and Nummedal, D. 1978. Proglacial braided outwash: a model for humid alluvial fan deposits. In A.D. Miall (Ed.), *Fluvial Sedimentology*. Canadian Society of Petroleum Geologists, Memoir 5, 641–668.

Bourke, M.C. 1994. Cyclical construction and destruction of flood dominated flood plains in semiarid Australia. In Proceedings of the Canberra Symposium. *Variability in Stream Erosion and Sediment Transport*, IAHS Publication No.224, 113–123.

Bowman, D. 1978. Determination of intersection points within a telescopic alluvial fan complex. *Earth Surface Processes*, **3**, 265–276.

Bull, W.B. 1977. The alluvial-fan environment. *Progress in Physical Geography*, **1**, 222–269.

Crouch, R.J. 1990. Rates and mechanisms of discontinuous gully erosion in a red-brown earth catchment, New South Wales, Australia. *Earth Surface Processes and Landforms*, **15**, 277–282.

Dubief, J. 1953. *Essai sur L'Hydrologie Superficielle Au Sahara*. Gouvernement Général de L'Algérie, Service des Études Scientifiques, Birmandreis.

Erskine, W. and Melville, M.D. 1983. Sedimentary properties and processes in a sandstone valley: Fernances Creek, Hunter Valley, New South Wales. In R.W. Young, and G.C. Nanson (Eds), *Aspects of Australian Sandstone Landscapes*. Australian and New Zealand Geomorphology Group Special Publication No.1, 94–105.

Erskine, W. and Melville, M.D. 1984. Sediment movement in a discontinuous gully system at Boro Creek, Southern Tablelands, N.S.W. In R.J. Loughran (Ed.), *Drainage Basin Erosion and Sedimentation: A Conference on Erosion, Transportation and Sedimentation in Australian Drainage Basins*. University of Newcastle and Soil Conservation Service of New South Wales, 197–204.

Freeman, M.J. 1986. *Huckitta SF53–11: 1:250 000 Geological Map Series – Explanatory Notes*. Department of Mines and Energy / Northern Territory Geological Survey, Darwin.

Friend, P.F. 1978. Distinctive features of some ancient river systems. In A.D. Miall (Ed.), *Fluvial Sedimentology*. Canadian Society of Petroleum Geologists, Memoir 5, 531–542.

Glennie, K.W. 1970. *Desert Sedimentary Environments*. Developments in Sedimentology 14, Elsevier, Amsterdam.

Gohain, K. and Parkash, G. 1990. Morphology of the Kosi Megafan. In A.H. Rachocki and M. Church (Eds), *Alluvial Fans: A Field Approach*. Wiley, Chichester, 151–178.

Goudie, A.S. and Wells, G.L. 1995. The nature, distribution and formation of pans in arid zones. *Earth-Science Reviews*, **38**, 1–69.

Graeme, D. and Dunkerley, D.L. 1993. Hydraulic resistance by the river red gum, *Eucalyptus camaldulensis*, in ephemeral desert streams. *Australian Geographical Studies*, **31**, 141–154.

Graf, W.L. 1988. Definition of flood plains along arid-region rivers. In V.R. Baker, R.C. Kochel and P.C. Patton (Eds), *Flood Geomorphology*. Wiley, New York, 231–242.

Hays, J. 1967. Land surfaces and laterites in the north of the Northern Territory. In J.N. Jennings and J.A. Mabbutt (Eds), *Landform Studies from Australia and New Guinea*. ANU Press, Canberra, 182–210.

Hogg, S.E. 1982. Sheetfloods, sheetwash, sheetflow or ... ? *Earth-Science Reviews*, **18**, 59–76.

Hooke, R. LeB. 1967. Processes on arid–region alluvial fans. *Journal of Geology*, **75**, 438–460.

Karcz, I. 1972. Sedimentary structures formed by flash floods in southern Israel. *Sedimentary Geology*, **7**, 161–182.

Kelly, S.B. and Olsen, H. 1993a. A terminal fan model. In *Keynote Addresses and Abstracts, Proceedings of the 5th International Conference on Fluvial Sedimentology*. University of Queensland, Brisbane, Australia, 62.

Kelly, S.B. and Olsen, H. 1993b. Terminal fans – a review with reference to Devonian examples. *Sedimentary Geology*, **85**, 339–374.

Knighton, A.D. and Nanson, G.C. 1994. Waterholes and their significance in the anastomosing channel system of Cooper Creek, Australia. *Geomorphology*, **9**, 311–324.

Lancaster, N. and Teller, J.T. 1988. Interdune deposits of the Namib Sand Sea. *Sedimentary Geology*, **55**, 91–107.

Langford, R.P. 1989. Fluvial–aeolian interactions: Part 1, modern systems. *Sedimentology*, **36**, 1023–1035.

Linton, P.L. and McCarthy, T.S. 1995. Fluvial morphology of the Mkuze River floodplain, Zululand, South Africa. In *Programme with Abstracts. International Association of Geomorphologists Southeast Asia Conference*, National University of Singapore/Nanyang Technological University, Singapore, 55.

Litchfield, W.H. 1969. *Soil Surfaces and Sedimentary History near the Macdonnell Ranges, N.T.* CSIRO Soil Publication No. 25, Melbourne.

Mabbutt, J.A. 1962. Part VII. Geomorphology of the Alice Springs area. In R.A. Perry (Ed.), *General Report on Lands of the Alice Springs Area, Northern Territory, 1956–57*. CSIRO Land Research Series, No. 6, 163–184.

Mabbutt, J.A. 1963. Wanderrie banks: micro-relief patterns in semi-arid Western Australia. *Geological Society of America, Bulletin*, **74**, 529–540.

Mabbutt, J.A. 1965. The weathered land surface in central Australia. *Zeitschrift für Geomorphologie NF*, **9**, 82–114.

Mabbutt, J.A. 1966. Landforms of the western Macdonnell Ranges: a study of inheritance and periodicity in the geomorphology of arid central Australia. In G.H. Dury (Ed.) *Essays in Geomorphology*. Heinemann, New York, 83–119.

Mabbutt, J.A. 1967. Denudation chronology in central Australia: structure, climate and landform inheritance in the Alice Springs area. In J.N. Jennings and J.A. Mabbutt (Eds), *Landform Studies from Australia and New Guinea*. ANU Press, Canberra, 144–181.

Mabbutt, J.A. 1977. *Desert Landforms*. ANU Press, Canberra.

Mabbutt, J.A. 1986. Desert lands. In D.N. Jeans (Ed.), *Australia – A Geography. Volume One: The Natural Environment*. Sydney University Press, Sydney, 180–202.

McCarthy, T.S. and Cadle, A.B. 1995. Aluvial fans and their natural distinction from rivers based on morphology, hydraulic processes, sedimentary processes and facies assemblages – discussion. *Journal of Sedimentary Research*, **A65**, 581–583.

McPherson, J.G. and Blair, T.C. 1993. Alluvial fans: fluvial or not? In *Keynote Addresses and Abstracts, Proceedings of the 5th International Conference on Fluvial Sedimentology*. University of Queensland, Brisbane, Australia, K33–K41.

Miall, A.D. 1978. Fluvial sedimentology: a historical review In A.D. Miall (Ed.), *Fluvial Sedimentology*. Canadian Society of Petroleum Geologists, Memoir 5, 1–47.

Mukerji, A.B. 1975. Geomorphic patterns and processes in the terminal triangular tract of inland streams in Sutlej-Yamuna Plain. *Journal of the Geological Society of India*, **16**, 450–459.

Mukerji, A.B. 1976. Terminal fans of inland streams in Sutlej-Yamuna Plain. India. *Zeitschrift für Geomorphologie NF*, **20**, 190–204.

Nanson, G.C. and Croke, J.C. 1992. A genetic classification of floodplains. In G.R. Brackenridge and J. Hagedorn (Eds), *Floodplain Evolution, Geomorphology*, **4**, 459–486.

Nanson, G.C., East, T.J., and Roberts, R.G. 1993. Quaternary stratigraphy, geochronology and evolution of the Magela Creek catchment in the monsoon tropics of northern Australia. *Sedimentary Geology*, **83**, 277–302.

Nanson, G.C., Chen, X.Y. and Price, D.M. 1995. Aeolian and fluvial evidence of changing climate and wind patterns during the past 100 ka in the western Simpson Desert, Australia. *Palaeogeography, Palaeoclimatology, Palaeoecology*, **113**, 87–102.

Nemec, W. and Steel, R.J. 1988. What is a fan delta and how do we recognize it? In W. Nemec and R.J. Steel, (Eds), *Fan Deltas: Sedimentology and Tectonic Settings*. Blackie, Glasgow, 3–13.

O'Brien, P.E. and Burne, R.V. 1994. The Great Cumbung Swamp – terminus of the low–gradient Lachlan River, eastern Australia. *AGSO Journal of Australian Geology and Geophysics*, **15**, 223–233.

Olsen, H. 1987. Ancient ephemeral stream deposits: a local terminal fan model from the Bunter Sandstone Formation (L. Triassic) in the Tønder-3, -4 and -5 wells, Denmark. In L. Frostick and I. Reid (Eds), *Desert Sediments: Ancient and Modern*. Geological Society, Special Publication No. 35, 69–86.

Ori, G.G. 1989. Terminal fluvial systems under different climatic conditions. *Program and Abstracts, 4th International Conference on Fluvial Sedimentology*, Sitges, Barcelona, 97.

Parkash, B., Awasthi, A.K. and Gohain, K. 1983. Lithofacies of the Markanda terminal fan, Kurukshetra district, Haryana, India. In J.D. Collinson and J. Lewin (Eds), *Modern and Ancient Fluvial Systems*. International Association of Sedimentologists, Special Publication No. 6, 337–344.

Patton, P.C., Pickup, G. and Price, D.M. 1993. Holocene paleofloods of the Ross River, central Australia. *Quaternary Research*, **40**, 201–212.

Pickup, G. 1986. Fluvial landforms. In D.N. Jeans (Ed.), *Australia – A Geography. Volume One: The Natural Environment*. Sydney University Press, Sydney, 148–179.

Pickup, G. 1991. Event frequency and landscape stability on the floodplain systems of arid central Australia. *Quaternary Science Reviews*, **10**, 463–473.

Pickup, G., Allan, G. and Baker, V.R. 1988. History, palaeochannels and palaeofloods of the Finke River, central Australia. In R.F. Warner (Ed.), *Fluvial Geomorphology of Australia*. Academic Press, Sydney, 105–127.

Rachocki, A.H. 1981. *Alluvial Fans: An Attempt at an Empirical Approach*. Wiley, Chichester.

Reineck, H.-E. and Singh, I.B. 1975. *Depositional Sedimentary Environments: With Reference to Terrigenous Clastics*. Springer-Verlag, Berlin.

Riley, S.J. and Taylor, G. 1978. The geomorphology of the upper Darling River system with special reference to the present fluvial system. *Proceedings of the Royal Society of Victoria*, **90**, 89–102.

Rodier, J.A. 1985. Aspects of arid zone hydrology. In J.C. Rodda (Ed.), *Facets of Hydrology Volume II*. Wiley, Chichester, 205–247.

Rust, B.R. and Gostin, V.A. 1981. Fossil transverse ribs in Holocene alluvial fan deposits, Depot Creek, South Australia. *Journal of Sedimentary Petrology*, **51**, 441–444.

Sadler, S.P. and Kelly, S.B. 1993. Fluvial processes and cyclicity in terminal fan deposits: an example from the Late Devonian of southwest Ireland. *Sedimentary Geology*, **85**, 375–86.

Schumm, S.A. 1961. The effect of sediment characteristics on erosion and deposition in ephemeral stream channels. *US Geological Survey Professional Paper*, 352C, 31–70.

Schumm, S.A. 1977. *The Fluvial System*. Wiley-Interscience, New York.

Shaw, R.D. and Warren, R.G. 1975. *Alcoota, Northern Territory: 1:250 000 Geological Series – Explanatory Notes*. Department of Minerals and Energy/Bureau of Mineral Resources, Geology and Geophysics, Australian Government Publishing Service, Canberra.

Slatyer, R.O. and Mabbutt, J.A. 1964. Hydrology of arid and semiarid regions. In V.T. Chow (Ed.), *Handbook of Applied Hydrology*. McGraw-Hill, New York, 24-1–24-46.

Sneh, A. 1983. Desert stream sequences in the Sinai Peninsula. *Journal of Sedimentary Petrology*, **53**, 1271–1279.

Stanistreet, I.G. and McCarthy, T.S. 1993. The Okavango Fan and the classification of subaerial fan systems. *Sedimentary Geology*, **85**, 115–133.

Stewart, G.A. and Perry, R.A. 1962. Part I. Introduction and summary description of the Alice

Springs area. In R.A. Perry (Ed.), *General Report on Lands of the Alice Springs Area, Northern Territory, 1956–57*. CSIRO Land Research Series, No. 6, 9–19.

Stidolph, P.A., Bagas, L., Donnellan, N., Walley, A.M., Morris, D.G. and Simons, B. 1988. *Elkedra SF53-7: 1:250 000 Geological Map Series – Explanatory Notes*, Department of Mines and Energy/Northern Territory Geological Survey, Darwin.

Sturt, C. 1849. *Narrative of an Expedition into Central Australia ... during the years 1844, 1845, 1846*, 2 Volumes. T. and W. Boone, London.

Sullivan, M.E. 1976. *Drainage Disorganisation in Arid Australia and its Measurement*. MSc Thesis, University of New South Wales.

Thompson, R.B. 1991. *A Guide to the Geology and Landforms of Central Australia*. Northern Territory Geological Survey, Alice Springs.

Tunbridge, I.P. 1984. Facies model for a sandy ephemeral stream and clay playa complex; the Middle Devonian Trentishoe Formation of North Devon, U.K. *Sedimentology*, **31**, 697–715.

Ward, J.D. 1988. Eolian, fluvial and pan (playa) facies of the Tertiary Tsondab Sandstone Formation in the central Namib Desert, Namibia. *Sedimentary Geology*, **55**, 143–162.

Wasson, R.J. 1974. Intersection point deposition on alluvial fans: an Australian example. *Geografiska Annaler*, **56A**, 83–92.

Wells, N.A. and Dorr, J.A. 1987a. Shifting of the Kosi River, northern India. *Geology*, **15**, 204–207.

Wells, N.A. and Dorr, J.A. 1987b. A reconnaissance of sedimentation on the Kosi alluvial fans of India. In F.G. Ethridge, R.M. Flores and M. Harvey (Eds), *Recent Developments in Fluvial Sedimentology*. Society of Economic Paleontologists and Mineralogists, 39, 51–62.

Williams, G.E. 1969. Flow conditions and estimated velocities of some central Australian stream floods. *Australian Journal of Science*, **31**, 367–369.

Williams, G.E. 1970a. The central Australian stream floods of February–March 1967. *Journal of Hydrology*, **11**, 185–200.

Williams, G.E. 1970b. Piedmont sedimentation and late Quaternary chronology in the Biskra region of the northern Sahara. *Zeitschrift für Geomorphologie*, Supplementband **10**, 40–63.

Williams, G.E. 1971. Flood deposits of the sand–bed ephemeral streams of central Australia. *Sedimentology*, **17**, 1–40.

10

Fluvial Form Variability in Arid Central Australia

MARY C. BOURKE[1] AND GEOFF PICKUP[2]

[1] *Center for Earth and Planetary Studies, Smithsonian Institution, Washington, DC, USA*
[2] *CSIRO, Land and Water, Canberra, Australia*

ABSTRACT

Many arid rivers are sensitive to variations in discharge and subject to extreme flows. These non-equilibrium rivers exhibit significant fluvial form variability. In the Todd River in arid central Australia, this variability is manifest in two ways. First, downstream variation in fluvial morphology occurs as a series of step changes rather than gradually. Abrupt morphological change at river confluences is pronounced. Discordant and barred junctions result from asynchronous tributary activity and differences in flow magnitude from contributing systems. Factors such as tributary spacing, tributary length, rainfall and runoff variability, hydrological lag and transmission losses are important in determining fluvial form variability. Additional influences include boundary roughness and resistance to erosion, and sediment supply. The pulsed movement of sediment in ephemeral systems creates a complex assemblage of landforms. Patterns of vegetation growth and stability are important influences on ephemeral sedimentation, storage time and associated landform distribution.

Second, there are three scales of landforms: small, medium and large. Extreme floods have deposited large-scale fluvial forms across the landscape. The resultant morphology and sedimentology of these events modulate the response of lower-magnitude flows by directing floodwaters and providing an abundant supply of sediment to the system. In this way form evolution occurs episodically and preservation is closely linked to flood magnitude.

INTRODUCTION

The geomorphology of arid and semi-arid rivers remains one of the least studied and understood aspects of fluvial systems (Reid and Frostick, 1997). Reviews of desert environments include descriptions of the variation in rainfall, the rapid runoff from contributing slopes, transmission losses, the flashy nature of flood hydrographs, asynchronous tributary activity, the generally high suspended sediment loads and the tendency for suspended sediment concentrations to increase downstream (e.g., Mabbutt, 1977; Graf, 1988a; Scoging, 1989, Reid and Frostick, 1997; Cooke et al., 1993; Thornes,

Varieties of Fluvial Form. Edited by A.J. Miller and A. Gupta
© 1999 John Wiley & Sons Ltd

1994a,b). What is not equally well documented is the morphological response of arid rivers to these dynamic processes. This paper addresses this issue by describing the geomorphology and aspects of form variability of a central Australian arid river.

Accepted models of river behaviour that have been predominantly developed in temperate and humid areas consider that fluvial systems develop a dynamic equilibrium between river form, discharge and sediment (Chorley and Kennedy, 1971). This concept is difficult to apply to arid systems dominated by variability that do not display an obvious quasi-equilibrium. Pickup and Rieger (1979) noted two types of non-equilibrium rivers: those sensitive to variations in discharge, and those subject to extreme flow. In this sense many arid-zone rivers are non-equilibrium rivers.

The discontinuous nature of flow in arid ephemeral systems limits the channel's ability to adjust. For example, Schumm (1961) found that channels in aggrading ephemeral rivers follow a pattern of alternate deposition and erosion (Schumm and Hadley, 1957). Such behaviour led Graf (1988a) to suggest that arid river systems essentially are not in equilibrium. Additionally, event-based changes in ephemeral rivers are common and channels respond to the variable discharge regime by dramatically adjusting their morphology and pattern. For example, the Salt and Gila compound channels in Arizona oscillate between a low-flow meandering channel and a larger braided flood channel (Graf, 1988b). Similar adjustments to variations in discharge were described along Cooper Creek in semi-arid Australia (Rust and Nanson, 1986) where braiding and anastomosing networks coexist.

The behaviour of arid-zone rivers is dominated by extreme events and the morphological response is typically step-like. Research in the Nahal Yael watershed (Schick, 1974) and in Central Australia (Bourke, 1994) indicates that intermittent discharges produce a channel unable to accommodate the occasional very large flood. The occurrence of low frequency "superfloods" completely alters the process–morphology relationship. Hence, ephemeral streams are best described as catastrophic whereby very large discharges drastically alter channel morphology. The Todd River is an example of a non-equilibrium river that is sensitive to variations in discharge and subject to extreme flows.

We describe the geomorphology of the Todd River in four zones: the headwaters, the piedmont zone, the alluvial plains and finally the longitudinal dunefield of the northern Simpson Desert. We discuss the impact of three factors on the patterns of fluvial form: the influence of tributaries, sediment transport dynamics and the variable discharge regime. In this ephemeral system, landform variability is expressed in two ways. First, fluvial morphology changes downstream as a series of steps rather than gradually. Factors such as asynchronous tributary activity and distances between points of tributary entry, variable boundary roughness and resistance have a dominant morphological and sedimentological signature. Second, landforms occur at three scales (small, medium and large).

CHARACTERISTICS OF THE TODD RIVER

The Todd River drains from the MacDonnell Ranges in central Australia (23°40′ S, 133°50′ E) passing through the town of Alice Springs (Figure 10.1). It is one of several internally draining ephemeral streams in the Lake Eyre Basin (>1 300 000 km^2), which

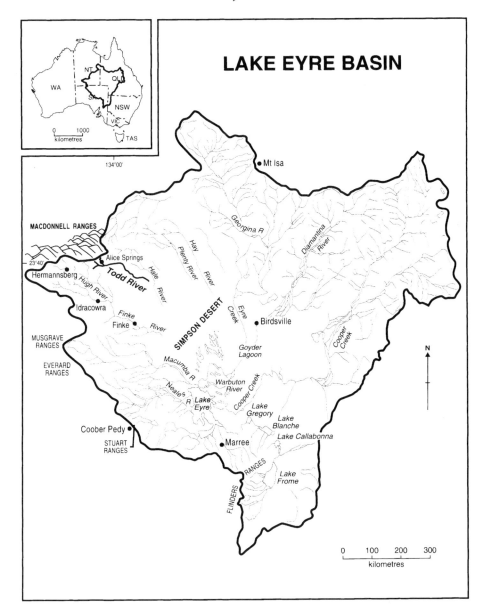

FIGURE 10.1   The Lake Eyre Basin, showing the location of the Todd and Finke Rivers

contains a wide range of river types and catchment sizes. These include the large braiding and anastomosing systems such as Cooper Creek (306 000 km^2) and the Diamantina River (365 000 km^2; Kotwicki, 1986) that rise in the tropical region of north Queensland (for detailed description see Rust and Nanson (1986) and Knighton and Nanson (1993)). Smaller systems include the Neales River (*c.* 35 000 km^2) rising in the more arid western basin (Croke et al., 1996). The Lake Eyre playa is the terminus for many of these

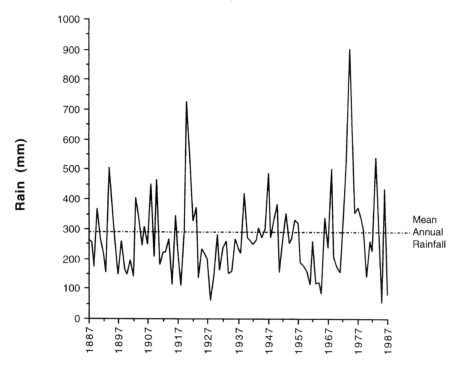

FIGURE 10.2   Total annual rainfall recorded at Alice Springs Post Office between 1887 and 1987. For location see Figure 10.3. Note the interannual variability in rainfall

internally drained river systems (Magee et al., 1995) but not the Todd which dissipates amongst longitudinal dunes in the northern Simpson Desert.

Rainfall in the Todd catchment is infrequent and most events occur between November and March during the Australian summer. Mean annual rainfall about the MacDonnell Ranges is 274.4 mm (median 238 mm) and displays high interannual variability (Figure 10.2). The spatial and temporal variation of rainfall is an important influence on fluvial landforms because it affects the magnitude and timing of drainage network activation. The asynchronous activity of channels results in a spatially variable distribution, and a temporally variable development of fluvial landforms. Rain typically reaches central Australia as discrete storms that move across the catchment. For example, a six-day rain event in March 1972 precipitated rainfall totals close to the annual average across the eastern Todd catchment but rainfall maxima occurred on different days at different gauges. Giles Creek (Ringwood gauge; Figure 10.3) recorded peak rainfall on the day prior to the peak at Ross River (Figure 10.4). Rainfall over the middle and upper Todd catchment during this period was much less than in the Ross and Giles that join the lower Todd.

The Todd River has few gauges downstream of the town of Alice Springs (Figure 10.3). The hydrological regime is best described as "flashy" as the river bed is dry 98% of the time and flow events rise and recede rapidly. The largest discharge event on record was measured in Alice Springs in 1988 and had a peak discharge of 1190 $m^3 s^{-1}$ for a catchment area of 450 km^2 (Barlow, 1988). This is regarded as a one-in-50 year event, on the basis of gauging since 1953.

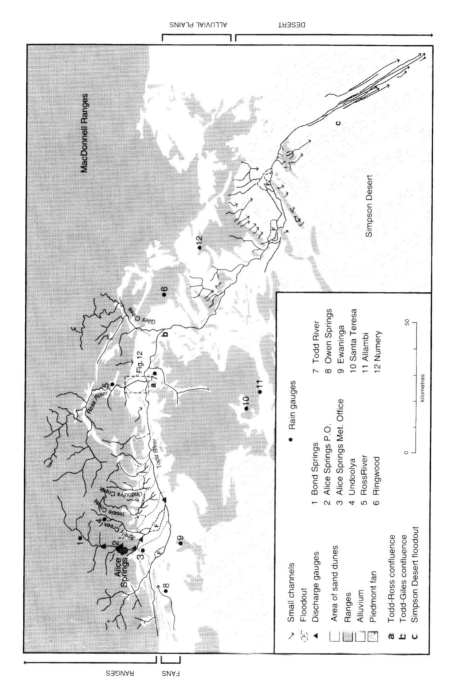

FIGURE 10.3 Todd River catchment, indicating the location of Alice Springs. rain and flow gauges, the main tributaries and the topographic domains: ranges, piedmont fans, alluvial plain and desert. The location of Figure 10.12 is also shown

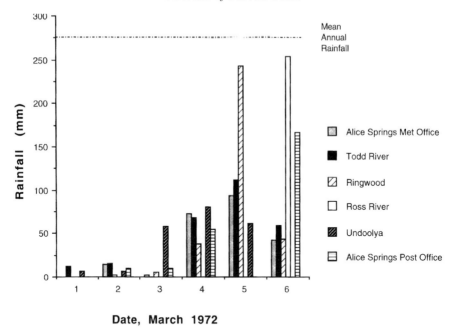

FIGURE 10.4   Temporal and spatial variation of rainfall, March 1972, Todd River catchment. For location of gauges see Figure 10.3

The catchment upstream of the Simpson Desert has an area of approximately 9300 km^2, with several large tributaries draining from the north (Figure 10.3). The headwaters of the Todd and its tributaries rise in Proterozoic crystalline and metamorphic rocks of the MacDonnell Ranges (Shaw and Wells, 1983). Soil in the ranges is predominantly shallow sandy lithosols with flanking outwash plains capped by red clays. South of the ranges, the soils are dominantly siliceous sands (Northcote and Wright, 1983). Since European settlement of the region over 100 years ago, land use within the catchment has become predominantly pastoral. Prior to this, nomadic hunting and gathering was practised by Aboriginal people (Spencer, 1896).

The Todd River essentially has a single, straight or gently winding channel, changing to a distributary pattern in floodouts (Figure 10.5b). Channel gradients, measured from 1:50 000 topographic maps, vary from approximately 0.00266 m m^{-1} in the ranges to

FIGURE 10.5   *(opposite)* Schematic diagrams of Todd River planform and cross-sections in each topographic domain (see Figure 10.3). Arrows indicate flow direction. (a) Gorge/strike valley. A–A′, cross-section through the gorge reach: a confined channel with coarse lateral deposits. B–B′, cross-section through the strike valley reach: a relatively unconfined channel inset in alluvium with well-defined channel and floodplain development. (b) Piedmont reach. A–A′, single-thread channel inset in piedmont fan alluvium. B–B′, multiple channel network with reduced channel capacity and dense vegetation. Note distributary channel pattern of floodout. (c) Alluvial plains with outcropping ridges. A–A′, single-thread channel inset into alluvium. B–B′, confined channel with asymmetrical channel cross-section indicating a deep bedrock pool and lateral sandy deposit. (d) Desert reach. A–A′, cross-section through terminal floodout as channel flows along aeolian dune swales. Note trellis-style channel pattern

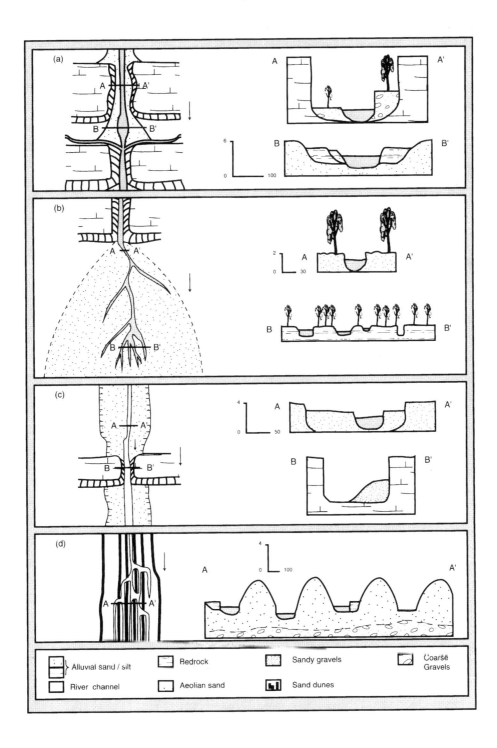

$0.00150 \, \text{m m}^{-1}$ further downstream. The channel cross-sections are wide and shallow (average width–depth ratio is 90). The channel is usually well defined and is incised into Pleistocene alluvium, palaeoflood deposits and modern alluvium. Banks are composed of erodible sandy material, but at some locations older cemented Pleistocene alluvium and aeolian deposits form resistant channel boundaries. The channel carries a coarse sandy load, deposited as large-scale ripples and tabular bars. These may be interspersed with fine-grained silt and clay deposits that settle out in localized channel lows.

Floodplains are well developed but infrequently inundated and, where the channel is confined, are laterally restricted. In these confined reaches floodplain morphology and sedimentology reflect a system subject to high-energy flows, characterized by "chaotic" sequences of inset fills (Figure 10.6) composed of vertically accreted layers of gravelly sands, sands and mud (Bourke, 1994). The youngest floodplain elements in these confined reaches are composed of sediments 1–2 m thick, deposited since 1950 (Bourke, 1998a). This recent phase of floodplain building is believed to relate to the period of above-average rainfall (1972–1979) following a severe drought (1958–1965) (Pickup, 1991; Bourke, 1998a).

The landscape through which the Todd River and its tributaries flow includes the strike ridge-dominated MacDonnell Ranges, Pleistocene piedmont fans, wide alluvial plains, and a terminal floodout in longitudinal dunes of the northern Simpson Desert. Below is a summary of the broad topographic domains (Figure 10.3) and a brief description of the landform assemblage in each domain (Figure 10.5a–d).

In the headwaters, the Todd River and its tributaries flow through narrow and sometimes meandering gorges, which pass south through low, sharp-crested, east–west trending ranges, formed through large-scale etching of folded Precambrian quartzite, sandstone and accessory carbonate rocks. These ridges are separated by narrow, strike valley plains, and the structurally controlled drainage pattern is trellised. Channel morphology in the steep-gradient gorges, through the strike ridges (Figure 10.5a), is similar to that of the upper Finke River described by Pickup et al. (1988) and includes an assemblage of coarse gravel bar deposits. Slackwater deposits sometimes occur in the narrower valleys and small gorges that follow structural lineaments.

The Todd River and its tributaries exit from the MacDonnell Ranges through a series of low-angle fans (approximate average gradient is $0.0036 \, \text{m m}^{-1}$). Channels of the larger tributaries are continuous, but those of smaller tributaries often terminate in distributaries (Figure 10.5b).

Downstream from the fans, the Todd River traverses a broad, strike-trending lowland, occupied by a wide alluvial plain bordering a broad sand plain interspersed with minor rounded hills and discontinuous strike valleys. These dissected plains and tablelands also contain small aeolian dunefields that are outliers of the Simpson Desert (Figure 10.3). In places, the Todd River has incised into carbonate-rich red earth, developed on the Pleistocene alluvial surface (Figure 10.6).

In its lower reaches, the Todd River channel occupies 300–500 m wide, interdune corridors between longitudinal northwest trending dunes of the northern Simpson Desert (Figure 10.5d). It maintains a well defined channel for 10 km, then the channel crosses a longitudinal dune and bifurcates. In this reach, incipient floodplains form discontinuous, bench-like insets composed of fine sandy material, unconformably overlying the aeolian sands. Approximately 20 km further downstream, the channel splits again and occupies

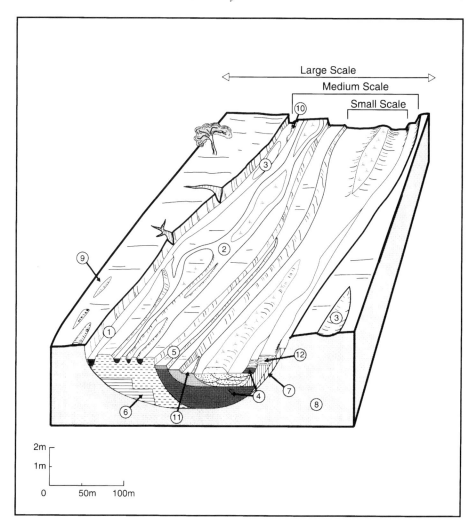

FIGURE 10.6   Schematic block diagram of the Todd River channel and floodplain. Note the variable surface morphology and complex subsurface fill units. 1, Back channel; 2, flood channel; 3, swirl pit; 4, channel inset; 5, floodplain inset; 6, buried floodplain remnant; 7, surface floodplain remnant; 8, Pleistocene red earth alluvium terrace; 9, overbank bars; 10, back channel inset; 11, stripped surface; 12, floodplain veneer deposit. For full description see Bourke (1994). The hierarchical scheme of flood forms is indicated as small, medium and large scale. Reproduced from Bourke (1994) by permission of IAHS Press

several parallel swales. Channels support dense stands of coolibah trees (*Eucalyptus microtheca*) and channel morphology is subdued. Where the channel passes through dunes, mud is deposited as backflood sediments in the dune swale. This pattern continues for a further 40 km until flow dissipates in a terminal floodout (Figure 10.3). The pattern of the longitudinal dunes controls the prevailing trellised, distributary channel pattern in this reach (Figure 10.5d).

PHYSIOGRAPHIC FACTORS INFLUENCING LANDFORM VARIABILITY

Downstream changes in channel morphology, planform and sediments are step-like in the Todd River. The channel bed in some places is cut into relatively indurated Pleistocene alluvium or bedrock, while further downstream it may be locally aggrading and contain thick sequences of sandy tabular bars. Many factors, such as the nature of sediment transport, sediment supply, channel boundary irregularities and knickpoint development, contribute to this pattern of variability. This section will focus on the step-change in channel morphology at tributary confluences.

In ephemeral streams, the pronounced variability of flow from confluent channels results in an exaggerated and often rapid change in channel morphology (Mabbutt, 1977; Reid et al., 1989), and channels often aggrade at or below junctions (Schumm, 1961; Everitt, 1993; Thornes, 1994b). It is the variability in the timing and magnitude of flow discharges between the contributing systems at confluences that influence the change in channel bed morphology and sediments above and below confluences in the Todd catchment. Two factors dominate the magnitude and direction of change: tributary spacing and tributary length.

Tributary spacing and tributary length

The channel distance between significant tributary inputs is an important influence on the behaviour of the floodwave downstream. In arid and semi-arid channels, transmission losses through drainage diffusion, infiltration and evaporation/evapotranspiration are significant (e.g., Schumm and Hadley, 1957; Schumm, 1961; Knighton and Nanson, 1994). While no transmission data exist for the Todd River, reports from the manager of the Todd River Station indicate that many large flows recorded at Alice Springs do not reach the Ross River confluence 78 km downstream (Figure 10.3) (I. Lovegrove, pers. comm.). Transmission losses into the channel bed result in a rapid downstream attenuation of flood magnitude. Flow is only augmented by a few large tributaries, the most significant of these being the Ross River and Giles Creek (Figure 10.3).

The proximity of the tributaries to sediment sources in the mountain ranges is an important factor affecting channel morphostratigraphy at confluences. The layout of the

TABLE 10.1   Channel distance from piedmont apex to confluence

Tributary	Channel distance to confluence from range outlet (km)	
	Tributary	Trunk stream
Emily	8	5
Jessie	10	28
Undoolya	13	35
Ross	10	78
Giles	13	100

Todd catchment is such that, south of Alice Springs, the trunk stream drains eastwards along a wide strike valley (Figure 10.3). As a consequence, tributaries travel relatively short distances from the ranges to confluence points (Table 10.1, Figure 10.3). Accordingly, the sediment loads, textures, channel gradients and flow magnitudes in the larger tributaries are often greater than those in the trunk stream, despite the observation that transmission losses probably increase between the mountain and alluvial fan reaches. As a consequence, several of the small tributaries have unchannelled junctions with the trunk stream.

RESPONSES OF A MAJOR CONFLUENCE TO RAINFALL EVENTS

The confluence dynamics of the Todd and Ross Rivers were monitored between 1993 and 1995 and are used here to illustrate the impact of tributary spacing and length upon confluence morphology and sediments (schematically represented in Figure 10.7). During this period, the geomorphic effects of a series of low flows down the trunk stream of the Todd River in conjunction with a series of moderate to high flows from the Ross River tributary were observed. The distinctive morphostratigraphy that developed in the Todd channel bed upstream of the confluence is illustrated (Figure 10.7a, Table 10.2). Upstream of the confluence, the Todd channel bed material fined downstream, from coarse and medium sands arranged in large tabular bars to horizontally laminated, thinly interbedded layers of fine sand and silty clay. This rapid downstream fining of bed material was accompanied by a decrease in channel gradient with a gradient reversal ($-0.0007 \, \text{m} \, \text{m}^{-1}$) near the confluence. Channel width decreased from 120 m to 40 m and the channel had a trench-like morphology with vertical channel banks. Floodplain sediments along this reach are vertically accreted, thickly bedded flood couplets and sand sheets.

At that time, bed elevation stepped upwards at the entrance point of the Ross River (Figure 10.7a), forming a barred confluence. Channel width increased from 40 m to 180 m (Figure 10.8), and channel bed material coarsened to gravelly sands. The channel bed gradient increased to $0.0032 \, \text{m} \, \text{m}^{-1}$ (Table 10.2). Channel bed morphology is typical of the wide braided reaches of central Australian streams described by Zwolinski (1985) with large bar forms and multiple channels. In addition there are a series of alluvial benches inset against the steep channel banks (Bourke, 1994).

Material deposited by the Ross River during the final phase of moderate to high flows, locally aggrades the bed, partially migrates up the Todd channel and effectively dams the lower-magnitude flows travelling down the Todd (Figure 10.7a), which deposit suspended load augmented by backflooding of fine sediment and plant detritus from the Ross.

However, this is a highly dynamic confluence, owing to uneven patterns of rainfall distribution described earlier. Hence, relative magnitude and timing of flows from Alice Springs and the Ross River vary through time and so too does the confluence morphostratigraphy. A recent flow (a one in five year event at Alice Springs) transformed the barred confluence to a discordant confluence, where both systems contain gravelly sands (Figure 10.7b).

During this event, total rainfall receipts at the Ross gauge measured 204.2 mm (peak rainfall 12–13 March) and 188.5 mm at the Bond Springs gauge (peak rainfall 16 March) north of Alice Springs (for locations see Figure 10.3). The Ross River flow peaked at the confluence before flow that travelled down the Todd and the Ross River aggraded its bed

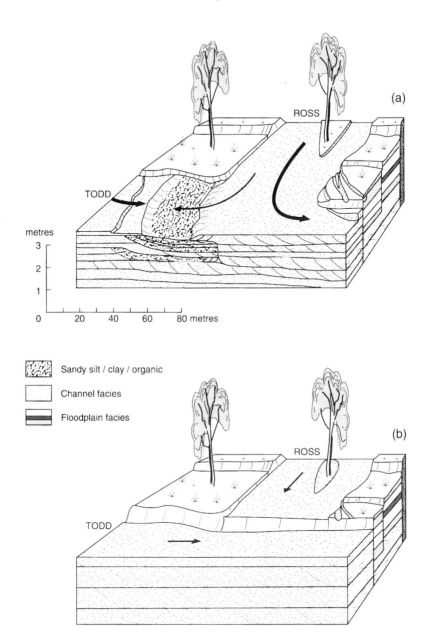

FIGURE 10.7 Todd/Ross confluence dynamics. Channel and floodplain morphostratigraphy following: (a) a sequence of low flows down the Todd channel and high flows down the Ross channel, resulting in an interfingering of coarse and fine units and a reversed local channel gradient. Sediment size coarsens significantly along the main channel through the confluence; (b) a high-energy flow down the Todd channel. The river has incised its bed and transformed the channel bed morphostratigraphy, forming a discordant junction in addition to eroding its floodplains

TABLE 10.2   Characteristics of the Todd channel upstream and downstream of the Ross River confluence in 1993

Variable	Upstream of confluence	Downstream of confluence
Gradient (m m^{-1})	−0.0007	0.0032
Width (m)	40	180
Bed material	Horizontally bedded clay-rich silt and fine sand, plant material	Cross-bedded gravelly sand
Channel pattern	Straight	Locally braiding

in a manner similar to that illustrated in Figure 10.7a. Arriving later, the Todd flood peak incised and truncated the alluvial dam from the Ross River, leaving the channel bed of the Ross hanging 2.6 m above the thalweg of the Todd River (Figure 10.7b). The discordant junction morphology persisted for several months until a lower-magnitude flow event on the Ross incised a small inset channel to local base level at the confluence.

These adjustments of confluence morphology illustrate three aspects of landform variability in ephemeral systems. First, during high-magnitude flows in ephemeral streams, it is the process that controls form, and forms created by such flows partly control processes during subsequent smaller events. In this way the persistent erosion and

FIGURE 10.8   An oblique aerial photograph of the confluence of the Todd and Ross River, taken in 1993. The Ross River enters the photograph on the lower right and the Todd on the upper right. Flow is towards the left of the photograph. Note the relatively narrow channel dimensions and fine sediment, as indicated by the darker colour, in the Todd channel bed upstream of the confluence. The Ross River tributary is clearly the dominant system at this time

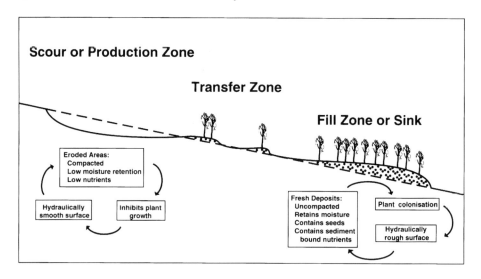

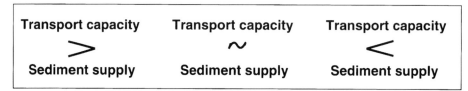

FIGURE 10.9 Schematic model of an erosion cell. The scour zone is an area of erosion and actively sheds both water and sediment. The transport zone occurs downslope and is occupied by discontinuous patches of sediment in transit or in temporary storage. In arid and semi-arid regions, storage may be for a considerable time as transport events are intermittent and short-lived. The fill zone lies downslope and is an area of sediment deposition. Accumulation rates are variable and are dependent on rates of sediment delivery and the propensity for the channel to change location. Reproduced from Pickup (1988a) by permission of Academic Press Ltd

depositional patterns generated by floods modulate the landscape response to subsequent and smaller events. Second, they highlight the importance of the order and magnitude of preceding events on the geomorphic response to flow events (cf. Pickup and Rieger, 1979). Third, it supports the view of Cooke et al. (1993) that forms and processes are rarely in equilibrium in ephemeral streams.

Similar effects from asynchronous flow peaks at confluences include the alternating damming and breaching of tributary confluences along the Wadi Watir in the

FIGURE 10.10 *(opposite)* The temporally variable landform assemblage of an erosion cell mosaic in a flat area such as an alluvial fan or floodplain. The shifting pattern is determined by the evolution of the erosion cell over time and the complex interaction of large- and small-scale erosion cells. Stage 1: erosion cell established as small production, transfer and sink zones. All three zones begin to extend upslope. Stage 2: upslope extension continues. The sink zone is elongated and no longer aggrades at the downstream end, but receives sediment-deficient flow. Gully initiation in distal reaches. Stage 3: development of new erosion cell. Primary sink may become inactive as new erosion cells capture diverted runoff. Reproduced from Pickup (1985) by permission of the Australian Rangeland Society

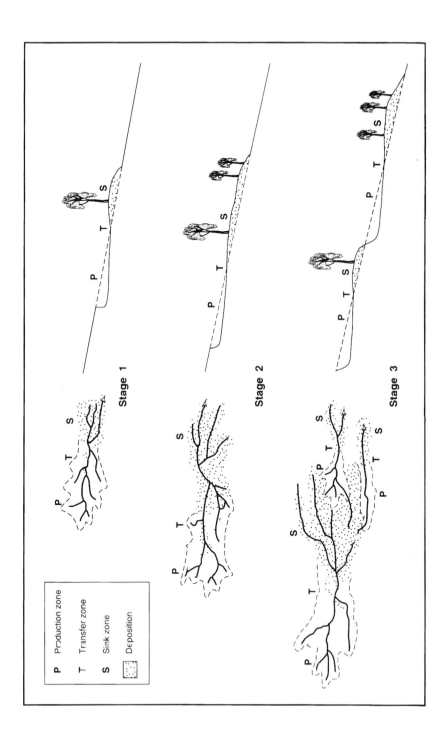

Stage 1

Stage 2

Stage 3

P Production zone
T Transfer zone
S Sink zone
    Deposition

southeastern Sinai (Schick and Lekach, 1987) and the distinct sedimentological signature of asynchronous tributary activity in channel bed sediments downstream of the confluence of the Il Warata and Il Naitiwa in Kenya (Reid et al., 1989).

PATTERNS OF EPHEMERAL SEDIMENTATION

Sediment supply in arid fluvial systems is episodic in time and variable in space and is strongly linked to patterns of vegetation cover. Pickup (1985, 1988a,b) has shown that patterns of shifting erosion and sedimentation, at different scales, in central Australia are modulated by colonization of newly deposited sediment by plants. The landscape can be considered in three zones: a sediment scour or production zone, a sediment transfer zone, and a sediment sink or fill zone (Figure 10.9). Newly deposited sediment in the fill zone (Figure 10.9) is uncompacted and potentially can hold more moisture than older compacted sediment, and may also contain seeds and a higher proportion of sediment-bound nutrients. Thus it tends to be readily colonized by plants, which stabilize the surface and promote further deposition. At the smallest scale these three elements can be identified within channels as scours, planar beds and shoals formed during smaller floods; at a larger scale, downcut channel reaches represent the sediment source zone, sandy channel reaches with transient bar features the transport zone, and aggrading channel reaches represent the fill zone. At the largest scale are extreme flood deposits, described later. These three elements shift around the landscape and scours become sediment traps while transfer channels and fill zones become new source zones (Figure 10.10).

At a relatively large scale the scour–transport–fill (STF) model is illustrated by floodouts located in the piedmont zone of the MacDonnell Ranges (Figures 10.3 and 10.5b). The bare, steep rock slopes in the steep-gradient headwaters of the Todd River constitute the source zone from which sand and angular weathered clasts pass through a series of small-scale fans into a network of gullies which feed into the channel of the Todd River. This detritus is then transported together with locally reworked alluvium to the broad piedmont where it is deposited, contributing to extended aprons of alluvial fill. These features are schematically represented in Figure 10.5b and are typically vegetated by coolibah trees (*Eucalyptus microtheca*).

The scales of STF sequences are not exclusive and may have complex spatial relationships; for example, the distal zone of one floodout may be the production zone for another STF sequence, or smaller cells may be embedded within larger ones. Erosion cells may not be fully developed or represent latent features in the landscape (Pickup, 1988b). The transition between linked cells tends to be morphologically abrupt. This patterning of arid-zone sediment transport and deposition produces a shifting mosaic of ground surface sediments known as an erosion cell mosaic (Pickup, 1985; Figure 10.10).

VARIABLE DISCHARGE REGIMES: A MULTISCALE APPROACH TO FLUVIAL LANDFORMS

Fluvial landforms that develop under variable flow regimes are best described using a multiscale approach (Gupta, 1988, 1995; Pickup, 1991). In these environments, landform assemblages are associated with discrete flow magnitudes. Examples include the nested channel systems described by Graf (1988b) and the nested fluvial forms of the Auranga River, India (Gupta, 1995). Under the present climatic regime, the Todd River is subject to flow at extremes of the magnitude/frequency spectrum, i.e. there are long periods with no flow in the channel interspersed by flood events including occasional superfloods (Bourke, 1998a). Hence, fluvial landforms evolve episodically rather than continuously and fall into discrete size classes.

Fluvial landforms of the Todd River fall into three categories based primarily on landform scale and associated flow magnitude: landforms generated by within-channel flows (small scale); those formed by flows which extend across floodplains (medium scale); and superflood landforms (large scale) (Figure 10.6). Dimensions across these three classes vary by an order of magnitude; for example, longitudinal bars in the three categories measure 10 m, 100 m and up to 2000 m, respectively. Approximate return periods for flows at each respective scale are less than one-in-10 year flood, one-in-100 year flood, and an estimated one-in-1000 year flood. Geomorphologically, these different scales of flow produce complex nested assemblages of fluvial landforms in the Todd River system. A description of the geomorphic effects of small- and medium-scale events is given, highlighting the importance of local factors on landform variability, followed by a description of large-scale flood features.

Small-scale fluvial landforms, formed by flows within the active channel banks, have a significant impact on channel side and channel bed. The main geomorphic effects include localized but sometimes substantial channel widening; erosion of within-channel benches in some reaches and aggradation in others; bank accretion; channel bed aggradation and incision (Figure 10.6).

FIGURE 10.11 Elliptical scour around river red gum (*Eucalyptus cameldulensis*) formed during medium-scale flow in Todd River channel. Person in right foreground of the scour gives scale. Flow is towards the camera. Note the partial re-excavation of buried trees along the right bank

Medium-scale fluvial landforms are formed by flows which spread across floodplains (Figure 10.6). The lateral extent of these floods is limited by event magnitude and by topographic barriers such as older, higher alluvial and aeolian surfaces. The geomorphic effects are more pronounced in areas where the active channel is entrenched into older, more resistant alluvium. Where floodplains are laterally extensive, flood effects decrease with distance from the channel. The largest gauged event on the Todd River occurred in 1988, with an estimated return interval of one-in-50 years. The geomorphic effects of this event include channel widening by 300%, the lateral and downstream extension of braided reaches, especially downstream of confluences (Pickup, 1991), and the deposition of ripples, longitudinal, transverse and linguoid bars in the channel bed throughout the catchment. Also observed was the removal of within-channel benches, lateral and vertical erosion of floodplains, erosion around large trees, and discontinuous deposition of sand sheets, splays and bars on floodplain surfaces.

Local factors strongly influence channel response to small- and medium-scale events. We have already mentioned the effects of asynchronous flooding at confluences and the importance of pre-existing channel configuration upon local changes at the Todd–Ross confluence (Figure 10.7). Other influential factors include channel roughness and variable boundary resistance. A spectacular example is the effect of vegetation on the development of asymmetric, elliptical scours which are formed by eddies around river red gums (*Eucalyptus cameldulensis*), both in the channel bed and on floodplain surfaces (Dunkerley, 1992; Bourke, 1994) (Figures 10.7b and 10.11). Large trees on elongated in-channel bars act as sediment traps at some flows and initiate substantial leeward scour at others. Variations in channel width are related to local bank resistance: extensive channel widening commonly occurs where channel banks are composed of unconsolidated sand, often associated with aggradation of the channel bed. In areas with more resistant boundary materials, for example where the river flows through indurated Pleistocene alluvium, large scour holes dominate the channel bed (similar to that shown in Figure 10.5c).

In central Australia, as in other arid and semi-arid regions, extreme events are geomorphologically the most important (Schick, 1974, 1995; Thornes, 1976; Wolman and Gerson, 1978; Patton and Baker, 1977; Harvey, 1984; Pickup, 1991; Patton et al., 1993). The largest-scale fluvial landforms in the Todd River (Figures 10.12 and 10.13) are formed by extreme flows, which occur rarely and inundate vast tracts of the landscape including the swale systems of aeolian dunefields. The occurrence of very large floods in the palaeohydrological record of central Australia was demonstrated by Baker et al. (1983) and Pickup et al. (1988), who reported late Holocene slackwater sequences in the Finke gorge. Pickup (1991) and Patton et al. (1993) described the geomorphic effects of very high-magnitude events on piedmont fans of the Todd and Ross, and Bourke (1998a) has detected the geomorphic effects of extreme flood events downstream to the terminal floodouts in the northern Simpson Desert.

Large-scale flood landforms include sand sheets, ripple fields, overflow channels (Pickup, 1991), large-scale palaeobraid channels, levee deposits and broad, low-relief bars (Patton et al., 1993) (Figures 10.12 and 10.13). In addition, large-scale channel avulsions have eroded longitudinal dunes of the northern Simpson Desert. The beds of these scoured channels are punctuated by large-scale swirl pits up to 2.5 m deep and 15 m

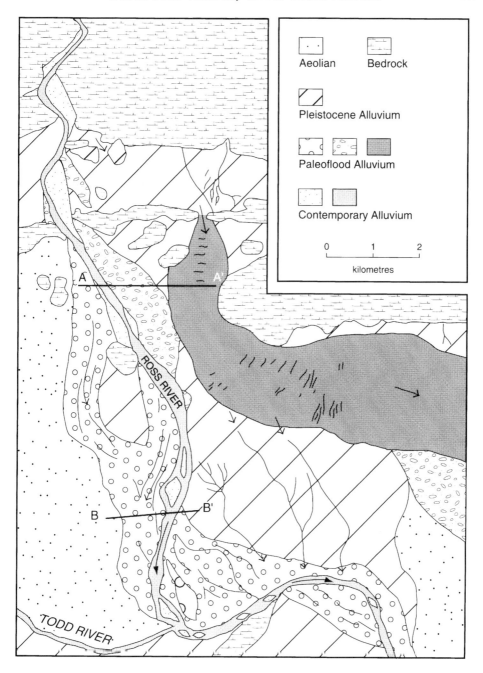

FIGURE 10.12   The Ross River palaeoflood channels. For detailed description see Patton et al. (1993). These channels cross the piedmont reach south of the MacDonnell Range. Note the higher Pleistocene surfaces around which the high-magnitude flows are deflected. Reproduced from Patton et al. (1993) by permission of Academic Press, Inc.

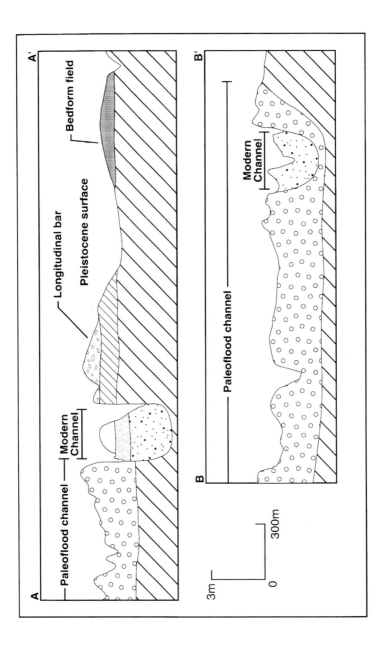

FIGURE 10.13 Schematic cross-sections of the superflood morphology of the Ross River palaeoflood channel. For location see Figure 10.12. The assemblage of large-scale palaeoflood channel forms extends long distances from the contemporary channel and displays a variety of morphologies. For full description see Patton et al. (1993). Reproduced from Patton et al (1993) by permission of Academic Press, Inc.

long. Selected cross-sections (Figure 10.13) illustrate three aspects of the superflood landscape: fluvial landforms extend several kilometres from the modern channel; the scale of fluvial landforms is more than 30 times that of the modern channel; and palaeofloods generate a variety of morphologies including palaeochannels, high-level bars and ripple fields. These large-scale landforms evolve episodically and remain inactive for long periods of time (Pickup, 1991), and may appear to be in disequilibrium with the contemporary channel.

In addition to the geomorphological imprint of a highly variable discharge regime, three other effects are noted. First, the scale of flows affects the frequency of reworking of alluvial deposits and the balance changes downstream, so that the headwaters and piedmont reaches of the system are more frequently reworked by small- and medium-scale flows than the downstream reaches, owing to transmission losses downstream. Second, the preservation of flood features is scale-dependent: forms created during large-scale flows potentially are preserved for longer than those generated during medium- and small-scale flows. Third, as channel recovery from extreme floods is ongoing, the landforms produced by smaller-scale flows cannot be adequately explained without reference to the more regional-scale events (Pickup, 1991; Bourke, 1998b).

CONCLUSION

To summarize, in this variable discharge regime, landforms develop episodically, and downstream transitions tend to be abrupt. At tributary confluences disequilibrium is expressed as discordant and barred junctions. Asynchronous tributary activity is influenced by the spatial and temporal patterns of rainfall and runoff, the tributary spacing and length, hydrological lags and transmission losses. Discharge events range from the extreme superflood to small flows which dissipate rapidly downstream. This variability produces an assemblage of fluvial forms that varies across a hierarchy of scales. Form preservation is related to distance downstream and distance from the active channel with the larger-scale forms preserved for longer periods. The older forms and sediments deposited during high-magnitude events modulate the geomorphic response to lower-magnitude flows. This assemblage of fluvial forms is the product of a system where form and process are rarely in equilibrium.

ACKNOWLEDGEMENTS

The authors thank John Chappell, Jacky Croke, Avijit Gupta, Andy Miller and Ellen Wohl for providing helpful reviews. The Coombs Cartography Unit assisted with diagrams and Coombs Photography produced the plates. An ANU postgraduate scholarship and field research funds from the National Greenhouse Advisory Committee and the Sydney Water Board are gratefully acknowledged. Dave Marshall and Phil Ward provided field assistance. The Manager of Todd River Station, I. Lovegrove, is thanked for his assistance during the challenging field conditions of the 1995 flood.

REFERENCES

Baker, V.R., Pickup, G. and Polach, H.A. 1983. Desert palaeofloods in central Australia. *Nature*, **301**(5900), 502–504.

Barlow, F.T.H. 1988. *Hydrology of the Todd River Flood of March 1988*. Power and Water Authority, Alice Springs, Australia, 26 pp.

Bourke, M.C. 1994. Cyclical construction and destruction of flood dominated floodplains in semiarid central Australia. In L.J. Olive, R.J. Loughlan and J.A. Kesby (Eds), *Variability in Stream Erosion and Sediment Transport*. International Association of Hydrological Sciences, Publication 224, 113–123.

Bourke, M.C. (1998a). *Fluvial Geomorphology and Paleofloods in Arid Central Australia*. PhD Thesis, Australian National University.

Bourke, M.C. (1998b). *Channel Adjustment to Extreme Events in Central Australia*. Proceedings of the Australian–New Zealand Geography Conference, Tasmania.

Chorley, R.K. and Kennedy, B.A. 1971. *Physical Geography: A Systems Approach*. Prentice-Hall, London, 370 pp.

Cooke, R., Warren, A. and Goudie, A. 1993. *Desert Geomorphology*. UCL Press, London, 526 pp.

Croke, J., Magee, J. and Price, D. 1996. Major episodes of Quaternary activity in the Lower Neales River, north west of Lake Eyre, central Australia. *Palaeogeography, Palaeoclimatology, Palaeoecology*, **124**, 1–15.

Dunkerley, D.L. 1992. Channel geometry, bed material, and inferred flow conditions in ephemeral stream systems, Barrier Range, Western N.S.W. Australia. *Hydrological Processes*, **6**, 417–433.

Everitt, B. 1993. Channel responses to declining flow on the Rio Grande between Ft. Quitman and Presidio, Texas. *Geomorphology*, **6**, 225–242.

Graf, W.L. 1988a *Fluvial Processes in Dryland Rivers*. Springer, New York, 343 pp.

Graf, W.L. 1988b. Definition of flood plains along arid-region rivers. In V.R. Baker, R.C. Kochel and P.C. Patton (Eds), *Flood Geomorphology*, Wiley, New York, 231–242.

Gupta, A. 1988. Large floods as geomorphic events in the humid tropics. In V.R. Baker, R.C. Kochel and P.C. Patton (Eds) *Flood Geomorphology*, Wiley, New York, 301–315.

Gupta, A. 1995. Magnitude, frequency, and special factors affecting channel form and processes in the seasonal tropics. In J.E. Costa, A.J. Miller, K.W. Potter and P.R. Wilcock (Eds), *Natural and Anthropogenic Influences in Fluvial Geomorphology*. American Geophysical Union, Washington, DC, Monograph 89, 125–136.

Harvey, A.M. 1984. Geomorphological response to an extreme flood: a case from southeast Spain. *Earth Surface Processes and Landforms*, **9**, 267–279.

Knighton, A.D. and Nanson, G.C. 1993. Anastomosis and the continuum of channel pattern. *Earth Surface Processes and Landforms*, **18**, 613–625.

Knighton, A.D. and Nanson, G.C. 1994. Flow transmission along an arid zone anastomosing river, Cooper Creek, Australia. *Hydrological Processes*, **8**, 137–154.

Kotwicki, V. 1986. *Floods of Lake Eyre*. Water Supply Department, Adelaide.

Mabbutt, J.A. 1977. *Desert Landforms*. ANU Press, Canberra. 340 pp.

Magee, J.W., Bowler, J.M., Miller, G.H., Williams, D.L.G. 1995. Stratigraphy, sedimentology, chronology and palaeohydrology of Quaternary lacustrine deposits at Madigan Gulf, Lake Eyre, South Australia. *Palaeogeography, Palaeoclimatology, Palaeoecology*, **113**, 3–42.

Northcote, K.H. and Wright, M.J. 1983. Sandy desert region (1). In *Soils: an Australian Viewpoint*. Division of Soils, CSIRO, Melbourne, 173–178.

Olive, L.J., Loughran, R.J. and Kesby, J.A. (Eds), 1994. *Variability in Stream Erosion and Sediment Transport*. IAHS Publication No. 224, 116.

Patton, P.C. and Baker, V. 1977. Geomorphic response of Central Texas stream channels to catastrophic rainfall and runoff. In D.O. Doehring (Ed.), *Geomorphology in Arid and Semi-Arid Regions*. Allen and Unwin, Boston, 189–217.

Patton, P.C., Pickup, G. and Price, D.M. 1993. Holocene palaeofloods of the Ross River, Central Australia. *Quaternary Research*, **40**, 201–212.

Pickup, G. 1985. The erosion cell – a geomorphic approach to landscape classification in range assessment. *Australian Rangeland Journal*, **7**(2), 114–121.

Pickup, G. 1988a. Modelling arid zone soil erosion at the regional scale. In R.F. Warner (Ed.), *Fluvial Geomorphology of Australia*. Academic Press, Sydney, 105–127.

Pickup, G. 1988b. Hydrology and sediment models. In M.G. Anderson (Ed.), *Modelling Geomorphological Systems*. Wiley, Chichester, 153–215.

Pickup, G. 1991. Event frequency and landscape stability on the floodplain systems of arid Central Australia. *Quaternary Science Reviews*, **10**, 463–473.

Pickup, G. and Rieger, W.A. 1979. A conceptual model of the relationship between channel characteristics and discharge. *Earth Surface Processes and Landforms*, **4**, 37–42.

Pickup, G., Allan, G. and Baker, V.R. 1988. History, paleochannels and palaeofloods of the Finke River, Central Australia. In R.F. Warner (Ed.), *Fluvial Geomorphology of Australia*. Academic Press, Sydney, 177–200.

Reid, I. and Frostick, L.E. 1997. Channel form, flow and sediments in deserts. In D.S.G. Thomas (Ed.), *Arid Zone Geomorphology*. Wiley, Chichester, 205–229.

Reid, I., Best, J.L. and Frostick, L.E. 1989. Floods and flood sediments at river confluences. In K. Bevan and P. Carling (Eds), *Floods: Hydrological, Sedimentological, and Geomorphological Implications*. Wiley, Chichester, 135–150.

Rust, B.R. and Nanson, G.C. 1986. Contemporary and palaeo channel patterns and the Late Quaternary stratigraphy of Cooper Creek, southwest Queensland, Australia. *Earth Surface Processes and Landforms*, **11**, 581–590.

Schick, A.P. 1974. Formation and obliteration of desert stream terraces – a conceptual analysis *Zeitschrift für Geomorphologie NF*, Supplementband **21**, 88–105.

Schick, A.P. 1995. Fluvial processes on an urbanizing alluvial fan: Eilat, Israel. In J.E. Costa, A.J. Miller, W. Potter and P.R. Wilcock (Eds), *Natural and Anthropogenic Influences in Fluvial Geomorphology*. American Geophysical Union, Washington, DC, Geophysical Monograph 89, 209–218.

Schick, A.P. and Lekach, J. 1987. A high magnitude flood in the Sinai Desert. In L. Mayer and D. Nash (Eds), *Catastrophic Flooding*. Allen and Unwin, Boston, 381–410.

Schumm, S.A. 1961. Effect of Sediment Characteristics on Erosion and Deposition in Ephemeral–Stream Channels. *US Geological Survey Professional Paper* 352 C.

Schumm, S.A. and Hadley, R.F. 1957. Arroyos in the semi arid cycle of erosion. *American Journal of Science*, **255**, 161–174.

Scoging, H. 1989. Runoff generation and sediment mobilisation by water. In D.S.G. Thomas (Ed.), *Arid Zone Geomorphology*. Wiley, Chichester, 87–116.

Shaw, R.D. and Wells, A.T. 1983. *1:250,000 Geological Series – Explanatory Notes, Alice Springs (second edition) Northern Territory, Sheet SF/53–14 International Index*. Department of Resources and Energy, Bureau of Mineral Resources, Geology and Geophysics, 44 pp.

Spencer, W.B. 1896. *A Report on the Work of the Horn Scientific Expedition to Central Australia*. Melville, Mullen and Slade, Melbourne.

Thornes J.B. 1976. *Semi-Arid Erosional Systems: Case Studies from Spain*. London School of Economics, Geography Department Papers, No. 7.

Thornes, J.B. 1994a. Catchment and channel hydrology. In A.D. Abrahams and A.J. Parsons (Eds), *Geomorphology of Desert Environments*. Chapman and Hall, London, 257–287.

Thornes, J.B. 1994b. Channel process, evolution, and history. In A.D. Abrahams and A.J. Parsons (Eds), *Geomorphology of Desert Environments*. Chapman and Hall, London, 288–317.

Wolman, M.G. and Gerson, R. 1978. Relative scales of time and effectiveness of climate in watershed geomorphic processes. *Earth Surface Processes and Landforms*, **3**, 189–208

Zwolinski, Z. 1985. Depositional model for desert creek channels: Lake Eyre Region, Central Australia. *Zeitschrift für Geomorphologie*, Supplementband **55**, 39–56.

11

The Uniab River Fan: an Unusual Alluvial Fan on the Hyper-arid Skeleton Coast, Namibia

A.C.T. Scheepers[1] and I.C. Rust[2]

[1] *Department of Earth Sciences, University of the Western Cape, Bellville, South Africa*
[2] *Department of Geology, University of Port Elizabeth, Port Elizabeth, South Africa*

ABSTRACT

The Uniab River alluvial fan, located on the hyper-arid wave-dominated Atlantic Ocean Skeleton Coast of Namibia, is characterized by: a very low longitudinal slope (less than 1:80); a suite of abandoned and incised channels with rapids, waterfalls, terraces and hanging mouths; evidence of fluvial incision during Late Quaternary rising sea level; a prominent regressive sea cliff; and indications of occasional mega-flood events caused by damming behind a major linear dunefield barrier up to 7 km wide across the main river channel.

The fan is not associated with any rising mountain front, and at present discharges directly into the Atlantic Ocean. Relative sea-level change, rather than tectonism, is the main control on the Uniab River fan architecture.

The Uniab River flows very infrequently, and seldom for longer than a few hours. The dunefield across the main channel of the Uniab River may, however, at times dam the discharge. It is postulated that breaching of the dunefield dam wall by overtopping or piping gives rise to occasional short-lived mega-floods capable of transporting boulders more than 1m in diameter.

Unconsolidated alluvial gravel fans in hyper-arid coastal regions are particularly susceptible to significant modifications due to marine erosion during relative rise in sea level. The undercut sea cliff produced by marine erosion during rising sea level serves to steepen channel gradients, leading to incision at a time when flooding and backfilling rather than fluvial erosion is expected.

INTRODUCTION

Alluvial fans occur in many environments (Leggett et al., 1966; Ryder, 1971; Kesel, 1985; Ritter and Ten Brink, 1986; Lecce, 1990), but they are common in arid settings where infrequent, intense, high-energy flash floods are the main driving force for transporting and depositing the debris that ultimately accumulates as a steep-sloped fan-

---

Varieties of Fluvial Form. Edited by A.J. Miller and A. Gupta
© 1999 John Wiley & Sons Ltd

shaped deposit (Blissenbach, 1954; Beaty, 1963; Hooke, 1967, Wells, 1977; Derbyshire and Owen, 1990; Abrahams and Parsons, 1994). Alluvial fans are in many cases associated with areas of rugged topography linked with tectonic activity (Bull, 1963; Blair, 1987; Miall, 1981), but not all (Carryer, 1966; Church and Ryder, 1972; Ono, 1990; Maizels, 1990; Lecce, 1990).

The Uniab River mouth is located in the Skeleton Coast National Park along the hyper-arid Namib Desert, Namibia. Information on the geomorphology of this desolate area and environs is scarce (Gevers, 1936; Jeppe, 1952; Mabutt, 1952; Hallam, 1964; Stengel, 1966; De Beer et al., 1981; Van Zyl and Scheepers, 1991, 1992, 1993; Scheepers, 1990, 1992; Jacobsen et al., 1995).

The deposits at the river mouths of the region have been variously called "deltas" (Stengel, 1964; Ward and Von Brunn, 1985; De Beer et al., 1981), "alluvial fans" (Hallam, 1964), "funnel-shaped" (Stengel, 1966) and "arid deltas" (Scheepers, 1992). In this paper we refer to the deposits in the lower reaches of the Uniab as "alluvial fan", accepting that they deviate in some respects from the diagnostic criteria listed by Blair and McPherson (1994).

The Uniab alluvial fan is an erosional remnant of a previously larger fan, which may originally have been a coarse-grained fan delta of the Hjulström type (Postma, 1990). The Uniab fan is unusual in several aspects, including: the very low radial gradient of the fan; the location of the fan at the shoreline and its dissection during recent rise in sea level; the apparently inconspicuous role of tectonics in the fan's development; and the control on the fan's morphology and operational processes by a linear dunefield that lies across its main channel.

Leckie (1994) describes features from the Canterbury Plains, New Zealand, that match many of those reported here for the Uniab River fan deposits, especially the relationship between changes in relative sea level, and channel incision in the coastal fluvial gravels. In many other respects, however, the two sites are unalike, particularly in terms of regional tectonic setting, climatic control and discharge dynamics.

This chapter is a progress report listing our initial findings and current ideas on the dynamics of the Uniab fan, our investigations being incomplete.

UNIAB RIVER DRAINAGE BASIN

The mouth of the Uniab River is located near Torra Bay in the desolate Skeleton Coast on the Atlantic Ocean, about 350 km north of Walvis Bay, the main commercial harbour of Namibia. The Uniab is one of the shorter rivers draining this sector of the generally hyper-arid coastal desert zone of Namibia. The length of the river is about 117 km, and its drainage area, located below the Great Southern African Escarpment, is about 4650 km^2 (Figure 11.1).

The topography of the Uniab drainage basin consists mostly of rolling hills, the hill tops being generally below 450–600 m above sea level, rising steadily inland towards the main watershed at about 1200 metres elevation. There is no indication whatsoever of a tectonically controlled rising scarp or mountain range to drive the development of the Uniab alluvial fan, as is the case, for example, for the Death Valley fans in southern California (Blair, 1987). Even the Great Southern African Escarpment, which forms the main landward watershed for the Uniab basin, is not a structural feature, but a post-

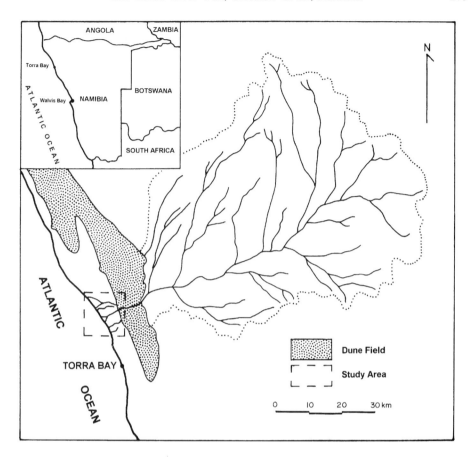

FIGURE 11.1   Locality map and drainage basin of the Uniab River, Skeleton Coast, Namibia. The southern limit of the Skeleton Coast Dunefield is indicated; it continues northwards for several hundred kilometres into Angola. The Skeleton Coast is a deserted area with no towns. Torra Bay is a camp site open for surf fishing a few weeks per year

Gondwana break-up erosional product. Regional tectonic uplift has been considered the reason for the prominent eroded sea cliffs of the main alluvial fans along the Skeleton Coast (Van Zyl and Scheepers, 1991; Scheepers, 1992), but although this may account for certain small-scale morphological features of the fan, our current thinking is that regional uplift does not appear to be the primary driving force for the formation of the Uniab fan.

Except for the area where the Uniab fan itself is developed, the basin is almost entirely underlain by basalt (Cretaceous Etendeka Formation). The fan lies on a wave-cut platform cut into high-grade gneisses of Precambrian age. In places lithified gravel and sandstone (?Late Pleistocene Oswater Formation; Ward, 1988), similar in composition to the Uniab fan gravel itself but clearly much older, form the floor of the lower reaches of the Uniab fan.

A significant feature of the lower drainage basin is the dunefield that lies across the main Uniab channel about 5 km upstream of the mouth. This dunefield is part of the

FIGURE 11.2   Looking south across the gravel-floored main Uniab River channel (foreground) to the Skeleton Coast Dunefield, here consisting mostly of barchanoidal forms. The dune sand has a distinctive yellow colour, and the dune surfaces are tinted red and black by a superficial lag of red garnet and black ilmenite and magnetite sand grains. The interdune areas in this vicinity consist of a peneplain rock floor and terraced slackwater flood deposits

Skeleton Coast Dunefield, an extensive linear dunefield corridor that originates near Torra Bay and continues more or less parallel to the coastline for several hundred kilometres into Angola (Figure 11.2). The dominant southerly gales drive the sand northwards in a series of separate barchans and linked barchanoid dunes. The width of the dunefield where it crosses the Uniab channel is about 7 km, and the height of the highest dune crests in the dunefield exceeds 50 m above the surrounding terrain of the Uniab drainage basin. During periods of prolonged drought, as during the past seven years, migrating dunes block the Uniab channel. Currently the main Uniab channel is open following recent flooding (March 1995), the highest dunes at present in the channel being generally less than 2 m high. The positions of five identified abandoned channels preserved on the eastern edge of the fan (Figure 11.3) indicate that at times the main channel of the Uniab River breached the dunefield in different places. The ages or causes of these avulsion events are not known at this stage.

FIGURE 11.3   *(opposite)* Channel systems of the Uniab River alluvial fan showing the currently active channel (channel 5) and its main distributary (channel 6); the remaining channels are at present abandoned. The main terrace scarps have been traced from aerial photographs of the area. Notable knickpoints in the channels are shown, the one in channel 5 being a 10 m waterfall. All the channel mouths, including the active mouth, hang between 1.5 m and 10 m above present sea level. The plan view of the fan is vaguely triangular but the lateral limits are difficult to recognize due to aeolian deflation and degradation. Only part of the Skeleton Coast Dunefield is shown. Its total width in this region, where it is crossed by the Uniab River, is about 7 km. Note the spread of the abandoned channels as they emerge from the dunefield cover. The history of avulsion is not known

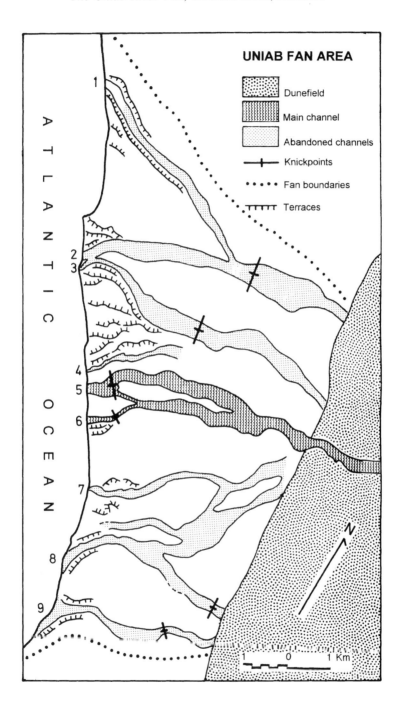

UNIAB FAN AREA

Dunefield

Main channel

Abandoned channels

Knickpoints

Fan boundaries

Terraces

ATLANTIC OCEAN

N

0 ···· 1 Km

RAINFALL AND DISCHARGE

Very little reliable information is available regarding rainfall and discharge in this desolate area. In general the annual rainfall is less than 100 mm, decreasing to less than 25 mm in the coastal zone. Heavy mist is common but rainfall is rare. During late summer (January to March), scattered thunderstorms locally produce runoff in the main drainage basin, not all of it necessarily reaching the sea.

The Uniab has occasionally been observed in flood. The depth of flow in the main channel more or less in mid-fan position is generally less than 40 cm, flowing over a width of a few hundred metres at the widest point. The peak stage seldom lasts more than a few hours, and the entire flood event rarely persists for more than a day or two. Geological information indicates, however, that the Uniab is capable of extreme flood events, during which gravel transport is the main mode. Boulders up to 4 m in maximum dimension are moved during such events.

UNIAB ALLUVIAL FAN

Morphology

The Uniab fan, an area of some 45 km^2, presumably has its apex hidden under the Skeleton Coast Dunefield (Figure 11.3). The lateral limits of the fan are vague. This may be due to degradation of the fluvial gravels by aeolian deflation. The plan view shows a poorly defined triangular pattern.

Along the Atlantic coastline the fan has been attenuated by vigorous wave attack. The coast is wave-dominated and micro-tidal. At the edge of the Uniab River fan a sea cliff some 14 km long forms a distinct convex scarp, rising up to 35 m elevation at its axis, and

FIGURE 11.4   View of the fan cliff face, looking north. Gravel lenses form the near-vertical outcrops. The rest of the slope is underlain by very weakly consolidated sand and silt. The silt layers are mostly less than 0.5 m thick and form subhorizontal drapes interbedded with the interfingered sand and gravel lenses. Desiccation structures are common in the silt layers

sloping gently down to sea level towards the north and south (Figures 11.4 and 11.5). The face of the cliff ranges in slope between about 35° and vertical. Active ongoing marine erosion of the fan front (Figure 11.6) indicates that the fan's observed radial length of some 5–6 km is not the original value. The inference is that the observed fan is a remnant of a larger fan, much of which has been removed by marine erosion, probably as a result of sea level rise of up to 130 m during the Flandrian Transgression (Miller, 1990; Van Andel, 1989).

The fan is unusual in that its regional radial gradient is generally between 1:110 and 1:80 (0.5–0.7°) (Figure 11.5). Most recognized alluvial fans have a radial slope along the surface of the fan cone at least double that value, and mostly much steeper (Blair and McPherson, 1994).

Apart from the currently active channel at least five other abandoned channels can be recognized. As a result of bifurcations of these channels eight or nine mouths reach the sea. All of them, including the active mouth, are hanging above sea level (Figure 11.5). The channel floor elevation of the mouths varies between about 1.5 m above spring high tide level for the active channel (Figures 11.7 and 11.8), to more than 12 m for some of the abandoned channels (Figure 11.9).

All the channels, except the oldest(?), most degraded channels, are marked by notably flat, sandy floors characterized by braided patterns (Figure 11.10). No conspicuous bedforms can be recognized. If the adjacent terraces are ignored, the channel floors are continuous and flat, channel-in-channel features being absent. Mass wasting of the channel walls produces a variety of mini-talus fans, slump debris fans and other small-scale features along the sides of the channel floors.

Longitudinal sections of the channels (Figure 11.5) reveal at least two generations of knickpoints in the channel floors (see also Figure 11.3). The more recent of the knickpoints is represented by a prominent waterfall in the main active channel (channel 5) and a rapids zone in the associated active distributary. The lip of the 10 m high waterfall marks in a spectacular manner the landward limit of headward incision in the active channel of the Uniab River. Locally the main channel is a narrow gorge only a few metres wide with a gradient of 1:33 (Figure 11.11); downstream the gorge widens rapidly to a steep-sided box canyon several scores of metres wide (Figure 11.7). The locality of both these knickpoints is controlled by outcrops of the Oswater Formation. The older generation of knickpoints occurs in the abandoned channels. The channel slope steepens locally to about 1:50, compared to the normal value of about 1:110, and marks the location of abandoned rapids. These knickpoints are controlled by resistant zones in the Precambrian floor rocks.

A most conspicuous feature of the Uniab fan is the large number of incised terraces (Figure 11.7), not only in the lower zone, where they are common, but also in the more inland area (Figure 11.12). The terraces occur as paired, single and complex terrace cusps. Two relationships with the parent channel have been recognized. narrow channel-parallel terraces (usually less than 50 m wide) that mark initial channel-in-channel incision; and fan-shaped terrace sets or cusps (100–600 m wide) that converge upstream, indicating systematic channel migration during episodic incision. The difference in elevation between adjacent terrace surfaces varies from a few centimetres to more than 5 m.

Most of the terrace surfaces are remarkably planar but some, mostly the larger (and older) terraces, reveal faint remnants of channel depressions and undulations. The terrace

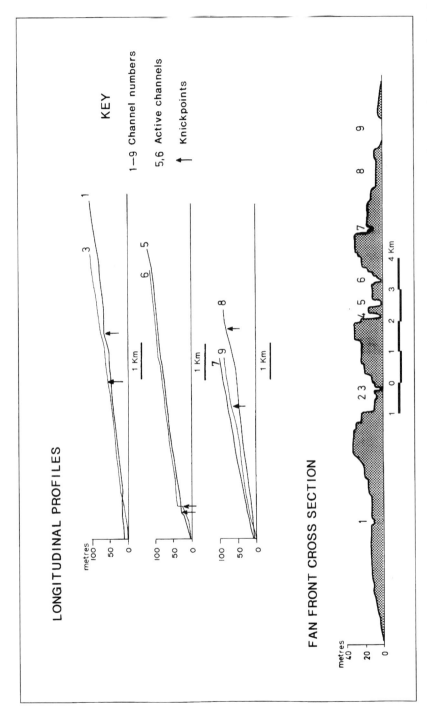

FIGURE 11.5   The upper part of the figure shows the longitudinal profile of all the main channels. The profiles are divided into three groups for clarity and comparison. Note the indicated knickpoints and the hanging mouths. The lower part of the figure displays the outline of the fan cliff face and shows the various channel mouths and terraces as exposed along the beach. The lenticular outline of the fan cliff face is conspicuous. Note that the currently active channel traverses the highest region of the fan cone

FIGURE 11.6   Mass wasting is a daily dynamic process at the Uniab fan cliff face, the foot of which is located at the level of storm wave run-up along the Atlantic Ocean. This coast is well known for its major storms. The cobbles on the gravel beach are almost exclusively basalt clasts weathered from the cliff face. This area is affected by a very strong northerly longshore drift. The combination of frequent storm events and the longshore current appears to sweep the base of the cliff face clean of slumped debris

FIGURE 11.7   The main active channel of the Uniab River near the mouth. Note the distinctive box-like channel with its flat floor. The channel in the near foreground (right) leads up to a 10 m high waterfall (upstream of the observer) at a notable knickpoint in the currently active thalweg. The skyline is formed by the typically flat terrace surfaces of the Uniab alluvial fan. The steps along the channel walls are remnants of previously more extensive paired terraces

FIGURE 11.8   Main active mouth of the Uniab River. The channel floor hangs above spring high tide level. The gravel berm is a small storm washover fan; this feature occurs in some of the other hanging channels as well

surfaces, though planar, show a notable gradient only slightly less than that of the associated channel.

The older terraces have been extensively deflated (Figure 11.12) by the southerly gales that commonly blow at more than 20 m s^{-1} during the windy season, leading to a well developed deflation armour and widespread development of ventifacts. The youngest terrace surfaces more clearly retain their original fluvial texture.

FIGURE 11.9   Hanging mouth of an abandoned distributary. The elevation of the degraded channel floor is about 10 m above sea level. Local thunderstorms raining on the fan area itself may cause short-lived discharge in this and other similar abandoned channels, but no water from the main active channel reaches these abandoned channels

FIGURE 11.10 Typical appearance of an active channel floor. This channel experienced minor discharge a few months before, hence the fairly well preserved in-channel braided pattern. In the background several minor terrace remnants are visible. The skyline marks one of the regional terraces in this area

Isolated terrace surfaces occur in the inland fan area on the seaward side of the Skeleton Coast Dunefield (Figure 11.13). They are prominent flat-topped hills, some with more than one terrace surface. They stand much higher above the regional marine platform than the other Uniab fan terraces but they consist of the same material as the rest of the Uniab fan. Their continuation with one another and their relationship with the Uniab fan terraces have not been clarified. They may not be part of the Uniab alluvial fan as described here; if they are, they represent a very large fan now almost entirely

FIGURE 11.11   The gorge immediately downstream of the main waterfall in the active channel of the Uniab River. During flood discharge conditions water may spill over the gorge wall as well

FIGURE 11.12   The very strong wind regime of the Skeleton Coast leads to a distinctive aeolian deflation armour forming on the main terrace surfaces. In addition, this terrace surface is marked by extra-large boulders of locally derived Precambrian granite–gneiss. The largest of these boulders may reach 2–3 m in diameter. In the far distance the even skyline is made by the next higher-lying terrace; the elevation difference between the two main terraces is about 4–5 m

FIGURE 11.13   In the foreground is exposed the rock pavement composed of Precambrian granite–gneiss that forms the floor of the Uniab alluvial fan. The floor is a marine-cut feature of Miocene age (or older). This rock peneplain continues underneath the Skeleton Coast Dunefield. The flat-topped hill is an erosional remnant of a probable older version of the Uniab fan; it consists of material identical to the present Uniab fan. The Skeleton Coast Dunefield barrier can be seen in the background. The active channel of the Uniab River crosses the dunefield a few kilometres north (to the left) of this site

destroyed by erosion. Thermoluminescence dating in progress may resolve some of these relationships.

Upstream of the Skeleton Coast dunefield barrier, extensive terrace surfaces occur in the rock-floored tributary valleys of the Uniab River and in the interdune valleys along the eastern boundary zone of the dunefield. These terraces are not part of the main Uniab alluvial fan, but they may have a significant relationship with the development of the fan itself. The terrace surfaces occur at several elevations, the highest terraces being several metres above the present main channel floor.

Fan composition

The material constituting the fan can be seen in detail along the dissected fan front (Figures 11.14 and 11.15). Other exposures occur in the sidewalls of the terrace incisions (Figure 11.15). The fan surface is generally degraded and extensively modified by deflation (Figure 11.12).

The fan sediment consists dominantly of unconsolidated sand and pebble- to boulder-sized gravel, interbedded in a complex arrangement of erosively based lenses. Mud-cracked silt layers commonly form thin blanket-like drapes that can be traced for several tens to scores of metres along the cliff face. A typical example of the fan's internal architecture can be illustrated by a measured section of the cliff face just south of the

FIGURE 11.14 Slackwater deposits in a minor tributary valley of the Uniab River upstream of the Skeleton Coast Dunefield. These distinctive fine-grained sand and silt deposits are distributed widely in the interdune valleys and other depressions and minor side valleys near the main Uniab channel east (upstream) of the barrier dunefield. The deposits occur up to several metres above the present active Uniab channel and record slackwater deposition during major flood events when the Uniab may have been dammed up against the dunes blocking flow down the main channel. In places, original depositional surfaces of the ponded deposits are exposed over large areas, showing the local distribution of the slackwater deposits. A variety of primary sedimentary structures can be recognized indicating depositional conditions ranging from lag gravel-bed deposition by fast-flowing streaming water, to suspension sedimentation in ponded water

FIGURE 11.15   Typical internal make-up of the Uniab fan. This outcrop is a side-wall of the main active channel. The composition varies considerably from place to place, and may range from near-massive gravel beds several metres thick, to clean sand with large cross-bed foresets up to 3 m high. Normally the beds are lenticular and interfingered. Not shown in this view are the common interbedded silt drapes with mud-crack structures

mouth of channel 3 (Figure 11.16). At this point the part of the fan exposed above low tide level is about 13 m thick. Locally, six main erosively based fining-up cycles can be recognized. Typically a cycle starts as coarse-grained, matrix-supported, poorly sorted gravel, and grades upwards into horizontally laminated or planar cross-bedded sand with interbedded pebbly layers. In other parts of the fan, drapes of mud-cracked silty lenses commonly terminate the cycles. The thickness of the cycles seen along the cliff face varies from slightly less than 1.5 m to about 4.0 m. Each of these cycles is interpreted as the depositional product of a single-event channel-fill process.

In the central part of the cliff face where the fan is at its thickest, lateral (across-stream) continuity of the bed sets ranges from a few tens of metres to about 150 m. The interbed relationship is complex, and interfingering, coarsening-up and fining-up cycles are common. Individual beds vary in thickness from a few centimetres to more than 4 m. Most of the thicker sand beds are devoid of internal structures but the thinner beds are commonly cross-bedded, usually planar cross-beds.

The gravel clasts are almost exclusively Etendeka basalt (Figure 11.6), the clasts being subrounded to rounded; minor rock types represented as clasts are Oswater Conglomerate blocks and Precambrian granite-gneiss, the larger clasts being subangular to subrounded. In places, especially towards the axis of the fan, massive, structureless pebble beds with rare cobbles form irregular, discontinuous, lenticular beds with faint indications of internal fabric. The basal contact is commonly erosive. Internal gravel–gravel contacts, not easily recognized, also display erosively based channel-like characteristics. Some internal layers, many inclined in the fashion of cross-beds, consist of somewhat better sorted pebbles and small cobbles, and show imbrication. Away from the axis of the fan, the gravel beds become better organized, better bedded and better sorted, and tend to form sheet-like, rather than lenticular, bodies.

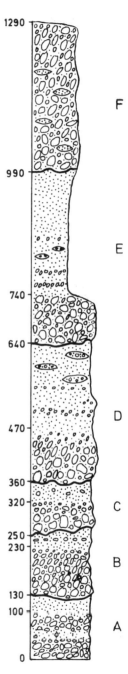

FIGURE 11.16 This typical measured section of the exposed cliff face just south of the mouth of channel 3 shows six main fining-upward cycles, labelled A–F, but further subsidiary units (not indicated on the diagram) can be recognized. The base of the 13 m thick section is at the beach, and the top is a local terrace surface. The numbers indicate stratigraphic thickness in centimetres above the base. The graphic symbols are mostly self-explanatory. In general the cycles are erosively based and start with coarse-grained (up to small boulders), poorly sorted, matrix-supported gravel indicative of debris flows and hyperconcentrated flows, grading upwards into sandy units with a variety of distinct stratification features indicative of traction and saltation sedimentation in normal low-concentration streams of flowing water. Lateral continuity of the individual cycles is commonly restricted to a few scores of metres along the cliff face. Not shown in this profile are the distinctive silt beds with desiccation features that elsewhere occur at the tops of the cycles

A minor but conspicuous component of the fan composition is blanket-shaped silt beds, mostly less than 50 cm thick, draped over the other sediments, but clearly filling extensive shallow depressions at the time of deposition; the terminal ends of the beds as exposed in the cliff face pinch out abruptly and unpredictably, and the beds are curved gently upwards. These silt beds are characterized by flat laminations, ripple laminations and desiccation cracks. Root impressions are rare.

The terrace surfaces are covered by angular basalt fragments that develop by *in situ* break-up of the original fluvial clasts through the action of salt crystal growth that takes place in the first few centimetres of the gravel. In places on the older terraces, large boulders up to 1 m in diameter, not only of basalt but also granite-gneiss, are now represented by an *in situ* mass of degraded rock particles that mark the original boulder exposed on the surface.

The terrace deposits in the area upstream of the Skeleton Coast dunefield are distinctive. They represent slackwater deposits, consisting mostly of fine-grained sand with flat laminations, and silt layers with distinctive ripple lamination, dewatering slump structures and desiccation cracks (Figure 11.14). Interbedded with the silt beds, but mostly on top of the terraces closest to the present channel, are comparatively thin beds of pebble-sized gravel, some showing imbrication of flattened clasts, and cross-bed laminations indicating gravel bar foresets.

In general the Uniab fan represents a relatively thin wedge-shaped veneer of sediment on the underlying gently sloping marine platform rock floor. No detailed measurements of sediment thickness have been made; observed maximum thickness is about 35 m and it seems unlikely that the maximum preserved thickness is more than 50 m.

The fan sediment is essentially unconsolidated and is freely undercut by vigorous surf runup at the base of the sea cliff (Figure 11.6). Mass wasting of the sea cliff surface continuously delivers debris to the beach below. We were impressed by the steady trickle of sand and clasts streaming off the cliff face simply as a result of surface moisture in the sand being driven off by the sun, the resulting cohesionless sand and gravel sliding and rolling downslope. The basalt pebbles and cobbles that arrive on the narrow beach below the cliff face are very quickly abraded and sorted by the vigorous surf action to produce an openwork gravel storm berm consisting of very well rounded clasts (Figures 11.4 and 11.6).

Some of the abandoned channels, for example channel 2, contain erratics up to several metres in dimension. These super-large boulders are, as expected, poorly rounded, and occur in isolation, or in clustered sets of boulders. The super-large boulders have not been seen in the cliff face exposures.

Upstream of the waterfall in the currently active main channel, a widespread waterline marked by plant debris flotsam and large boulders (up to 1.5 m in diameter) occurs at an elevation of about 2–3 m above the local channel floor (Figures 11.17 and 11.18). Locally the Oswater Formation forms an outcrop in the channel and surrounding area where the waterline material is deposited, and many of the large boulders appear to be derived locally. It appears as if floodwaters tend to dam up locally upstream of the waterfall knickpoint, leading to a notable fall zone for large clasts and flotsam debris along the waterline of the "dam". The plant flotsam consists of woody material, sticks and small trunks of as yet unidentified species. We speculate that the age of this material is of the order of a few decades at most, but no dating has been done to confirm this idea.

FIGURE 11.17   Channel of the Uniab River immediately upstream of the main waterfall knickpoint. Note the very large boulders scattered on the fan surface along the skyline; some are more than 2 m in maximum dimension. Driftwood flotsam is scattered all over this area along an apparent waterline located several metres above the thalweg. It appears as if these extra-large boulders and the flotsam driftwood represent deposits of one or more very large flood events. The organic material has not yet been identified or dated

FIGURE 11.18   Close-up view of one of the many super-large boulders found in the currently active channel. It is not known, however, when last these large boulders were moved, but plant flotsam debris that appears to be not older than a few decades at most, is associated with these large boulders

Fan age

Preliminary thermoluminescence dating on samples collected at the cliff face indicates that the base of the fan may be about 175 000 years old (Council for Scientific and Industrial Research (CSIR) dating laboratory reference number Pta 1-0045), and the top about 18 000 years (CSIR reference number Pta C4719). More analyses are in progress to improve the reliability of the dating. The plant material on the high waterline of the present active channel has not been dated but is likely to return a comparatively recent date (decades rather than centuries).

DISCUSSION

The hyper-arid Uniab River fan is controlled by two main forces: (1) the dunefield barrier across the main channel; and (2) the change in sea level during the Late Quaternary.

The very low gradient of the Uniab fan is atypical of alluvial fans in general, and it may be argued that the deposit is not a fan but simply a complex channel deposit. However, the regional triangular planview of the deposit, the nature of the sediment and its internal fabric, and the prominently convex outline of the fan surface as displayed by the sea cliff indicate that the deposit is a true alluvial fan. It is possible that its low slope is a combination of four effects: (1) the generally low gradient of the main drainage basin; (2) the discharge pattern in the hyper-arid drainage basin; (3) the absence of a tectonically rising front at the apex of the fan; and (4) the effect of the gently sloping marine platform on which the fan lies.

The role of the dunefield in the dynamics of the fan development may be pivotal. We suggest that the dunefield acts as a type of switch or filter in regulating the frequency and/or intensity of the floods. During times of prolonged drought, the migration of individual dunes in the 7 km wide dunefield eventually blocks the main Uniab River channel over a distance of scores to hundreds of metres, if not more. Depending on the extent of the blockage, flood discharge from the Uniab basin may be prevented from passing beyond the dunefield. A small discharge faced by a major dunefield blockage would probably simply deposit the sediment load in the flood lake dammed up behind the dunefield "dam wall". A major discharge could conceivably manage to breach the barrier by eventual overtopping, or more likely by piping, and, once flow is initiated, the unconsolidated sandy "dam wall" would be destroyed rapidly, leading to a mega-flood event magnified by the accumulated discharge stored in the floodwater lake.

The consequences of such a scenario are several.

(1) While the main channel passage is open, most flood events would reach the fan area relatively unimpeded, and, depending on the actual discharge, the action on the fan itself would vary from almost no effect (low discharge) to sedimentation and/or erosion (high discharge). The most recent known flood events appear to be mostly of the low discharge type. During this phase, fine-grained sedimentation may be more prevalent than coarse-grained deposition, leading to playa development rather than conventional channel and overbank deposition of gravel and sand.

(2) While the main channel passage is blocked, small discharge runoff from the main basin will have no effect on the fan itself, but will serve to build up slackwater lake

deposits on the upstream side of the dunefield barrier, now seen as the interdune and tributary valley terraces. During this phase the fan itself will be dormant, and subjected to a variety of degradation processes.

(3) The effect of a major runoff discharge from the basin will be magnified several times if the dunefield barrier across the main channel is sufficiently substantial to retard break-through long enough to accumulate a large volume of water behind the "dam wall". We suggest that such mega-flood events are responsible, amongst other things, for transporting the macro-boulders observed on the fan, and that they may also be responsible for excavating some of the terrace incisions. During this phase we envisage that the fan will be subjected to catastrophic changes, channel switching and avulsion, channel bifurcation, channel-in-channel erosion, overbank sedimentation, in-channel sedimentation, etc. The lake and slackwater deposits in the vicinity of the main channel upstream of the dunefield barrier will be extensively eroded during the accelerated drainage of the "dam"; hyper-concentrated streams and debris flows may form as a result.

The rise in sea level during the past 18 000 years or so, and the presence of an extensive set of terraces and other evidence of incision in the fan, appear to be in conflict. After all, a rise in sea level effectively raises the erosion base level and reduces fluvial gradient, leading normally to valley backfill, not to erosion and incision. The conspicuous sea cliff may, however, provide the key to resolving the conflict. We suggest that the marine surf environment of the Atlantic Ocean is so vigorous that the rising sea level managed to undercut the unconsolidated sediments of the originally larger Uniab fan at a much faster rate than the hyper-arid river system could manage to keep up with changing base level. In effect, therefore, the channel gradient in the lower sectors of the fan was increased, not reduced. The hanging river mouths are dramatic proof. Each of the now-abandoned channels was at one time the main active channel of the day, and each in turn developed a hanging mouth, the hyper-arid river losing the battle with the surf erosion driven by the rising sea level. This caused the channel to excavate the fan gravel in its vicinity as the channel strived to re-establish grade depending on the extent of undercutting and retreat of the fan front. The erodibility of the fan gravels, therefore, allowed the fan to be incised while sea level was actually rising. Pulsed mega-flood events of the type described above may also have been instrumental in excavating some of the terraces.

The knickpoints are indicators of the respective landward limits of retreat of fluvial incision initiated at the hanging river mouths as a result of the effective increase in gradient in the channels. The waterfall in the currently active channel is a particularly spectacular demonstration of the most recent incision.

The role of tectonic uplift in the evolution of the Uniab fan is unclear. It has been suggested (Van Zyl and Scheepers, 1991) that local uplift of the Uniab fan area can be deduced from the presence of a number of small faults noted in the Oswater Conglomerate that underlies part of the fan. This uplift, if contemporaneous with the deposition of the fan, would have caused incision of the channels, as well as the abnormal number of distributaries of the lower Uniab River, and the hanging mouths. However, we emphasize that in a conventional sense (the Death Valley alluvial fan model, for example) tectonic control is not significant in the formation of the main Uniab fan itself. There is no evidence in the Uniab drainage basin of a rising tectonic block that drives the slope differential between mountainland and adjacent lowland, a scenario distinctive of many other alluvial fans. We do not exclude, however, the possibility of minor modifications to the Uniab fan landscape under the influence of crustal flexures. Work continues to resolve the issue of tectonism versus sea level changes as the main driving force behind the configuration of the Uniab fan.

Finally, we postulate that hyper-arid fans located in the coastal zone are particularly susceptible to significant modifications due to marine effects associated with relative changes in sea level. This is because of the comparative disparity in energy levels and frequency of driving events in the hyper-arid fluvial environment and the contiguous oceanic littoral environment. In the case of the Uniab fan we suggest further that because of this disparity the fan was actively incised during a relative rise in sea level. Also, a unique condition applicable to the Uniab River, namely a dunefield barrier across its main channel, gives rise to the possibility of either starving the fan from time to time, or subjecting it to unusually magnified mega-flood events.

ACKNOWLEDGEMENTS

Our work is funded by the University of the Western Cape, the University of Port Elizabeth and the Foundation for Research Development. Dating was done by J.C. Vogel, CSIR, Pretoria. The rangers of the Skeleton Coast Park have provided generous permission for off-route entry into the park.

REFERENCES

Abrahams, A.D. and Parsons, A.J. (Eds) 1994. *Geomorphology of Desert Environs*. Chapman & Hall, London.

Beaty, C.B. 1963. Origin of alluvial fans, White Mountains, California and Nevada. *Association American Geographers, Annals*, **53**, 516–535.

Blair, T.C. 1987. Hydrologic and tectonic controls on alluvial fan/fan-delta flood basin sedimentation, southwestern United States, and their implications to the development of cyclic marginal fan and flood-basin sequences. *International Symposium on Fan Deltas*, Geological Institute, Bergen, Norway, 121–122.

Blair, T.C. and McPherson, J.G. 1994. Alluvial fans and their natural distinction from rivers based on morphology, hydraulic processes, sedimentary processes, and facies assemblages. *Journal of Sedimentary Research*, **A64**(4), 450–489.

Blissenbach, E. 1954. Geology of alluvial fans in semi-arid regions. *Geological Society of America, Bulletin*, **65**, 175–190.

Bull, W.B. 1963. Alluvial fan deposits of western Fresno County, California. *Journal of Geology*, **71**, 243–251.

Carryer, S.J. 1966. A note on the formation of alluvial fans, New Zealand. *Journal of Geology and Geophysics*, **9**, 91–94.

Church, M. and Ryder, J.M. 1972. Paraglacial sedimentation: a consideration of fluvial processes conditioned by glaciation. *Geological Society of America, Bulletin*, **83**, 3059–3072.

De Beer, J.H., Joubert, S.J. and Van Zyl, J.S.V. 1981. Resistivity studies of an alluvial aquifer in the Omaruru delta, South West Africa/Namibia. *Transactions of the Geological Society of South Africa*, **84**, 115–122.

Derbyshire, E. and Owen, L.A. 1990. Quaternary alluvial fans in the Karakoram Mountains. In A.H. Rachocki and M. Church (Eds), *Alluvial Fans – a field approach*. Wiley, New York, 27–54.

Gevers, T.W. 1936. The morphology of the western Damaraland and the adjoining Namib Desert. *South African Geography Journal*, **19**, 61–79.

Hallam, C.D. 1964. The geology of of the coastal diamond deposits of southern Africa. In S.H.

Haughton (Ed.), *Some Ore Deposits in Southern Africa, Vol. 2*. Geological Society of South Africa, Johannesburg, 671–728.

Hooke, R.L. 1967. Processes on arid-region alluvial fans. *Journal of Geology*, **75**, 438–460.

Jacobsen, P.J., Jacobsen, K.M. and Seely, M.K. 1995. *Ephemeral rivers and their catchments – sustaining people and development in Western Namibia*. Desert Research Foundation of Namibia and the Department of Water Affairs, Namibia, Windhoek.

Jeppe, J.F.B. 1952. *The geology of the area along the Ugab River west of Brandberg*. PhD Thesis, University of the Witwatersrand, Johannesburg, South Africa.

Kesel, R.H. 1985. Alluvial fan systems in a wet-tropical environment, Costa Rica. *National Geographic Research*, **1**, 450–469.

Lecce, S.A. 1990. The alluvial fan problem. In A.H. Rachocki and M. Church (Eds), *Alluvial Fans – a field approach*. Wiley, New York, 3–26.

Leckie, D.A. 1994. Canterbury Plains, New Zealand – Implications for sequence stratigraphic models. *American Association of Petroleum Geologists, Bulletin*, **78**, 1240–1256.

Leggett, R.F., Brown, R.J.E. and Johnson, G.H. 1966. Alluvial fan formation near Aklavik, NWT, Canada. *Geological Society of America, Bulletin*, **77**, 15–30.

Mabbutt, J.A. 1952. The evolution of the middle Ugab Valley, Damaraland, South West Africa/Namibia. *Transactions of the Royal Society of South Africa*, **33**, 334–366.

Maizels, J. 1990. Long-term paleochannel evolution during episodic growth of an exhumed Plio-Pleistocene alluvial fan, Oman. In A.H. Rachocki and M. Church (Eds), *Alluvial Fans – a field approach*. Wiley, New York, 271–304.

Miall, A.D. 1981. Alluvial sedimentary basins: tectonic setting and basin architecture. In A.D. Miall (Ed.), *Sedimentation and tectonics in alluvial basins*. Geological Association of Canada, Special Paper 23, 1–34.

Miller, D.E. 1990. A South African Late Quaternary sea-level curve. *South African Journal of Science*, **86**, 456–458.

Ono, Y. 1990. Alluvial fans in Japan and South Korea. In A.H. Rachocki and M. Church (Eds), *Alluvial Fans – a field approach*. Wiley, New York, 91–108.

Postma, G. 1990. Depositional architecture and facies of river and fan deltas: a synthesis. In A. Colella and D.B. Prior (Eds), *Coarse-grained deltas*. International Association of Sedimentologists, Special Publication 10, 13–27.

Ritter, D.F. and Ten Brink, N.W. 1986. Alluvial fan development in the glaciofluvial cycle, Nenana Valley, Alaska. *Journal of Geology*, **94**, 613–625.

Ryder, J.M. 1971. The stratigraphy and morphology of para-glacial alluvial fans in south–central British Columbia. *Canadian Journal of Earth Science*, **8**, 279–298.

Scheepers, A.C.T. 1990. *Landvorme en sandverplasing oor die Kuiseb delta* [*Landforms and sand movement of the Kuiseb delta, Namibia*]. MSc Dissertation, University of Stellenbosch, Stellenbosch.

Scheepers, A.C.T. 1992. Waaiervorme langs die Namibkuslyn [Fan-shaped forms along the Namib coastline]. *South African Geographer*, **19**, 91–105.

Stengel, H.W. 1964. *The rivers of the Namib and their discharge into the Atlantic I: The Kuiseb and Swakop*. Scientific Papers of the Namib Desert Research Station, Namibia, 22.

Stengel, H.W. 1966. *The rivers of the Namib and their discharge into the Atlantic II: Omaruru and Ugab*. Scientific Papers of the Namib Desert Research Station, Namibia, 30.

Van Andel, J.H. 1989. Late Pleistocene sea-levels and the human exploitation of the shore and shelf of southern Africa. *Journal of Field Archaeology*, **16**, 133–155.

Van Zyl, J.A. and Scheepers, A.C.T. 1991. Landforms and the geomorphic processes of the Uniab River mouth area, Namibia. *South African Geographer*, **18**, 31–45.

Van Zyl, J.A. and Scheepers, A.C.T. 1992. Quaternary sediments and the depositional environment of the lower Uniab area, Skeleton Coast, Namibia. *South African Journal of Geology*, **95**, 108–115.

Van Zyl, J.A. and Scheepers, A.C.T. 1993. Geomorphic history and landforms of the lower Koichab River, Namibia. *South African Geographer*, **20**, 12–23.

Ward, J.D. 1988. On an interpretation of the Oswater Conglomerate Formation, Kuiseb Valley, Namib Desert. *Palaeoecology of Africa*, **19**, 119–125.

Ward, J.D. and Von Brunn, V. 1985. Sand dynamics along the lower Kuiseb River. In B.J. Huntley (Ed.), *The Kuiseb Environment; the development of a monitoring baseline*. South African National Scientific Programmes Report 106, CSIR, Pretoria, 51–72.

Wells, S.G. 1977. Geomorphic controls on alluvial fan deposition in the Sonoran Desert, southwestern Arizona. In D.O. Doering (Ed.), *Geomorphology in Arid Regions*. Publications in Geomorphology, State University of New York, Binghamton, 27–50.

NOTE ADDED IN PROOF

A poster displayed at the 6th International Conference on Fluvial Sedimentology, Cape Town, 1997, by J.D. Ward, R. Swart and A. Bloem entitled "The Hunkab River flood of 1995: an example of dune–river interaction from northwestern Namibia" showed dramatic photographic proof of the "dune dam wall" hypothesis presented in our paper; the Hunkab River is the next river north of the Uniab River.

12

Downstream Adjustments in Allochthonous Rivers: Western Deccan Trap Upland Region, India

Leena A. Deodhar[1] and Vishwas S. Kale[2]

[1] *Department of Geography, S. P. College, Pune, India*
[2] *Department of Geography, University of Poona, Pune, India*

ABSTRACT

The rivers of the Deccan Trap region of India have some distinct characteristics with respect to flood hydrology and channel forms, owing to their monsoonal and allochthonous nature of flow. This paper describes the morphological and geomorphic characteristics of these allochthonous rivers. An attempt has been made to evaluate the at-a-station hydraulic geometry, and to discuss the downstream changes in the width–depth ratio at high and low flows. In view of the limited data, some generalizations have been made. It is suggested that the Trap rivers are decreasing their width–depth ratio during large floods, which could be one way of compensating for the downstream decrease in discharge.

INTRODUCTION

A river that flows across one major climatic zone for much of its course is described as autochthonous. In contrast, a river that is initiated in a high-rainfall area and flows through a dry area for much of its course has been referred to as allogenous, allogenic or allochthonous (Fairbridge, 1968; Kale, 1990a; Iriondo, 1993). In allochthonous rivers a large part of the drainage area is in the rainshadow and/or dry zone and so there is very little contribution to discharge from that part of the drainage basin. Such rivers adjust their hydraulic geometry in the upstream and downstream reaches to distinctly separate regime conditions. Many examples are known of basins which are dominated by the fluvial dynamics of allochthonous rivers: the Nile (Fairbridge, 1968); the western Deccan Upland rivers of India (Kale, 1990a); and the rivers of western Chaco, Argentina (Iriondo, 1993).

Relatively few studies consider the nature of channel adjustments in allochthonous rivers. The emphasis has been on the river channel morphologies of the perennial and

Varieties of Fluvial Form. Edited by A.J. Miller and A. Gupta
© 1999 John Wiley & Sons Ltd

seasonal rivers (Baker, 1977; Graf, 1983; Knighton, 1987; Kale, 1990a; Gupta, 1993, 1995). The major objectives of this paper are: (i) to describe the morphological and geomorphic characteristics of the allochthonous rivers of the western Deccan Trap Upland region; (ii) to evaluate the at-a-station hydraulic geometry; and (iii) to discuss the downstream changes in the form ratio at high and low flows.

THE STUDY AREA

The upper Krishna, upper Godavari and Bhima River basins form a part of the Deccan Peninsula of India. These rivers and their major tributaries emerge from the high-rainfall zone of the Western Ghats of India and traverse the semi-arid parts of the east (Figure 12.1). The rivers are deeply cut into the Cretaceous–Eocene basalts and/or late Quaternary alluvium. The source region and the extreme eastern parts of the study area exhibit relatively thin alluvial cover, but the area between the two is a belt of relatively thick deposits (Kale, 1990a). Therefore, bedrock control on river channels is stronger upstream as well as downstream, but less pronounced in the middle. The most striking features of the Trap rivers are their low-gradient, large valleys, markedly elongated basins (Table 12.1), and confined nature (Kale, 1990a). The upper reaches of the valleys of many rivers are unusually wide (Figure 12.2A). Each channel is cut deeply in bedrock (Figure 12.2B) or alluvium (Figure 12.3A), with gravel beds (Figure 12.3B) and localized basalt outcrops. On the basis of the channel boundary composition, the river channels can be classified as gravel-bed channels, with composite or cohesive banks (Figures 12.4A, B). Alluvial banks consist of calcareous sandy–silty alluvium with sufficient cohesion to stand for years in vertical cuts (Figure 12.4B). The rivers have well defined and stable courses, and usually have adequate channel capacity to accommodate large monsoon flood discharges.

The study region has a monsoonal-type climate, with the majority of precipitation falling in the wet southwest monsoon season (June–September), after a long dry season of about eight months. In the Western Ghat zone the rainfall exceeds 5000 mm annually. Towards the east, rainfall decreases rapidly, dropping to about 500 mm within a distance of 200 km (Figure 12.1). On occasions, the rainfall is heavy and vigorous in association with the cyclonic storms and depressions. Maximum one-day rainfall is about 12%, two-day about 18% and three-day about 23% of the mean annual rainfall (Dhar and Mhaiskar, 1970).

The hydrological record is generally short and discontinuous. However, gauge data for some sites are available for the last 10 to 40 years. Historical and gauge records show that very severe floods on different rivers of the Trap region were recorded in 1817, 1825, 1840, 1851, 1882, 1914, 1938, 1939, 1941, 1953, 1958, 1961, 1969, 1976, 1983 and 1994. In other words, one major high-magnitude flood occurs at least once every 10–20 years on one of the rivers. The average monsoon daily discharge on the rivers is highly variable, with the greatest flow following heavy monsoon spells. The daily mean monsoon discharge of 14–$434\,\mathrm{m^3\,s^{-1}}$ (unit discharges = 0.002 to $0.72\,\mathrm{m^3\,s^{-1}\,km^{-2}}$) increases by an order of magnitude during large floods. For such large floods the unit discharge ranges between 0.03 and $9.55\,\mathrm{m^3\,s^{-1}\,km^{-2}}$ and in exceptional cases increases to $23.04\,\mathrm{m^3\,s^{-1}\,km^{-2}}$. In the post-monsoon season daily

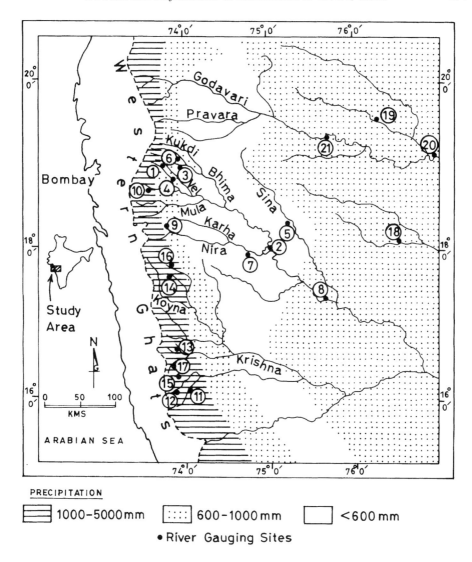

FIGURE 12.1   Location map of the study area. Numbers refer to gauging sites listed in Table 12.4

average flow drops to $<35\,\mathrm{m}^3\,\mathrm{s}^{-1}$ in the principal rivers (Figure 12.5A) and to zero in the lower-order streams (Figures 12.4A, B).

Over 95% of the annual suspension load and bedload is carried in three to four months of the monsoon season (Gole and McManus, 1988; Kale, 1990a). For the Trap rivers draining the Deccan Trap basalt region, the total suspended particulate matter of the wash load is primarily silt and clay. Available data indicate that in the Ghat zone the annual suspended sediment transport ranges between 36 and $275\,\mathrm{t\,km}^{-2}\,\mathrm{a}^{-1}$ and it occurs entirely within the four-month monsoon period (Gole and McManus, 1988). However, due to the allochthonous nature of the major rivers, the channel floor pebbles and cobbles become

TABLE 12.1 Hydrogeomorphic parameters of the Upland rivers (after Kale and Rajaguru, 1988)

River basins	Absolute relief (m a.s.l.) (source)	Basin relief (m)	Channel length (km)	Channel slope (%)	Valley floor slope (%)	Basin circularity (%)	Catchment area (km^2)
Godavari*	1277	831	321	0.24	0.32	65.3	18956
Pravara	1468	998	236	0.42	0.55	44.4	6220
Mula	1367	897	224	0.40	0.56	14.5	2896
Kukdi	856	294	105	0.28	0.37	38.2	1953
Ghod	1002	490	183	0.26	0.35	42.7	4704
Vel	1069	533	56	0.94	1.19	21.3	504
Bhima†	1060	621	353	0.27	0.33	69.6	22369
Indrayani	622	63	96	0.07	0.09	29.1	983
Mula-Mutha	1139	620	153	0.39	0.49	33.7	2925
Karha	989	418	83	0.51	0.65	32.4	1549
Nira	1189	657	206	0.31	0.39	34.9	6875
Krishna‡	1371	853	308	0.35	0.38	48.7	21609
Koyna	1372	798	156	0.50	0.69	34.2	2385
Venna	1371	770	56	1.36	1.66	16.8	214
Yerala	952	404	113	0.36	0.41	47.4	3236

* Up to Paithan;
† Up to confluence with Nira;
‡ Up to 75°E long

FIGURE 12.2 (A) View of the source of Puspavati River, a tributary of Kukdi River. Flow is towards the right. The Western Ghat scarp, which is the continental water divide, is clearly visible in the centre. Note the unusually flat valley floor right in the source of the river. A similar situation exists in many other valleys. (B) Rocky channel of Krishna River. Note the box-shaped nature of the channel. The width of the channel is roughly 22 m

FIGURE 12.3   (A) River Venna close to the source. Note the wide and open nature of the valley as well as the river channel. The channel width is about 56 m. The drainage area is about 195 km². A 3 m high terrace is visible on the left bank of the river. Flow is from left to right. (B) View downstream showing the wide and shallow gravel-bed channel of Kukdi River, with large channel bars and narrow thalweg. The average channel width at this reach is about 115 m and the catchment area is slightly less than 700 km². Alluvial banks are visible on the right

FIGURE 12.4 (A) View upstream of the wide, shallow channel of Vel River, with coarse bed material. The width of the channel is about 55 m and the drainage area is about 312 km². (B) View of the Markandi River, deeply cut into late Quaternary alluvium. The width of the channel is about 45 m and the upstream drainage area is approximately 310 km². The thalweg is indicated by light-coloured surface (due to deposition of calcium carbonate) and the point bar deposits are represented by dark gravels. Flow is from left to right

FIGURE 12.5   (A) The Krishna River at Sangli during low water flow. Average depth of water about 2 m. (B) High-water monsoon flow. Depth of flow is about 26 m. The view of high flow was taken about 20 m away and about 4 m higher than the low-flow position. Flow is from left to right. Channel width is about 240 m and catchment area about 11 125 km^2

important throughout the study area. The coarse sediment is brought in by the large number of tributaries rising from the divides in the semi-arid zone. An association between the channel type and the sediment size exists (Kale, 1990a). Bedrock channel reaches are characterized by coarser sediments, and alluvial channel reaches are dominated by relatively finer sediments.

THE ALLOCHTHONOUS RIVERS: RELATIONSHIPS BETWEEN DISCHARGE AND CONTROL VARIABLES

Downstream change in channel discharge is commonly associated with increasing drainage area (Knighton, 1987) and channel length (Kale, 1990a). A conservative relationship exists between the catchment area and discharge, that is, enlargement of drainage area is accompanied by a corresponding or proportional increase in discharge (Hack, 1957). The relationship between discharge (Q, in $m^3 s^{-1}$) and drainage area (A, in km^2) is expressed as (Alexander, 1972):

$$Q = cA^k \tag{12.1}$$

where c and k are regression constants. The exponent k usually ranges between 0.5 and 0.9 (Thomas and Benson, 1970; Alexander, 1972; Mutreja, 1986; Knighton, 1987). Since the ability of a stream to perform geomorphic work, expressed in terms of stream power, is the product of discharge, slope and water density, increasing catchment area should also reflect a corresponding or proportional increase in geomorphic effectiveness.

A series of bivariate relations was determined which describes the trend in discharge parameters as a function of catchment area and river length. Mean annual discharge (Q_m) and the peak discharge on record (Q_{max}) are used as indices of the hydrology. To describe and compare the downstream channel adjustments, the relations between Q_{max} and Q_m and the control variables were evaluated separately for the rivers of the western Deccan Trap Upland region (Trap), other rivers of the Indian subcontinent (INR) and other tropical rivers (OTR). Discharge data for INR and OTR were obtained from the UNESCO records of large floods, and from Stoddart (1969) and Sinha and Friend (1994). For the Trap rivers, annual peak discharge and 10-day discharge data were obtained for 21 gauging stations. The 10-day discharge is the average discharge from 1 to 10, 11 to 20 and 21 to the last day of every month, according to the definition of the Irrigation Department of Maharashtra State. The discharge data are available for at least the last 10 years, and in certain cases for more than 20 years.

Table 12.2 presents some comparative data for the Trap, INR and OTR rivers. The exponent values (k) for the Trap rivers are smaller than the rate of change for INR and OTR rivers (Figure 12.6). An identical, but inversed trend of relationship is indicated by the unit discharge–area relationship. This is to say that as more and more tributaries drain into the Ghat rivers the catchment area increases, but the discharge does not increase at the same rate. This further suggests that for a given catchment area the Trap rivers have lower discharges than the OTR and INR rivers. One can, therefore, infer that, other things being the same, the Trap rivers are less effective in terms of geomorphic work for a comparable drainage area.

Alluvial and semi-controlled rivers flow in self-formed channels and reveal a consistency of form in the downstream direction (Knighton, 1987; Kale, 1990a). The hydraulic geometry approach expresses the dependent variables, width, depth and

TABLE 12.2   Relation between discharge parameters and catchment area, for Trap, INR and OTR rivers.

Rivers	Discharge parameters $(m^3 s^{-1})$	Rate of change (k)	R^2
Deccan Trap rivers	Q_{max}	0.48	0.370**
	Q_m	0.55	0.240
	Unit Q_{max}	−0.57	0.379**
	Unit Q_m	−0.47	0.168
Other Indian rivers	Q_{max}	0.76	0.740*
	Q_m	0.85	0.720*
	Unit Q_{max}	−0.23	0.230*
	Unit Q_m	−0.14	0.059
Other tropical rivers	Q_{max}	0.64	0.613*
	Q_m	0.80	0.770*
	Unit Q_{max}	−0.36	0.330*
	Unit Q_m	−0.19	0.123*

* Significant at 0.01 level
** Significant at 0.05 level

velocity, as simple power functions of discharge (Knighton, 1987). In the absence of discharge data, catchment area and channel length are considered as surrogates for discharge (Hack, 1957) in order to understand the downstream adjustments in the channel parameters (Richards and Greenhalgh, 1984; Kale, 1990a).

Since the relationship between drainage area and Q_{max} is statistically significant (Table 12.2), the downstream adjustments in channel morphological properties can be evaluated. Table 12.3 and Figure 12.7 illustrate the nature of increase of width, depth, form ratio and channel gradient downstream. The rate of increase in width with channel length and catchment area is not statistically significant (Table 12.3). The poor correlation reflects the high longitudinal variability in the channel width. According to Kale (1990a), tributaries rising from the divides in the semi-arid zone create deviations in the channel form. Field observations at several locations show that tributary junctions are marked by fan-like deposits and a sudden change in the channel width and slope in the semi-arid parts (Kale, 1990a; Saxena, 1993) as well as in the upper reaches (Unde, 1996). Other factors responsible for the variability in width are: random variation in perimeter lithology, and abrupt changes in gradient in association with knickpoints and incised channels. For example, field investigations reveal the occurrence of large, shallow channels over resistant basalt flows. Similarly, coarse, unconsolidated sediments in the banks can also contribute to channel widening (Figure 12.3A).

Regression analysis also indicates that the unit stream power and channel boundary shear stress rapidly decrease in the downstream direction (Table 12.3), probably because the rapid decrease in channel slope is not compensated by a corresponding increase in discharge. Since the ability of a stream to erode and transport is directly related to the product of discharge and channel slope (and the unit weight of water, assumed constant),

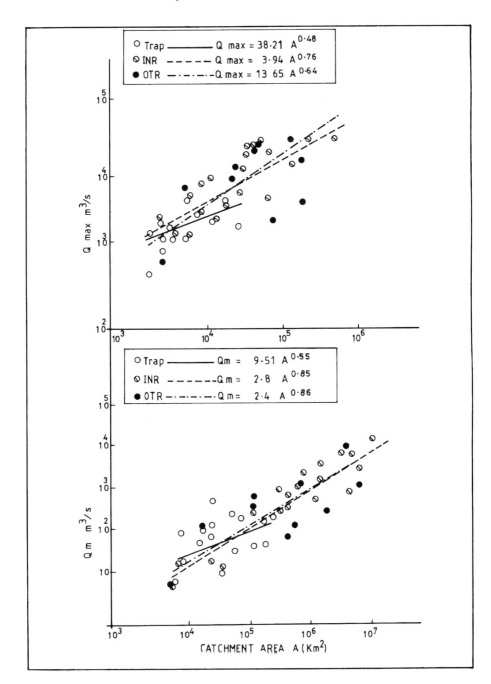

FIGURE 12.6 Relationship between catchment area and peak discharge for Deccan Trap rivers (Trap), other Indian rivers (INR) and other tropical rivers (OTR)

Varieties of Fluvial Form

TABLE 12.3   Relation between catchment area, distance downstream and channel parameters for Trap rivers

Channel parameter	Rate of change of drainage area vs other channel parameters	Rate of change of channel length vs other channel parameters
Width	0.312	0.21
Depth	0.099*	0.11*
W/D	0.199*	0.08
Slope	−0.708*	−0.75*
Unit stream power	−0.558*	−0.89*
Boundary shear stress	−0.630*	−0.49*

* Significant at 0.01 level

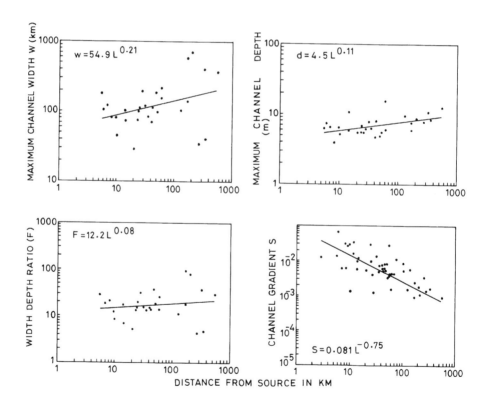

FIGURE 12.7   Plot of various channel parameters against channel length

any decrease in one parameter should normally be compensated by a corresponding increase in the other. This is to say that in order to maintain the unit stream power at approximately constant rate, these rivers should increase their channel slope or decrease their width–depth ratio. In the case of the Trap rivers there is no evidence to suggest that there is any tendency to steepen their slopes downstream. However, as discussed in the next section, it seems that the rivers are compensating for the decrease in discharge by decreasing their width–depth ratio downstream.

Catchment area and discharge are usually related, but in the study area the correlations are not stronger because the contribution from the tributaries to discharge downstream is not proportionate. Nevertheless, the moderate to high correlations between catchment area, channel length and some channel parameters indicate that the type and degree of adjustment to the current hydrological regime are not totally random, but semi-deterministic in nature.

AT-A-STATION HYDRAULIC GEOMETRY

At-a-station hydraulic geometry describes the manner in which the channel characteristics of width (b), depth (f) and velocity (m) adjust to changes in discharge (Leopold and Maddock, 1953). Therefore, to provide a simple summary of the relations among channel and flow characteristics for these allochthonous rivers, the hydraulic geometry equations for 21 gauging sites were calculated. The results presented in Table 12.4 reveal that for a

TABLE 12.4 Values of exponents in the relations for at-a-station hydraulic geometry of Trap river channels

Gauging site*	Width, b	Depth, f	Velocity, m
1. Chaskaman	0.120	0.431	0.439
2. Hingangaon	0.082	0.577	0.341
3. Kalamb	0.627	0.254	0.119
4. Khed	0.088	0.578	0.333
5. Kolegaon	0.186	0.466	0.347
6. Manikdoh	0.638	0.281	0.079
7. Sarati	0.110	0.481	0.408
8. Takali	0.130	0.431	0.438
9. Umbre	0.133	0.451	0.414
10. Velholi	0.364	0.341	0.294
11. Ajra	0.309	0.239	0.451
12. Anapwadi	0.417	0.286	0.212
13. Awarde	0.490	0.208	0.301
14. Gundhe	0.174	0.416	0.409
15. Mandukali	0.354	0.455	0.191
16. Parali	0.243	0.351	0.405
17. Patraychiwadi	0.158	0.295	0.546
18. Arudshahajani	0.216	0.585	0.198
19. Rajewadi	0.403	0.359	0.238
20. Nanded	0.243	0.634	0.123
21. Shagad	0.315	0.330	0.355

* See Figure 12.1 for location of gauging sites

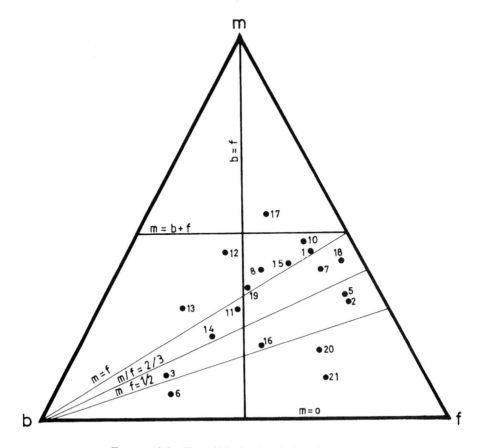

FIGURE 12.8   The width–depth–velocity (*b*-*f*-*m*) diagram

majority of stations, the rate of increase in mean depth (*f*) is usually greater than the corresponding rate of increase in width (*b*). Only four sites have significantly higher *b/f* values. Field observations at these sites indicate that the channels are unusually wide and shallow, and the entire channel is used only during large floods. Unusually wide channels and broad valleys (Figure 12.2A) in the source areas seem to be a common characteristic of many large rivers (e.g. Krishna, Indrayani, Kukdi, Godavari and Puspavati). Since there is insufficient geological evidence to suggest that basalt flows are more resistant in the source area, the only plausible explanation is that the rivers have been beheaded during the elevation and subsequent recession of the Western Ghats (Kale and Rajaguru, 1988).

Rhodes (1977) proposed a *b–f–m* diagram for achieving an empirical classification of channel cross-sections and for the interpretation of the at-a-station hydraulic geometry. Since the majority of points are clustered towards the '*f*-corner' on the ternary diagram (Figure 12.8), it can be inferred that the principal adjustment in channel form with increase in discharge is in the mean depth. Hence higher *f*-values imply greater values of shear stress (depth–slope product) and unit stream power (depth–slope times velocity). This in turn implies that there is a definite increase in the transportational and erosional capacity of the Trap rivers during large floods.

Another useful concept for evaluating the hydraulic efficiency of the channel is the width–depth ratio. Rivers of the seasonal tropics show a rapid decrease in the width–depth ratio with the arrival of high-magnitude floods, when the entire channel is used

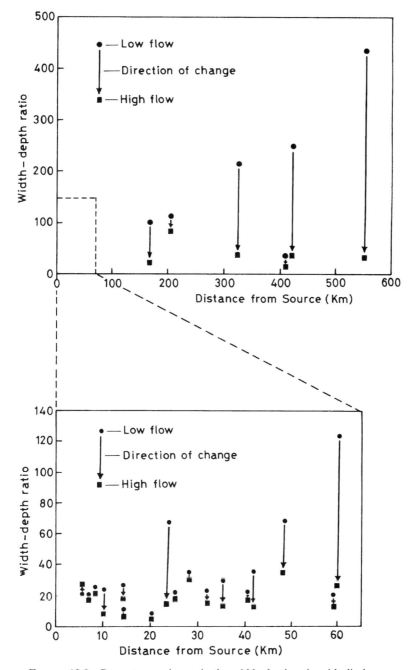

FIGURE 12.9 Downstream change in the width–depth ratio with discharge

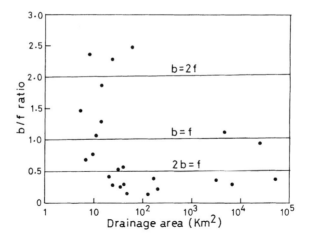

FIGURE 12.10   Plot of *b/f* (width–depth exponents) ratio against drainage area

(Gupta, 1995). Width–depth ratios were calculated separately for different peak discharges and were plotted against channel length and catchment area. The plot given in Figure 12.9 illustrates that, in the majority of cases, there is a significant decrease in the width–depth ratio with increasing discharge. The decrease is of a higher magnitude than shown by Gupta (1995) for some other Indian and Jamaican rivers. Rough estimates show that during large floods, a dramatic decrease in the width–depth ratio could raise the shear stress and power per unit area by a factor of 3 to 10. This in turn suggests that during high-magnitude floods the flows are capable of accomplishing a variety of geomorphic work, including transportation of coarse sediments and perhaps some erosion of bedrock. The rapid decrease in the width–depth ratio can be attributed to the incised and confined nature, and to the nested pattern (Gupta, 1995) displayed by the Trap channels.

The *b/f* ratios (ratio of width–depth exponents) presented in Figure 12.10 show that for larger catchment areas (>100 km^2), by and large, the ratio is equal to or less than unity. But for small catchment areas (<100 km^2) the ratio ranges from slightly above zero to 2.5. This is to say that whereas channels with smaller drainage areas have a higher rate of change in width, channels with larger catchment areas are characterized by a rapid increase in mean depth. Downstream, since the channels are deeply cut in rock or alluvium, as the discharge increases the depth increases more rapidly than width. The misfit nature of the channels in the headwaters has already been alluded to, and may account for the unexpected and anomalous relationship.

CHANNEL PROCESSES

Although an allochthonous flow regime characterizes the Trap rivers, they behave similarly to normal meandering rivers (Figure 12.11), but lack classic features associated with channel migration. All the rivers are generally sinuous and exhibit a pool–riffle bed

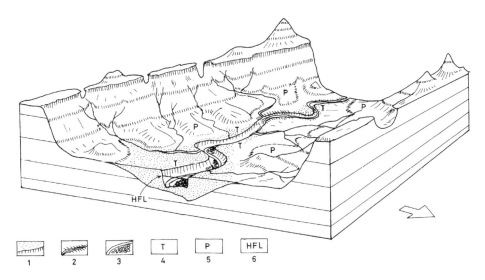

FIGURE 12.11   Schematic representation of the Trap rivers showing the setting of the incised river channel within an open and wide valley (not to scale). 1 = late Quaternary alluvium; 2 = bedrock channel; 3 = point bars; 4 = river terraces in Quaternary deposits; 5 = pediments or pediment terraces; 6 = high monsoon flood level

topography. Their sinuosity is highly variable, ranging between 1.1 and 1.3. Point bars consisting of coarse gravel occur along the concave banks (Figure 12.11) and bed armouring is common. There is no evidence that the rivers are cutting their outer banks or are shifting their courses. Whereas in the alluvial sections high cliffs in the late Quaternary deposits on both banks confine the channels, in the rocky stretches bedrock exposures restrict the movement of the channels (Figure 12.11). The adjacent terraces in late Pleistocene or mid-Holocene alluvium are generally 3–8 m high but sometimes they rise up to 10 m or more above the channel bed. The common high flows during the monsoon are incapable of filling the entire channel. Since the rivers can neither freely shift their channels nor overtop their banks, the normal floodplain formation processes do not operate. This situation is similar to other rivers of the seasonal tropics described by Gupta (1995). During the dry post-monsoon season, little or no erosion or transportation of sediments takes place.

   The channel forms and sinuosity appear to be a response to the seasonally high rate of water and sediment supply and variable bed and bank erodibility (Kale, 1990a). The channel bed sediments consist predominantly of pebble- and cobble-sized basalt clasts (Figure 12.3 and 12.4). These large sediments have a relatively high threshold of motion and control the bed morphology. During the height of the wet season, the channels are filled above point bar levels. The rivers submerge all the bedforms and carry large amounts of suspended sediments as well as bedload (pebbles and cobbles). Because of the box-shaped nature of channels (Figures 12.2B and 12.4A, B), as the flood level rises, width does not markedly increase as depth increases rapidly. Lower width–depth ratios imply greater values of flow velocity, boundary shear stress and unit stream power. The consequent increase in velocity and shear stress greatly increases the equilibrium rate of sediment transport, and cobble-sized sediments are also moved. In the absence of data regarding the water surface slopes, it is not possible to estimate precisely the flood power associated with different width–depth ratio values. Rough estimates indicate that the values of bed shear stress and unit stream power associated with normal high monsoon flows range between 4 and 250 N m^{-2}, and 11 and <800 W m^{-2}, respectively. Flows with unit stream power in this range can move coarse sediments. The rate of movement or residence time of these large particles is unknown. However, some experiments carried out by Rajaguru (1970) show that pebbles (*c.* 10 cm) move at the rate of *c.* 20 mm per day during the monsoon season. Observations of marked gravels in two rivers by the authors indicate that during peak monsoon floods, coarse gravel, measuring 12–20 cm in intermediate axis, is carried a maximum of 10 to 60 m downstream. Calculations based on Williams' (1983) empirical equations also suggest that the measured maximum velocities (usually between 1.5 and 2.5 but in exceptional cases as high as 3.0 m s^{-1}) are certainly sufficient to transport pebble- and cobble-sized sediments. The mean velocity, power per unit area and bed shear stress values, however, are not high enough to cause erosion of bedrock.

In bedrock reaches, which are essentially knickpoints or knickzones or simply discontinuities in gradient, the rivers flow through deep bedrock channels or gorges with cataracts and rapids. At some places the channels flowing over bedrock are very wide, with multiple channels and rocky islands. It seems likely that the rocky segments, because of their relatively greater thalweg gradients, are the sites of most efficient and intense bedrock erosion. The morphological conditions are favourable for the development of secondary circulation, flow separation, and the formation of vortices during large floods. The effects of such intense erosion are indicated by the presence of erosional features such as potholes, grooves, inner channels, scablands and boulder berms as well as by the near-absence of sediments on the channel floor. Data on bedrock erosion are lacking, but it is interesting to note that erosional features are conspicuously absent on the floor of artificially carved channels downstream of dams which were constructed in the last few decades. This implies that the process of bedrock erosion is very slow and the rate is negligible over a period of a few decades.

In normal alluvial rivers, the major sources of sediments are tributaries and the sediments on the bed and banks. In such rivers, bank erosion occurs by shearing through hydraulic action or through liquefaction of bank sediments. Although data on bank erosion are not available, it may be assumed that, as there is no evidence of significant bank erosion and channel widening, these processes are ineffective. It therefore appears that the common monsoon high flows are only capable of moving large volumes of silt and clay in suspension, and pebble- and cobble-sized sediments as bedload, over short distances, but are totally incapable of eroding the alluvial banks.

Cross-sectional surveys during the pre- and post-monsoon season in two rivers – Urmodi (humid region) and the Vel (semi-arid region) – show evidence of marginal to noteworthy scouring and deposition on the channel bed (Ghodke, 1989; Kale et al, in prep.). Similarly, longitudinal surveys of Mutha River carried out in 1966/67 and again in 1993/94, show noteworthy changes in the channel bed topography (CWPRS, 1994). Though systematic data on bed and bank erosion on the regional scale are lacking, it may be assumed that major changes in the bed morphology occur only during rare, high-magnitude floods that occur once every 10–20 years or so.

In spite of the allochthonous nature of the rivers, there are no significant changes in the channel forms, channel patterns and channel bed material type in the semi-arid parts. Throughout the area the rivers are characterized by gravel-bed channels. The rivers are not truly braided in any reach; however, downstream of tributary confluences, some degree of braiding is observed at moderate flows. At very low discharges the flows follow the lowest points of the thalweg and at high flows all the bedforms are submerged.

Given the decreasing nature of discharge, the Trap rivers have to either increase their channel slope or decrease their width–depth ratio to maintain the unit stream power at a constant rate. Since there is no tendency to increase the channel slope, it is suggested that the rivers have adjusted to the diminishing discharge by decreasing their width–depth ratio (Figure 12.10), thus maintaining their hydraulic efficiency and geomorphic effectiveness at the level required to sustain water and sediment flux. However, the occurrence of fan-like deposits at the junction of tributaries in the semi-arid parts may reflect the fact that the increase in efficiency during common monsoon floods is insufficient to cause substantial transportation of coarse sediments and erosion of banks.

There is enough evidence from the study area to suggest variability in stream behaviour and morphology in the past. Kale and Rajaguru (1987) have summarized the fluvial records and have inferred aggradation during a period of reduced runoff between >30 ka and 10 ka BP. From 17 ka to 10 ka BP the aggrading rivers changed from coarse-bedload streams to suspended-load streams. Early Holocene incisional and excavational activity persisted until about 4.5 ka BP, when a phase of overbank sedimentation generated a lower inset fill, and after 3 ka BP the streams were characterized by general incision. This phase of incision has continued until the present. There is no known evidence of palaeo-channels. However, recently a palaeochannel of the Kukdi River was located near Bori, which is clearly defined by volcanic ash (Kale, 1990b). The volcanic ash is believed to be 74 ka in age (Acharyya and Basu, 1993). The palaeochannel is highly sinuous and narrow (17–24 m in width). In comparison, the modern channel is three to four times wider. It suggests a low-energy environment and a very different hydraulic regime from the present, as evident from unusually thick clayey deposits associated with the ash and the palaeochannel (Kale, 1990b). Similar evidence has also been found in Karha Valley (Kale et al., 1993).

Several field visits in the last 15 years or so have not revealed any evidence of notable changes in the channel form, dimensions and sinuosity. This is suggestive of the general stability of the channels. One of several reasons is a general decrease in the magnitude of large floods due to the construction of dams in the source areas of almost all the large rivers. For instance, statistical analysis using annual maximum series data from 1940 to 1992 shows that there are significant differences between the pre- and post-dam flood magnitudes recorded on the Mutha River (CWPRS, 1994). Therefore, it appears that the primary function of the modern channels is to convey sediment and water discharge during the monsoon season, and under the present hydrological regime they are incapable of significantly modifying or altering their channel forms and dimensions.

CONCLUSIONS

The rivers of the Deccan Trap region of India have some distinct behavioural characteristics with respect to flood hydrology, channel forms and denudational capabilities, owing to their monsoonal and allochthonous nature of flow. Since the rivers flow through low-rainfall areas for much of their courses, there is very little contribution to discharge downstream. Therefore, for a given catchment area the Trap rivers could be geomorphologically less effective than the non-allochthonous rivers that are character-ized by a corresponding or proportional increase in discharge downstream. However, it appears that the rivers are decreasing their width–depth ratio during large floods, which could be one way of compensating for the decrease in discharge. This suggests that large flows are the dominant control on many aspects of channel morphology in the seasonal tropics (Gupta, 1995). Furthermore, the low to moderate correlations between channel morphological parameters and controlling factors suggest the influence of several other factors: random variations in perimeter lithology, temporal ordering of flood events and imposition of coarse sediments by tributaries (Kale, 1990a). This has significant implications for Quaternary geomorphology, applied hydrology and palaeoflood hydrology.

Considering the fact that the unit stream power, which expresses the rate of doing geomorphic work, decreases downstream more rapidly than expected for perennial rivers, one can possibly account for the preservation of relatively thick Quaternary deposits in the middle reaches. The ineffectiveness of the higher-order streams to transport sediments deposited by the tributaries heading in the semi-arid parts also lends support to the inference regarding the decreasing competence of flows along the rivers. As for channel competence, the channel gradient is also an important controlling variable, but in the case of the study area the lower rate of increase in discharge with catchment area has a more important bearing on the channel's capacity to do erosional and transportational work. It is of utmost importance, therefore, that besides the fluvial geomorphologists, the hydrology and engineering communities in India should recognize the distinct behavioural characteristics of the Deccan Trap rivers while estimating the flood parameters. The results of the analyses presented should provide a basis for the design of stable, allochthonous channels with gravel beds.

ACKNOWLEDGEMENTS

The authors thank the Irrigation Department of Maharashtra State and the officials of the Central Water Commission for providing all the necessary data. Avijit Gupta and Andy Miller are especially thanked for reading earlier versions of the paper and for making helpful comments. Their suggestions significantly improved the paper.

REFERENCES

Acharyya, S.K. and Basu, P.K. 1993. Toba ash on the Indian subcontinent and its implications for correlation of late Pleistocene alluvium. *Quaternary Research*, **10**, 10–19.

Alexander, G.N. 1972. Effect of catchment area on flood magnitude. *Journal of Hydrology*, **16**, 225–240.

Baker, V.R. 1977. Stream channel response to floods, with examples from central Texas, *Geological Society of America, Bulletin*, **88**, 1057–1071.

CWPRS. 1994. *Mutha River Improvement Project Water Surface Profiles*. Central Water and Power Research Station, Pune, Technical Report No. 3159, 7–8.

Dhar, O.N. and Mhaiskar, P.R. 1970. A study of maximum point and areal rainfall in the Bhima basin up to Ujjani dam site. *Indian Journal of Meteorology and Geophysics*, **21**, 451–458.

Fairbridge, R.W. 1968. *The Encyclopaedia of Geomorphology*, Vol. III. Reinhold, New York, 284 pp.

Ghodke, B.D. 1989. *Variation in Bed-forms and Microrelief along Channel Bed: A case study of Urmodi Channel at Parali, Satara*. Unpublished MPhil thesis, University of Pune, Pune.

Gole, S.P. and McManus, J. 1988. Sediment yield in the upper Krishna Basin, Maharashtra, India. *Earth Surface Processes and Landforms*, **13**, 19–25.

Graf, W.L. 1983. Flood related channel changes in an arid region river. *Earth Surface Processes and Landforms*, **8**, 125–139.

Gupta, A. 1993. The changing geomorphology of the humid tropics. *Geomorphology*, **7**, 165–186.

Gupta, A. 1995. Magnitude, frequency, and special factors affecting channel form and processes in the seasonal tropics. In J.E. Costa, A.J. Miller, K.W. Potter and P. Wilcock (Eds), *Natural and Anthropogenic Influences in Fluvial Geomorphology*. American Geophysical Union, Washington, DC, Monograph 89, 125–136.

Hack, J.T. 1957. Studies of longitudinal stream profiles in Virginia and Maryland. *US Geological Survey Professional Paper*, 294–B.

Iriondo, M. 1993. Geomorphology and late Quaternary of the Chaco, South America. *Geomorphology*, **7**, 289–303.

Kale, V.S. 1990a. Morphological and hydrological characteristics of some allochthonous river channels, Western Deccan Trap Upland Region, India. *Geomorphology*, **3**, 31–43.

Kale, V.S. 1990b. *Flood Geomorphology of the Allochthonous River Channels of Western Upland Maharashtra, India*. Geological Society of India, Memoir, 18, 94–104.

Kale, V.S. and Rajaguru, S.N. 1987. Late Quaternary alluvial history of the northwestern Deccan upland region. *Nature*, **325**, 612–614.

Kale, V.S. and Rajaguru, S.N. 1988. Morphology and denudation chronology of the coastal and upland river basins of western Deccan Trappean landscape (India): a collation. *Zeitschrift für Geomorphologie*, **32**, 311–327.

Kale, V.S., Patil, D.N., Pawar, N.J. and Rajaguru, S.N. 1993. Discovery of a volcanic ash bed in the alluvial sediments at Morgaon, Maharashtra. *Man and Environment*, **18**, 141–143.

Kale, V.S., Gupta, A., Hire, P. and Kirkute, A. (in prep.). Monsoon floods and changes in the channel characteristics of the Vel River, Bhima basin, India.

Knighton, A.D. 1987. River channel adjustment – the downstream dimension. In K. Richards (Ed.), *River Channels Environment and Processes*. Blackwell, Oxford, 95–128.

Leopold, L.B and Maddock, T. 1953. The hydraulic geometry of stream channels and some physiographic implications. *US Geological Survey Professional Paper* 252. 1–57.

Mutreja, K.N. 1986. *Applied Hydrology*. Tata McGraw Hill, New Delhi, 676 pp.

Rajaguru, S.N. 1970. *Studies in Late Pleistocene of Mula-Mutha Valley*. PhD Thesis, University of Pune, Pune.

Rhodes, D. 1977. The b-f-m diagram: representation and interpretation of at-a-station hydraulic geometry. *American Journal of Science*, **277**, 73–96.

Richards, K. and Greenhalgh, C. 1984. River channel change: problems of interpretation illustrated by the river Dervent, North Yorkshire. *Earth Surface Processes and Landforms*, **9**, 175–180.

Saxena, H. 1993. *Channel morphology and hydrological characteristics of the Kukdi River, Maharashtra*. PhD Thesis, University of Pune, Pune.

Sinha, R. and Friend, P.F. 1994. River systems and their sediment flux, Indo-Gangetic plains, North Bihar, India. *Sedimentology*, **41**, 825–845.

Stoddart, D.R. 1969. World erosion and sedimentation. In R.J Chorley (Ed.), *Water, Earth and Man*. Methuen, London, 43–64.

Thomas, D.M. and Benson, M.A. 1970. Generalization of streamflow characteristics from drainage-basin characteristics. *US Geological Survey Water-Supply Paper*, 1975, 1–55.

Unde, M. 1996. *The problem of within channel siltation downstream of Koyna Dam*. PhD Thesis, University of Pune, Pune.

Williams, G.P. 1983. Palaeohydrological methods and some examples from Swedish fluvial environments. I. Cobble and boulder deposits. *Geografiska Annaler*, **64A**, 227–243.

13

Debris Flow and Sheetflood Fans of the Northern Prince Charles Mountains, East Antarctica

JOHN A. WEBB[1] AND CHRISTOPHER R. FIELDING[2]

[1] *School of Earth Sciences, La Trobe University, Bundoora, Australia*
[2] *Department of Earth Sciences, University of Queensland, Australia*

ABSTRACT

Small fans in the Beaver Lake area, northern Prince Charles Mountains, have formed in the very cold, arid Antarctic climate by both debris flow and sheetflood processes active during the summer thaw (rainfall is completely lacking). Individual fans are dominated by only one of these processes. Thick, coarse debris flows and thinner, finer-grained, more dilute flows generate fans in areas where there is an abundance of till. The relatively low slope of the debris flow fans (5°) reflects the low catchment topography (120 m). None of the debris flow fans shows fan-head incision, indicating that this feature is not always part of the development of small alluvial fans.

The single sheetflood fan has a relatively small area given the size of its catchment, reflecting low sediment availability compared to the supply of water during flooding events. Debris flow lobes at the fan apex do not indicate a general downfan progression from debris flow to sheetflood deposits; instead their position is due to the fortuitous presence of a source of poorly sorted sediment (till) on either side of the fan apex.

All fan surfaces have experienced a small amount of post-depositional modification. The most obvious changes are evident on the toes of the debris flow fans, where wetting–drying cycles caused by the tidal waters of Beaver Lake have disintegrated the sandstone clasts of the very poorly sorted debris flows, giving a sandy sediment completely unlike the unmodified fan deposits.

The Beaver Lake fans have some characteristics that reflect the cold, arid climate in which they have formed. Occasional freeze–thaw cracks disrupt the surface pavements, and till is a component of all fans. It can be identified by its mixture of clast sizes and types (many not locally derived). Biological activity is absent and rates of chemical weathering are low, so there is minimal soil development. The resultant low availability of sediment may be responsible for the small size of the fans. However, the processes that constructed the Beaver Lake fans (debris flows and sheetfloods) are not diagnostic of the cold, arid climate.

Varieties of Fluvial Form. Edited by A.J. Miller and A. Gupta
© 1999 John Wiley & Sons Ltd

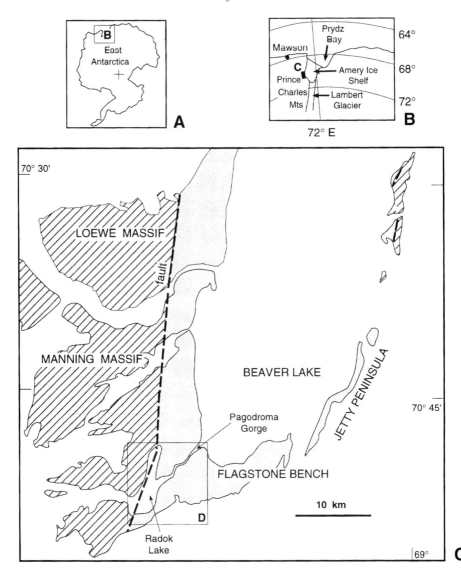

FIGURE 13.1   Location of debris flow and sheetflood fans. Debris flow fans are numbered 1 to 8 from east to west; those fans identified by number are described in the text. For detailed maps of representative debris flow fan (fan 4) and sheetflood fan (Glossopteris Gully fan) see Figures 13.3 and 13.10, respectively

INTRODUCTION

Alluvial fans have been studied in a variety of climatic regimes: hot arid and semi-arid climates (e.g. Bull, 1964; Denny, 1965; Hooke, 1967; Wasson, 1979; Harvey, 1989; Ritter et al., 1995), humid–temperate climates (e.g. Schumm, 1977; Pierson, 1980; Blair, 1987; Clarke, 1987; Wells and Harvey, 1987; Kochel, 1992), humid–tropical climates

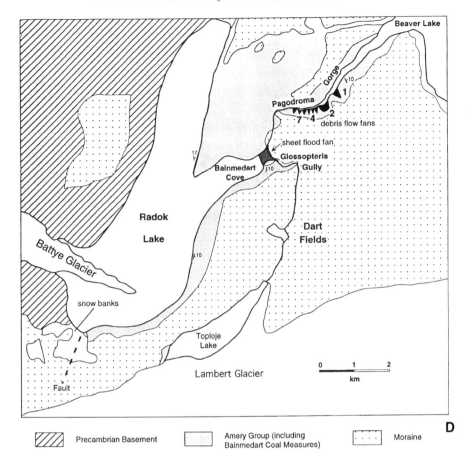

FIGURE 13.1   (*continued*)

(e.g. Brierley et al., 1993), and humid–glacial and paraglacial climates (e.g. Ryder, 1971; Brazier et al., 1988; Eyles and Kocsis, 1988; Krainer, 1988; Catto, 1993). However, the influence of climate on fan formation is hotly debated; some authors believe that changes in climate will be preserved as differences in fan facies and morphology (e.g. Nemec and Postma, 1993; Dorn, 1994; Ritter et al., 1995), whereas others have suggested that climate is rarely the main factor governing alluvial fan characteristics (Blair and McPherson, 1994).

Thus it is important to have a good understanding of fans that have formed under all climatic regimes, and detailed descriptions of alluvial fans from cold arid regions like Antarctica are lacking. Antarctica is a cold desert; the climate lacks rainfall and over most of Antarctica the annual snow balance is less than $10 \, \mathrm{g \, cm^{-2}}$, making the continent one of the most arid regions in the world (Budd, 1981). Studies on fluvial processes in the Dry Valleys of the Trans-Antarctic Mountains have shown that sediment transport rates are two orders of magnitude less than in humid–glacial Arctic and Alpine regions (Rains et al., 1980; Mosley, 1988). Alluvial fans have been recorded in ice-free areas of Antarctica (Calkin and Nichols, 1972; Mosley, 1988; Adamson, 1989), but have not previously been studied in detail. Fan deposition is due to the annual summer thaw, and the fan surfaces are frozen and preserved intact for the remainder of the year. There is, as a result, relatively little of the reworking by secondary processes that dominates fan surfaces elsewhere in the world, masking the primary fan-building mechanisms. In addition, there is little biological activity (mosses and lichens are the only plants throughout most of Antarctica), so weathering and soil development are minimal. Thus the Antarctic fans offer an opportunity to assess the imprint of a cold, arid climate on fan morphology and sedimentology.

Alluvial fans in the northern Prince Charles Mountains of East Antarctica were studied during an Australian National Antarctic Research Expedition over the summer of 1989/90, based at a temporary field camp on the southeastern edge of Glossopteris Gully fan (Figure 13.1). The surface morphology of the fans was surveyed by compass, clinometer and tape. Changes in the fans were observed over a four-week interval in January 1990, corresponding approximately to the summer thaw. Sediments exposed on the fan surfaces were described and sampled for textural analysis, and clast macro-fabrics were measured at selected locations.

REGIONAL SETTING

Geomorphology

The study area is located close to Beaver Lake in the northern Prince Charles Mountains, East Antarctica (Figure 13.1), about 350 km southeast of Mawson research base. Beaver Lake lies on the western side of the Lambert Glacier, generally recognized as the largest valley glacier in the world, as much as 80 km wide and more than 500 km long.

Beaver Lake is ice-covered and tidal; it is connected beneath the Amery Ice Shelf to Prydz Bay. On the southwestern edge of Beaver Lake is the smaller, land-locked Radok Lake (Figure 13.1), which is steep-sided and over 300 m deep (Adamson and Darragh, 1990). The two lakes are connected by Pagodroma Gorge (Figure 13.2); on the southern side of the central part of the gorge are several small debris flow fans.

On the eastern side of Radok Lake is a large inlet, Bainmedart Cove (Figure 13.1), notably shallower than the main north–south part of the lake. Flowing into the eastern end of this inlet is a meltwater stream, fed by two lakes to the south. The stream has incised a small gorge, Glossopteris Gully, at the mouth of which an extensive alluvial fan has developed.

Geology

Late Permian and Triassic strata of the Amery Group crop out extensively around the shores of Beaver Lake. The Amery Group comprises, in ascending stratigraphic order, the Radok Conglomerate, Bainmedart Coal Measures, and Flagstone Bench Formation (McKelvey and Stephenson, 1990; Fielding and Webb, 1996; McLoughlin et al., 1997). The Late Permian Bainmedart Coal Measures, which form the cliffs along the western side of Beaver Lake and eastern side of Radok Lake, are dominated by coarse-grained light yellow–brown quartzose and feldspathic sandstone, with minor proportions of siltstone, coal and claystone.

FIGURE 13.2   View eastwards down Pagodroma Gorge, which is flooded by the tidal waters of Beaver Lake. To the right (south) are buttresses of Bainmedart Coal Measures overlain by Pagodroma Tillite; debris flow fans (partly obscured) lie between the buttresses, below the headwall snowfield

The Amery Group overlies and is faulted against granulite facies Proterozoic metamorphics, mostly felsic granulite and gneiss (McKelvey and Stephenson, 1990). These crop out to the west and south of Beaver Lake (Figure 13.1).

Unconformably overlying both Amery Group strata and Precambrian metamorphics is an irregularly distributed blanket of semi-lithified tillite up to 100 m thick, called the Pagodroma Tillite (McKelvey and Stephenson, 1990). The tillite contains Late Miocene and Middle Pliocene micro-fossils, and is probably Late Pliocene or Early Pleistocene in age. Younger glacial moraines are of relatively limited extent around Radok Lake.

Climate

In the Prince Charles Mountains, precipitation (snowfall) is very low, and typical of Antarctica as a whole (less than 5 cm per year). South-southwesterly katabatic winds blow all year round; they originate over the centre of the continent as temperature inversions, and blow more or less directly down the topographic gradient to the coast. Even in summer, strong winds (>20 knots) are recorded on most days, and gusts can reach 60 knots. In midsummer, temperatures at low elevations can rise above 0°C during the day, melting the ice at the edges of Radok and Beaver Lakes, and generating surface streams down Glossopteris Gully and Pagodroma Gorge. This period of peak melt activity does not usually exceed two months, and in colder summers the streams and lakes remain frozen. Humidity in summer ranges from 40 to 60%, so evaporation is significant despite the low temperatures.

PAGODROMA GORGE FANS

Pagodroma Gorge

Pagodroma Gorge is a narrow, steep-sided valley which runs northeast from Radok Lake (7 m above sea level) to Beaver Lake (at sea level; Figure 13.1), and extends underwater beneath Beaver Lake (Adamson, 1989). The V-shaped gorge (Figure 13.2) was clearly cut by a river, but there is little active fluvial erosion now. The northeastern portion of the gorge is flooded by Beaver Lake (Figure 13.2), which has a measured tidal range here of about 1.5 m; the water is only slightly saline.

The central east–west part of the gorge is cliff-lined on the northern side (Figure 13.2), with scree slopes and occasional debris flows along the foot of the cliffs. Mass-wasting processes are dominated by rockfall, due largely to frost wedging. Exposed sandstone faces frequently exhibit honeycomb weathering, perhaps reflecting saline groundwaters due to the low humidity.

In contrast to the northern side, the steep southern wall, which rises about 120 m, is lined by eight small alluvial fans (Figure 13.2). Extensive permanent snowbanks cover the uppermost 10–20 m of this wall; the snow has been blown from the plateau to the south by the katabatic winds. Rock buttresses 40–50 m high separate the alluvial fans; these are composed of gently dipping sandstones of the Bainmedart Coal Measures. Overlying the sandstones is the Pagodroma Tillite, which reaches a maximum thickness of about 60 m at the eastern end of this part of the gorge; the till weathers to form relatively gentle slopes on top of the sandstone cliffs.

Morphology of debris flow fans

Most of the debris flow fans are single bodies 70–80 m in width and 40–50 m from apex to toe (Figures 13.3 and 13.4A), i.e. total areas of 3000–4000 m^2. Some form arc-shaped projections into the waters of the gorge, but most end in a more-or-less straight line (Figure 13.3). Towards the east is a larger fan, 80–90 m from apex to toe and over 150 m wide (fan 2, Figure 13.1). The apex of each fan marks a substantial change in slope (Figures 13.3 and 13.4A); above the apex the slope averages 25–30°, until it finally steepens at the snowbanks, which may have overhanging cornices. On the fans themselves the slope averages 5–6°; they have very irregular surfaces, with a variation in surface elevation of up to 2 m. The cross-fan profiles are shallowly convex-up. The longitudinal profiles are broadly concave-up; superimposed on this are the convex-up mounds of large debris flows (Figure 13.3). None of the fans shows fan-head incision (Figures 13.4A,B), so there are no intersection points, and the long profiles are not divided into segments. No macroscopic biological activity is evident; even lichens are absent. Freeze–thaw cracks disrupt the fan surfaces in places.

The catchment areas for the fans are difficult to define precisely, because there are no clearly marked divides between adjacent fan catchments. The snowfields at the head of the southern gorge wall are more-or-less continuous (Figures 13.2 and 13.4A). Visual estimates from aerial photographs indicate that the catchment areas are at least three times as large as the fans themselves, taking into account the entire sloping area above each fan.

The fans are all composed of overlapping debris flow lobes (Figures 13.3 and 13.4A, B). These are tongue-shaped bodies that range in width from 7 to 20 m, and in length

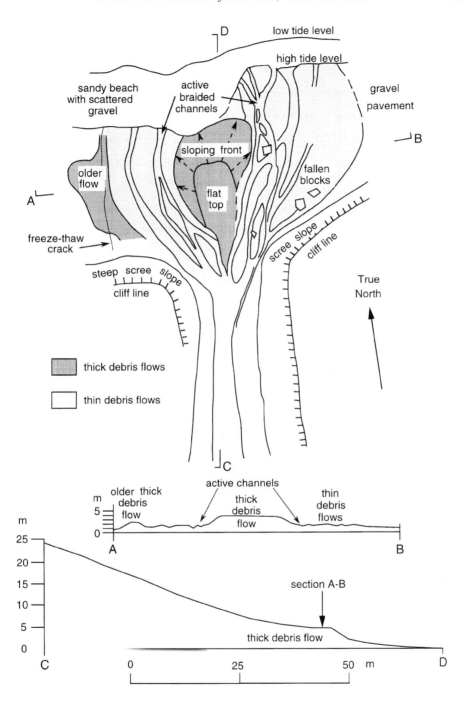

FIGURE 13.3 Map and cross-sections of debris flow fan 4 in Pagodroma Gorge; no vertical exaggeration (see Figure 13.1 for location and Figure 13.4A, E for photographs)

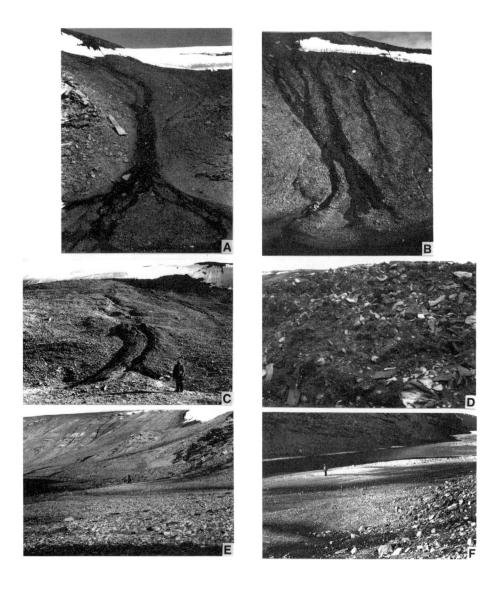

FIGURE 13.4   Debris flow fans of Pagodroma Gorge (see Figure 13.1 for location). (A) Oblique view of fan 4, showing headwall snowfield, buttress of Bainmedart Coal Measures to the left (east), thick debris flow lobe in the centre of the fan, and meltwater streams diverted either side (see Figure 13.3 for map). (B) Series of small recent debris flows on the eastern side of fan 7, which terminates in a sandy beach. (C) Recent debris flow with prominent levees, which show longitudinal cracking. In the foreground, below the change in slope, is the sediment lobe deposited by the debris flow. (D) Coarse matrix-supported sediment of thick debris flow; small shovel on left for scale. (E) Eastern flanks of fan 4, showing low relief, finer-grained debris flows (figure in centre for scale). (F) Beach in front of fan 7; size of sandstone clasts rapidly diminishes below high-tide level, which is marked by the change in slope in the right foreground (see Figure 13.7 for measurements)

from 12 to 35 m; length–width ratios vary from one to two. Two types of lobes are present: thicker lobes with very coarse clasts, and thinner lobes with finer clasts.

The thicker debris flow lobes have flat tops and steeply sloping fronts and sides (Figure 13.3). In longitudinal profile they form convex-up mounds of sediment from less than 1 m to over 3 m thick. They tend to dominate the gross morphology of the fans; thinner debris flows are diverted either side of them, as are meltwater streams (Figure 13.4A).

The thicker debris flows are composed of massive, poorly sorted clast- or matrix-supported sediment with a dark grey muddy matrix (Figure 13.4D). Grain size analysis of a representative sample (Figure 13.5) shows that the mud content is less than 4%, with about 1% clay. X-ray diffraction identified the clay as predominantly kaolinite, with a substantial amount of mixed layer illite–smectite. The high proportion of coarse sand largely represents grains released by disintegration of sandstones from the Bainmedart Coal Measures. There is an excess of coarse material (phi quartile skewness = -1.2), which comprises subangular, tabular clasts of sandstone (from the till and the scree slopes under the sandstone cliffs) and subrounded to well rounded cobbles to boulders of Precambrian basement (mostly felsic granulite, derived from the till). Few of the clasts from the till show any evidence of glacial action, e.g. facetting or striations. The Trask sorting coefficient ($(d_{75}/d_{25})^{0.5}$, d in mm) is 4.0, typical of debris flows (Costa, 1984), and the sediment plots within the debris flow fields on graphs of both d_{50} against quartile deviation (Pe and Piper, 1975) and d_1 against d_{50} (Bull, 1964; Costa, 1984).

Maximum clast size of individual thick debris flows (mean of 10 largest clasts) ranges from 20 to 80 cm. The surface tabular sandstone clasts on the flat tops of a few flow lobes are imbricated; the percentage of imbricated clasts increases towards the apex. Most of the clasts dip into the debris flow away from the flow termination, and have long axes oriented subperpendicular to the sides of the flow and subparallel to the flow front (Figure 13.6). On the steeply sloping front and sides of the lobe, imbrication is lacking and the tabular clasts mostly lie flat, parallel to the surface.

The thinner debris flow lobes have a lower relief than the thicker lobes, and are composed of smaller clasts (Figure 13.4E). The lobes may be only tens of centimetres thick, and the maximum clast size (mean of 10 largest clasts) ranges from 10 to 40 cm. On fan 4 (Figure 13.3), a thin debris flow on the eastern side of the fan has a maximum clast size of 27 cm, whereas the thick central debris flow lobe has a maximum clast size of 76 cm. On fan 2, a similar comparison is 13 cm to 43 cm. The majority of thinner debris flows are composed predominantly of sandstone clasts; few Precambrian basement clasts are present. The tabular pebbles and cobbles of sandstone generally lie parallel to the surface of the debris flow, and form a clast-supported surface armour about 10 cm thick across the entire surface of the flow (Figure 13.4E). Beneath this is the original dark grey, matrix-rich debris flow material. Grain size analysis of the latter material (Figure 13.5) shows that it is very poorly sorted and similar to the sediment of the thicker, coarser lobes, with abundant coarse material and little mud (less than 4%), and is typical of debris flow sediments (e.g. Trask sorting coefficient = 5.1). The clay is again mostly kaolinite, with some mixed layer illite–smectite.

Leveed channels extend upslope from recent debris flows on the fans (Figures 13.4B, C), often to headwall scars. The change from flat-topped toe to leveed channel corresponds to the increase in slope at the apex of the fan. The levees, which may be longitudinally cracked (Figure 13.4C), typically rise 20–50 cm above the channel floor,

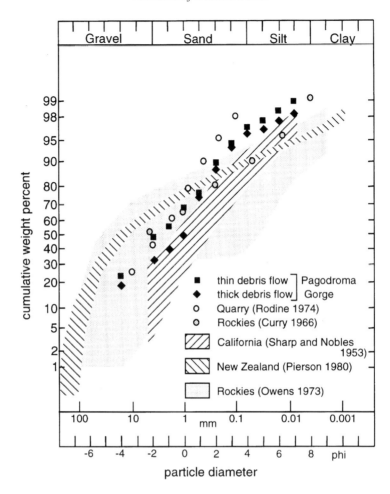

FIGURE 13.5 Cumulative distribution of particle sizes for representative samples of thick and thin debris flows; maximum clast size of thick debris flow sample is 68.5 mm, and that of thin debris flow sample is 44.3 mm. For comparison, distribution curves and envelopes from debris flows elsewhere are shown (modified after Pierson, 1980)

and are up to 1 m wide, with the central channel a similar width. Levees are less obvious or absent upslope of the older debris flows, indicating that they are readily destroyed by reworking processes.

Debris flow processes on the fans

Debris flows can be understood using the Bingham plastic model; before flow will occur, a certain yield stress or shear strength must be exceeded (Leeder, 1982). The shear strength of the debris is provided by matrix cohesion and internal friction between the larger clasts. Addition of water elevates the hydrostatic pore pressure and forces the grains apart; increasing water content dramatically decreases the matrix strength

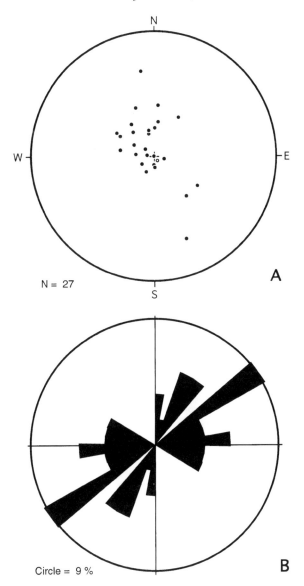

FIGURE 13.6   Imbrication of the surface clasts on top of a thick debris flow lobe on fan 7. (A) Equal-area projection of poles to a–b planes of the sandstone clasts. The debris flow surface (open circle) is flat and dips shallowly towards the front of the flow; most of the clasts dip more steeply than the lobe surface and in the opposite direction, i.e. away from the flow front. (B) Rose diagram of the long axes of clasts, which are oriented subperpendicular to the sides of the debris flow (which trend north-northwest) and subparallel to the front of the flow

(Johnson, 1970; Rodine and Johnson, 1976; Hooke, 1987), and also increases the weight of the material. Thus debris flows are generally triggered by a sudden influx of water, either rare, high-magnitude storm events (Clarke, 1987; Harvey, 1989) or melting snow and ground ice (Sharp, 1942; Harris and Gustafson, 1993; Catto, 1993). Along the southern wall of Pagodroma Gorge, the snowbanks on the upper slopes (Figures 13.4A, B) partially melt during warm summers, because they face north and receive sunshine for 24 h a day. The meltwaters infiltrate the colluvium on the steep gorge slopes (25–30°). In addition, the permafrost within the colluvium progressively melts, thickening the unstable layer of debris, and making it more prone to failure.

The debris flows in Pagodroma Gorge begin as slumps; headwall scars mark where sediment has been removed as an intact mass sliding on a detachment plane, which is probably the upper limit of frozen colluvium (permafrost), 0.5–1 m below the ground surface. The slumps then appear to have transformed into debris flows as a result of dilation and jostling between clasts (Costa, 1984; Blair and McPherson, 1994).

Downslope movement of debris flows is by laminar plug flow (Leeder, 1982); the high mobility of the flow is maintained by the presence of clay in the matrix, which reduces permeability and increases pore pressure (Rodine and Johnson, 1976; Pierson, 1980; Costa, 1984). The amount of muddy matrix required is small; in most debris flows it is usually 10–20%, and the percentage of clay may be only a few per cent (Costa, 1984). The Pagodroma Gorge debris flows contain very little mud (<4%) and clay (<1%), showing clearly that even though matrix clay is considered an essential part of every debris flow, the amount present may be extremely small.

Debris flow levees represent the margins of the frontal lobe of the flow, where the velocity is too low to overcome the inherent shear strength of the debris, which effectively "freezes". As the debris flow advances, these frontal lobe accumulations are progressively pushed aside (Sharp, 1942; Harris and Gustafson, 1993); there may also be some lateral forcing and internal sorting involved (Pierson, 1980). In Pagodroma Gorge, the levees are cracked (Figure 13.4C); the levee material collapses backwards into the channel area after the bulk of the debris flow has continued downslope and lateral support has been withdrawn (Pierson, 1980).

The debris flow processes change from net transport to net deposition when the gravitational forces driving the flow are no longer sufficient to overcome the shear strength of the matrix. This most commonly occurs where the slope suddenly decreases, and lateral spreading allows the debris thickness to decrease below that necessary for flow (Pierson, 1980). Debris flows typically require slopes in excess of 8–10° to maintain significant flow (Kochel, 1992), so in Pagodroma Gorge the debris flows quickly stop when they encounter the low gradients (<5°) at the foot of the gorge wall. The abrupt dumping of the sediment at one spot creates a convex-up mound of material, the debris flow lobe. In arid-zone fans, debris flows may terminate when the water in the matrix infiltrates into the underlying sediment (Hooke, 1967; Mabbutt, 1977), but this is unlikely to affect the Pagodroma Gorge fans, which are underlain at a shallow depth by permafrost.

Sediment flow fabrics in the main part of a debris flow may be present as imbricated clasts with their long axes parallel to the flow direction, and their a–b planes dipping into the flow centre (Leeder, 1982; Wells and Harvey, 1987). In the head region of a debris flow, long axes of clasts may be aligned transverse to the flow direction, and a–b planes dip upslope (Drewry, 1986). The latter fabric is well illustrated by some Pagodroma Gorge debris flows with abundant tabular clasts, which dip into the flow and are oriented subparallel to the flow front (Figure 13.6). The fabric probably originates when longitudinal compressive strain becomes dominant at the flow head, due to successive pulses or surges pushing into the pre-existing mass of debris (Pierson, 1980; Harris and Gustafson, 1993). This is more likely if the flow is particularly viscous (Bull, 1977).

Debris flows can vary greatly in strength and viscosity, from fairly slow-moving, high-viscosity events to rapidly moving flows transitional to streamflow (Lawson, 1982). The higher water content of the less viscous flows may be an original feature, or may result from increasing dilution of a debris flow with time, due to increasing water content or decreasing sediment availability (Wells and Harvey, 1987). Transitional flows may also be generated as 'runouts" during the last phase of a classic debris flow cycle, with a fluid slurry following a slow-moving debris flow (Johnson, 1970; Pierson, 1980).

Slower, thicker debris flows have a more viscous matrix and can carry larger clasts; thinner, more dilute flows can support only smaller clasts (Lawson, 1982), because matrix strength decreases greatly as the moisture content increases (Johnson, 1970). Thus the maximum particle size of a debris flow (mean of 10 largest clasts) often correlates with bed thickness (Bluck, 1967), and may be regarded as diagnostic of a mass flow origin (Nemec and Steel, 1984). Apart from being finer grained, the deposits of dilute transitional flows are often distinguished by a surface armour of clast-supported gravel, the fine-grained matrix having been transported downwards as water drained out of the flow (Wells and Harvey, 1987).

The Pagodroma Gorge fans probably contain deposits of both viscous and dilute transitional debris flows. The thicker flows with coarser clasts (Figure 13.4D) are certainly deposits of viscous debris flows. These flows generally have a greater percentage of Precambrian basement clasts than the thinner flows, indicating that they originated high up on the walls of the gorge, where till overlies the Bainmedart Coal Measures.

The thinner flows, which have smaller clasts and lower relief (Figure 13.4E), are probably the deposits of dilute (transitional) debris flows. The common presence of a surface gravel armour suggests that they originated as transitional flows. Fluvial reworking of the flow surfaces will also produce a surface gravel layer, but it is difficult to see how this process can evenly rework the entire surface of each flow, affecting channelled and unchannelled areas alike. The thinner debris flows could represent "runouts" following more viscous flows, able to transport only smaller sandstone clasts. Alternatively, they may be completely separate flows that originated in more water-rich parts of the colluvium on the gorge wall.

Reworking processes

Secondary, non-catastrophic processes inevitably remould fan surfaces during the periods between aggradational episodes (Blair and McPherson, 1994), but have had relatively little impact on the Pagodroma Gorge fans. Wind ablation removes silt and clay from the surfaces of freshly deposited debris flows, and fluvial action washes away the fine-grained matrix of the debris flows along stream channels. The central channel of the debris flows usually contains a small stream (Figure 13.4A), fed by the melting snowbanks above, with a gently incised U-shaped cross-section floored by open-framework gravel. At the toe of the flow the stream is diverted to one side by the convex-up sediment lobe. Sometimes the stream splits more or less equally either side of the toe

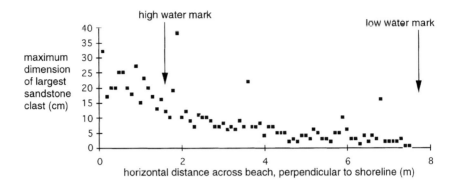

FIGURE 13.7   Maximum individual clast size on a traverse perpendicular to the shoreline below the scree slope west of fan 7 (see Figure 13.4F for photograph). All the clasts are composed of sandstone derived from the adjacent cliffs of Bainmedart Coal Measures

(Figure 13.4A), and bifurcating and braiding of stream channels is characteristic of the mid-fan area. The stream channels terminate at the high-water mark, where sandy or gravelly flats separate channel mouths.

The most extensive post-depositional modification of the fans has occurred where the debris flows enter the tidal waters of Pagodroma Gorge. Between low- and high-water marks are wide, gently sloping sandy beaches (Figures 13.3 and 13.4F), and small sandbanks or gravel bars, elongated east–west and up to several metres long, may be present just offshore. Because the katabatic winds blow down the gorge from the southwest, there is some northeasterly movement of sand along the beaches, so the fans are slightly asymmetrical.

The beaches have much less gravel than the immediately adjacent debris flows (Figure 13.4F), and the large clasts on the beaches are almost all composed of Precambrian basement, in contrast to the abundance of sandstone fragments in the debris flows. The size and shape of sandstone clasts change dramatically from above high-water mark (angular, >30 cm diameter) to low-tide mark (often subrounded, <1 cm; Figure 13.7). As the tidal waters of the gorge rise and fall each day, the alternate wetting and drying of the porous sandstone clasts causes them to gradually disintegrate. Presumably freeze–thaw processes are largely responsible rather than salt crystallization, because salt crusts are not evident on the sandstone clasts and the water in the gorge has low salinity. Clay expansion may be a factor, since some mixed layer smectite–illite is present in the sandstone.

There is no evidence of low-velocity gelifluction affecting the fan surface; subparallel ridges and stepped terracettes, as described from alluvial fans in the permafrost regions of Canada (Catto, 1993), are absent.

Comparison with debris flow fans from other regions

Most alluvial fans in both arid and humid environments are longer than they are wide (e.g. Heward, 1978; Kochel, 1992). By contrast, the Pagodroma Gorge fans are substantially wider than they are long. The thick debris flows in the centre of many of the fans divert smaller flows and meltwater streams to either side, so the fans build laterally. In addition, the debris flows may not have been able to run out to their full extent, because they terminate in the tidal waters of the gorge. There has also been some reworking of the fan fronts.

None of the Pagodroma Gorge fans shows fan-head incision or intersection points, as have been described commonly elsewhere (e.g. Bull, 1964; Denny, 1967; Hooke, 1967). Fan-head trenching has been attributed to external causes such as climatic change (Ritter et al., 1995) or faulting (Bull, 1964). It may also be an intrinsic part of fan development, as continued natural downcutting in the source terrain leads to incision at the fan apex (Wasson, 1977; Mabbutt, 1977; Nilsen, 1982). Experimental studies by Weaver and Schumm (1974) showed that fan-head trenching was due to downcutting within the oversteepened fan apex when its slope exceeded a threshold of stability, and was therefore inherent in the natural growth of an alluvial fan. The trench may be eroded by the more dilute, lower viscosity flow directly following a debris flow (Pierson, 1980; Hooke, 1987), or by subsequent stream action.

The lack of incision at the apices of the Pagodroma Gorge fans reflects the absence of tectonism (there is no evidence of recent fault activity) and apparently low climatic variability. Pagodroma Gorge has not been affected by major glaciation since at least the Late Tertiary (Mabin, 1990; Adamson and Darragh, 1990), and the climate was presumably periglacial throughout that time, because the small area of the fans argues that the rate of sediment supply, and therefore also the climate, have never been substantially different from the present. The absence of fan-head entrenchment may also reflect the small size of the fans; they have not prograded far enough to oversteepen the fan apex, and the toes cannot build out into the gorge because they are continually removed by wave and current reworking. In addition, there has been no significant retreat of the gorge wall behind the fans, and therefore no reduction in the gradient of the source area. The fact that fluvial incision can occur for only a few weeks a year may also be significant. Thus there are no external or internal causes for fan-head trenching, which is therefore not always part of the development of small fans.

Alluvial fan gradients have been related to the area, relief and lithology of the catchment, as well as the dominant depositional process on the fan. Steeper, higher-relief catchment slopes supply coarser debris, mainly as debris flows, and form alluvial fans with steeper slopes (Bull, 1964; Mabbutt, 1977). In addition, fans derived from more erodible lithologies are steeper than those on resistant rock types, because of more frequent debris flows and more rapid aggradation (Hooke, 1968), but may be smaller, either because the fine-grained sediment is more easily transported away (Harvey, 1989), or because more sediment is stored within the catchment rather than being delivered to the fan (Lecce, 1991). Fan slope also increases with decrease in catchment area (Mabbutt, 1977). The average gradient of the Pagodroma Gorge fans (about 5–6°) represents the average slope of alluvial fans worldwide (Nilsen, 1982), and the lower limit of debris flow fan gradients (5–15°; Blair and McPherson, 1994). Debris cones are normally steeper (12–25°; Brazier et al., 1988). Given the coarse sediment supply, erodible catchment lithology (till), small catchment areas and predominance of debris flows, the gradient of the Pagodroma Gorge fans might be expected to be higher. It is most likely that the low catchment relief along Pagodroma

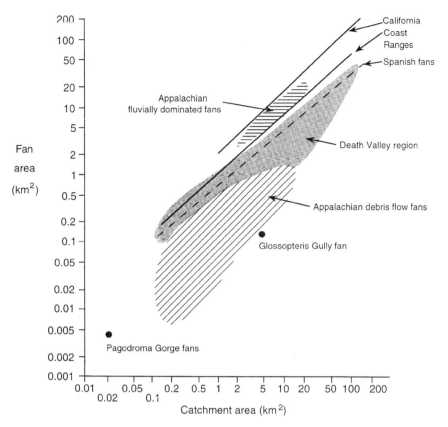

FIGURE 13.8   Fan area–catchment area plot for the Beaver Lake fans, compared with data from humid temperate fans in the Appalachian region of the eastern USA (from Kochel, 1992), semi-arid fans in Spain (from Harvey, 1989) and California (from Bull, 1977), and arid fans in Death Valley (from Hooke, 1968). Diagram modified after Harvey (1989)

Gorge (*c.* 120 m) has overridden the other factors and caused the low slope of the alluvial fans there; most alluvial fans worldwide have at least an order of magnitude greater catchment relief.

Alluvial fan area is directly related to catchment area, by an empirical relationship of the form (Hooke, 1968):

$$\text{fan area} = \text{constant} \times (\text{catchment area})^n$$

where the exponent is the gradient on a log–log plot of fan area against catchment area, and the constant $= 10^{y-\text{intercept}}$ on this graph (Figure 13.8). The exponent can vary from 0.6 to 1.2, but averages 0.9 (Harvey, 1989; Lecce, 1991). The Pagodroma Gorge fans have a constant of approximately 0.13, assuming $n = 0.9$ and a maximum catchment area of 0.02 km^2 (Figure 13.8). This is within the values for humid eastern USA fans (Kochel, 1992), but outside or at the limit of the trends derived for arid and semi-arid Death Valley and Californian fans in the USA (Hooke, 1968; Bull, 1977) and Spanish fans (Harvey, 1989).

GLOSSOPTERIS GULLY FAN

Glossopteris Gully

Glossopteris Gully enters the eastern end of Bainmedart Cove (Figure 13.1) as a narrow cliffed gorge incised into shallowly dipping sandstones, siltstones and coal seams of the Bainmedart Coal Measures (Figure 13.9A). There are extensive outcrops of Pagodroma Tillite immediately to the north and south of the mouth of the gorge. The 60 m cliffs lining the gorge are strongly jointed, and coarse, angular rockfall mantles their bases; they decrease in height upstream, as the floor of the gorge rises steeply. Where the gorge swings to the south, it is blocked by an ice wall, from which a waterfall emerges in summer. Upstream is a broad, shallow stream valley running southwards across a till-covered plateau (Dart Fields of McKelvey and Stephenson, 1990) to two lakes (Figure

FIGURE 13.9   Glossopteris Gully fan (see Figure 13.1 for location and Figure 13.10 for detailed map). (A) General view, showing the till-covered plateau (Dart Fields) on the skyline and the mouth of Glossopteris Gully gorge, cut through strata of the Bainmedart Coal Measures; Pagodroma Tillite mantles the steep slopes either side of the gorge. Braided stream channels cross the centre of the fan and Radok Lake lies in the foreground; note tent for scale (arrowed). (B) Reworked debris flow in the central southern part of the fan, showing large, scattered boulders, many composed of Precambrian basement. (C) Imbricated gravel pavement on the central part of the fan; the sandstone clasts dip upslope to the right (east), towards the fan apex (note hammer in foreground for scale). (D) Fine gravel pavement on the distal part of the fan, overlying a sand unit exposed in the channel bank in the foreground

13.1). The larger (Lake Toploje) abuts the edge of Lambert Glacier, and is largely frozen even in midsummer.

The small stream running through the gorge in summer periodically carries large flows, as shown by the rounding of the sandstone cobbles and boulders in the stream bed, and the benches eroded into the colluvium along the sides of the gorge, at heights up to 1 m above present stream level. Near the head of the gorge, a large rockfall temporarily dammed the stream, probably in the 1988/89 summer; freshly broken sandstone blocks up to 3 m across were washed over 30 m downstream when the rockfall dam catastrophically failed.

Morphology of Glossopteris Gully fan

Glossopteris Gully fan (Figures 13.9A and 13.10) is a single alluvial fan about 550 m across and 450 m from apex to furthest extent. Total area is approximately 0.12 km^2. The long profile is broadly concave-up (Figure 13.10), flattening from a maximum gradient of around 5° at the apex to 1–2° distally; average gradient across most of the fan is around 2–3°. The cross-fan profile is shallowly convex-up. The fan has prograded into Radok Lake (Figures 13.1 and 13.9A), which is freshwater and non-tidal.

Flanking the southern and northeastern sides of the fan are irregular slumps and debris flows (Figures 13.9A and 13.10), derived from the extensive areas of Pagodroma Tillite present along the eastern side of Pagodroma Gorge. The southern debris flows have a steep margin 2–3 m high where they abut the fan, and have forced the southern stream across the fan into an arc-shaped course. The debris flow material is mostly matrix-supported, has a maximum clast size (mean of 10 largest clasts) of 81 cm, and consists of an approximately equal mixture of clasts of sandstone (from the Bainmedart Coal Measures) and Precambrian basement.

The fan is crossed radially by three separate downstream-bifurcating, low-sinuousity stream systems (Figures 13.9A and 13.10), which diverge from the apex. The central braided stream carries the largest flow. At the fan apex it has a channel up to 2 m wide incised over 1 m into the surrounding sediments. Downslope the channel shallows and splits into numerous small distributaries, separated by low banks of gravel. There is a small area of active sand deposition where the largest distributary enters Radok Lake. In the centre of the fan some channels have diffuse terminations, and the flow spreads across the fan surface as shallow sheets of water. During the peak melt season, large parts of the fan surface are covered by meltwater sheetflow.

The entire fan is frozen at a depth of less than 1 m, even in summer. Large freeze–thaw cracks cross the surface almost everywhere, except in the active channel areas. Small increases in stream flow cause rapid rises in the water table across the whole fan, because the permafrost is so shallow.

The central part of the fan, south of the main stream channel, is a distinct convex-up sediment mound in the middle of the overall concave-up long profile (Figure 13.10), and is probably an old, reworked debris flow lobe. It is covered by large numbers of boulders, separated by cobbles and pebbles (Figure 13.9B). The maximum clast size is 145 cm, and many of the boulders are composed of Precambrian basement. This boulder gravel is a surface layer overlying matrix-supported material of the same maximum grain size.

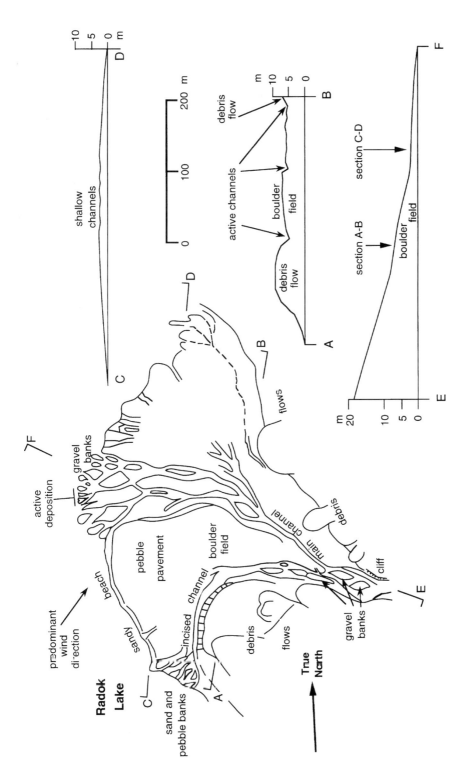

FIGURE 13.10 Map and cross-sections of Glossopteris Gully fan; vertical exaggeration × 5 (see Figure 13.1 for location and Figure 13.9 for photographs)

Downslope from the boulder area, and across most of the remainder of the fan, the surface is covered by a clast supported gravel pavement composed of moderately well sorted pebbles and cobbles, predominantly of sandstone (Figure 13.9C). Maximum grain size ranges from 45 cm in the centre of the fan to 20–30 cm on the distal areas, to 10–15 cm on the gravel banks on the lake shore. Imbrication of tabular clasts is common (Figure 13.9C), with the a–b planes of the clasts dipping upslope. Areas of finer-grained gravel may have tabular pebbles lying parallel to the surface (Figure 13.9D). The sediment beneath the surface pavement on the distal part of the fan is plane-laminated gravelly sand (Figure 13.9D).

The main sandy areas are along the lake shore (Figure 13.10). The active stream channels deposit small amounts of sand-sized sediment as low banks within the channels, or as thin gravelly sand lobes at the lake edge. Along the southern side of the fan, facing the prevailing katabatic wind direction, there are narrow sandy beaches, with a wave-eroded notch 40–50 cm above lake level.

Sheetflooding

Sheetflooding may be the most important fluid gravity process in the construction of many alluvial fans (Blair and McPherson, 1994). Flashy runoff from snowmelt or high-intensity rainfall entrains colluvial sediment, often as a result of slumping from slopes; sudden melting of interstitial ice in a permafrost area may also be important (Wells and Harvey, 1987; Catto, 1993). When the sediment-charged flow reaches the fan apex, it expands into a more or less uniform sheet as the flow becomes instantaneously unconfined; velocities may reach $10\,\mathrm{m\,s^{-1}}$ (Hogg, 1982). This attenuation of the flow rapidly increases downfan, depositing sediment of decreasing grain size under supercritical flow conditions, as a result of antidune migration and washout (Blair and McPherson, 1994). Sheetflooding deposits planar beds containing couplets of coarse gravel interstratified with fine gravel or coarse plane-laminated sand; imbrication is common (Wasson, 1979; Nemec and Steel, 1984; Blair and McPherson, 1994).

Alluvial fans dominated by sheetflooding have low slopes (typically 2–8°), concave-up long profiles, high-capacity feeder channels and large, low-gradient catchment areas often floored by bedrock (Wells and Harvey, 1987; Blair and McPherson, 1994). The greater discharge resulting from the catchment area characteristics produces more dilute sediment gravity flows, i.e. sheetfloods rather than debris flows. The fan slope is low because the greater discharge transports alluvium further across the fan (Mabbutt, 1977).

Deposition on the Glossopteris Gully fan is most likely dominated by sheetflooding. The imbricated gravel pavements, distally overlying plane-laminated sand layers, are probably sheetflood deposits (Figures 13.9C, D), and the overall low slope of the fan and downfan decrease in maximum clast size also fit this mode of deposition. There are several mechanisms that could generate sheetflooding in Glossopteris Gully: sudden failure of a rockfall dam across the gorge (there is evidence that this has occurred relatively recently; see above); melting of the ice wall at the head of the gorge; or sudden melting of either or both of the upstream lakes. Any of these mechanisms would generate a large volume of water which would entrain sediment from the till cover on Dart Fields and scree within the gorge.

Fluvial modification

The surface of Glossopteris Gully fan has been little modified by non-catastrophic streamflow. The gravel pavement covering the fan largely retains its original sheetflood imbricated fabric (Figure 13.9C), although there has been some removal of fines by diffuse and channelized meltwater flow. Fluvial incision is limited to about 1 m of downcutting at the fan apex, decreasing to 10 cm on the distal fan (Figure 13.9D).

Comparison with other sheetflood fans

As previously mentioned, alluvial fan area is approximately related to catchment area by the equation:

$$\text{fan area} = \text{constant} \times (\text{catchment area})^{0.9}$$

For the Glossopteris Gully fan, the constant is 0.045 (Figure 13.8), which falls below or at the lower limit of trends derived for arid and humid fans elsewhere (e.g. Hooke, 1968; Harvey, 1989; Kochel, 1992). The constant for the Pagodroma Gorge fans is greater (Figure 13.8), but also less than values in most other regions of the world. The very low value for the Glossopteris Gully fan reflects the small area of this fan relative to its large catchment, which includes a substantial lake. The small size of the fan may be due to the relatively low availability of sediment in the area; the till cover on Dart Fields is thin, and the amount of scree in the gorge is limited. This lack of sediment may be a general characteristic of cold, arid climatic regimes, where the thin regolith reflects the absence of vegetation and low rates of chemical weathering.

It is notable that the Glossopteris Gully fan is affected by both debris flow and sheetflood processes, with the former confined to the apex of the fan, and the latter probably responsible for most sediment deposition. Although this matches early models of fan facies (e.g. Hooke, 1967), recent studies (Blair and McPherson, 1994) have shown that most fans around the world are dominated by one process or the other, and there is usually no downfan change in facies. The fortuitous presence of till deposits next to a gully with a large catchment area explains why both processes are active on Glossopteris Gully fan.

Intrinsic geomorphic thresholds, like size and gradient of catchment, determine whether debris flows or sheetfloods are dominant processes on alluvial fans, by controlling the water/sediment ratio (Wells and Harvey, 1987; Harvey, 1989). On Glossopteris Gully fan, the water/sediment ratio largely reflects sediment availability. Debris flows on this fan are only found where there is an abundant supply of matrix-rich sediment (till); sheetfloods are generated when there is limited sediment but a large volume of water.

RECOGNITION OF ALLUVIAL FANS DEPOSITED UNDER COLD, ARID CONDITIONS

Despite the fact that the Beaver Lake fans are small on the global scale and have a limited chance of preservation in the stratigraphic record, they offer some interesting insights into fan processes. At Beaver Lake, debris flow and sheetflood fans exist almost side by side, and both processes can be found on the same fan. Clearly, there is no climatic significance in the dominant fan process (Blair and McPherson, 1994), although changes in the process could have climatic origins (Nemec and Postma, 1993; Dorn, 1994; Ritter et al., 1995).

Alluvial fans associated with glaciation derive their deposits from glacially eroded detritus, in contrast to the concurrent nature of denudation and sedimentation in other fan types (Ryder, 1971). Since till usually contains a significant component of clay-rich matrix, it may be more likely to generate debris flows rather than sheetfloods, and many alluvial fans in glacial and paraglacial areas are dominated by debris flows (e.g. Catto, 1993; Harris and Gustafson, 1993). Nevertheless, sheetfloods can also occur in these regions, given favourable circumstances (Dorn, 1994), as shown by the Glossopteris Gully fan.

Fans and debris flows in permafrost regions are not inherently different from those in other climates (Catto, 1993; Harris and Gustafson, 1993), and it has proved very difficult to generate a general model for climate control of alluvial fan development (Dorn, 1994). In the case of the Beaver Lake fans, the climate under which they formed is evident from several lines of evidence. The small size of the fans may be due to the relatively low availability of sediment, reflecting the absence of vegetation and low rates of chemical weathering characteristic of cold, arid climatic regimes. The mixed nature of the debris flow clasts (sandstone and high-grade metamorphics) strongly indicates the influence of glaciation, particularly as outcrops of Precambrian metamorphics are at least several kilometres away (Figure 13.1). However, glacial facetting and striations on pebbles are uncommon. The surface gravels that mantle the fans lack any evidence of biological activity or soil development, and are disrupted by freeze–thaw cracks, again indicating the cold arid climate. The frozen state of the fans for much of the year has relatively little effect on the fan surfaces and deposits; its main effect is to temporarily suspend activity.

As a final note, it should be pointed out that sediment reworking along the shoreline of Pagodroma Gorge has the potential to dramatically change the characteristics of the alluvial fan sediments (Figures 13.4F and 13.7), without leaving an obvious marine signature. This is another example of the difficulty of distinguishing marine deposits in alluvial conglomerate sequences (Nemec and Steel, 1984). Destruction of sandstone clasts by marine processes in Pagodroma Gorge means that a rise in sea level would result in a layer of sand with scattered clasts of Precambrian basement mantling a debris flow sequence containing little sand and abundant sandstone clasts, and would most likely be interpreted as a change in source material.

CONCLUSIONS

The alluvial fans in the Beaver Lake area show the typical features of alluvial fans worldwide, but they have particular characteristics that reflect the climate, topography and sediment supply of the area in which they are found. They have formed by both debris flow and sheetflood processes, but individual fans are dominated by only one of these processes. The debris flow fans are found in areas where there is an abundance of colluvium that can be mobilized each summer when it is saturated by meltwater from nearby snowfields. Both thick, coarse debris flows and thinner, finer-grained debris flows are represented; the latter were more dilute. The relatively low slope of these fans (5°)

reflects the low catchment topography (120 m). None of the debris flow fans shows fan-head incision, indicating that this feature is not always part of the development of small fans.

The sheetflood fan has a relatively small area given its large catchment; this reflects the low sediment availability compared to the supply of water during flooding events. Debris flow lobes at the fan apex are due to the fortuitous presence of till deposits nearby, and do not indicate a general downfan progression from debris flow to sheetflood deposits.

Because they are frozen for at least 10 months of the year, fan surfaces have been modified to only a limited extent, as shown by the presence of intact debris flow or sheetflood surfaces on portions of all fans. However, the wetting–drying cycles to which the toes of the Pagodroma Gorge fans are subjected by the tidal waters of Beaver Lake result in major sediment modification. Sandstone clast sizes are diminished and the clasts may disintegrate completely, giving a sandy sediment completely unlike the unmodified debris flow deposits.

The Beaver Lake fans contain some features that show the imprint of the climate in which they were deposited. First, there is the ubiquitous presence of till, identified by its mixture of clast sizes and types (many not locally derived), rather than the presence of striated, facetted clasts. Second, biological activity is absent and rates of chemical weathering are low, so there is minimal soil development. The resultant low availability of sediment may be responsible for the small size of the Beaver Lake fans. Third, there are occasional freeze–thaw cracks which disrupt the surface pavement. However, the processes that constructed the fans (debris flows and sheetfloods) are not diagnostic of the cold, arid climate.

ACKNOWLEDGEMENTS

This work was carried out as part of the Australian National Antarctic Research Expedition's summer programme of 1989/90 in the northern Prince Charles Mountains. The financial and logistical assistance of the Australian Antarctic Division is gratefully acknowledged. In particular, field leader Martin Betts and senior pilot Pip Turner are thanked for their considerable help. Sam Ross and Tom Bernecker assisted with preparation of the maps, and Derek Fabel carried out the grain size analyses and clay identifications. The comments of two anonymous referees and Andrew Miller greatly improved this paper.

REFERENCES

Adamson, D. 1989. Quaternary geomorphological and botanical studies in the Northern Prince Charles Mountains 1988–89. In R. Ledingham (Ed.), *Prince Charles Mountains Field Report Nov 1988–Jan 1989*, 32–37. ANARE, Kingston, Tasmania.

Adamson, D. and Darragh, A. 1990. Field evidence on Cainozoic history and landforms in the Northern Prince Charles Mountains, East Antarctica. In D. Gillieson and S. Fitzsimons (Eds), *Quaternary Research in Australian Antarctica: Future Directions*. Department of Geography and Oceanography, Australian Defence Force Academy, Special Publication 3, 5–14.

Blair, T.C. 1987. Sedimentary processes, vertical stratification sequences, and geomorphology of the Roaring River alluvial fan, Rocky Mountains National Park, Colorado. *Journal of Sedimentary Petrology*, **57**, 1–18.

Blair, T.C. and McPherson, J.G. 1994. Alluvial fans and their natural distinction from rivers based on morphology, hydraulic processes, sedimentary processes, and facies assemblages. *Journal of Sedimentary Research*, **A64**, 450–489.

Bluck, B.J. 1967. Deposition of some upper Old Red Sandstone conglomerates in the Clyde area; a study in the significance of bedding. *Scottish Journal of Geology*, **3**, 139–167.

Brazier, V., Whittington, G. and Ballantyne, C.K. 1988. Holocene debris cone evolution in Glen Etive, western Grampian Highlands, Scotland. *Earth Surface Processes and Landforms*, **13**, 525–531.

Brierley, G.J., Liu, K. and Crook, K.A.W. 1993. Sedimentology of coarse-grained alluvial fans in the Markham Valley, Papua New Guinea. *Sedimentary Geology*, **86**, 297–324.

Budd, W.F. 1981. The role of Antarctica in Southern Hemisphere weather and climate. In N.W. Young (Ed.), *Antarctica: Weather and Climate*. Royal Meteorological Society, Melbourne, 29–48.

Bull, W.B. 1964. History and causes of channel trenching in western Fresno County, California. *American Journal of Science*, **262**, 249–258.

Bull, W.B. 1977. The alluvial fan environment. *Progress in Physical Geography,* **1**, 222–270.

Calkin, P.E. and Nichols, R.L. 1972. Quaternary studies in Antarctica. In R.J. Adie (Ed.), *Antarctic Geology and Geophysics*. Universitetsforlaget, Oslo, 625–643.

Catto, N.R. 1993. Morphology and development of an alluvial fan in a permafrost region, Aklavik Range, Canada. *Geografiska Annaler*, **75A**, 83–93.

Clarke, G.M. 1987. Debris slide and debris flow historical events in the Appalachians south of the glacial border. *Geological Society of America Reviews in Engineering Geology*, **7**, 125–138.

Costa, J.E. 1984. Physical geomorphology of debris flows. In J.E. Costa and P.J. Fleischer (Eds), *Developments and Applications of Geomorphology*. Springer-Verlag, Berlin, 268–317.

Curry, R.R. 1966. Observations of alpine mudflows in the Tenmile Range, central Colorado. *Geological Society of America, Bulletin*, **77**, 771–776.

Denny, C.S. 1965. Alluvial fans in the Death Valley region, California and Nevada. *US Geological Survey Professional Paper* 466, 1–62.

Denny, C.S. 1967. Fans and pediments. *American Journal of Science*, **265**, 81–105.

Dorn, R.I. 1994. The role of climatic change in alluvial fan development. In A.D. Abrahams and A.J. Parsons (Eds), *Geomorphology of Desert Environments*. Chapman and Hall, London, 593–615.

Drewry, D. 1986. *Glacial Geological Processes*. Edward Arnold, London, 276 pp.

Eyles, N. and Kocsis, S. 1988. Sedimentology and clast fabric of subaerial debris flow facies in a glacially influenced alluvial fan. *Sedimentary Geology*, **59**, 15–28.

Fielding, C.R. and Webb, J.A. 1996. Facies and cyclicity of the Late Permian Bainmedart Coal Measures in the Northern Prince Charles Mountains, MacRobertson Land, Antarctica. *Sedimentology*, **43**, 295–322.

Harris, S.A. and Gustafson, C.A. 1993. Debris flow characteristics in an area of continuous permafrost, St Elias Range, Yukon Territory. *Zeitschrift für Geomorphologie N.F.*, **37**, 41–56.

Harvey, A.M. 1989. The occurrence and role of arid zone alluvial fans. In D.S.G. Thomas (Ed.), *Arid Zone Geomorphology*. Wiley, New York, 136–158.

Heward, A.P. 1978. Alluvial fan sequence and megasequence models: with examples from Westphalian D – Stephanian B coalfields, northern Spain. In A.D. Miall (Ed.), *Fluvial Sedimentology*. Canadian Society of Petroleum Geologists, Memoir 5, 669–702.

Hogg, S.E. 1982. Sheetfloods, sheetwash, sheetflow, or ..? *Earth Science Reviews*, **18**, 59–76.

Hooke, R.LeB. 1967. Processes on arid region alluvial fans. *Journal of Geology* **75**, 438–460.

Hooke, R.LeB. 1968. Steady-state relationships on arid-region alluvial fans in closed basins. *American Journal of Science*, **266**, 609–629.

Hooke, R.LeB. 1987. Mass movement in semi–arid environments and the morphology of alluvial fans. In M.G. Anderson and K.S. Richards (Eds), *Slope Stability*. Wiley, Chichester, 505–529.

Johnson, A.M. 1970. *Physical Processes in Geology*. Freeman, San Francisco, 577 pp.

Kochel, R.C. 1992. Geomorphology of alluvial fans in west-central Virginia. In G.R. Whittlecar (Ed.), *Alluvial fans and boulder streams of the Blue Ridge Mountains, west-central Virginia*. Southeastern Friends of the Pleistocene, 1992 Field trip, 79–128.

Krainer, K. 1988. Sieve deposition on a small modern alluvial fan in the Lechtal Alps (Tyrol, Austria). *Zeitschrift für Geomorphologie N.F.*, **32**, 289–298.

Lawson, D.E. 1982. Mobilization, movement and deposition of active subaerial sediment flows, Matanuska Glacier, Alaska. *Journal of Geology*, **90**, 279–300.

Lecce, S.A. 1991. Influence of lithological erodibility on alluvial fan area, western White Mountains, California and Nevada. *Earth Surface Processes and Landforms*, **16**, 11–18.

Leeder, M.R. 1982. *Sedimentology – process and product*. Allen and Unwin, London, 344 pp.

Mabbutt, J.A. 1977. *Desert Landforms*, ANU Press, Canberra, 340 pp.

Mabin, M.C.G. 1990. The glacial history of the Lambert Glacier–Prince Charles Mountains area and comparisons with the record from the Transantarctic Mountains. In D. Gillieson and S. Fitzsimons (Eds), *Quaternary Research in Australian Antarctica: Future Directions*. Department of Geography and Oceanography, Australian Defence Force Academy, Special Publication, 3, 15–23.

McKelvey, B.C. and Stephenson, N.C.N. 1990. A geological reconnaissance of the Radok Lake area, Amery Oasis, Prince Charles Mountains. *Antarctic Science*, **2**, 53–66.

McLoughlin, S., Lindstrom, S. and Drinnan, A.N. 1997. Gondwanan floristic and sedimentological trends during the Permian–Triassic transition: new evidence from the Amery Group, northern Prince Charles Mountains, Antarctica. *Antarctic Science*, **9**, 281–298.

Mosley, M.P. 1988. Bedload transport and sediment yield in the Onyx River, Antarctica. *Earth Surface Processes and Landforms*, **13**, 51–67.

Nemec, W. and Postma, G. 1993. Quaternary alluvial fans in southwestern Crete: sedimentation processes and geomorphic evolution. In M. Marzo and C. Puigdefabregas (Eds), *Alluvial Sedimentation*. International Association of Sedimentologists, Special Publication 17, 235–276.

Nemec, W. and Steel, R.J. 1984. Alluvial and coastal conglomerates: their significant features and some comments on gravelly mass-flow deposits. In E.H. Koster and R.J. Steel (Eds), *Sedimentology of Gravels and Conglomerates*. Canadian Society of Petroleum Geologists, Memoir 10, 1–31.

Nilsen, T.H. 1982. Alluvial fan deposits. In P.A. Scholle and D. Spearing (Eds), *Sandstone Depositional Environments*. American Association of Petroleum Geologists, Memoir 31, 49–86.

Owens, I.F. 1973. *Alpine mudflows in the Nigel Pass area, Canadian Rocky Mountains*. PhD Thesis, University of Toronto (quoted in Pierson, 1980).

Pe, G.G. and Piper, D.J.W. 1975. Textural recognition of mudflow deposits. *Sedimentology*, **13**, 303–306.

Pierson, T.C. 1980. Erosion and deposition by debris flows at Mt Thomas. North Canterbury, New Zealand. *Earth Surface Processes*, **5**, 227–247.

Rains, R.B., Selby, M.J. and Smith, C.J.R. 1980. Polar desert sandar, Antarctica. *New Zealand Journal of Geology and Geophysics*, **23**, 595–604.

Ritter, J.B., Miller, J.R., Enzel, Y. and Wells, S.G. 1995. Reconciling the roles of tectonism and climate in Quaternary alluvial fan evolution. *Geology*, **23**, 245–248.

Rodine, J.D. 1974. *Analysis of the mobilization of debris flows*. Final report to US Army Research Office, Durham, NC, Report ARO 9973.1–EN (quoted in Pierson, 1980).

Rodine, J.D. and Johnson, A.M. 1976. The ability of debris heavily freighted with coarse clastic materials to flow on gentle slopes. *Sedimentology*, **23**, 213–234.

Ryder, J.M. 1971. The stratigraphy and morphology of para-glacial alluvial fans in south-central British Columbia. *Canadian Journal of Earth Sciences*, **8**, 279–298.

Schumm, S.A. 1977. *The Fluvial System*, Wiley, New York, 338 pp.

Sharp, R.P. 1942. Mudflow levees. *Journal of Geomorphology*, **5**, 222–227.

Sharp, R.P. and Nobles, L.H. 1953. Mudflow of 1941 at Wrightwood, southern California. *Geological Society of America, Bulletin*, **64**, 547–560.

Wasson, R.J. 1977. Last–glacial alluvial fan sedimentation in the lower Derwent Valley, Tasmania. *Sedimentology*, **24**, 781–799.

Wasson, R.J. 1979. Sedimentation history of the Mundi Mundi alluvial fans, western New South Wales. *Sedimentary Geology*, **22**, 21–51.

Weaver, W.E. and Schumm, S.A. 1974. Fan-head trenching – an example of a geomorphic threshold. *Geological Society of America Abstracts with Programs*, **6**, 481.

Wells, S.G. and Harvey, A.M. 1987. Sedimentologic and geomorphic variations in storm-generated alluvial fans, Howgill Fells, northwest England. *Geological Society of America, Bulletin*, **98**, 182–198.

Part 4

Patterns of Alluvial Deposition

14

The Fly River, Papua New Guinea: Inferences about River Dynamics, Floodplain Sedimentation and Fate of Sediment

WILLIAM E. DIETRICH[1], GEOFF DAY[1] AND GARY PARKER[2]

[1] Department of Geology and Geophysics, University of California, Berkeley, USA
[2] St Anthony Falls Laboratory, University of Minnesota, Minneapolis, USA

ABSTRACT

The Fly River drains the Southern Fold Mountains of Papua New Guinea where it descends down steep valleys bordered by peaks of up to 4000 m and then crosses a low-relief, nearly flat plain to the Gulf of Papua. Only about 30% of the 75 000 km^2 basin is in the rapidly eroding uplands. The basin receives up to 10 m a^{-1} rainfall in the uplands and still exceeds 2 m a^{-1} in the lowlands. Here we focus on the 450 km long Middle Fly River and its floodplain. Along this reach, the slope decreases by a factor of more than three (from 6.6 × 10^{-5} to 2 × 10^{-5}), the median grain size declines by a comparable amount (to about 0.1 mm) and duration of flooding greatly increases, causing the rainforest vegetation of the upper reach floodplain to give way to a swamp grass-dominated lower reach. Our analysis of observations on the Fly sheds some light on three questions: (1) what controls the rate of channel migration? (2) what controls the rate of floodplain sediment deposition? (3) what controls downstream fining in a large sand-bedded river? Previous studies have interpreted the well preserved floodplain features of former channel positions (scroll bar complexes, filled oxbows and oxbow lakes) as indicating a rapidly migrating, unstable river which frequently reworks its floodplain. Simple assumptions regarding frequency of loop cutoff, comparison of available topographic maps, and recent high resolution aerial photographs all indicate that, instead, the Fly River migrates slowly, typically less than 0.0045 channel width per year in the forested floodplain reach, and appears to be virtually stagnant in the swamp grass reach. The well preserved floodplain features in the roughly 200 km long swamp grass reach suggest that channel migration rates here were higher in the recent past. One mechanism which might allow a meander to slow or cease movement is for the river slope to decrease, causing the boundary shear stress on the bank to drop below a critical value. Such a change may also be responsible for the unusually large-wavelength bends in the swamp grass reach which have long, nearly straight limbs. Slope reduction may have been brought about by tectonics, hydro-isostatic effects, eustatic effects or backwater influence from evolution of a large downstream tributary (the Strickland River). Rather than indicating highly active channel migration, the well preserved floodplain features instead give evidence of low rates

Varieties of Fluvial Form. Edited by A.J. Miller and A. Gupta

of overbank deposition, probably less than 1 mm a^{-1} in the forested reach and of the order of 0.1 mm a^{-1} in the swamp grass reach during the late Holocene. Floodplain deposition occurs by advection of sediment with overbank flows, by lateral diffusion from the sediment-rich Fly main channel, and by transport up tributaries or small channels connecting the river with off-river water bodies (tie channels). Low rates of sedimentation in the swamp grass reach (despite long periods of inundation) result from: (1) low concentration of sediment in the overbank flows; (2) dense swamp grass vegetation which may damp turbulence of advecting overbank flows and reduce diffusion; and (3) inhibition of river overbank flows across the floodplain due to pre-existing standing water resulting from local rainfall and low drainage rates. Prevention of tie channel plugging with sediment may depend on rather rare hydrological events, perhaps linked to drought-producing El Niño events which cause large pressure head differences between the river and the off-river water bodies and enable the strong currents in the tie channels to scour. The observed downstream fining of the Fly can result from only about a 15% reduction in the total sand load which may occur by preferential net deposition of the coarser sand in the channel bed and surrounding floodplain.

INTRODUCTION

Large river systems may show significant lags in response to changes in controls due to long travel distances. Church and Slaymaker (1989), for example, have argued that rivers draining western Canada are still passing the increased sediment load arising from glaciation which terminated over 10 000 years ago. Fluctuating sea level has forced waves of incision and aggradation on rivers, and the current morphology of large lowland rivers may strongly reflect responses that are still evolving to the current sea level high stand (e.g. Ikeda, 1989). An analysis of any large lowland river draining to the sea, then, must face the issue of sorting out the evolutionary relics of the river's response to past sediment load, water discharge and sea level rise from the current dynamic linkages between morphology and current conditions. Importantly, such rivers may be responding simultaneously to large changes created in the Holocene and to recent effects associated with land use, and these effects may force opposing or reinforcing morphologic adjustments on the river.

The Fly River in Papua New Guinea (Figures 14.1 and 14.2) is currently the subject of intense investigations directed towards documenting and predicting the geomorphic and ecological effects of sediment waste from a large copper mine that has been operating since 1985 in the headwaters of the river (e.g. Eagle et al., 1986; Higgins et al., 1987; Pernetta, 1988; Higgins, 1990; Smith and Hortle, 1991; Wolanski and Eagle, 1991; Day et al., 1993; Smith and Bakowa, 1994). Of particular concern with regard to the geomorphology have been the effects of increased sediment load on the long-term planform stability of the river, the rate of channel aggradation, and the rate of floodplain sedimentation. A monitoring programme has been established to assess hypotheses developed from earlier and ongoing studies and to guide ongoing theoretical development directed towards predicting long-term effects of the mine discharges. In this paper we re-examine previously collected field data, report new findings and, through some simple estimations, reach conclusions that in some cases differ significantly from those of previous studies. Here we will focus on the pre-mine behaviour of the system which had been negligibly influenced by human activity. The river geomorphology has been strongly influenced by Holocene sea level rise, therefore distinguishing historical relics from ongoing tendencies will be an important issue in our analysis. We specifically

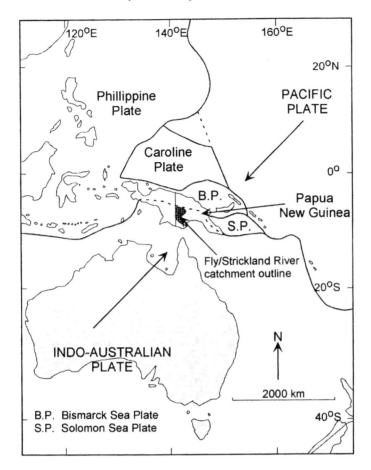

FIGURE 14.1 Plate tectonic framework of the Fly River catchment in Papua New Guinea showing relative plate motion (modified from Hill, 1991)

examine three general questions: (1) what controls the rate of channel lateral migration? (2) what controls the rate of deposition of sediment on the floodplain? and (3) what controls downstream fining of bed material in lowland sandy rivers?

THE FLY RIVER SYSTEM

The overall character of the Fly River system owes much to its tectonic setting. According to Pigram et al. (1989), middle Oligocene collision of the northern carbonate shelf margin of the Australian craton with a subduction system led to the formation of a foreland basin. This caused a broad flexure of the margin; as the northerly migration of the plate pushed into a tropical climate, an extensive carbonate platform formed. By the late Pliocene and Quaternary, sediment shed from the mountains overwhelmed the foredeep and clastic deposition, and rapidly spread southward burying the carbonate platform. Continued tectonic activity has caused previously buried Quaternary sediments

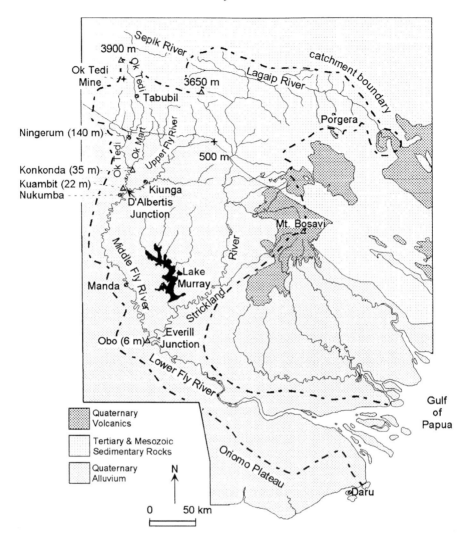

FIGURE 14.2   Fly River catchment showing very generalized geology (modified from Bureau of Mineral Resources, Geology and Geophysics, Canberra, ACT, 1976). Numbers in parentheses next to site names are elevations of the bank. (Note that "Ok" means "river" in the local dialect)

to be uplifted and, consequently, dissected, such that while the topography across the former carbonate platform is subdued, it is locally highly dissected (Loffler, 1977). Based on exploratory drilling, about 1000 m of Quaternary sediments mantle the carbonate platform at Ningerum (on the Ok Tedi, Figure 14.2) progressively thinning to about 280 m at Lake Murray (Taylor, 1979). This suggests that the deposition rate has been of the order of 1 mm a^{-1} near the mountains, decreasing to about 0.1 mm a^{-1} towards the Gulf of Papua.

This tectonic setting leads to: (1) high sediment loads in the headwaters; (2) steep, massively unstable topography which can shed large pulses of sediment to rivers; (3)

mechanically weak sediment which quickly breaks downs upon transport in the rivers (i.e. Parker, 1991); (4) orographically induced high rainfall rates; (5) a rapid decline in river slope and an overall very low gradient in the lowland bordering the mountains; (6) possible tectonic influence on river orientation (Blake and Ollier, 1969, 1971; Blake, 1971); and (7) limited net storage of sediment on land upstream of the Gulf of Papua.

The Fly River system consists of three major tributaries, the Ok Tedi, Fly and Strickland, all of which originate in the steep, rapidly uplifting subduction zone complexes and marine sediments of the Southern Fold Mountains where peaks reach up to 4000 m in elevation (Figure 14.2). These tributaries quickly descend through narrow canyons, crossing the uplifted Quaternary sediments at about 500 m, and join in the lowland area enroute to the delta. Where the Ok Tedi meets the Fly, the elevation of the water surface at bankfull is about 18 m (the local bank height for Kuambit reported in Figure 14.2 is on a terrace), and the river is 846 km from the delta mouth. At this junction (known as D'Albertis Junction), the drainage area of the Ok Tedi is 4350 km^2 and that of the Fly is 7600 km^2. The Fly then follows a 412 km sinuous path to its junction (Everill Junction) with the Strickland River. Here the Strickland drains 36 740 km^2 while the Fly drains only 18 400 km^2; this notwithstanding, the main stem below the junction is called the Fly River. Downstream of Everill Junction the Fly River gains an additional 19 860 km^2 before discharging into the Gulf of Papua. Of the 75 000 km^2 of land that drains via the Fly to the Gulf, only about 30% occupies the steep headwaters area.

Rainfall depends strongly on elevation, with annual rainfall in excess of 10 m occurring at the mine site on the Ok Tedi (elevation about 1500 to 2000 m), and decreasing from 8 m at Tabubil (600 m) to about 2 m at Obo and Daru near sea level (Markham, 1995; Harris et al., 1993). A distinct seasonality occurs in the southern half of the catchment with June to August tending to be drier. Despite heavy rainfall, droughts still strike, particularly in the lowlands, causing the Fly to fall and drain its normally swampy floodplain.

River flows increase downstream with increasing drainage area and become less flashy such that along the Middle Fly (the reach of the Fly between D'Albertis and Everill Junctions), flood stage may remain approximately constant for many months during which time the floodplain will be under several metres of water. We have observed continuous standing water in the swamp reach of the Fly floodplain for more than 18 months, and according to Higgins (1990) much of that water may originate from local rainfall and runoff. Average annual flow on the Fly below D'Albertis Junction is 1930 $m^3 s^{-1}$ (Higgins, 1990), about 2244 $m^3 s^{-1}$ at Obo just above Everill Junction and about 3110 $m^3 s^{-1}$ on the Strickland. Hence, the average annual flow at the delta is less than 7000 $m^3 s^{-1}$ (not the 15 000 $m^3 s^{-1}$ proposed by Blake and Ollier, 1971). Tidal influence on stage extends upstream of Everill Junction, and perhaps, during low river stage, as far as Manda, 570 km upstream from the Gulf.

Before the start of mining on the Ok Tedi, the sediment discharge from the Ok Tedi and Upper Fly was about 5 $\times 10^6$ t a^{-1} each (based on early sediment monitoring records), but was subject to large temporal variations due to occasional massive landsliding (Pickup et al., 1981). Little sediment appeared to be added between D'Albertis Junction and Everill Junction, despite the drainage area increasing from 11 950 to 18 400 km^2 along this reach, because this low-lying portion of the catchment has gentle topography, is densely vegetated, and the tributaries are mostly blocked and empty into lakes as they enter the floodplain area. Instead, net sediment loss may have dominated at this time. According to various estimates and direct monitoring, the sediment load from the Strickland is about 70 to 80 $\times 10^6 \, t \, a^{-1}$, and while there is no doubt that the Strickland sediment yield was much higher than the Middle Fly before mining began, the uncertainty on this estimate is large. Perhaps less than 30% of this load is coarser than silt. The natural sediment discharge from the Fly above Everill Junction is relatively high compared to other rivers of comparable drainage area, but low compared to other rivers with comparable runoff (from data reported by Milliman and Syvitski, 1992). By implication, this means the sediment concentration in the Fly must have been fairly low, even though it is chronically turbid. On the Fly just below D'Albertis Junction the mean sediment concentration was about 100 $mg \, l^{-1}$, whereas on the Strickland upstream of Everill Junction it was about 770 $mg \, l^{-1}$ (average of 53 depth-integrated samples collected during 1991 to 1996 by personnel working for Ok Tedi Mining Ltd). Figure 14.3 shows the downstream variation in runoff and sediment yield throughout the Ok Tedi–Fly system. On a per unit area basis, sediment yield and runoff decrease, illustrating how the uplands area is the dominant source for both water and sediment.

While bed material size on the Ok Tedi–Fly River declines downstream with decreasing slope, it does not do so uniformly (figure 2 in Pickup, 1984; Parker, 1991). Pickup proposed that five bed material zones, bounded by sharp transitions, make up the downstream variation in bed sediment characteristics. They are: (1) source (boulders and gravel, poorly sorted, close to sediment supply); (2) armoured (gravel over sand-rich

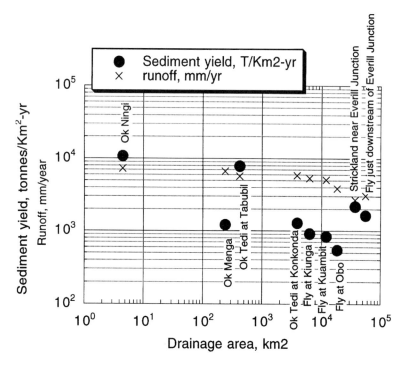

FIGURE 14.3   Downstream variation in sediment yield and runoff along the Ok Tedi and Fly River. Data are from Pickup et al. (1981) and OTML hydrology reports

substrate); (3) gravel–sand transition (bimodal); (4) sand (mean size about 0.2 to 0.3 mm); and (5) backwater (mean size less than 0.16 mm, but fluctuates with stage). Although he did not specifically locate these zones, they roughly correspond to: (1) the Ok Tedi above Ningerum (see Figure 14.2 for location); (2) Ningerum to above the Ok Mart Junction (at Konkonda); (3) a short reach just upstream of the Ok Mart; (4) Ok Mart to roughly half-way to Everill Junction; and (5) the remainder of the Fly to Everill Junction. Pickup suggested that slope, sediment supply and geomorphic history enforce these sharp bed material breaks. The backwater zone was proposed to be due to Strickland water level at its confluence with the Fly River. Based on a simple mass balance, Pickup estimated that net deposition of all sand at the distal end of the zone would cause the sand zone to reach the junction with the Strickland between 1400 years and 22 000 years if the bed level of the Strickland were to remain constant.

Blake and Ollier (1971), in a pioneering examination of the Fly River system, recognized several important attributes of the floodplain of the Fly. They distinguished five distinct floodplain environments: scroll complexes; back swamps (similar to Allen's (1965) floodbasins); alluvial plains of minor tributaries; blocked valley swamp; and lakes (Figure 14.4). Blake and Ollier proposed that the sediment-laden Fly and Strickland Rivers aggraded during Holocene sea level rise, blocking all small tributaries draining from the lowlands, and creating blocked valley lakes (including Lake Murray) and swamps. The scroll complexes were inferred to be aggraded relative to surrounding plains (Blake and Ollier, 1971, p.5) giving rise to the back swamp areas. They identified numerous geomorphic features on aerial photographs and introduced the term "tie channel" to describe the small channels that almost invariably connect the main stem Fly River to cutoff and blocked valley lakes and through which the water may flow in either direction. They conclude that the well formed, numerous scroll bar complexes on the Fly indicate that "the alluvial plains are highly unstable and that lateral migration and accretion dominate over vertical accretion". Loffler (1977), Pickup et al. (1979) and Pickup and Warner (1984, p.39) agree with this inference of active, rapid channel migration on the Fly. We will re-evaluate this interpretation below.

Blake (1971) reported a radiocarbon age date of 27 000 years BP in sediments found near Kiunga (Fly River above D'Albertis Junction) which he believes are overlain by a regionally extensive alluvial unit referred to as the Lake Murray beds. Blake and Ollier (1969) further proposed that the land mass south of the Fly River (the Oriomo Plateau, Figure 14.2) was uplifted 15 m after deposition of the Lake Murray beds and that this uplift contributed to forcing an ancestral Fly River (which they assumed headed southwestwards to the sea) to head eastwards to the Gulf of Papua. Subsequently Loffler (1977, p.20, 98) questioned whether the regional stratigraphy was sufficiently understood that a single date can be widely applied. He argued that the fans from the volcano Mount Bosavi (Figure 14.2) overlie the broad alluvial plain proposed by Blake to have been dissected in the last 27 000 years, and yet the Bosavi volcano construction must extend well into the Pleistocene. Finally he argued that the rate of erosion required in 27 000 years to generate the fine-scale ridge and valley topography that borders the Fly–Strickland north of the Oriomo Plateau would greatly exceed that expected from a low-lying alluvial plain. Instead, he proposed that the Fly River entrenchment and dissection of the surrounding hills occurred repeatedly during the Pleistocene as sea level fluctuated with the glacial cycles. Because other rivers turn from a southerly direction to an easterly

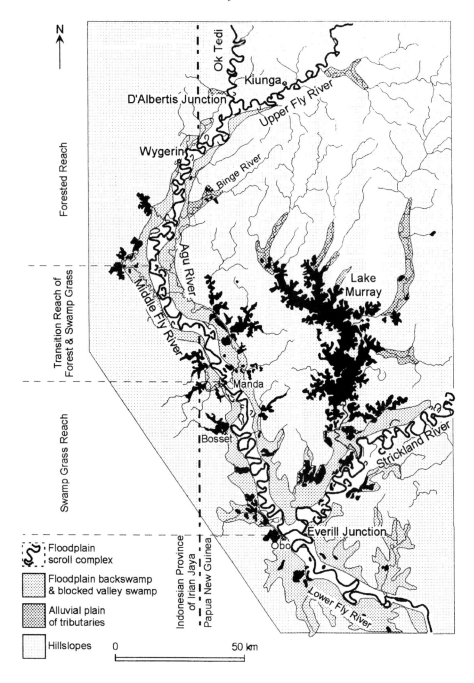

FIGURE 14.4   Generalized depositional environments of the Fly River and lower Strickland area
(modified from Blake and Ollier, 1971)

direction to the sea in this area, he also reasoned that the Fly River was not turned by recent uplift of the Oriomo Plateau, but instead has simply followed the "predominant slope of the depositional surface" of the alluvial plains (Loffler, 1977, p.90). Harris et al. (1996) reported that deep incised valleys, up to 120 m deep, extend across the continental shelf into the Gulf of Papua, indicating that the Fly River has flowed eastwards since at least the last glacial maximum.

Pickup et al. (1979) used the Blake (1971) observation and the presence of weathered alluvium in the bed of the Fly River near Kiunga to infer that the Fly River cut down approximately 10 m between 27 000 and 17 000 BP and then aggraded to its present level by 5000 BP. They concluded that the river aggraded at about $1 \, \text{mm} \, \text{a}^{-1}$ during sea level rise (a rate similar to the longer time scale deposition recorded in the Quaternary sediments). Taylor (1979) also relied on the Blake (1971) observation as well as data that indicate cessation of volcanism at Mount Bosavi about 30 000 to 50 000 BP to estimate an average aggradation rate in the Middle Fly River area of about $1 \, \text{mm} \, \text{a}^{-1}$. Dietrich (1988), using these data and other inferences, reasoned that the aggradation rate decreased from about $1 \, \text{mm} \, \text{a}^{-1}$ to about $0.1 \, \text{mm} \, \text{a}^{-1}$ down the Middle Fly. Hettler and Lehmann (1995) reported a similar conclusion.

Harris et al. (1996) reviewed current understanding of Late Quaternary sea level changes in this area, pointing out that over much of the last 100 000 years eustatic sea level has been in the range of 40 to 70 m below its present position. They also suggested that the Holocene transgression may have experienced brief periods of rapid rise and that current eustatic sea level was reached about 6500 years ago. Isostatic adjustments due to rising water levels (i.e. Chappell et al., 1982) and sediment loading on the shelf probably have occurred, but their magnitudes are not yet established.

In the following, we examine more closely the spatial and temporal patterns of channel migration, floodplain deposition and grain size variation along the Middle Fly observed by earlier workers.

RIVER PLANFORM DYNAMICS OF THE MIDDLE FLY RIVER

Along the Middle Fly River from D'Albertis to Everill Junction, the river has a sinuosity of 2.1, and within the distinct meander belt there is abundant evidence of past channel position, including complex scroll patterns and filled cutoffs and cutoff lakes connected to the main channel via tie channels (Figures 14.5 and 14.6). The floodplain and meander belt progressively widen downstream from each being about 4 km at Kiunga to 14 and 8 km, respectively, by Everill Junction (Pickup et al., 1979). Channel width, on the other hand, is about 200 m at Kiunga, increases to an average width of about 350 m within 90 km downstream of D'Albertis, and then narrows downstream thereafter to about 250 m before reaching Everill Junction (Pickup et al., 1979). Vegetation is an uninterrupted rainforest from D'Albertis Junction through the reach bordering the Indonesian province of Irian Jaya. This gradually gives way to a swamp grass reach where the Agu River joins the Fly (near Manda) and continues as such until after the Strickland enters the Fly (Figure 14.4).

Throughout the Middle Fly reach, the meanders are not entirely free; occasionally, along the outer banks of bends, higher banks composed of intensely weathered, bright red alluvium are exposed which clearly offer greater resistance to channel migration. Pickup

FIGURE 14.5 Scroll bars and oxbows of the swamp
grass reach of the lower Middle Fly. Not all scroll bars
shown. Based on aerial photographs take in 1982. Flow
direction is from top to bottom of figure. Arrow is
located in the first bend downstream from where the
Agu River joins the Fly (at a bend apex on the Fly)

et al. (1979) proposed these banks to be composed of Pleistocene sediment, which seems
very likely. There is also a distinct change in the meander morphology along the river. In
the forested upper Middle Fly, meanders have typical shapes and wavelength-to-width
ratios comparable to those of most rivers (about 11), but in the swamp grass reach this
ratio grows to 20 and the bends tend to consist of long straight limbs attached to short,
highly curved bend apices reminiscent of meanders in muskeg (Figures 14.4–14.6).

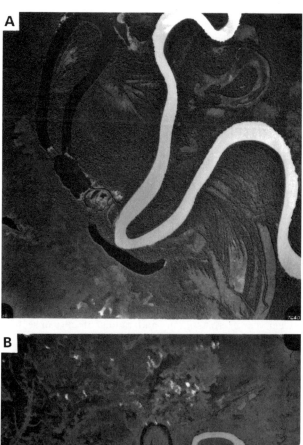

FIGURE 14.6   Aerial photographs of the Fly in the forested reach (A) and swamp grass reach (B). In (A), which is downstream of the Binge River (Figure 14.4), oxbows in various stages of infilling are visible and scroll bars are recorded by long curved rows of trees. A distinct tie channel joins the Fly with the oxbow lake. In (B), which is just upstream of the Obo gauging station (at bottom of Figure 14.5), a double oxbow with typical bends lies in contrast to the dogleg pattern of the current Fly River. A tie channel connects the oxbows to the Fly and scroll bars dominate the fine-scale topography within the meander belt

In order to determine migration rates and assess the inference that the river "meanders vigorously", three different measurements were made. First, a crude assumption, but one for which there is evidence on other rivers (e.g. Fisk, 1944), is that the sinuosity of the river and resulting slope are relatively constant over time. In this case, the number of meander bends for a given reach would stay approximately constant, with reaches that experienced cutoff quickly recovering their sinuosity and other bends increasing in size. Hence, if a cutoff occurs, the loop will be replaced, and if we know the total number of bends and the rate of cutoff, we can calculate the mean migration rate.

On the Middle Fly, according to Pickup et al. (1979, p.104), three loops were cut off since 1899, suggesting a cutoff rate of three per 90 years or one every 30 years. Between D'Albertis and Everill Junctions there are 111 individual bends (over 400 km). If these cut off at the rate of one in 30 years and are immediately replaced with new loops such that this number stays constant, then on average an individual loop forms and is then cut off in 111×30 years or 3330 years. Cutoff loops typically have an amplitude of about 1.5 km at the time of cutoff. Hence if the typical loop grows to 1.5 km in 3330 years, its maximum lateral migration rate at the bend apex is on average about $0.5 \, \mathrm{m \, a^{-1}}$. Studies of meander loop evolution typically show that they tend to first grow and migrate downstream and then enlarge laterally, slowing down as the loops become large (e.g. Hooke (1984), Larson (1995) and as implied in the Nanson and Hickin (1986) relationship between bank erosion rate and radius of curvature to width ratio). The average migration rate of $0.5 \, \mathrm{m \, a^{-1}}$, then, is probably low compared to the early stages of actively migrating loops, and gives a minimum average rate of about 0.002 channel widths per year.

The second method of estimating channel migration rates consisted of comparing topographic maps. Two high quality maps were available: 1:100 000 topographic maps produced by the Royal Australian Survey Corp in 1969 and based on aerial photography from 1963 to 1966, and 1:50 000 navigational charts produced by the Snowy Mountains Engineering Corporation (SMEC) in 1981 and modified from the Australian Survey Corp maps using 1980/81 Landsat Imagery and ship radar observations during surveying. Although of relatively high quality, these data have unknown positional inaccuracies due to the lack of extensive ground control. The strength of our findings reported below relies on the comparison among the three methods. Older, less reliable maps included: 1:250 000 US Army Map Service maps from aerial photography and field observations made in 1953–1963, and 1:500 000 US Army Map Service dated 1942 and compiled from many sources including the Netherlands New Guinea and Papua maps at 1:250 000 and US Hydrographic charts. We also found a 1:500 000 1943 map captured from the Japanese after World War Two that was modified from an Australian 1:1 000 000 map printed in 1941. This map was particularly important for examining the swamp grass reach. Aerial photographs dating from the 1930s apparently exist but we have not yet been able to gain access to them.

Channel migration analysis was divided into three distinct reaches: Kiunga to D'Albertis Junction; D'Albertis Junction to the upstream end of the swamp reach at the southern end of the Indonesian border; and from the upstream end of the swamp reach to Everill Junction. Of the 18 bends between Kiunga and D'Albertis only 30% had measurable displacement. Average shifting in these active bends over an approximate 20 year period was typically about 1 to 1.5% of channel width (or about 2 to $4 \, \mathrm{m \, a^{-1}}$).

Comparisons using the older Japanese maps were less certain because of clear mapping errors; nonetheless, these maps also support the interpretation that migration is occurring at a measurable pace in only a fraction of the bends and the rate is less than 2% of channel width per year in these bends. Hence, average channel migration rate overall in this reach is estimated to be less than $0.02 \times 0.30 = 0.006$ channel width per year and may be as low as 0.003 channel width per year.

Between D'Albertis Junction and the end of the Indonesian border upstream of Manda, US Army maps were compared to the SMEC maps to determine migration rates. Of the 62 bends along this international border, about 30% showed evidence of measurable displacement, with most of these bends in the southern part of the reach. Average shift rate where it occurred was about 1.5% of channel width per year (or about 4 to $5 \, \mathrm{m \, a^{-1}}$). This pre-mine estimate for channel migration rate is about $0.015 \times 0.30 = 0.0045$ channel width per year.

In the swamp grass reach, we found the Japanese maps to be reasonably accurate and only minor differences were found between these maps made in the early 1940s with those published in 1981. Comparison of the Australian Royal Survey Corp maps with the SMEC navigation charts showed only one bend other than a short reach affected by a cutoff to have any measurable displacement. Nearly the entire reach has shown no perceptible migration for over 50 years.

The third method of analysis became possible when aerial photographs were taken in 1992 and compared to 1982 aerial photographs obtained by BHP Engineering. This work was accomplished by Barr Engineering Co. (1995) by digitizing channel centrelines and analysing channel shift using a program called MEANDER, originally described by MacDonald et al. (1991). Barr Engineering divided the distance from Kiunga to D'Albertis Junction into three reaches, and found the average shift normal to the downstream direction similar in each, averaging 0.0036 channel width per year. Downstream of D'Albertis, they divided the 200 km distance to the start of the swamp reach into eight separate analyses of migration. Figure 14.7 summarizes their findings, showing that migration rate was found to be highest just downstream of D'Albertis Junction and declined to values of about 0.0047 channel width per year at the downstream end. Shortly below their lowest reach, the swamp reach begins and the migration rates drop to very low values. Barr Engineering interpreted the elevated migration rate downstream of D'Albertis Junction as resulting from bar growth due to introduction of mine waste to the Ok Tedi, which began in earnest in 1985. Whereas this is probably correct, this pattern of accelerated bank migration downstream of a tributary that introduces significant bedload can occur naturally. This is precisely what happens downstream of Everill Junction, where significant channel migration has been previously noted (SMEC, 1981, p.52, figure 17). In contrast, in the swamp reach upstream of Everill Junction, bars are normally absent and channel migration rates are presently nearly zero. Hence, the pattern of downstream decreasing migration rates below D'Albertis probably existed before, but has been exaggerated by the introduction of mine-related sediment to the system.

Comparison of these three estimates of migration rates gives remarkably similar results (especially given the assumptions employed and inaccuracies of the first two methods). Cutoff rates give a minimum estimate of 0.002 channel width per year. Comparison of topographic maps up to 1981 indicates migration rates between 0.003 and 0.006 for the

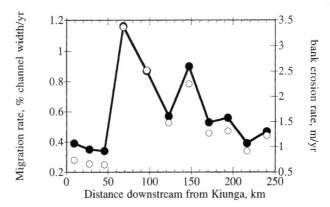

FIGURE 14.7   Channel migration rate downstream of Kiunga shown as metres per year (open circles) and percentage of channel width per year (solid circles) (data from Barr Engineering Co., 1995)

reach upstream of D'Albertis, 0.0045 below D'Albertis to the swamp reach, and no migration in the swamp reach. Repeated recent aerial photographs give values of 0.0036 channel width per year above Kiunga, and an accelerated rate (due to mine-related sediment) below D'Albertis Junction which declines to 0.0047 channel width per year in the unaffected lower reach above the swamp reach.

Despite the well preserved evidence of lateral migration of the Fly, the documented migration rates do not reveal "intensive meandering activity" that could cause "rapid floodplain destruction" (Pickup and Warner, 1984, p.39) as many have suggested. The migration rates are low, or, in the case of the swamp reach, essentially zero. An analysis of migration rates of 18 rivers in western Canada by Nanson and Hickin (1986) gave a median migration rate of about 0.015 channel width per year. A compilation by MacDonald et al. (1991) of 16 streams in Minnesota, USA, gave values of migration per unit channel width per year of 0.0044 to 0.066, with an average value of 0.0187. Most of the rivers in the two studies were narrower rivers conveying smaller discharges. Two reaches on the Mississippi River, however, which had widths of 56 and 67 m, had migration rates of 0.005. Larsen (1995) reviewed migration rates of a number of rivers (besides the MacDonald et al. sites) including several that had widths greater than 100 m. None of the rivers greater than 100 m wide had migration rates less than 0.013 channel width per year, a value much higher than the mean pre-mining rates on the Fly River. Empirically, then, the migration rates for the Fly fall within the low range of values reported for other rivers.

If the strong preservation of scroll bars and old channel paths in the floodplain cannot be explained by rapid or vigorous lateral migrations, then their preservation instead suggests that the rate of overbank deposition is sufficiently low that floodplain features are not obliterated in spite of the slow channel migration. On the Fly above the swamp grass reach, if the average lateral migration is 0.004 channel width per year, the mean channel width is 240 m and the meander belt width is 6 km, then to traverse half that distance at this rate (assuming that alternate bends migrate in opposite directions) would take about 3000 years, roughly the time estimated for a loop to form and cut off. The traverse time would be shorter if the peak migration rate during loop expansion is higher, as it probably is. If cutoff loops do represent 3000 years of floodplain development, then historical rates of floodplain overbank deposition must be less than $1 \, \text{mm} \, \text{a}^{-1}$ as $3 \, \text{m}$ would probably eradicate much of the scroll bar relief so prominent on the floodplains.

This issue is particularly pertinent to the swamp reach upstream of Everill Junction where very little bank migration was detected for the past 50 years, yet scroll bar features, cutoff loops and old channel paths etch the floodplain (Figure 14.5). This seeming contradiction suggests that the rate of channel migration has reduced significantly.

What would cause lateral migration in a large lowland river to nearly cease? Howard (1992) reasoned that four factors constrain bank erosion rates: (1) rate of deposition on the point bar; (2) ability of the stream to remove the bedload component from eroded bank deposits; (3) ability of the stream to entrain cohesive bank slump deposits; and (4) rate of weathering of cohesive bank materials. It appears that constraint 1 would at least partially explain the accelerated bank migration rate below D'Albertis Junction. In the swamp grass reach there are no well developed point bars, but the channel is relatively narrow, having a width/depth ratio of 15 or less, hence local flow velocities are not significantly diminished due to a lack of point bar confinement. Constraint 4 is unlikely to undergo the apparently relative rapid transition suggested by the preservation of floodplain features. So this points to a decreased ability of the river to entrain (constraint 3) and/or transport (constraint 2) in-place or fallen bank material.

Howard (1992) reviewed the basis for the widely used bank erosion law that assumes the erosion rate (E) is proportional to the near-bank velocity perturbation (u_b), an assumption originally proposed by Ikeda et al. (1981), i.e

$$E = Ku_b \qquad (14.1)$$

$$u = U(1 + u_b) \qquad (14.2)$$

in which K is a proportionality constant, u is the local vertically averaged velocity and U is the average velocity for the channel cross-section. While field observations support this assumption, Howard also pointed out that Odegaard (1989) and Hasegawa (1989) argue that depth perturbation near the outer bank may be a better or at least an equally important predictor of bank erosion rates. Furthermore, Howard reasoned that there is likely to be a critical near-bank shear stress below which bank erosion ceases, hence the simple velocity perturbation assumption may be incomplete. Alternatively, then, one might formulate Equation (14.1) using a critical bank velocity, u_c:

$$E = K(u_b - u_c) \qquad (14.3)$$

Such a model would provide a direct mechanism for slowing or even halting channel migration through either increases in bank resistance or decrease in the perturbation velocity. The most likely case is reduction in perturbation velocity due to reduction in mean slope. If a river was actively migrating and experienced a slope decline (due to local base level rise, for example), then mean velocity might decline and with that the perturbation velocity. If the slope declined progressively, the perturbation velocity would remain in excess of critical for progressively smaller lengths of channel until the entire bend ceased to migrate. Such a decline could lead to a morphological change as well, from the typical bends of actively migrating rivers, to the dogleg bends with large wavelength-to-width ratio of the present swamp reach.

On the Fly River, average slope decreases from about 6.6×10^{-5} near D'Albertis Junction to about 2×10^{-5} in the swamp reach (based on SMEC (1981) data). Correspondingly, the bankfull boundary shear stress declines from about 80 to $30 \, dyn \, cm^{-2}$. In addition, for a given discharge (or for bankfull discharge), average velocity declines from Kuambit to Obo (Figures 14.2 and 14.8). It is possible, then, that earlier in the Holocene, the slope was steeper in the lower Middle Fly and then reduced to its current value. Slope reduction could have resulted from sea level rise, build-up of the Strickland and backwater development, or tilting due to tectonics. We briefly consider each of these controls.

After deglaciation at the end of the Pleistocene, sea level quickly rose to close to its current level by about 6500 years ago (as reviewed by Harris et al., 1996). We infer that the change in slope along the Fly is more recent than 6500 years ago because the floodplain features are well preserved (some preliminary dating of oxbow lakes appears to support this inference). Hence, sea level rise seems an unlikely mechanism. Build-up of the Strickland and development of a backwater up the lower Middle Fly was proposed by Pickup and Warner (1984) to explain the swampy conditions; they did not, however, comment on swamp reach migration rates. At Everill Junction, the Strickland drains twice the drainage area of the Fly and probably carries 10 times the sediment load. The Strickland floodplain near Lake Murray is frequently flooded, but forest lines the channel and spreads out into the surrounding floodplain, suggesting that it is better drained than the lower Middle Fly. A build-up of sediment in the Strickland would presumably occur progressively, and the backwater development could just be the result of ongoing

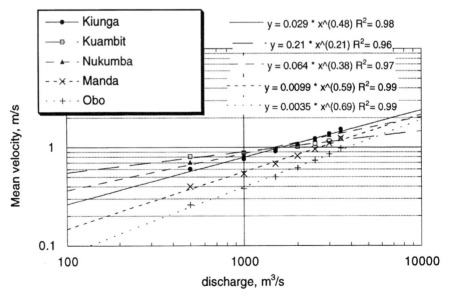

FIGURE 14.8 Variation in mean velocity with discharge at gauging stations along the Fly. Nukumba is about 5 km downstream of Kuambit and Manda is at the head to the swamp grass reach just upstream of where the Agu River joins the Fly

response of the Strickland to Holocene sea level rise. The bank elevation of the Fly falls only about 3 m over the roughly 152 km swamp reach, hence relatively minor aggradation of the Strickland could greatly influence the average slope.

One indication of backwater effects on the swamp grass reach is the stage dependent slope that can be calculated from hydraulic geometry relationships at gauging stations. Using Manning's relationship:

$$u = d^{0.67} S^{0.5}/n \tag{14.4}$$

solving for the slope, S, and replacing the mean velocity, u, and average depth, d, with the hydraulic geometry relationships (e.g. Leopold et al., 1964):

$$u = kQ^m \quad \text{and} \quad d = cQ^f$$

gives

$$S = (kQ^m)^2 n^2 / (cQ^f)^{1.33}$$

or

$$S = k^2 c^{-1.33} Q^{(2m-1.33f)} n^2 \tag{14.5}$$

For the Obo gauging station, Equation 14.5 is

$$S = 3.8 \times 10^{-7} Q^{1.35} n^2$$

showing that the slope is strongly discharge-dependent. A Manning's n of 0.03 gives reasonable estimates (ignoring stage dependency of resistance) of slope in this reach increasing from 3.8×10^{-6} to 1.7×10^{-5} as discharge increases from 1000 to 3000 $m^3 s^{-1}$. This strong stage dependency may reflect the varying influence of the Strickland on the Fly, whereby the effect is strongest at low flow, causing the swamp grass reach to have an exceptionally small slope. This analysis is consistent with field observations made during low flow in which we saw a dramatic decrease in suspended sediment concentration as the flow entered the swamp reach. While there is strong evidence that the Strickland creates a backwater on the Fly, it does not necessarily follow that this backwater is responsible for the apparent change in migration character of the swamp reach.

Given the location of the Fly River on the leading edge of the Australian plate, it is also not unreasonable to suggest that recent tectonics may have tilted or lowered the lower Middle Fly, causing it to become swampy and reducing its slope to the point where bank migration has nearly ceased. On the *Geology of Papua New Guinea* map (Bureau of Mineral Resources, Geology and Geophysics, Canberra, ACT, 1976), a fault is depicted as starting just upstream of the swamp reach and running parallel to the axis of the Fly valley, passing just downstream of Everill Junction and curving eastward following the Fly valley to where the river begins to widen rapidly as it enters the delta. How this fault, if it is correctly shown, affects the Fly River is not clear, but it does suggest that tectonics could affect this reach.

By whatever mechanism, if the slope has been reduced at the transition from the forested to the swamp grass reach, there should be a tendency for aggradation. We have recognized no obvious evidence for this. Furthermore, preliminary results of high resolution surveying of bank and floodplain heights using a global positioning system fail to show a significant slope break at this transition: a nearly constant gradient extends from nearly 100 km upstream of the transition to Obo at the lower end of this reach. This is consistent with the lack of a distinct zone of aggradation. The lack of aggradation at the transition reach raises the possibility that the higher channel migration in the past in the current swamp reach was due to historically higher sediment loads. According to Blong (1991), about 8800 BP a 7 km^3 landslide swamped the Ok Tedi near Tabubil. However, the majority of this pulse of sediment had probably passed through the system, however, well before the bank migration slowed in the current swamp reach.

At this time, although we can make a strong case that the pace of current channel migration is too slow to develop the complex floodplain features of the swamp reach, we cannot yet assign causality.

SEDIMENTATION RATES AND PROCESSES

Floodplain deposition

Geological evidence cited above suggests that long-term rates of floodplain deposition decline downstream along the Middle Fly from about 1 mm a^{-1} to 0.1 mm a^{-1}. These low rates are compatible with preservation of scroll bar features formed by the slowly migrating Fly River. Futhermore, preliminary analysis of radiocarbon dating in the swamp reach supports the inference that sedimentation rates are on the order of 0.1 mm a^{-1} (Hettler and Lehmann, 1995). Higgins (1990) estimated that 3% of the mine-derived sediment currently carried by the Fly River will be deposited on the floodplain. If this deposition rate is representative of the fate of the entire sediment load, then we can estimate a floodplain deposition rate. Mean sediment load of the river due to mine waste addition is about 50×10^6 tonnes annually and if 3% of that is spread out over the roughly 3300 km^2 floodplain then for a bulk density of 1 t m^{-3} the deposition rate is 0.5 mm a^{-1}; for a bulk density of 1.7 t m^{-3}, it is 0.3 mm a^{-1}. Sampling of floodplain copper concentrations confirm these estimates and show that the aggradation rate is higher in the forested reach than the swamp reach (Day et al., 1993). These estimates are for the high sediment load associated with the mine which is roughly five times natural load.

Inferences regarding preservation of floodplain features suggest that the swamp grass reach, despite being chronically in flood, actually has the lowest floodplain deposition rate of the Middle Fly. Even before mining began, the sediment discharge of the Fly River above Everill Junction was higher than average for a basin of its size (Milliman and Syvitski, 1992, figure 2), and now it is very high. Why is the sedimentation rate of the floodplain not higher? What controls the rate of floodplain sedimentation?

Floodplain deposition is controlled by the frequency and duration of river flow onto the floodplain and the concentration and size of sediment carried by the river. Three distinct processes transport sediment from the Fly to the surrounding plain: advection by overbank flow; diffusion during flood stage; and transport up tie channels and tributaries. Advection by overbank flows depends on the relative heights of water in the river and on the floodplain. On the Fly, the strongest advective transport probably occurs when the water level across the floodplain is low and the river experiences a large flow generated from storms in the upper catchment. The probability of this occurring is much greater in the forested upper Middle Fly than in the poorly drained, chronically flooded swamp grass reach.

Once the swamp reach becomes flooded, there is probably little pressure gradient between the river and the flooded plain. Water is lost by evapotranspiration from the plain, but can be completely replenished by rainfall on the floodplain itself. In fact, during the flooded state in the swamp reach, it is common to see black, sediment-free floodplain waters bleeding into the river after a few hours of intense local rainfall has elevated the floodplain waters relative to that of the river. Figure 14.9A shows the river just at the transition from flooding onto and draining of the floodplain. On the far bank (flow is from left to right), light-coloured sediment-rich water has spilled behind the tree-lined levee a distance of about one-half the channel width across flooded swamp grass. On the right bank, an irregular boundary between light-coloured river water and black floodplain waters reveals the Fly River partly spilling sediment-rich water onto the plain and partly draining sediment-free floodplain waters.

Higgins (1990, p. 406) estimates that on average for only one month per year does "mainstream flow rather than local runoff contribute significantly to water stored in the floodplain". Hence, despite being flooded for periods in excess of 18 months, the swamp grass reach is not receiving a large advective transport of sediment. This must be a significant contributor to the low overbank deposition rates in this reach. It may also explain why the forested, better drained reach upstream has higher overbank deposition rates; frequency and duration of overbank flows onto the floodplain in this reach are probably greater than in the swamp reach.

A simple calculation emphasizes this point. Let us assume that all the water on the floodplain originates from the channel and has a concentration, C, the same as that in the channel. We can ask how deep, h, must this water be to provide enough sediment to deposit a specified thickness, d for a given bulk density of sediment, ρ_s:

$$C \times (1/\rho_s) \times h \times (\text{unit area of the floodplain}) = d \times (\text{unit area})$$

$$h = d \times \rho_s/(C) \tag{14.6}$$

If $C = 100 \, \text{mg} \, l^{-1}$, $\rho_s = 1 \, \text{g} \, \text{cm}^{-3}$, and $d = 0.1 \, \text{mm} \, a^{-1}$, then $h = 1 \, \text{m}$. Clearly, then, if the Fly River were to spread across the entire floodplain just 1 m deep carrying sediment at the same mean concentration as in the channel, it could cause $0.1 \, \text{mm} \, a^{-1}$ of sediment deposition. Certainly the concentration near the surface will be less than the mean for the flow as a whole and the bulk density will increase well above 1.0 with deep burial; both of these effects require deepening of the standing water needed to cause the required deposition. However, the floodplain waters are typically much deeper in the swamp grass reach, often being 3 to 5 m deep, yet the deposition rate is nonetheless probably much closer to $0.1 \, \text{mm} \, a^{-1}$. For the swamp grass reach to have such a low deposition rate it seems reasonable to conclude that rainwater contributes significantly to flooding here and that this inhibits overbank sediment-rich flows.

Dietrich and Parker (1992), in an unpublished report, describe the findings of the first three years of an intensive sampling programme being conducted by G. Day to monitor the spread of mine-derived sediment across the Fly River floodplain. These preliminary data gave floodplain deposition rates identical to that estimated by Higgins (1990) of 3%

FIGURE 14.9 Photographs showing the penetration of light-coloured, sediment-rich Fly River waters onto the already flooded sediment-poor floodplain. (A) View of the Fly in the swamp reach taken in 1988 (flow from left to right). River-derived waters have reached a distance of about one-half the channel width onto the floodplain from the left (top of picture) bank. Along the right bank the boundary between dark floodplain-derived waters and light-coloured river waters is irregular, with limited overbank deposition in some areas and spilling of floodplain waters to the river elsewhere. The river is about 250 m wide. (B) View of the Agu River about 10 km upstream from the Fly junction. Here the Fly River water is travelling up the Agu and is spilling via a network of distributary tie channels into lakes on the adjacent swampy floodplain. The Agu is roughly 70 m wide

of the total load, and indicated that the deposition rate was roughly 1 to $2\,\text{mm}\,\text{a}^{-1}$ in the forested reach and less than $0.5\,\text{mm}\,\text{a}^{-1}$ in the swamp grass reach. They pointed out that the chronically high water level of the swamp grass reach would prevent the development of pressure gradients from the main channel to the floodplain and thus inhibit floodplain deposition there.

Recent analysis of remote sensing data in several large river systems around the world by Mertes (1997) suggests that the inhibition of sediment-laden overbank flows across the floodplain due to pre-existing water on the floodplain is a widespread phenomenon. Based on our experience on the Fly, this tendency for a given floodplain system may vary widely depending on antecedent conditions on the floodplain. Following a drought in which the swamp grass reach of the Fly completely drains, if rainfall occurs primarily in the uplands then flooding in the lowlands may arise from overbank advection rather than from local precipitation. It may be these rare events that contribute the most important sediment load to the floodplain. An intensive field monitoring programme is now underway on the Fly to document controls on flooding.

The second process of sediment delivery to the floodplain, diffusive transport, occurs because of concentration gradients between the river source and the floodplain sink (e.g. Pizzuto, 1987). It will be strongly influenced by grain size, as coarser particles, once out of the turbulence of the channel flow, will tend to settle out (according to grain size). Presumably advective transport dominates over diffusive transport in the forested reach. Diffusive transport may take on greater significance in the swamp grass reach if advective transport is inhibited. The dense swamp grass may reduce large-scale turbulence and thereby retard diffusive and advective transport. At this time we cannot quantify the relative importance of these processes.

The third process, tie channel and up-tributary flow, follows discrete pathways into the floodplain environment and can cause significant sedimentation across it. Sediment-laden flow up tie channels and tributaries occurs more frequently than does overbank flow across the floodplain. Flow up tie channels may also be an important pathway for rising river flow to flood the surrounding plains before overbank flow develops. We have seen Fly River water travel 40 km up the Agu River (Figure 14.4), and along the way spill out of the Agu (Figure 14.9B), depositing sediment in bordering valleys and plains. Tie channels commonly have distinct deltas oriented away from the Fly, where they enter adjoining oxbows or blocked valley lakes (as noted by Blake and Ollier, 1971), recording the transport of the Fly River sediment often many kilometres across the floodplain. On a tie channel which connects two oxbows to the Fly just upstream of Everill Junction (Figure 14.6B), Markham and Day (1994) monitored flow direction and water level, and found that in this tidally influenced reach, flow could alternate direction in the tie channel many times a day.

The high sedimentation generated by transport through the tie channels raises the question as to why the channels do not quickly plug with sediment. We have witnessed two distinct stage conditions that maintain the channels. The river reaches a low stage during rapid drawdown of the Fly after a period of low rainfall in the uplands or during chronic drawdown during El Niño droughts. The forested reach of the river has been observed to drop over 4 m in less than 48 h following reduced rain in the mountains. This creates a strong pressure gradient from the off-river water bodies towards the mainstem along the tie channels. During one such event we observed flow cascading down the tie channel from an oxbow lake, cutting a trough into the underlying sediments. We have also observed very strong flow up the tie channels during a period when the river stage rose more quickly than rainwater flooded the adjacent plains. This suggests that maintenance of the tie channels depends on particular, relatively rare hydrological events.

A very crude estimate of oxbow infilling, largely due to tie channel sediment contribution, can be made by assuming that production rate of oxbows is balanced by infilling rate such that there is no net increase through the late Holocene of the number of oxbow lakes. The rate of oxbow production may be about one in 30 years (as discussed above) and there are about 29 oxbows between D'Albertis and Everill Junction (about 25 have tie channels). To keep pace with production an individual oxbow would have to fill in 870 years (29 multiplied by 30). For an initially 10 m deep lagoon, this implies a sedimentation rate of $1 \, cm \, a^{-1}$. We can estimate the number of times per year flow must enter the oxbow to generate this deposition through a simple mass balance. If deposition is assumed to occur by settling of sediment at concentration C at N times per year, then when the depth of the lake is H with a fluid density of ρ_w, the deposited sediment will have a grain density of ρ_s and a porosity of p, and deposition rate D is simply:

$$D = NHC(\rho_w/\rho_s)(1/1 - p) \qquad (14.7)$$

If H equals 1000 cm, $\rho_w = 1.0 \, g \, cm^{-3}$, $\rho_s = 2.65 \, g \, cm^{-3}$, $p = 0.35$, $C = 100$ ppm, and $D = 1 \, cm \, a^{-1}$, then N is 17 times per year. Clearly, as the oxbow shallows with deposition, the frequency of flow from the Fly into the lake would have to increase to maintain this deposition rate. Hence it is likely that deposition rates decline through time; to average $1 \, cm \, a^{-1}$, early stage deposition rate would have to be higher.

Alternatively, if we assume instead that the oldest oxbows (which tend to be completely filled) were formed shortly after eustatic sea level stabilized (about 6500 years BP), then the average aggradation rate is $1.5 \, mm \, a^{-1}$. The actual sedimentation rate probably falls between these two estimates of 1 to $10 \, mm \, a^{-1}$, and is much higher than the overbank deposition rate on the floodplain.

One other distribution system deserves special attention. Along the eastern side of the Fly River floodplain between the Binge River and the upstream end of the swamp reach, the Agu River drains southward parallel to the general trend of the Fly (Figure 14.4). It drains at least $1800 \, km^2$ of lowland landscape and passes through a seasonal lake system before it joins the Fly at the head of the swamp reach just above Manda. Various topographic maps indicate that the Agu is connected to the Fly via tie channels at a minimum of five locations (Figure 14.4). Recent field observations during an extended period of flooding confirm that significant flow is draining from the Fly via these channels into the Agu. During high stage, the Agu is also connected to the Binge River upslope, which means that nearly all of the eastern side of the Fly River floodplain could at times drain via the Agu River. As previously mentioned, under certain conditions, flow on the Fly will also travel up the Agu from the tributary junction near Manda. The channels which convey flow from the Fly River to the Agu also convey sediment, hence the Agu system must play a major role in the lateral distribution of sediment, in this case taking sediment beyond what might otherwise be considered the edge of the floodplain.

Downstream fining

The bed material of the pre-mine Fly River systematically fines downstream from Kiunga to Everill Junction, where it coarsens for a short distance with additional sediment from the Strickland, and then fines out into the delta (Figures 14.10, 14.11 and 14.12). This downstream fining raises the possibility that significant deposition is required along the Fly system, which would seem to be at odds with the inferences gained from preservation of floodplain features and the slow meander migration rates. Furthermore, the data do not reveal a sharp grain size reduction at the start of the swamp reach (about 240 km from Kiunga), despite the apparent backwater effects that should reduce sediment transport capacity here.

We examine this problem in two ways. First, we analyse downstream changes in potential mode of sediment transport, then we perform a simple mass balance analysis. The downstream decline in slope of the Fly, if not accompanied by some combination of

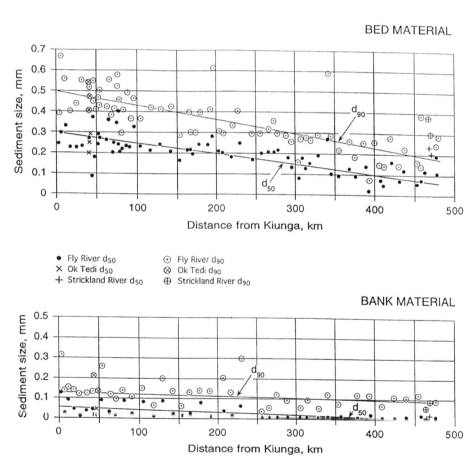

FIGURE 14.10 Downstream change in bed material and bank material (top of bank adjacent to channel). Symbols d_{50} and d_{90} refer to the median grain size and the size for which 90% of the bed is finer, respectively (slightly modified from Pickup et al., 1979)

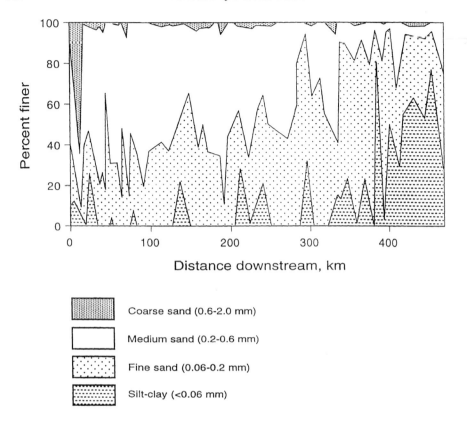

FIGURE 14.11 Bed material grain size variation from Kiunga to Everill Junction (modified from Pickup et al., 1979).

increasing depth, decreasing grain size or declining form drag, should lead to diminished sediment transport capacity of bed material load.

In Figure 14.12, we compare the downstream varying bankfull shear velocity with the downstream decline in bed material size. In order to estimate the shear velocity, the total boundary shear stress was estimated from slope values reported by SMEC (1981) and mean depths determined from cross-sections reported by SMEC and Pickup et al. (1979). Due to the influence of channel sinuosity, bars and finer scale bedforms (dunes and ripples), this total boundary shear stress was reduced by a factor of three (based on experience elsewhere; Dietrich et al., 1984) to estimate the average boundary shear stress responsible for sediment transport. We assumed that for suspension, the local shear velocity must equal or exceed the particle settling velocity (e.g. Raudkivi, 1990); we used the empirical curves of Dietrich (1982) to convert grain size to settling velocity. The vertical axis labelled "maximum grain size suspended" was scaled to correspond to the opposing values of shear velocity using an empirical relationship derived from Dietrich's curves. Hence, it is possible to read on the left axis the shear velocity, and on the right axis the corresponding grain size that can be suspended by that shear velocity.

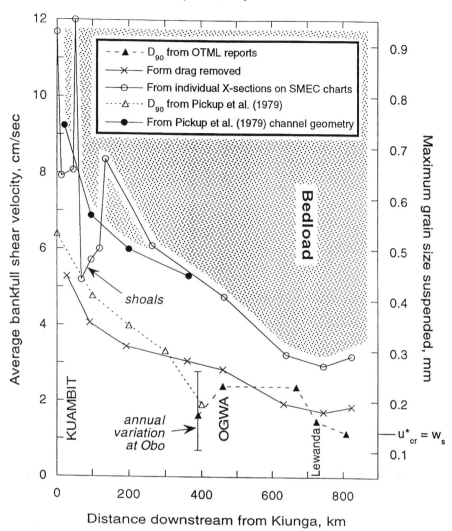

FIGURE 14.12 Downstream variation in average bankfull shear velocity (defined as the $(gdS)^{0.5}$), form drag corrected shear velocity, median and d_{90} of bed material, and maximum grain size that can be suspended by the calculated shear velocity. Note that we use the criteria that suspension occurs when the shear velocity equals or exceeds the settling velocity. If the settling velocity exceeds the shear velocity it will travel primarily as bedload and this area is shown as shaded for the shear velocity values not corrected for form drag. The point where the critical shear velocity for initial motion is equal to settling velocity of the particle is shown on the left vertical axis. Annual variation at Obo is shown because there is large temporal variance of bed material grain size, apparently due to periodic backwater effects of the Strickland.

In Figure 14.12 the downstream varying shear velocity is compared with the downstream varying grain size for which 90% of the particles are finer. The two curves closely follow each other. If the form drag correction of the total boundary shear stress is correct, then this plot also indicates that as the shear velocity declines so does the grain size, and the shear velocity is close to values capable of suspending the entire bed. Apparently, virtually all bed material can be carried in suspension through the Fly despite the large downstream decrease in shear velocity. While Figure 14.12 shows that downstream fining compensates downstream diminished shear velocity, it does not indicate how this fining occurs and whether it requires significant net deposition of sediment.

In a sand-bedded river, downstream grain size reduction due to breakdown of coarser sand during transport is unlikely. Some breakdown undoubtedly occurs during weathering and soil formation when sediment is temporarily stored in the floodplain. If the average bank erosion for the 250 km of active Middle Fly is just $1 \, \text{m} \, \text{a}^{-1}$ and that contributes a bank of material 10 m high, with a bulk density of $1.7 \, \text{t} \, \text{m}^{-3}$, then this introduces 4.3×10^6 tonnes into the river annually, a value equivalent to about half of the annual total load below D'Albertis Junction. Residence of this sediment in the floodplain probably is on average several thousand years, and in this environment weathering does clearly (as seen in bank exposures) alter sediments. However, weathering is inhibited by ground saturation and this probably significantly reduces weathering of deep sediments. One clear indication, however, that it is not only weathering that causes downstream fining, is the fact that the downstream fining has persisted (albeit somewhat muted) despite an increase in sediment load by over a factor of five times due to mine waste dumping.

The bed could fine downstream if a significant portion of the coarser sand is lost to net deposition in the bed and floodplain contributing to aggradation. Alternatively, a greater portion of the fine sand carried in suspension upstream could reside on the bed and travel as bedload or weak suspended load, causing dilution of the sand.

Simple mass balance assumptions can be used to test which of these two mechanisms could explain downstream fining. If grain size reduction occurs without net deposition of sediment due to dilution of coarser sediment with the fine sediment coming out of suspension, then there must be an increase of bedload transport in proportion to the addition of fines. The amount of total sand load coarser than a specified size that travels as bedload at section i is:

$$f_i p_i a_i T_i$$

where T_i = total sediment load at i, a_i = proportion of total load that is sand at i, p_i = proportion of sand load that travels as bedload at i and f_i = proportion of bed material coarser than a specified size i.

In this equilibrium case, if we compare grain sizes at two sections, 1 and 2, then

$$T_1 = T_2; \quad a_1 = a_2; \quad a_1 T_1 = a_2 T_2$$

and

$$f_1 p_1 a_1 = f_2 p_2 a_2$$

hence

$$p_2 = p_1 f_1 / f_2 \qquad (14.8)$$

Note also that $p_1 a_1 T_1$ = bedload discharge at section 1, $(p_1 a_1 T_1)/T_1 = p_1 a_1$ = proportion of total load that is bedload at section 1, and $p_2 a_1$ = proportion of total load that is bedload at section 2.

According to Figure 14.11, at Wygerin 60% of the bed is coarser than 0.2 mm, whereas at Bosset Lagoon only about 20% of the bed material is coarser. This is just above where significant silt and clay were found in the bed. The proportion of the total load that travels as bedload just downstream of D'Albertis Junction (at Kuambit) has been estimated from both measurement and theoretical analysis to be less than 10% (Ok Tedi Mining Ltd, 1987, p.25); here we will use 5 to 10%. Limited data at Kuambit suggest that about 20% of the suspended load is sand, hence $a_1 = 0.95(0.2) + 0.05(1)$ to $0.9(0.2) + 0.1(1)$ or 0.24 to 0.28. This also implies that the proportion of the sand load that travels as bedload, p_1, is 0.21 to 0.36. In Equation 14.8 with $f_1 = 0.6$ and $f_2 = 0.2$, $p_2 = 3p_1$, that is, the proportion of sand load that travels as bedload would increase three times between Wygerin and Bosset. Consequently p_2 would equal 0.63 to 1.0 by Bosset; that is most of the sand would be travelling as bedload by Bosset and the proportion of the total load travelling as bedload would be 0.15 to 0.28.

These simple calculations argue that fining downstream is probably *not* caused by downstream addition of fine sand residing as bed material and travelling as bedload. Note, too, that there is probably little sediment contributed from the few tributaries between D'Albertis and Everill Junctions. No field evidence suggests that bedload transport is more significant at Bosset; in fact, the reduction in the size of bars in this lower reach may indicate an opposite trend. A large increase in downstream bedload transport also seems unlikely under the observed condition of downstream declining boundary shear stress. The relatively large proportion of the total load required to travel as bedload here is also very unlikely. Furthermore, Pickup et al. (1979) found that top-of-channel bank deposits adjacent to the channel still contained up to 30% sand in this reach.

If decrease in grain size is due to net deposition of the coarser fraction from the bedload only, then:

$$a_1 T_1 \neq a_2 T_2$$

but

$$f_1 p_1 a_1 T_1 - f_2 p_2 a_2 T_2 = a_1 T_1 - a_2 T_2 \tag{14.9}$$

i.e., the amount of total sand coarser than a specified size in the bedload at section 1 minus the amount at section 2 downstream equals the difference in the total sand load. Rearranging Equation 14.9, assuming $p_1 = p_2$, and using the values from the previous examples gives:

$$a_2 T_2 = a_1 T_1 (1 - f_1 p_1)/(1 - f_2 p_2)$$

$$= 0.91 a_1 T_1 \text{ to } 0.85 a_1 T_1$$

This surprising result states that the dramatic decrease in medium sand content in the bed material from Wygerin to Bosset and the corresponding decrease in median size from about 0.23 mm to 0.13 mm could occur if less than 15% of the total sand load is lost to deposition. The reason this number is small is the relatively small proportion of the total load that is bedload and the assumption that the medium sand travels only as bedload. This latter assumption is well supported by the expected mode of transport based on shear stress values (Figure 14.12). Also, some medium sand is found as overbank deposits. However, as long as the percentage of the suspended load coarser than 0.2 mm and the total suspended load discharge remain constant between Wygerin and Bosset, the results are the same.

The selective deposition of coarser sand that leads to downstream fining probably occurs by several mechanisms. Slow aggradation of the bed of the river can account for some. For example, if the sediment load before the mine below D'Albertis Junction was 10×10^6 tonnes annually of which 28% was sand, and of this total sand 15% were deposited in the bed between Wygerin and Bosset (about 220 km), then assuming a bulk density of 1.9 (upon deeper burial), the annual aggradation of this reach for a 200 m wide channel bed would be $(10 \times 10^6 \times 0.28 \times 0.15)/ (1.9 \times 220\,000 \times 200)$ or about $5\,\mathrm{mm\,a^{-1}}$, a rate much greater than inferred to be the long-term floodplain deposition. This suggests that significant portions of the medium sand are also deposited on levees and in off-river water bodies.

DISCUSSION AND CONCLUSION

Our analysis suggests that some inferences that have previously been made about the Fly were correct, and others, despite seemingly clear evidence, were not. Despite appearances, the Fly River is currently not an unstable river rapidly crossing its floodplain. On the other hand, estimates of long-term floodplain deposition rates are supported by further analyses presented here. These estimates and other observations, however, seem to contradict a proposal by Harris et al. (1996) regarding sediment storage in the Fly River floodplain. In essence, they report detailed analyses of delta and offshore sedimentation patterns for the Holocene and conclude that the large sediment input from the uplands areas cannot be accounted for in deltaic and near-shore deposition. To solve this missing mass problem, they propose that in the Holocene 30 m of aggradation took place in the Fly and Strickland floodplains. This deposition, they suggest, may have occurred in as little as 3000 years, implying an aggradation rate of $1\,\mathrm{cm\,a^{-1}}$. While Early to Mid-Holocene sedimentation rate was probably much higher than present, our estimates and those of previous workers do not support this higher sedimentation rate. Furthermore, Pickup et al. (1979) report encountering weathered (probably Pleistocene) sediments in the bed of the Fly River near Kiunga, implying limited incision during glaciation.

It seems likely that the swampy conditions with reduced channel migration in the roughly 200 km reach of the lower Middle Fly River formed relatively recently, perhaps due to slope reduction caused by backwater of the Strickland or possibly tectonic effects. A key to understanding what happened is to find out when slope reduction occurred, and an effort is underway to date channel migration history throughout the Middle Fly.

Qualitative observations suggest that sediment is delivered to the floodplain by at least three processes: overbank advection, diffusion, and transport via tie channels and tributaries. The most dramatic case of tributary transport is the drainage of the Middle Fly into the parallel-draining Agu River and the upriver transport of sediment from the downstream tributary junction. Very few data are available on rates of sedimentation caused by these various processes on large lowland floodplains. We can infer that rates of sedimentation will be controlled by near-surface river flow sediment concentration and grain size, frequency of overbank flow and frequency of flow up tie channels and tributaries. These latter frequencies appear to be tied to timing of upland versus lowland rainfall, which dictates how wet the floodplain already is when the flow on the Fly rises.

Recent analysis of remote sensing data in several large river systems around the world by Mertes (1997) suggests that the inhibition of sediment-laden overbank flows across the floodplain due to pre-existing water on the floodplain is a widespread phenomenon. Observations on the Fly suggest there will be annual variation in the relative timing of upland versus lowland rainfall and there will be longer time variations, for example El Niño drought events, that play a significant role in determining when important depositional events occur across the floodplain. A programme is now underway to try to monitor the floodplain hydrology directly so as to understand the controls on timing and magnitude of flooding.

Downstream fining of bed material on large sandy rivers has received little attention compared to that given to the phenomenon on gravel-bedded rivers (e.g. Paola et al., 1992). Nonetheless, dramatic fining can occur. Leopold et al. (1964) report data on the Mississippi River collected before 1935 when the river was not converted to a network of dams and lakes. These data show the mean grain diameter decreasing in size at a constant rate over 1500 km from about 0.65 mm just below New Madrid, Missouri, to less than 0.2 mm below New Orleans, Louisiana. This fining results from systematic loss of coarser sediment (including a small amount of gravel in the upstream reaches). Based on our analysis of the Fly, rather than reason that this reflects a large net loss of sediment and a significant decline in the river's transport capacity, we suggest that such fining can occur with a relatively minor amount of net aggradation of the river bed or net discharge to the floodplain environment. Alternatively, there could have been a dilution effect caused by addition of significant amounts of finer sediment from the large tributaries which, unlike the Fly, line the Mississippi.

Through reanalysis of existing data and addition of some new data on the Fly River, we have been able to reject some inferences, confirm or at least support others, and offer a few new interpretations. Field and modelling research is now underway that should allow us to investigate much more quantitatively the issues raised here, no doubt introducing new ones.

ACKNOWLEDGEMENTS

We thank Alan Howard and Geoff Pickup for valuable reviews of the manuscript. Martin Trso assisted in drafting some of the figures. Through the years this work has been supported by Murray Eagle, Ian Wood and Marshall Lee and the staff of the Environment Department of Ok Tedi Mining Ltd, Papua New Guinea.

REFERENCES

Allen, J.R.L. 1965. A review of the origin and characteristics of recent alluvial sediments. *Sedimentology*, **5**, 89–191.

Barr Engineering Co. 1995. *Investigation of the effect of mine tailings on meander rate of the Fly River*. Report to Ok Tedi Mining Ltd, February 1995.

Blake, D.H. 1971. *Geology and geomorphology of the Morehead-Kiunga area*. CSIRO Australian Land Research Series No. 29; 56–68.

Blake, D.H. and Ollier, C.D. 1969. Geomorphological evidence of Quaternary tectonics in southwestern Papua. *Revue Geomorphologie Dynamique*, **19**, 28–32.

Blake, D.H. and Ollier, C.D. 1971. Alluvial plains of the Fly River, Papua. *Zeitschrift für Geomorphologie*, Suppl. bd **12**, 1–17.

Blong, R.J. 1991. *The magnitude and frequency of large landslides in the Ok Tedi catchment.* Report to Ok Tedi Mining Ltd, May 1991.

Bureau of Mineral Resources, Geology and Geophysics, Canberra, ACT. 1976. *Geology of Papua New Guinea*, 1:2 500 000 map.

Chappell, J., Rhodes, E.G., Thom, B.G. and Wallensky, E. 1982. Hydro-isostasy and the sea-level isobase of 5500 BP in North Queensland, Australia. *Marine Geology*, **49**, 81–90.

Church, M. and Slaymaker, O. 1989. Disequilibrium of Holocene sediment yield in glaciated British Columbia. *Nature*, **337**, 452–454.

Day, G.M., Dietrich, W.E., Apte, S.C., Batley, G.E. and Markham, A. J. 1993. The fate of mine-derived sediments deposited on the middle Fly River flood-plain of Papua New Guinea. In R.J. Allan and J.O. Nriagu (Eds), *International Conference on Heavy Metals in the Environment*, Volume 1. CEP Consultants, Ltd, Edinburgh, 423–426.

Dietrich, W.E., 1982. Settling velocity of natural particles. *Water Resources Research*, **18**(6), 1615–1626.

Dietrich, W.E., 1988. *Effects of mined sediment discharges on the channel and floodplain of the Fly River: a geomorphic perspective.* Report to Ok Tedi Mining Ltd, October 1988.

Dietrich, W.E. and Parker, G. 1992. *Analysis of copper contamination of the Fly River floodplain.* Report to Ok Tedi Mining Limited, November 1992.

Dietrich, W.E., Smith, J.D. and Dunne, T. 1984. Boundary shear stress, sediment transport and bed morphology in a sand-bedded river meander during high and low flow. In *Rivers '83: Proceedings of a Specialty Conference on River Meandering*, October 1983 American Society of Civil Engineers, 632–639.

Eagle, A.M, Cloke, P.S. and Hortle, K.G. 1986. Environmental management, monitoring and assessment: Ok Tedi mining project, Papua New Guinea. *National Environmental Engineering Conference*, Melbourne 17–18 March 1986, 75–80.

Fisk, N.H. 1944. *Geological investigation of the alluvial valley of the lower Mississippi River.* Mississippi River Commission, Vicksburg, Miss, 78 pp.

Harris, P.T., Baker, E.K., Cole, A.R. and Short, S.A. 1993. A preliminary study of sedimentation in the tidally dominated Fly River Delta. Gulf of Papua, *Continental Shelf Research*. **13**, 441–472.

Harris, P.T., Pattiaratchi, C.B., Keene, J.B., Dalrymple, R.W., Bardner, J.V., Baker, E.K., Cole, A.R., Mitchell, D., Gibbs, P. and Schroeder, W.E. 1996. Late Quaternary deltaic and carbonate sedimentation in the Gulf of Papua foreland basin: response to sea-level change. *Journal of Sedimentary Research*, **66**(4), 801–819.

Hasegawa, J. 1989. Studies on qualitative and quantitative prediction of meander channel shift. In S. Ikeda and G. Parker (Eds), *River Meandering*. American Geophysical Union Water Resources Monograph 12, 215–236.

Hettler, J. and Lehmann, B. 1995. Environmental impact of large-scale mining in Papua New Guinea: Mining residue disposal by the Ok Tedi Copper-Gold Mine. *Berliner Geowissenschaftliche Abhandlungen*, Reihe A, Band, p. 49.

Higgins, R.J. 1990. Off-river storages as sources and sinks for environmental contaminants. *Regulated Rivers*, **5**, 401–412.

Higgins, R.J., Pickup, G., and Cloke, P.S. 1987. Estimating the transport and deposition of mining waste at Ok Tedi. In C.R. Thorne, J.C. Bathurst and R.D. Hey (Eds), *Sediment Transport in Gravel-bed Rivers.* Wiley, Chichester, 949–976.

Hill, K.C. 1991. Structure of the Papua fold belt, Papua New Guinea. *American Association of Petroleum Geologists, Bulletin*, **75**(5), 857–872.

Hooke, J.M. 1984. Changes in river meanders: A review of techniques and results of analysis. *Progress in Physical Geography*, **8**, 473–508.

Howard, A.D. 1992. Modeling channel migration and floodplain sedimentation in meandering streams. In P.A. Carling and G.E. Petts (Eds) *Lowland Floodplain Rivers.* Wiley, Chichester, 165–183.

Ikeda, H. 1989. Sedimentary control on channel migration and origin of point bars in sand-bedded meandering rivers. In S. Ikeda and G. Parker (Eds), *River Meandering*. American Geophysical Union Water Resources Monograph 12, 51–68.

Ikeda, S, Parker, G. and Sawai, K., 1981. Bend theory of river meanders, Part 1. Linear development. *Journal of Fluid Mechanics*, **112**, 363–377.

Larsen, E.W. 1995. *Mechanics and modeling of river meander migration* PhD. Dissertation, University of California, Berkeley, 342 pp.

Leopold, L.B., Wolman, M.G. and Miller, J.P. 1964. *Fluvial Processes in Geomorphology.* Freeman, San Francisco, 522 pp.

Loffler, E. 1977. *Geomorphology of Papua New Guinea.* Australian National University Press, Canberra, 258 pp.

MacDonald, T.E., Parker, G. and Leuthe, D.P. 1991. *Inventory and analysis of stream meander problems in Minnesota, Minneapolis.* St Anthony Falls Hydraulic Laboratory, University of Minnesota, 37 pp.

Markham, A.J. 1995. *Hydrology Annual Report, 1993–1994.* ENV95 06, OTML, 28 pp..

Markham, A.J. and Day, G. 1994. *Sediment transport in the Fly River basin, Papua New Guinea.* No. 224, 233–239.

Mertes, L.A.K. 1997. Documentation and significance of the perirheic zone on inundated floodplains. *Water Resources Research,* **33**, 1749–1762.

Milliman, J.D. and Syvitski, J.P.M. 1992. Geomorphic/tectonic control of sediment discharge to the ocean: the importance of small mountainous rivers. *Journal of Geology,* **100**, 525–544.

Nanson, G.C. and Hickin, E. J. 1986. A statistical analysis of bank erosion and channel migration in western Canada. *Geological Society of America, Bulletin,* **97**, 497–504.

Odegaard, A.J. 1989. River Meander model. II. Application. *Journal of Hydraulic Engineering,* **115**, 1451–1464.

Ok Tedi Mining Ltd. 1987. *Environmental Study Progress Report* No. 3, 174 pp.

Paola, C., Parker, G., Seal, R., Sinha, S., Southard, J.B. and Wilcock, P.R. 1992. Downstream fining by selective deposition in a laboratory flume. *Science,* **245**, 393–396.

Parker, G. 1991. Selective sorting and abaison of river gravel. I: Theory. *Journal of Hydraulic Engineering,* **117**(2), 131–149.

Pernetta, J.C. 1988. *Potential impacts of mining on the Fly River.* United Nations Environment Program, Regional Seas Reports and Studies No. 99, SREP Topic Review No. 33, 119 pp.

Pickup, G. 1984. Geomorphology of tropical rivers: I. Landforms, hydrology, and sedimentation in the Fly and lower Purari, Papua New Guinea. In A. Schick (Ed.), *Catena Supplement* **5**. 1–18.

Pickup, G. and Warner, R.F. 1984. Geomorphology of tropical rivers II: Channel adjustment to sediment load and discharge in the Fly and Lower Purari, Papua New Guinea., In A. Schick (Ed.), *Catena Supplement* **5**, 18–41.

Pickup, G., Higgins, R. J. and Warner, R. F. 1979. *Impact of waste rock disposal from the proposed Ok Tedi mine on sedimentation processes in the Fly River and its tributaries, Papua New Guinea.* Department of Minerals and Energy and Office of Environment and Conservation, 138 pp.

Pickup, G., Higgins, R.J. and Warner, R. F. 1981. *Erosion and sediment yield in Fly River drainage basins, Papua New Guinea.* International Association of Hydrological Science, Publication No. 132, 438–456.

Pigram, C.J., Davies, P.J., Feary, D.A. and Symonds, P.A. 1989. Tectonic controls on carbonate platform evolution in southern Papua New Guinea: Passive margin to foreland basin. *Geology,* **17**, 199–202.

Pizzuto, J.E. 1987. Sediment diffusion during overbank flow. *Sedimentology,* **34**, 301–317.

Raudkivi, A.J. 1990. *Loose Boundary Hydraulics.* Pergamon Press, Exeter, 53 pp.

Smith, R.E.W. and Bakowa, K.A. 1994. Utilization of floodplain water bodies by fishes of the Fly River Papua New Guinea. *Mitteilungen Societar Internationalis Limnologiae,* **24**, 187–196.

Smith, R.E.W. and Hortle, K.G. 1991. Assessment and prediction of the impacts of the Ok Tedi copper mine on fish catches in the Fly River system. *Environmental Monitoring and Assessment,* **18**, 41–68.

Snowy Mountains Engineering Corporation (SMEC). 1981. *Hydrographic Survey of the Fly River, Papua New Guinea – Western Province.*

Taylor, D. 1979. *Sediment sources and quantities in Fly drainage system.* Report by Paltech Pty. Ltd, 18 pp.

Wolanski, E. and Eagle, A.M. 1991. Oceanography and sediment transport, Fly River Estuary and Gulf of Papua. *10th Australasian Conference on Coastal and Ocean Engineering,* Auckland, New Zealand, 2–6 December 1991, 6 pp.

15

Downstream Changes in Valley Confinement as a Control on Floodplain Morphology, Lower Tuross River, New South Wales, Australia: A Constructivist Approach to Floodplain Analysis

ROB J. FERGUSON AND GARY J. BRIERLEY

School of Earth Sciences, Macquarie University, North Ryde, NSW, Australia

ABSTRACT

Tuross River, on the south coast of New South Wales, is a confined sandy meandering river. Although channel width remains remarkably uniform along the lower 35 km of the river, there are marked downstream changes in floodplain character. Assemblages of element-scale geomorphic units on reaches of discontinuous floodplain reflect differing processes of floodplain accretion and modification associated primarily with downstream changes in valley confinement. In order of decreasing stream power, three primary floodplain styles have been discerned: *vertically accreted sandy* floodplains, some of which are stripped or scoured; *laterally accreted sandy* floodplains, characterized by ridge and swale topography; and *vertically accreted silty* floodplains which are found in open valley reaches or in areas protected by bedrock. Floodplain styles are associated closely with downstream changes in the type of valley floor trough, differentiated into inset, incised and backwater settings. Element assemblages on floodplains in the inset and incised segments reflect high stream power values characteristic of confined valleys, with floodplain modification evidenced through floodchannel scour and catastrophic stripping. Floodplains in backwater reaches at the downstream end of the system reflect lower stream power values. However, the juxtaposition of laterally accreted sandy and vertically accreted silty floodplains in the open reaches of the lowermost Tuross indicates that stream power–floodplain morphology relationships break down in valley widths over 700 m, as hydraulic conditions become more diverse, and autocyclic processes such as channel migration become the key determinant of floodplain style and distribution.

INTRODUCTION

While considerable attention has been placed in the geomorphic literature on channel style (e.g. Rosgen, 1994), significantly less emphasis has been placed on variability in

Varieties of Fluvial Form. Edited by A.J. Miller and A. Gupta
© 1999 John Wiley & Sons Ltd

floodplain style, and the association between channel- and floodplain-forming processes. This dilemma, reported by Bridge (1985), is reflected in the scant regard given to floodplain depositional units in the approach to river sediment inventory compiled by Miall (1985), wherein one architectural element characterizes "overbank" deposits. Advances in our understanding of floodplain-forming processes were recently summarized in a genetic classification scheme for floodplains developed by Nanson and Croke (1992). Appreciation of the variability in floodplain character has come a long way since the simple differentiation of within-channel and overbank deposits (or lateral versus vertical accretion deposits, bottom-stratum versus top-stratum deposits; *sensu* Fisk, 1944, 1947; Wolman and Leopold, 1957). Nanson and Croke's approach to floodplain classification, based on stream power relations and resistance to erosion of the floodplain (viewed largely in terms of sediment texture), works on the principle that the distribution of flood energy within fluvial systems is a key control over processes of channel and floodplain formation, modification and preservation. As acknowledged by Nanson and Croke (1992), such classifications do not necessarily apply to entire systems, as individual reaches may demonstrate a differing floodplain style. For example, Miller (1995) illustrated the control that valley morphology exerts over the spatial distribution of stream power during a given flood event, thereby influencing floodplain-forming processes. In this paper, within-reach variability in floodplain style is characterized and explained for the Tuross Valley, on the south coast of New South Wales (NSW), Australia.

Analysis of river style is performed through assessment of assemblages of geomorphic units that comprise differing sections, or reaches, of Tuross River (Brierley, 1991a, b, 1996). The complex floodplains of the lower Tuross River, with their mosaics of geomorphic units, require the systematic analysis and characterization provided by the constructivist approach (Brierley, 1996). Classification of floodplains into genetically related types can only follow thorough characterization.

CONSTRUCTIVIST METHODOLOGY

The overall floodplain study is framed in terms of the hypothesis that valley confinement is the fundamental control over the observed floodplain assemblages along the lower Tuross River.

Geomorphic assessment of floodplains took place in the following steps:

(1) photo-interpretive mapping of channel and floodplain features at *c.* 1:6 000 scale;
(2) field survey and mapping of channel and floodplain features at 1:10 000 scale; qualitative assessment of form/process associations;
(3) analysis of spatial patterns of geomorphic units, and establishment of possible genetic relationships.

The *constructivist* approach outlined by Brierley (1996) provides a flexible and powerful framework to approach analysis of the extremely complex and variable floodplains found on the lower 35 km of the Tuross River above the estuary. Detailed geomorphic mapping (steps 1 and 2) provided the initial data set for analysis. The first outcome was an appreciation of every geomorphic unit (or element) present on the channel margin and floodplains, and the degree of variability within any given element. Individual elements

TABLE 15.1   Genetic floodplain style, element assemblages and assemblage packages of the lower Tuross River

Genetic floodplain style	Element assemblages include combinations of:	Assemblage packages Include combinations of:
Vertically accreted sandy	Levee Floodchannel Flat floodplain Crevasse splay Concave bank bench Bench	Vertically accreted sandy and vertically accreted silty
Laterally accreted sandy	Ridge and swale topography Levee Crevasse splay Flat floodplain Point bench Floodchannel	Laterally accreted sandy and vertically accreted silty
Vertically accreted silty	Levee Distal floodplain Lake Backswamp Concave bank bench	

"are viewed simply as the geomorphic building blocks of a river system" (Brierley, 1996, p.277).

Individual elements are then placed together in their naturally occurring combinations as step 3 (see Table 15.1) to give element assemblages. For example, levees and floodchannels are often found in association, as are ridge and swale topography and flat floodplain, and levees and distal floodplain. Element assemblages are combined to form assemblage packages, an essential step on some of the highly complex floodplains of the lower Tuross.

Elements that are generally recognized to be diagnostic of a particular style of floodplain accretion (Table 15.1) form the basis of a genetic classification that will subsequently be tested by detailed sedimentological investigation. The key assumption is that a particular geomorphic unit is indeed diagnostic of a floodplain accretion process (see Table 15.2).

REGIONAL SETTING

The Tuross River drains a catchment area of 2180 km^2 in southeastern NSW (Figure 15.1). Elevations in the rugged western half of the catchment exceed 1000 m, and over 80% of the catchment has slopes greater than 15°. The geology of the catchment is characterized by Devonian granites in the upstream half, and Ordovician metasediments in the downstream half. The granites typically weather to produce sand-sized quartz grains, whereas the metasediments produce fines and boulder-sized particles. In common with many other coastal valleys of NSW, the Tuross River is considered to be sediment supply-limited (cf. Nanson and Erskine, 1988).

TABLE 15.2   Relationship of floodplain elements to floodplain accretionary style on the lower Tuross River

Geomorphic unit	Inferred primary process	Source
Levee	Vertical accretion	Fisk (1944), Allen (1965), Nanson (1986)
Crevasse splay	Vertical accretion	Fisk (1944), Allen (1965)
Concave bank bench	Vertical accretion	Nanson and Page (1983)
Ridge and swale topography	Lateral accretion	Nanson (1980), Iseya and Ikeda (1989)
Flat floodplain (sandy)	Vertical accretion (sand sheet deposition)	Burkham (1972), Schumm and Lichty (1963)
Distal floodplain (silty)	Vertical accretion (fines settling out from suspension)	Fisk (1944), Allen (1965)
Floodchannel	Scour	Burkham (1972), Nanson (1986)

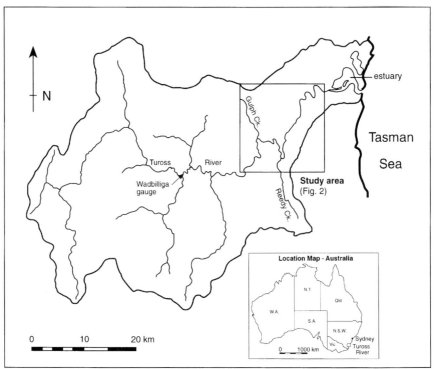

FIGURE 15.1   Map of the Tuross River, showing its location on the Australian continent, and the position of the study reach. The upstream boundary of the study reach is defined by valley widths too narrow to allow floodplain formation. The downstream limit is where the valley abruptly widens to around 5 km, where estuarine processes and landforms become dominant

Average annual rainfall ranges from 800 mm on the coast to 1100 mm on the elevated west of the catchment, with an average runoff of 280 mm (Millar, 1994). Logging has occurred over much of the catchment, but most upper catchment areas have returned to bush cover. A large part of the upper catchment is national park. Vegetation clearance has been restricted to floodplains and small pockets within 30 km of the coast with only minor exceptions.

The discharge record for the Tuross River is poor. Data from Wadbilliga station (located on Figure 15.1; record from 1964 onwards) indicate a mean annual flood of around 300 m^3 s^{-1}, while estimates of the one-in-10 and one-in-50 year events are around 2350 and 4750 m^3 s^{-1}, respectively. Annual variability in total flow (as expressed as a percentage of the mean annual discharge) ranges from 6% of the average during droughts (such as the mid-1960s and early 1980s) to 260% of average during wet years (such as the late 1980s to early 1990s). Flood events appear to be clustered (e.g. three events > *c.* 2000 m^3 s^{-1} within 18 months in 1991–92). Anecdotal reports suggest that most floods are flashy, resulting from high rainfall in the western half of the catchment. For example, the December 1992 event of approximately 2000 m^3 s^{-1} rose and fell approximately 10 m overnight (J. Simpson, pers. comm., 1994).

The study reach is the 35 km of the river valley immediately upstream of the estuary (Figures 15.1 and 15.2). Upstream of this reach, the valley is bedrock-confined and not wide enough to allow floodplain deposition. Examination of aerial photographs from 1944 to 1994 indicates negligible changes in channel position along the lower course of the river.

Upper, middle and lower valley segments have been demarcated in the study reach (Figures 15.2 and 15.3). Over the lower segment, channel bank height gradually increases from 4 to 8 m moving upstream, while channel width decreases from approximately 130 m to around 80–100 m (Table 15.3). Upstream of here, channel dimensions vary little in the study reach, with channel width generally between 90 and 110 m and bank height increasing to a maximum of 13 m.

In contrast to the relatively consistent downstream pattern in channel width, valley width is highly variable. In the uppermost segment of the study reach, valley widths are generally less than 500 m, and frequently less than 250 m. Discrete sections of discontinuous floodplain are up to 1600 m long, but are generally less than 1000 m, with width varying between *c.* 50 m and 400 m. Valley sides are steep (> 30°) and in many reaches the channel abuts on the valley wall. In the middle segment valley width is generally wider, averaging around 500 m, with exceptions at Tyronne (wider) and Rewlee (narrower) (see Figures 15.2 and 15.3). The channel is gently sinuous and switches from one side of the valley to the other, with less direct contact with bedrock than upstream. Valley sides slope at between 10 and 30°. The lowermost segment of the valley is characterized by a series of bedrock spurs which create embayments where floodplains up to 2000 m long have formed. At the end of the spurs, valley width varies between 350 and 600 m, and the more open sections of valley are generally up to 1200 m across (though with a maximum of 2400 m). Valley margins are typically around 10°, but can be up to 30° where the channel abuts bedrock.

The valley floor at the lower end of the study reach has been eroded to a depth of at least 42 m below sea level (NSW Department of Water Resources drill hole 39041). The fill of the valley to at least 12 km up the study reach is dominated by cohesive estuarine muds, which underlie fluvial deposits. The lower 23 km of the study reach has a gradient of between 0.0005 and 0.001 (see Figure 15.4a). The upper 10 km of the study reach has a mean channel gradient of around 0.0015. Lateral and point bars are common along the

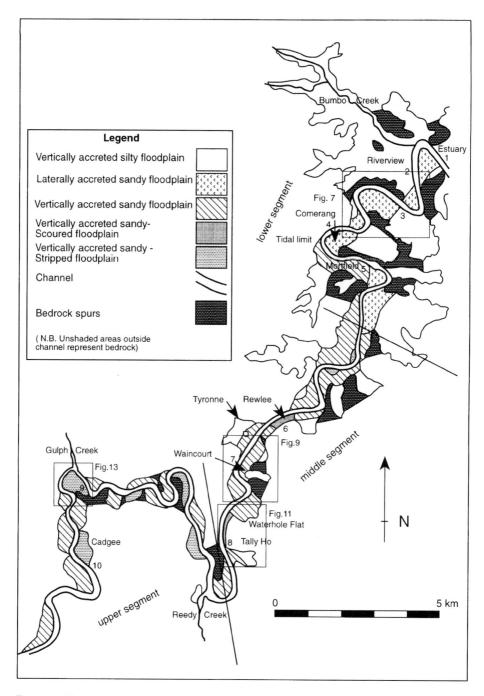

FIGURE 15.2   Distribution of floodplain styles throughout the study reach and the three main valley segments. Note the numerous small tributary valleys that abruptly increase valley width, and also the bedrock spurs (highlighted) downstream of Mortfield that control valley width and shape. Boxed areas are shown in detail in Figures 15.7, 15.9, 15.11 and 15.13

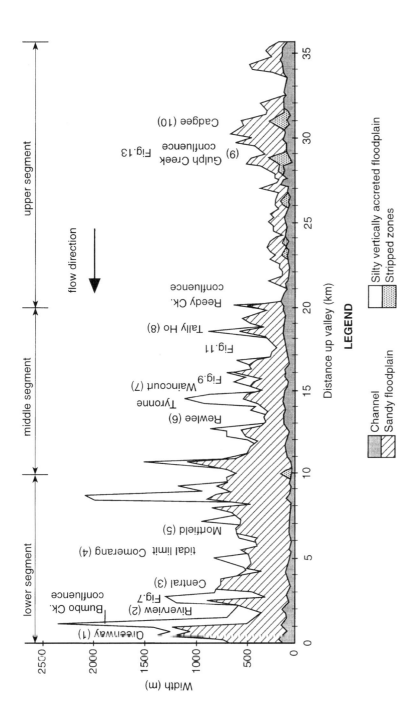

FIGURE 15.3 Comparison of valley width and the length and type of floodplain present in the 36 km long study reach. Place names are of sites listed in the text, and figure numbers refer to the locations of sites shown in detail. Numbers in parentheses (1–10) refer to locations where hydraulic parameters have been measured and bankfull discharge and within-channel unit stream power calculated (refer to Table 15.3). Channel, floodplain and valley dimensions were recorded at 250 m intervals down the length of the valley, perpendicular to approximate valley "centre". Sandy floodplain refers to both laterally and vertically accreted sandy floodplains

TABLE 15.3 Channel parameters and estimated discharge characteristics

Location*	Manning's n	Water surface gradient	Maximum bankfull depth (m)	Derived velocity† (ms^{-1})	Bankfull width (m)	Q_{bf}‡ ($m^3 s^{-1}$)	Max. unit stream power§ ($W\,m^{-2}$)
Cadgee 10	0.04	0.0015	13	5.35	100	6260	1022
Gulph Creek confl. 9	0.04	0.0013	13	4.97	90	5298	823
Tally Ho 8	0.04	0.001	12	4.14	85	3726	486
Waincourt 7	0.04	0.001	10	3.67	90	2061	360
Rewlee 6	0.04	0.001	9	3.42	90	1860	301
Mortfield 5	0.04	0.001	8	3.16	70	1517	247
Comerang 4	0.04	0.0005	8	2.24	70	1075	87
Central 3	0.04	0.0005	7	2.05	110	1435	70
Riverview Greenway 2	0.04	0.0005	5	1.64	175	1370	40
Greenway, 200 m upstream of bridge 1	0.04	0.0005	4.5	1.53	200	1306	33

* Numbers refer to locations shown in Figures 15.2, 15.3 and 15.4
† utilizing the equation: $v = d^{0.666} s^{0.5} / n$
‡ $Q_{bf} = A \times v$
§ Calculated using maximum depth and velocity, utilizing the equation:
$(d_{max} \times v) \times$ gravitational constant $\times$ gradient
d_{max} = depth of flow over the thalweg at Q_{bf}

(a)

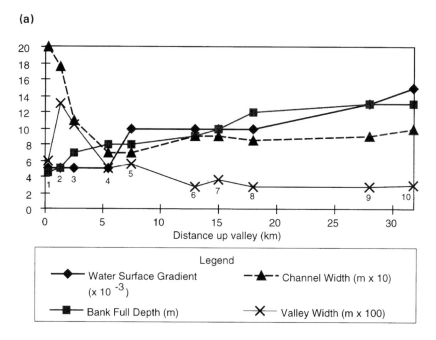

(b)

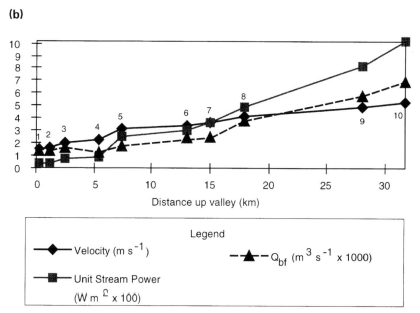

FIGURE 15.4　(a) Measurements of bankfull channel width, bankfull channel depth, water surface gradient and valley width at 10 sites within the lower Tuross Valley (see Figure 15.2 for locations). (b) Estimated velocity, bankfull discharge (Q_{bf}) and within-channel unit stream power for 10 sites within the lower Tuross Valley. See Table 15.3 for details of equations used. Note the dramatic decrease in Q_{bf} going downstream, a result of decreased gradient and depth that is not compensated for by increases in channel width

entire length of the channel, with chute channels up to 2 m above low flow level evident at a number of sites in the middle and upper reaches. Bars in the lowermost 7 km are exclusively sandy (coarse to very coarse) but gravel (b_{max} up to 100 mm) becomes increasingly common upstream.

The pronounced downstream changes in channel depth and slope within the study reach result in dramatic downstream changes in estimates for bankfull velocity and discharge, as well as unit stream power (Figure 15.4b). The implications of these changes on floodplain character are discussed later in the paper.

FLOODPLAIN CHARACTERIZATION ALONG THE LOWER TUROSS RIVER

Element description

Lower Tuross floodplains and channel margins are made up of 12 primary geomorphic elements (see Table 15.4 and Figure 15.5a,b).

Small benches are common throughout the river, occurring as two main types. *Point benches* on broad bends are up to 100 m wide at the bend apex, with their surfaces generally 1–2 m lower than the adjacent floodplain (see Figure 15.5).

Benches on straight reaches are typically only a few metres wide and less than 100 m long. One notable exception is found at Rewlee in the middle segment (see Figure 15.2), where a large bench (40–50 m wide, over 1000 m long and raised 5–6 m above the channel; see Figure 15.5) is inset 4–5 m below a prominent levee. Scour holes around trees are common on bench surfaces and at the entrance to floodchannels which cut sections of the floodplain. Ridges also occur in a number of places within the channel, and are up to 5 m high, 30 m wide and several hundreds of metres long. Chute channels, raised approximately 1–2 m above the low flow channel, occur between these ridges and the channel margin.

Levees are common throughout the lower Tuross and result from loss of flow competence when sediment-charged floodwaters overtop the channel banks (Figure 15.5 and Table 15.4). Levees with floodchannels on their distal margins (as opposed to the association with distal floodplain depicted in Figure 15.5) show direct evidence of scour, indicating that both erosion and deposition can control levee form. These levees aggrade during non-erosive and/or waning flood stages. In backwater reaches, levees on vertically accreted silty floodplains are approximately the same height as adjacent or opposite laterally accreted sandy floodplain surfaces.

Crevasse splays are narrow, elongate features oblique to the channel (Figure 15.5 and Table 15.4). They cut levee crests and are commonly deposited in floodchannels. Crevasse scours on the levee crest are only 5–10 m wide and 1–2 m deep. The depositional lobe is characterized by hummocky relief, and can be up to 50 m long.

Concave bank benches on the lower Tuross are found exclusively at the downstream end of floodplains (Figure 15.5 and Table 15.4), and result from the development of a secondary circulation cell, formed when flood flow abuts against the bedrock of the valley margin (Nanson and Page, 1983). The centre of the bench contains a low, gentle ridge. During a flood, the distal depression becomes a channel in which flood waters flow upstream, having been deflected off the bedrock at the downstream end of the bench.

Ridge and swale topography on the lower Tuross has up to 3 m of relief, with wavelengths of 30–50 m. Patterns are generally arcuate and subparallel to the channel bend (Figure 15.5 and Table 15.4), but on at least one floodplain swales bifurcate over short distances to form patterns generally not seen in "classic" ridge and swale topography (cf. Jackson, 1976; Nanson, 1980).

Numerous forms of *floodchannels* are evident along the lower Tuross River. In general, they are intimately associated with levees in high (usually > 300 W m^{-2}) stream power situations, where the distal levee forms one channel bank, and bedrock or a terrace edge the other (Figure 15.5 and Table 15.4). Although these floodchannels are found in high energy environments, they are fine grained at the surface and do not presently show evidence of bedload deposition. Floodchannels characteristic of backwater reaches (Figure 15.5) do presently transport bedload material, and have sandy floors and banks, in contrast to the well vegetated, levee-associated floodchannels.

There are two floodplain elements on the lower Tuross: flat and distal. *Flat floodplains* are sandy, flat-topped surfaces that are several metres higher than *distal floodplains* (Figure 15.5 and Table 15.4). When adjacent, the boundary between the two floodplain units is a bank 2–4 m high which slopes at gradients up to 20°. Distal floodplains are fine grained, are very gently sloping to flat, and frequently include lakes and backswamps in bedrock-protected margins.

Element assemblages

The ultimate aim of establishing floodplain assemblages is to create a classification scheme that reflects processes of floodplain genesis and/or modification. Analysis of the spatial assemblages of floodplain elements along the lower Tuross River (see Table 15.2 and Figures 15.2 and 15.6) led to the identification of the categories listed below.

(1) *Vertically accreted silty* floodplains are dominated by the distal floodplain–levee assemblage (see Figure 15.6). An example is illustrated in Figures 15.7 and 15.8. Lakes and/or backswamps are extremely common on these floodplains. A concave bank bench is found on one small floodplain of this type at the lowermost end of the study reach.

(2) *Laterally accreted sandy* floodplains are characterized by assemblages of ridge and swale topography, levees, floodchannels, flat floodplains, point benches and, rarely, lakes (Figure 15.6), and appear to represent the classic model of point bar migration on the inside of bends (Fisk, 1944, 1947; Jackson, 1976). Ridge and swale topography, and flat floodplain constitute the majority of these floodplains, with levees found on some proximal margins (see Figures 15.7 and 15.8). Floodchannels take the most direct course possible across the floodplain.

(3a) *Vertically accreted sandy* floodplains are dominated by assemblages consisting of flat floodplains, point benches, distal floodplain and floodchannels (Figure 15.6). The vast majority of the floodplain area is flat floodplain with point bench found on curved sections of channel, possibly representing channel contraction. Floodchannels are small (<10 m wide) and generally cut across from the upstream end of the floodplain into small areas of distal floodplain, which are always protected by bedrock or located in tributary valleys (see Figures 15.9 and 15.10). These floodplains generally show no evidence of "classic" proximal–distal change in geomorphic units (i.e. no backswamps or lakes) or lateral variability in material texture, reflecting the uniformity of floodplain accreting processes (possibly vertical accretion of sand sheets deposited under high energy conditions).

TABLE 15.4 Morphometric analysis of lower Tuross River geomorphic elements

Geomorphic element	Geometry	Scale	Position/associated units/assemblage	Floodplain style	Estimated stream power (W m^{-2})
Ridge and swale topography	Elongate/arcuate series of ridges and depressions, subparallel to channel, can bifurcate	Height 0.5–3 m; width 20–40 m; length 150–500+ m	Inside of bends, flat floodplain, levee	Laterally accreted sandy	Up to 250
Levee	Raised elongate ridge, steeper on proximal than distal side	Height above floodplain 1.5–5 m; width 10–100 m (crest to distal)	Channel margin; floodchannel, distal floodplain, ridge and swale topography	Lateral sandy; Vertical silty; Vertical sandy, scoured	Up to 250; Up to c.100; 300–500
Major floodchannel	Gently curved channel	Up to 4 m deep; 30 m wide; 700 m long	Immediately downstream of bedrock spurs; flat floodplain, ridge and swale	Laterally accreted sandy; Vertically accreted sandy	Up to 100; Up to 250
Minor floodchannel	Straight to gently sinuous	Up to 2 m deep; 5–15 m wide; 500 m long	Flat floodplain, vertically accreted sandy floodplain	Laterally accreted sandy; Vertically accreted sandy	Up to 100; Up to 500
Distal floodplain	Flat expanse	Up to 1.5 km long; 800 m wide	Lakes, backswamps, levee	Vertically accreted silty	Up to 100

(continues)

TABLE 15.4 (*continued*)

Lake	Circular to irregular	Up to 300 m long	Distal floodplain, backswamp, bedrock valley margin	Vertically accreted silty; Vertically accreted sandy	Up to 100
Backswamp	Irregular	Up to 500 m	Distal floodplain, lake bedrock valley margin	Vertically accreted silty	Up to 100
Flat floodplain	Irregular	Up to 1 km	Ridge and swale, major and minor floodchannels, levee	Laterally accreted sandy / Vertically accreted sandy	Up to 250 / Up to 1000
Flat–distal floodplain bank	Straight to gently curved (concave)	300–800 m long; 2–4 m high	Lateral or vertically accreted sandy; vertical silty		Up to 100
Bench	Elongate, straight to gently curved	Up to 50 m wide; 1 km long; 4–6 m above channel	Scour holes around trees	Vertically accreted sandy, scoured	Up to 300
Point bench	Arcuate	Up to 100 m wide; up to 300 m long; 4–8 m above channel	Scour holes around trees	Laterally accreted sandy; Vertically accreted sandy	Up to 360
Crevasse splay	Elongate, oblique to channel	Up to 20 m wide; 100 m long; 1–2 m high	Hummocky topography	Vertically accreted sandy and Vertically accreted sandy, scoured	Up to 500

(a)

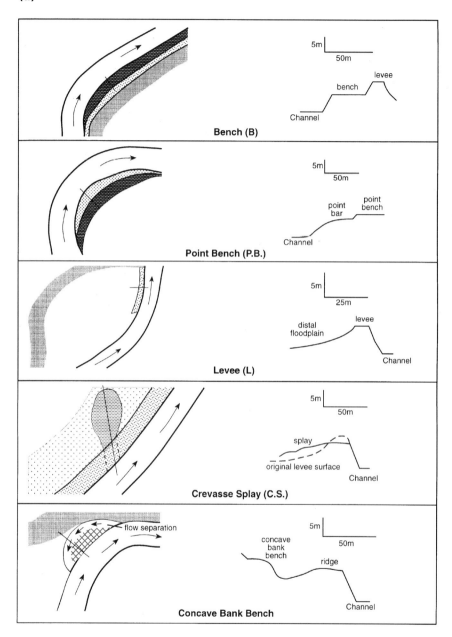

FIGURE 15.5   Schematic plan and cross-section views of the main geomorphic elements found on the lower Tuross River. All are based on real examples. Scale is approximate. See text for further discussion of each example and refer to Figures 15.7, 15.9, 15.11 and 15.13 for site-specific maps

(b)

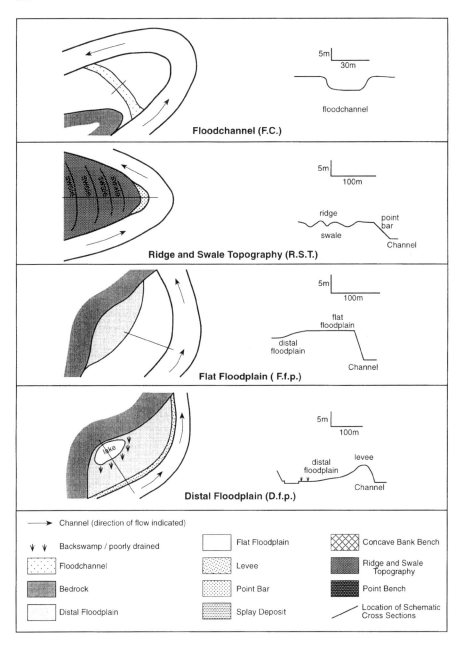

FIGURE 15.5   *(continued)*

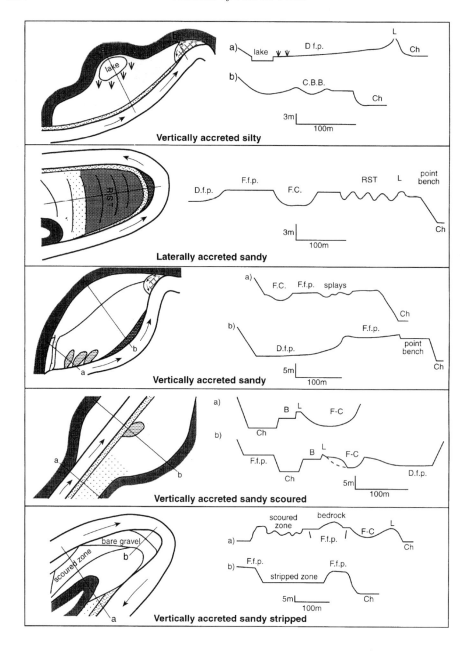

FIGURE 15.6   Schematic plan and cross-section views of the five main floodplain styles found on the lower Tuross River. For explanation of abbreviations see Figure 15.5. Scale is approximate. See text for further discussion of each example and refer to Figures 15.7, 15.9, 15.11 and 15.13 for site-specific maps

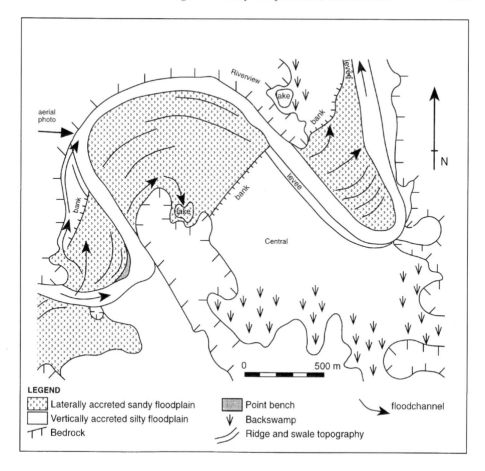

FIGURE 15.7   The complex floodplain packages that form in the backwater reaches are shown, with laterally accreted sandy and vertically accreted silty floodplains adjacent to each other, separated by a bank 3–4 m high. Laterally accreted sandy floodplains are always found at the upstream end of these floodplains, with a floodchannel cutting across the end of the bedrock spur, generally taking the most direct route to the main channel. The floodchannel that flows into the lake (centre left) is an exception to this rule

(3b) *Vertically accreted sandy, scoured* floodplains contain the simplest element assemblage, comprising a simple levee and floodchannel association (Figure 15.6). At one site (Rewlee), a large bench is present, and crevasse splays are common. An example is illustrated in Figures 15.11 and 15.12.

(3c) *Vertically accreted sandy, stripped* floodplains always contain flat floodplain. Depending on the type of stripping, levees, floodchannels and bars may be present (Figure 15.6), presenting a range of floodplain styles. Inset stripped surfaces are up to 7 m lower than the surrounding higher floodplain surface, and comprise a levee–floodchannel assemblage. Completely stripped surfaces are similar in origin to the stripped basal lag reported by Nanson (1986). Remnant floodplains are always flat floodplains, and occur in pockets up to 50 m wide, 200 m long and up to 5 m above the adjacent stripped zone

FIGURE 15.8 Low-level oblique aerial photograph of laterally accreted sandy and vertically accreted silty floodplains in the Riverview and Central area (see Figure 15.7 for map), looking east-southeast. Note the extensive bedrock confinement on the outside of the bend in the foreground. The patch of trees right of centre hides the floodchannel, which flows into the lake at the end of the bedrock spur (far right centre)

(Figures 15.13 and 15.14). Partially stripped surfaces, with extensive scour pits and chute channels, are similar to those reported by Bourke (1994) and Gardner (1977), but can still be called flat floodplain. The extensive scour holes are closely related to large trees (see Miller and Parkinson (1993) for a catalogue of floodplain erosion features).

VALLEY CONFINEMENT AS A CONTROL ON FLOODPLAIN CHARACTER

The highly variable nature of confinement along the Tuross Valley (see Figures 15.2, 15.3 and 15.4), in addition to the variable discharge regime and limited sediment availability, have produced distinctive sets of floodplain-forming and reworking processes along different sections of the valley. There is a strong relationship between valley width, stream power (which is in part a function of valley width), and resulting floodplain morphology. Hence, valley confinement is a primary determinant of floodplain style along the Tuross River. This accords with the role of the valley floor trough as a control on the pattern of fluvial sedimentation and sediment storage proposed by Warner (1992).

The terms incised, inset and backwater valley segments are taken from Warner (1992). *Incised* segments refer to narrow (< 500 m) valley trough profiles that are incised into bedrock. In contrast, Warner's use of the term *inset* reflects a channel that is constrained within a wider valley by Pleistocene deposits. The term *backwater* relates to tidal backwaters in protected areas of estuaries, which were extensive in coastal NSW systems during the mid-Holocene, but which have increasingly been infilled with fluvial sediments (Roy, 1984).

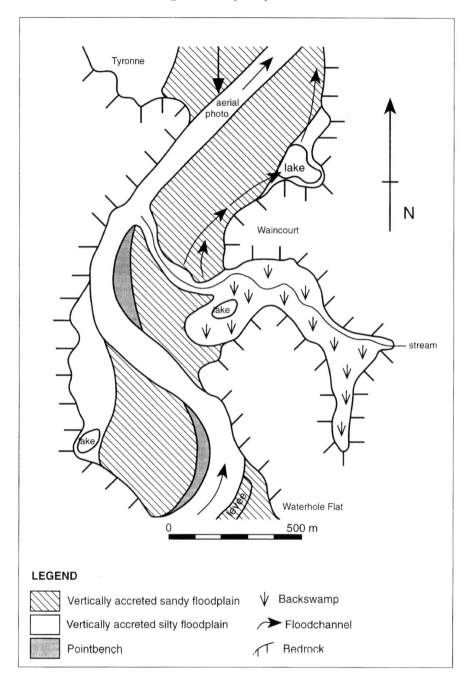

FIGURE 15.9 The role of tributary valleys in controlling floodplain style distribution is illustrated here, with the narrow side valley coming in from the east forming a sheltered environment allowing fine-grained deposition. In close proximity in the centre of the valley, high energy sand sheets are probably the main floodplain accreting mechanism, illustrating a highly effective mechanism for textural partitioning

FIGURE 15.10 Low-level oblique aerial photograph of the Waincourt floodplain, looking due south (see Figure 15.9). Trees cover the point bench on the inside of the foreground bend. Note the small lake, which is in an ampitheatre-shaped depression bounded by bedrock hillslopes and vertically accreted sandy floodplain

Incised reaches

The uppermost or incised reaches of the Tuross are characterized by either stripped or scoured sections of vertically accreted sandy floodplains. Scoured sections occur in narrow valley reaches (> 150 m but < 300 m; see Figures 15.11 and 15.12) and are equivalent to the "confined vertical accretion" category (A2) defined by Nanson and Croke (1992). In highly confined, steep-sided valleys, large discharges can attain great depths and the entire valley floor acts as a channel. All stream power estimates cited for the lower Tuross are based on estimated bankfull discharge (Q_{bf}; see Table 15.3 for methodology). Within-channel unit stream power estimates for these narrow reaches are in the range of 300-500 W m^{-2}.

Stripped sections occur in valley reaches with widths of between 200 and 500 m (see Figures 15.13 and 15.14). Remnant pockets of floodplain on the channel margin support the inference that stripping is an active process. Estimated specific stream power values range up to 1000 W m^{-2} (Table 15.3; Figure 15.4b). Stripped floodplains correspond to the "unconfined vertical accretion sandy floodplain" category (A3) of Nanson and Croke (1992), with good agreement in terms of stream power estimates. However, Nanson and Croke associate this type of floodplain with an unconfined situation (Cimarron River as reported in Schumm and Lichty (1963), and the Gila River as described by Burkham (1972)), which is clearly not the case on the Tuross.

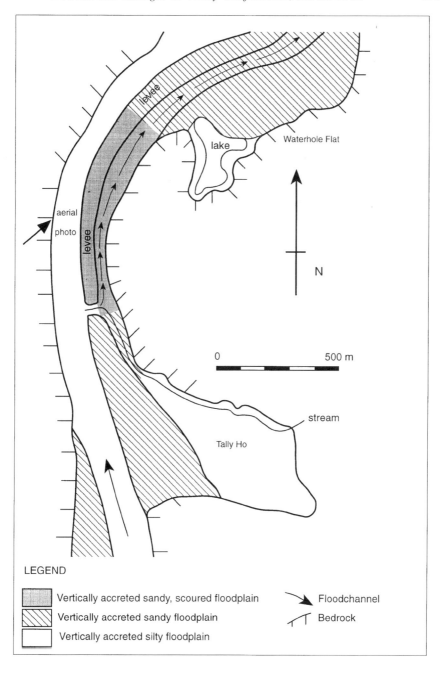

FIGURE 15.11   The vertically accreted scoured floodplain occurs only in the narrowest part of the valley, less than 250 m wide in this instance. The levee–floodchannel assemblage that characterizes the scoured section extends through the downstream vertically accreted sandy floodplain, with flat floodplain taking up the additional space. A lake has formed in the protected embayment behind the bedrock spur

FIGURE 15.12   Low-level oblique aerial photograph looking northeast over the constricted valley reach (< 250 m wide) that runs between Tally Ho and Waterhole Flat (see Figure 15.11). Flow is towards the upper left corner. Note the prominent floodchannel running between the levee crest and the road (on bedrock of the valley margin). This is a prime example of a vertically accreted sandy, scoured floodplain

Floodplain modification generally occurs in a dramatic fashion through floodplain stripping, or else as relatively minor scour by floodchannels. However, analysis of valley width and relation to stripping location shows no direct link in the lower Tuross Valley. A number of stripped areas have remnant patches of floodplain bordering the channel, and at these sites the channel width is much narrower than average, generally less than 50 m. It therefore seems likely that stripping is a response to high stream power values generated by channel constriction. Catastrophic stripping in 1991 at Cadgee (located on Figure 15.2) saw the removal of up to 3 m of floodplain sediment over an area up to 200 m wide and 500 m long. This large-scale erosion was not associated with an extremely high magnitude event (2200 m^3 s^{-1}, or a less than one-in-

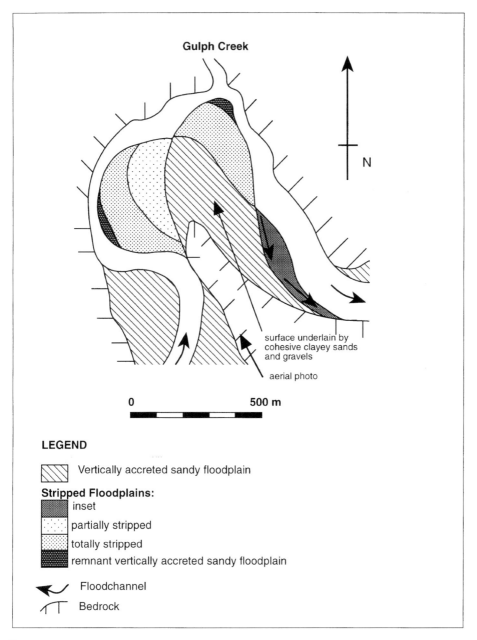

FIGURE 15.13 This one site shows enormous complexity, with three different stripping styles present in close proximity. The two areas of totally stripped floodplain are only 1–3 m above low flow level, yet the small remnant vertically accreted sandy floodplains associated with them are 3–5 m above low flow, indicating the removal of large volumes of material. The partially stripped floodplain surface is 5–7 m above low flow and 3–5 m below the central section of unstripped vertically accreted sandy floodplain, which itself is underlain by cohesive clayey sands and gravels. The inset floodplain has a levee with its crest 7–9 m above low flow, which drops down 3–4 m into a floodchannel

FIGURE 15.14   Low-level oblique aerial photograph of the Gulph Creek confluence area, looking just west of north. Gulph Creek enters at the top right corner. The prominent paddocks are on vertically accreted sandy floodplain, which is 2–3 m thick, and in turn rests on cohesive clayey sands and gravels. This surface is raised approximately 13 m above the channel. Note the contrast between the totally stripped surfaces (see Figure 15.13) and the partially stripped surfaces: the latter are covered with far more vegetation. The two areas of remnant vertically accreted sandy floodplain are also characterized by dense vegetation

10 year recurrence interval and less than bankfull), a fact in agreement with Nanson's (1986) assertion that catastrophic stripping is essentially random when related to discharge and cannot be directly linked to very high magnitude flood events. Miller and Parkinson (1993) caution that "longitudinal trends in the severity of erosion along a single valley are heavily influenced by local conditions, and not readily explained by a single parameter". Inset segment stripping sites are found well down the valley in the lower middle reach (see Figure 15.2), and are directly associated with a reach where the channel is narrow (< 50 m). However, valley widths are similar to those found at the major stripping sites in the uppermost reaches (Figure 15.13). Stripping is not found in any valley reach wider than 700 m, and they are usually under 500 m wide.

Flat floodplain surfaces, underlain by cohesive, clayey sands and gravels, play an important role in the upper Tuross reaches by constricting the valley floor trough. The zone of varied and large-scale stripping (Gulph Creek confluence; Figures 15.13 and 15.14) contains a prominent high floodplain burying a clayey sand and gravel unit 10 m above the channel bed. This cohesive unit greatly constricts the effective valley width (from 330 to 130 m wide), leading to stream power values higher than would be expected from an analysis of overall valley width alone. Miller and Parkinson (1993) report similar observations from confined valleys in Appalachian valleys following catastrophic flooding in 1985.

Inset reaches

These reaches are dominated by vertically accreted sandy floodplains, with a minor component of vertically accreted silty floodplain found only in protected environments, typically tributary valleys (Figures 15.2, 15.9 and 15.10). Valley widths average 400–700 m. Vertically accreted sandy floodplains in the open valley grade over distances of < 50 m into vertically accreted silty floodplains typical of protected tributary valley environments, reflecting the abrupt energy partitioning at the mouths of tributary valleys (Figures 15.9 and 15.10). This segment is split into two sections either side of a central constriction at Rewlee, where the valley is only 230 m wide. The upstream limit of the segment is around Reedy Creek confluence. Estimated bankfull stream power values are in the range of 300–400 W m^{-2} (Table 15.3).

The characteristically flat, vertically accreted floodplains of this valley segment probably result from sand sheet accumulation (cf. Schumm and Lichty, 1963; Burkham, 1972). Major floods in these relatively narrow reaches must be capable of maintaining competent flows for hundreds of metres across these floodplains. If flows were less competent, then levee features would be observed. Valley morphology plays a key role in determining the extent of sand sheet deposition, as the mouths of tributary valleys of all sizes always correspond to abrupt textural transitions from sand to fines and geomorphic transitions from higher, sandy floodplains to lower, silty floodplains with backswamps and lakes (Figures 15.9 and 15.10).

Wherever present, cohesive clayey sand or gravel units constrict the effective valley width. This is particularly significant in smaller flood events. For example, at Mortfield (see Figures 15.2, 15.3 and 15.4 and Table 15.3) the valley is around 600 m wide, but half of this width comprises a buried cohesive unit, underpinning a flat floodplain 8–10 m above the channel. The remaining half of the valley has a morphology similar to the scoured floodplains observed upstream, with a large levee–floodchannel–crevasse splay assemblage. The levee crest is 1 m lower than the edge of the high flat floodplain, suggesting that flood events contained within the "lower" floodplain only can generate stream power values to produce element associations typical of highly confined valley locations.

The closest fit to this style of floodplain in the Nanson and Croke (1992) scheme is their "unconfined vertical accretion, sandy" category (A3). However, this is by no means an ideal match. In Tuross Valley this floodplain style is associated with much greater confinement (< 700 m wide) than in other studies (i.e. Schumm and Lichty, 1963; Burkham, 1972), but stream power estimates are generally consistent (Tables 15.3 and 15.4).

Between- and within-reach transitions among floodplain styles are pronounced and abrupt along the Tuross River. When narrow reaches abruptly open up, the prominent levee–floodchannel association (vertically accreted sandy, scoured floodplain) characteristic of the narrow valley reach tends to be retained for the bulk of the length of the wider valley segment on the same side of the valley (see Figure 15.11). The levee tends to scale to the same height as the vertically accreted sandy floodplain on the other side of the channel in the wider segment. The extension of the levee–floodchannel association probably reflects maintenance of the efficient, low width:depth (around 10:1) channel characteristic of the narrow segment into the wider segment. In all other situations, floodplains terminate against bedrock constrictions.

Backwater reaches

Lowermost Tuross floodplains are dominated by two different floodplain styles (Figures 15.2, 15.3, 15.7 and 15.8): laterally accreted sandy and vertically accreted silty. These styles are directly adjacent, separated by a bank 2–4 m high bank (Figure 15.7). Laterally accreted sandy floodplains characteristically have levees and ridge and swale topography, possibly reflecting channel migration from an initial lowest possible sinuosity course to the present pattern of arcuate forms reflecting maximum possible sinuosity (cf. Jackson, 1976; Nanson, 1980; Iseya and Ikeda, 1989). This interpretation is supported by ^{14}C dating of charcoal extracted from ridge and swale deposits at Riverview and Comerang (located on Figure 15.2; see Table 15.5). At the Riverview site, charcoal from an auger hole in the distal ridge and swale floodplain gives a date of 2780 ± 70 radiocarbon years BP, while charcoal from a more proximal hole gives a date of 1190 ± 70 radiocarbon years BP. At the Comerang site, a distal ridge and swale floodplain sample gives an age of 3610 ± 220 radiocarbon years BP, while a proximal sample gives an age of 2320 ± 50 radiocarbon years BP. On both floodplains, the younger age closer to the channel supports the hypothesis of an initially low-sinuosity channel migrating outwards until reaching the valley margin.

Laterally accreted sandy floodplains correlate well with Nanson and Croke's (1992) "laterally migrated, scrolled floodplain" (B3b), while vertically accreted silty floodplains directly correlate with their "laterally stable, single channel" (C1). However, estimated specific stream power values for the lower Tuross are not in agreement with Nanson and Croke's values for the two different floodplain styles. The values for laterally accreted sandy fit well, but Tuross estimates are significantly higher than those stated for the

TABLE 15.5 Carbon-14 dates from laterally accreted sandy floodplains

Site	Auger hole location and code	Sample depth (cm)	ANSTO code	Age (^{14}C years)	1 sigma error (years)
Comerang	109Co12 195 m distal of bank top*	470	OZC304	3610	220
Comerang	116Co19 50m distal of bank top*	690	OZC305	2320	50
Riverview	32R15 281 m distal of bank top at apex of bend†	450	OZB918	2780	70
Riverview	20R3 41 m distal of bank top at apex of bend†	430	OZB917	1190	70

* On same transect, running perpendicular to the channel
† On same transect, bisecting the elongate floodplain

"laterally stable, single channel" class (< 100 W m^{-2} compared with < 10 W m^{-2} as stated by Nanson and Croke).

The presence of these two floodplain styles adjacent to each other reflects remarkably efficient sediment partitioning and depositional processes, possibly related to separation zones and reverse eddies during floods (A. Miller, pers. comm., 1997). Distal floodplain surfaces are only up to 1–2 m above sea level, indicating low rates of vertical accretion (up to 30 cm per 1000 years) since sea level stabilization in the mid-Holocene, in marked contrast to laterally accreted sandy floodplains.

Where these two types of floodplain occur on opposite banks, laterally accreted sandy floodplains are always on the inside of the bend immediately downstream of the bedrock at the head of the floodplain (Figures 15.7 and 15.8). This suggests that the highest stream power during overbank flow is concentrated on the inside of bends downstream of bedrock. This is supported by the distribution of major floodchannels and extensive scour around trees on the banks. Specific stream power estimates for the lower Tuross give values of < 100 W m^{-2}. Computer modelling work in similar valley layouts by Miller (1995) supports the presence of high velocity flows over floodplains where valley widths abruptly increase and flow is forced to step up 5–10 m onto the floodplain surface. Gardner (1977) notes that during high flood stages flows are aligned downvalley rather than downchannel. Furthermore, Miller and Parkinson (1993) and Miller (1995) note that valley width, orientation of the channel to the valley, within-valley restrictions such as terraces, and local obstacles such as trees and buildings, all need to be taken into account when assessing stream power distribution in confined valleys.

There is not a direct relationship between valley width and floodplain type in the backwater section of the valley. Hence, it must be concluded that some other process is controlling the distribution of floodplain types in the lowermost valley. During the mid-Holocene, most of the valley floor trough probably comprised a vertically accreted silty floodplain with large occluded lakes, but this has undergone a metamorphosis as the channel has migrated and increased its sinuosity. Therefore the present-day relationship of laterally accreted sandy and vertically accreted silty floodplain types is probably more a reflection of spatially and temporally variable hydraulic conditions and thus longer-term floodplain evolution, than contemporary stream power. Estimates of stream power based on the width of the valley floor trough in this part of the valley will be relatively little changed since the mid-Holocene (due to the high valley width), yet the floodplain assemblage has changed considerably.

Sea level change as a control on floodplain character

The post-glacial marine transgression has played a significant role in altering the valley floor trough in the lowermost reaches of the Tuross River. Drill hole-based valley cross-sections (Stone, 1993) suggest a slightly asymmetric V-shaped valley. During the last glacial maximum the channel would have been confined in a narrow valley, allowing little room for floodplain formation. The increase in height of base level has provided an enormous increase in accommodation space and area for floodplains to form above estuarine and deltaic deposits. Models of estuarine infilling on the NSW coast, developed by Roy (1984), suggest that backwater areas of estuaries gradually accrete vertically with fine-grained overbank sediments, eventually becoming distal floodplains.

Reaches upstream of any marine influence have not undergone this radical change in local base level, and have always been confined in relatively narrow valley floor troughs. As such, stream power values will always have been relatively high, leading to higher energy element assemblages (such as levee and floodchannel) than those characteristic of the lower end of the study reach (levee with broad distal floodplains).

FLOODPLAIN MODIFICATION

The relationship between within-channel and overbank unit stream power must play a significant role in determining floodplain erosion processes. The stream power estimates reported in this study are based on simple calculations (Table 15.3). It is likely that the consistent downstream trends seen in these values are valid, and estimates generally accord well with those reported in Nanson and Croke (1992). Furthermore, estimates agree with figures presented by Miller (1990) and Magilligan (1992), which suggest that within-channel unit stream power values greater than 300 $W m^{-2}$ are required to produce catastrophic floodplain stripping. Estimates derived indicate that completely stripped floodplain surfaces can reflect operation of the entire valley floor as a channel, and there is no need to invoke channel widening as the sole agent of floodplain destruction. For example, the June 1991 flood event (2200 $m^3 s^{-1}$, less than bankfull), which induced significant floodplain stripping at Cadgee had an estimated stream power of 245 $W m^{-2}$ assuming a velocity of 2 $m s^{-1}$. If a velocity of 4 $m s^{-1}$ is substituted, the estimated stream power becomes 360 $W m^{-2}$. However, no other significant erosion was reported along the river at this time, yet these stream power estimates would certainly have been exceeded in other, narrower reaches. This suggests that a range of other factors influence stripping potential, such as the nature of vegetation cover. Indeed, in most instances, partially stripped floodplain sites are associated with flood flows which short-cut the channel, with high velocity filaments aligned downvalley (cf. Gardner, 1977).

Post- or syn-depositional modification of floodplains has not been applied to "classic" models of floodplain formation, which tend to consider only accretionary processes. As noted by Brierley and Hickin (1992), selective preservation of floodplain deposits reflects reworking of sediments at different flow stages (or during differing flood events) and at differing positions in the floodplain. In confined valleys (where the entire valley may act as a channel bed during large-magnitude events) long-term floodplain preservation is considered unlikely. One consequence of this set of circumstances is the non-planar character of floodplains in valleys such as the Tuross: flat sections are the exception rather than the rule. In large part this pattern reflects the erosive impacts of floods in this confined valley setting, as floodplain modification is pronounced in upper sections of the study area (Figures 15.1 and 15.5).

SUMMARY

Although the lower Tuross River maintains a near-uniform channel morphology along the lower 35 km of its course, there is marked downstream variability in valley width. Element assemblages that make up the floodplain of the Tuross River can be differentiated into a range of floodplain styles. The spatial association of these floodplain styles reflects the downstream increase in valley width and associated decreasing stream power.

In summary, floodplain morphology of the lower Tuross Valley is characterized by the following attributes:

(1) Vertically accreted sandy, scoured floodplains, with a levee–floodchannel assemblage. High within-channel unit stream powers, and evidence from similar floodplains elsewhere (Nanson, 1986), suggest that these floodplains may be likely locations for catastrophic stripping to occur in the future.

(2) Vertically accreted sandy, stripped floodplains display geomorphic evidence of catastrophic stripping. They contain flat floodplains, with inset stripped surfaces. Channel constriction, combined with high unit stream power values, cause floodplain stripping.

(3) Vertically accreted sandy floodplains are dominated by flat floodplain, with no substantial modification. This indicates overbank stream powers high enough to deposit sand sheets but not generally high enough to cause substantial erosion.

(4) Laterally accreted sandy floodplains contain ridge and swale topography indicative of lateral channel migration. Valley morphology commonly promotes floodchannels at localized points of high stream power.

(5) Vertically accreted silty floodplains reflect slow accretion rates by depositon of fines from suspension, reflecting low stream power values.

The uniformity of vertically accreted sandy floodplains (regardless of extent of modification) is in marked contrast to traditional notions of proximal–distal floodplain trends. This reflects high stream power values in narrow valley floor troughs. These settings are categorically different from low energy alluvial systems where classic floodplain models have been derived (e.g. Fisk, 1944; Wolman and Leopold, 1957).

In general, the stream power associations for floodplain style summarized by Nanson and Croke (1992) work well within the Tuross Valley. Floodplain styles typically reflect the pronounced variability in stream power estimates associated with valley confinement (i.e. the morphology of the valley floor trough, *sensu* Warner, 1992). However, the juxtaposition of different floodplain styles within a given reach suggests that in some situations stream power ceases to be a useful indicator of floodplain style. In these situations, autocyclic processes, such as channel migration leading to floodplain metamorphosis, and complex flow hydraulics at flood stage, are the dominant controls. A more complex assessment of the spatial distribution of flood stage flow, through finite element computer modelling, is required to understand the complex hydraulics that doubtless occur in the lower Tuross Valley.

Other confined valley settings, such as the glacially carved Squamish (Brierley 1991a,b) and structurally controlled Appalachian valleys (Miller, 1990, 1995; Miller and Parkinson, 1993) generally do not exhibit such an extreme diversity of fluvial forms. Complexity occurs at a range of scales throughout the lower Tuross: the valley scale, between floodplains and within floodplains. While stressing variability, it must also be noted that similar assemblages are found in a range of valley reach settings, reflecting the fact that a limited range of fundamental processes is operative.

Estimated stream power values at floodplain stripping sites on the Tuross River accord well with the estimate of 300 W m^{-2} as a minimum threshold for the onset of catastrophic flooding (Miller, 1990; Magilligan, 1992). Such floodplains need to be considered as disequilibrium floodplains (Nanson, 1986), as they almost certainly undergo cycles of accretion followed by catastrophic stripping.

The primary overall control on the diversity of fluvial forms on the lower Tuross floodplains is the irregularity of the valley floor trough, which is essentially a reflection of numerous, abrupt changes in valley width and shape. Floodplain modification and the Holocene marine transgression have added further diversity of geomorphic form. Antecedent controls on landscape morphology, reflected at valley scale, are key controls on river–floodplain relationships along the lower Tuross River.

ACKNOWLEDGEMENTS

Primary funding for this project came from an Australian Research Council grant held by G.J.B. and Ken Woolfe (James Cook University, Queensland) which R.J.F. gratefully acknowledges for providing both research expenses and a stipend. Additional funds have been supplied by the Macquarie University Postgraduate Research Fund, Macquarie University School of Earth Sciences, Australian Institute of Nuclear Science and Engineering, and J. and S. Ferguson. All (the numerous) farmers along the lower Tuross are thanked for allowing access to their properties; however, Susan Tocchini is especially thanked for providing accommodation, logistic and moral support. The many undergraduate students, friends and colleagues who have helped out in the field, too many to name here individually, are also thanked. Andrew Beattie is thanked for rapidly drafting initial versions of most figures. Reviews by Gerald Nanson and an unknown reviewer, as well as detailed editorial and technical comments by Andy Miller, have greatly improved this paper. This paper stems from an ongoing doctoral study by R.J.F., supervised by G.J.B.

REFERENCES

Allen, J.R.L. 1965. A review of the origin and characteristics of recent alluvial sediments. *Sedimentology*, **5**, 89–191.

Bourke, M.C. 1994. Cyclical construction and destruction of flood dominated flood plains in semiarid Australia. In L.J. Olive, R. Loughran and J.A. Kesby (Eds), *Variability in Stream Erosion and Sediment Transport*. IAHS Publication No. 221, 113–123.

Bridge, J.S. 1985. Paleochannel patterns inferred from alluvial deposits: a critical evaluation. *Journal of Sedimentary Petrology*, **55**(4), 579–589.

Brierley, G. J. 1991a. Bar sedimentology of the Squamish River, British Columbia: definition and application of morphostratigraphic units. *Journal of Sedimentary Petrology*, **61**(2), 211–225.

Brierley, G. J. 1991b. Floodplain sedimentology of the Squamish River, British Columbia: relevance of element analysis. *Sedimentology*, **38**, 735–750.

Brierley, G. J. 1996. Channel morphology and element assemblages: a constructivist approach to facies modelling. In P.A. Carling and M.R. Dawson (Eds), *Advances in Fluvial Dynamics and Stratigraphy*. Wiley, Chichester, 263–298.

Brierley, G.J. and Hickin, E.J. 1992. Floodplain development based on selective preservation of sediments, Squamish River, British Columbia. *Geomorphology*, **4**, 381–391.

Burkham, D.E. 1972. Channel changes of the Gila River in Safford Valley, Arizona, 1846–1970. *US Geological Survey Professional Paper* 655–G, 1–24.

Fisk, H.N. 1944. *Geological investigation of the alluvial valley of the lower Mississippi River*. Mississippi River Commission Waterways Experiment Station, Vicksburg, Mississippi, 82 pp.

Fisk, H. N. 1947. *Fine–grained alluvial deposits and their effects on Mississippi River activity*. Mississippi River Commission, Vicksburg, Mississippi.

Gardner, J.S. 1977. Some geomorphic effects of a catastrophic flood on the Grand River, Ontario. *Canadian Journal of Earth Science*, **14**, 2294–2300.

Iseya, F. and Ikeda, H. 1989. Sedimentation in coarse-grained sand-bedded meanders: distinctive deposition of suspended sediment. In A. Taira and F. Masuda (Eds), *Sedimentary Facies in the Active Plate Margin*. Terra Scientific Publishing, Tokyo, 81–112.

Jackson, R. G. 1976. Depositional model of point bars in the lower Wabash River. *Journal of Sedimentary Petrology*, **46**(3), 579–594.

Magilligan, F. J. 1992. Thresholds and the spatial variability of flood power during extreme floods. *Geomorphology*, **5**, 373–390.

Miall, A.D. 1985. Architectural-element analysis: a new method of facies analysis applied to fluvial deposits. *Earth-Science Reviews*, **22**, 261–308.

Millar, K. 1994. *Surface Water, New South Wales South East Region Water Management Strategy – Water Planning for the Future*. New South Wales Water Resources Council.

Miller, A.J. 1990. Flood hydrology and geomorphic effectiveness in the central Appalachians. *Earth Surface Processes and Landforms*, **15**, 119–134.

Miller, A.J. 1995. Valley morphology and boundary conditions influencing spatial patterns of flood flow. In J.E. Costa, A.J. Miller, K.W. Potter and P.R. Wilcock (Eds), *Natural and Anthropogenic Influences in Fluvial Geomorphology*. American Geophysical Union, Washington, DC, 57–81.

Miller, A.J. and Parkinson, D.J. 1993. Flood hydrology and geomorphic effects on river channels and flood plains: the flood of November 4–5, 1985, in the South Branch Potomac River Basin of West Virginia. In R.B. Jacobson (Ed.), Geomorphic Studies of the Storm and Flood of November 3–5, 1985, in the Upper Potomac and Cheat River Basins in West Virginia and Virginia. *US Geological Survey Bulletin* 1981, E1–E96.

Nanson, G.C. 1980. Point bar and floodplain formation of the meandering Beatton River, northeastern British Columbia, Canada. *Sedimentology*, **27**, 3–29.

Nanson, G.C. 1986. Episodes of vertical accretion and catastrophic stripping: A model of disequilibrium flood–plain development. *Geological Society of America, Bulletin*, **97**, 1467–1475.

Nanson, G.C. and Croke, J.C. 1992. A genetic classification of floodplains. *Geomorphology*, **4**, 459–486.

Nanson, G.C. and Erskine, W.D. 1988. Episodic changes of channels and floodplains on coastal rivers in New South Wales. In R.F. Warner (Ed.), *Fluvial Geomorphology of Australia*. Academic Press, Sydney, 201–221.

Nanson, G.C. and Page, K. 1983. Lateral accretion of fine-grained concave benches on meandering rivers. In J.D. Collinson and J. Lewin (Eds), *Modern and Ancient Fluvial Systems*. Blackwell, Oxford, 133–143.

Rosgen, D.L. 1994. A classification of natural rivers. *Catena*, **22**, 169–199.

Roy, P.S. 1984. New South Wales estuaries: their origin and evolution. In B.G. Thom (Ed.), *Coastal Geomorphology in Australia*. Academic Press Australia, Sydney, 99–121.

Schumm, S.A. and Lichty, R.W. 1963. Channel widening and flood–plain construction along Cimarron River in southwestern Kansas. *US Geological Survey Professional Paper* 352-D, 71–88.

Stone, M.D. 1993. *Evolution of the Tuross River fluvial delta*. Unpublished BSc Honours Thesis, School of Geography, University of New South Wales.

Warner, R.F. 1992. Floodplain evolution in a New South Wales coastal valley, Australia: spatial process variations. *Geomorphology*, **4**, 447–458.

Wolman, M.G. and Leopold, L.B. 1957. River Flood Plains. Some Observations On Their Formation. *US Geological Survey Professional Paper* 282-C, 87–107.

16

Flow and Sedimentation at Tributary River Mouths: A Comparison with Mesotidal Estuaries

BARBARA A. KENNEDY

School of Geography, University of Oxford, UK

ABSTRACT

Studies of river–sea (estuary) and river–river junctions have largely been conducted in isolation from each other. For fluvial junctions, most is known about: flow and sedimentation in the confluence zone of streams of roughly equal discharge; and the evidence for flow reversal and slackwater sedimentation in small tributaries as a result of mainstream flood peaks. There is little published information about the behaviour of the lowest reaches of tributaries which are small relative to the trunk stream, especially at non-flood discharges. Observations of two small (Shreve magnitudes 1 and 23) right-bank tributaries of the Mahaveli Ganga, Peradeniya (Sri Lanka) give evidence of frequent disruption of tributary water and sediment movement at Mahaveli stages well below the annual monsoon flood peak. Comparisons of hydraulic geometry, water and sediment movements and tributary mouth sedimentary structures with those described for mesotidal estuaries suggest that the mouths of these small tributaries behave like estuaries over a wide range of flows. Whilst the irregular timing of peak flows and the likely absence of strong density contrasts between the two fluvial water bodies would seem to indicate that tributary mouths should not be viewed as strict homologues of tidal estuaries, observations reported here suggest a continuum of confluence sedimentary environments from alluvial fans to tidal estuaries.

INTRODUCTION

Many major rivers have estuarine sections. In those reaches, flow of water and sediment will reverse on a more-or-less regular basis determined by tidal regime, although the pattern will be complicated both by the river's flood peaks and the morphology of the estuary, which may lead to the development of tidal bores (Postma, 1967; Wright, 1978; Chen et al., 1994). There are several published studies of tidal channel systems from the perspective of standard (i.e. non-tidal) fluvial morphology (notably Myrick and Leopold, 1963; Bridges and Leeder, 1976), but these investigations largely avoided the

Varieties of Fluvial Form. Edited by A.J. Miller and A. Gupta
© 1999 John Wiley & Sons Ltd

TABLE 16.1 Water quality of Mahaveli Ganga and tributaries

Stream	Discharge $(m^3 s^{-1})$	pH	TDS $(mg\,l^{-1})$	SS $(mg\,l^{-1})$
Mahaveli Ganga	54	5.2	100	25
Ma Oya, confluence	1.14	7.0	40	10
Lovers' Lane, confluence	0.05	6.2	80	15

TDS, total dissolved solid; SS, suspended sediment
Data kindly analysed by Mr Weerasooria, Department of Geology, University of Peradeniya

complications of substantial freshwater runoff contributions to the basic tidal circulation (though the case described by Knighton et al. (1992) implies inputs from freshwater networks). Conversely, there have been some investigations of freshwater river mouths where flow blocking or reversal in tributaries induced by high-water stages on trunk streams has been inferred from slackwater deposits (Bretz, 1929; Baker, 1973; Kochel and Baker, 1982, 1988; Nanson et al., 1993). There is also a growing number of studies of the effects of combining the flow of two river channels: interest in the morphology, flow dynamics and sediment movement in the confluence area is exemplified by the work of Mosley (1976), Ashmore and Parker (1983), Best (1986, 1988), Best and Reid (1984), Roy and Roy (1988), Reid et al. (1989), Biron et al. (1993), Bristow et al. (1993), Kenworthy and Rhoads (1995), and Rhoads and Kenworthy (1995). It is, however, worth noting that both experimental and field studies of the confluence zone, such as those listed, have tended to focus upon tributaries of roughly equivalent catchments and discharges. There is also very limited consideration given to the effects of asynchronous flood peaks (but see Reid et al. (1989) for an exception). An examination of most channel networks shows that, on balance, there are many more junctions between streams whose Shreve magnitudes vary at least twofold (cf. Figure 16.1) than between those of approximately equal size (Kennedy, 1984).

The present literature seems, then, to lack extensive discussion of two cases which appear, a priori, to be common: first, the meeting of a small channel with a larger one; second, the non-synchroneity of discharge peaks and troughs at confluences.

What follows is a description of observations made on two small tropical tributaries during the declining stage of the monsoon flood peak on the main river. These observations are then compared with information on the behaviour of tidal passes/mesotidal estuaries (especially Battjes, 1967; Boothroyd, 1978; Carter, 1988; Hume and Bell, 1993; Pethick, 1984). This comparison leads to the suggestion that the lower reaches of many small streams may behave far more like tidal estuaries than has generally been supposed.

FIELD OBSERVATIONS

Cross-sections were surveyed with tape and ranging rods and then monitored over a two-month period at the mouths of two small (Shreve magnitudes 1 and 23) right-bank tributaries of the Mahaveli Ganga, Sri Lanka, during the falling stages of the autumn monsoon, 1984 (Figures 16.1 and 16.2). Flow was estimated at the tributary mouths and also at sites upstream, where bridges gave fixed cross-sections. The pattern of discharge

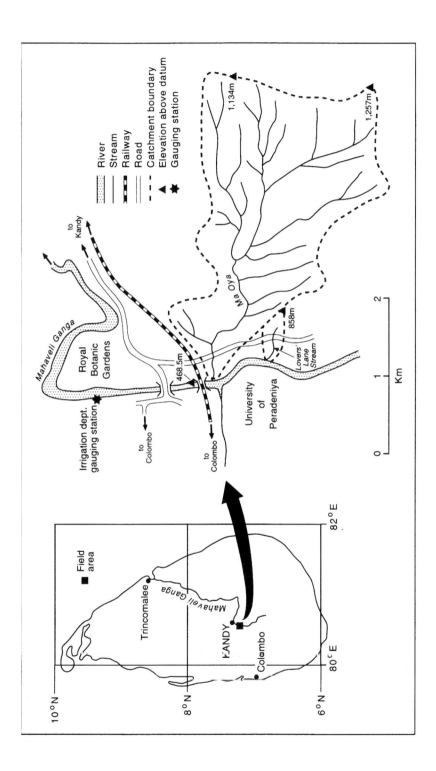

FIGURE 16.1   Location of the study sites

412

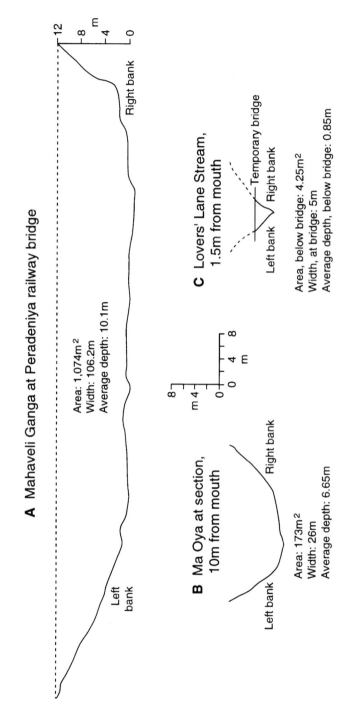

FIGURE 16.2 Cross-sections of three streams studied (the Mahaveli section is taken from the gauging records)

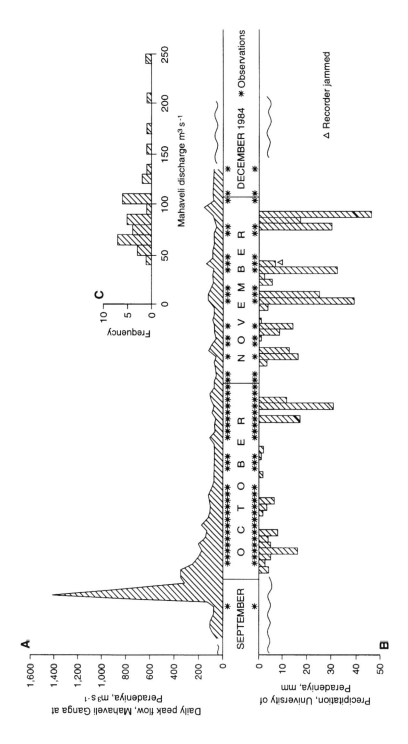

FIGURE 16.3 Discharge (A) and precipitation (B) at the study site, September–December 1984. (C) Range of gauged flows on Mahaveli at which measurements were made on Ma Oya and/or Lovers' Lane streams

on the Mahaveli at the nearby gauging station, and of precipitation measured during the study period, is shown in Figure 16.3. After the passage of the 1400 m^3 s^{-1} flood peak, 42 days of observations were made at the tributary mouths, covering a range of 40–240 m^3 s^{-1} discharges on the trunk stream (Figure 16.3C). The discharge upstream and downstream on each tributary was estimated, using surface floats. The downstream cross-sections, one on the smaller stream and two on the Ma Oya (that shown in Figure 16.2 and a second section at the line of confluence, 10 m downchannel, shown in Figure 16.9) were resurveyed, at 0.5 m intervals, from fixed points, to give estimates of scour and fill. Measurements at the Ma Oya mouth were impossible until the Mahaveli flow fell below 100 m^3 s^{-1} (since the Ma Oya was too deep to wade), but 27 detailed surveys of the confluence cross-section and 21 of that 10 m upchannel were obtained. The "Lovers' Lane" section was vandalized after seven sets of readings, but a further 31 were taken at a location 1 m from the confluence. The directions of surface water flow and of sediment transport were also observed, as were the depositional and erosional forms at the tributary mouths. The Mahaveli bed is of cobbles and sand, whilst the two tributaries flow on gravels and sand, with a surface covering of silt. There is a cobble deposit in the bed of the Ma Oya in a pool some 25 m above its confluence with the Mahaveli. Both tributaries show pronounced pool and riffle sequences, with those on the Ma Oya exhibiting as much as 1 m difference in bed elevation.

A total of 22 pairs of upstream/downstream discharge estimations for the larger tributary (Ma Oya), and 18 for the smaller tributary (Lovers' Lane) were obtained and are shown in Figure 16.4A. In the latter case, the absence of a relationship is certainly due to the impact of the hydraulic barrier imposed by the Mahaveli's flow: even at low flows, the main channel is plainly creating backwater effects on this small tributary. (Silt lines on vegetation suggested that, at the monsoon flood peak in September, backwater effects raised the level of the Lovers' Lane stream some 8 m above its low-flow channel.) When the estimated tributary discharges are plotted against that of the Mahaveli for the same time (obtained from the gauging record), the disruption to the small tributary's flow is evident (Figure 16.4C). For the larger Ma Oya, there is, in contrast, a straightforward positive relationship between the Ma Oya outflow and the Mahaveli stage (although the absence of Ma Oya data at Mahaveli flows in excess of 100 m^3 s^{-1} should be noted). However, when the hydraulic geometry relationships for channel width and depth against discharge are plotted for the Ma Oya (Figure 16.5), the relationships are non-standard. There is an absence of association between depth and discharge (Figure 16.5A) and a negative one between width and discharge (Figure 16.5B). These anomalies seem, in part, to be related to the stage and, hence, the discharge of the Mahaveli. Inspection of the data suggests that there may be a threshold level of main channel flow, in the region of 78 m^3 s^{-1}, which creates a minor, but real, hydraulic barrier to the Ma Oya outflow. In particular, there is some suggestion (in Figure 16.5A) that the Ma Oya discharge is higher at intermediate depths. If this relation is real, it would coincide with Myrick and Leopold's (1963) finding for their purely tidal system (figure 13, pp.13–14) and Pethick's (1984) statement that peak outflow in tidal estuaries characteristically occurs at mid-tide, i.e. at intermediate channel depth, rather than "bankfull", when water motion is halted. Lessa and Masselink (1995) describe a more complex tidal cycle, with peak flows on certain flood tides; interestingly, they ascribe this effect, in part, to the existence of a sill at the estuary mouth (see below).

415

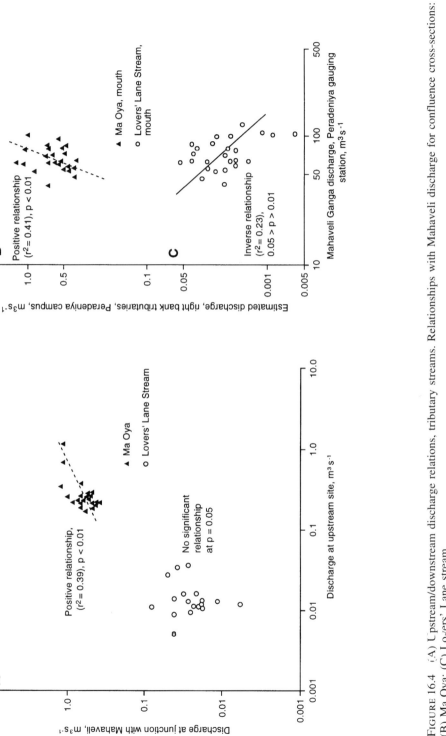

FIGURE 16.4 (A) Upstream/downstream discharge relations, tributary streams. Relationships with Mahaveli discharge for confluence cross-sections: (B) Ma Oya; (C) Lovers' Lane stream

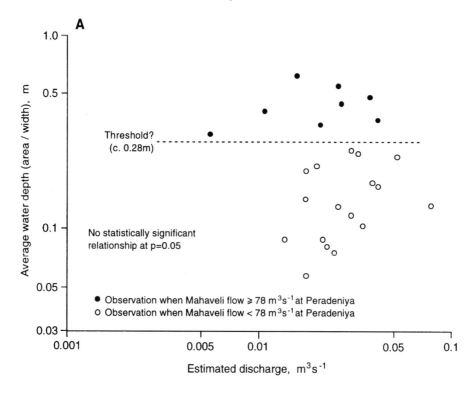

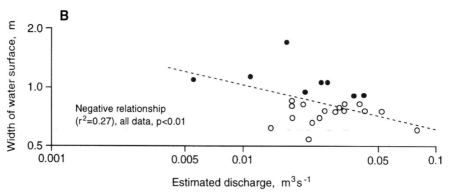

FIGURE 16.5   Hydraulic geometry, Ma Oya mouth: (A) depth; (B) width

The movement of water and sediment at both tributary mouths was carefully observed from 5 October 1984 to 5 December 1984 (Mahaveli discharge ranged from 175 to $42 \, \text{m}^3 \, \text{s}^{-1}$; Ma Oya gauged when Mahaveli flow dropped below $100 \text{m}^3 \, \text{s}^{-1}$); it became clear in each case that reversed flows were relatively common. In the 30 days of observations at the Ma Oya mouth, silt from the Mahaveli was observed to move up the tributary as much as 10 m as a subsurface "cloud" at least once each day (see Figure 16.6A), and there were nine instances where the surface water flow was reversed. This

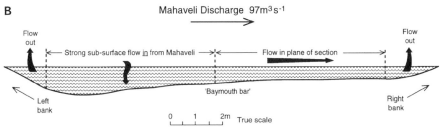

FIGURE 16.6   (A) Influx of Mahaveli silt into Ma Oya, 10 November 1984. Mahaveli discharge $42 \, m^3 \, s^{-1}$, Ma Oya *c.* $0.7 \, m^3 \, s^{-1}$ (units on staff, 0.1 m). (B) Section showing flow patterns observed at Ma Oya confluence, 15 November 1984

occurred with Ma Oya discharges from 1.1 to $0.41 \, m^3 \, s^1$ and mean water depths of 0.67 to 0.26 m. There were nine instances where the surface water flow was reversed: this occurred with Mahaveli flows from 170 down to $64.5 \, m^3 \, s^{-1}$. In some cases, reversals were clearly assisted by wind blowing up-tributary. Overall, the flow patterns at the Ma Oya mouth were often complex, as shown in Figure 16.6B. In the 31 observations at the Lovers' Lane outlet, 12 cases of subsurface silt clouds and two of surface flow reversal were noted. the corresponding Mahaveli discharges for the latter were 128 and $88 \, m^3 \, s^{-1}$.

The subsurface silt cloud irruptions noted on both tributaries are presumably more-or-less analogous to the case of salt-wedge intrusions into tidal estuaries, although the density differences for river waters are likely to be minor in comparison with tidal cases (Postma, 1967; Wright, 1978). This seems to be supported by the analysis of water samples obtained by submerging polythene bottles in the flow on 30 November 1984, shown in Table 16.1.

The reversed surface flows on both small tributaries are unexpected at such relatively low stages on the trunk stream (recall Figure 16.3), but may be comparable with the general impact of tides on freshwater inlets (cf. Postma, 1967). It is also possible that the Ma Oya, with a broad pool located 25 m upstream of the mouth, may experience the development of a lens of still water in the depths of the pool which (cf. Guilcher, 1967) acts as a weir to trap coarse bed material moving downvalley, whilst the Mahaveli silt moves upchannel over the top (cf. Van Rijn, 1993, figure 11.41): certainly, the pool was noted to have a cobble floor.

The net erosion and deposition at the two tributary mouths over the study period were rather different (Figure 16.7). The smaller stream showed both net erosion (Figure 16.7A)

A

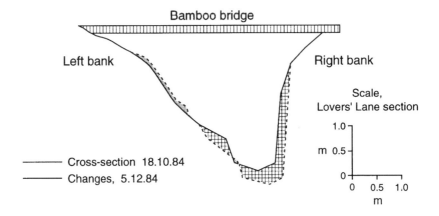

B

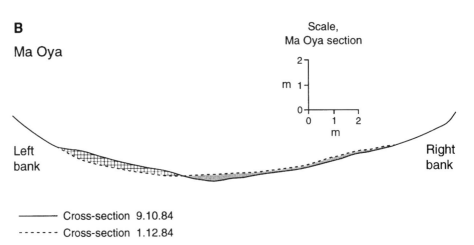

FIGURE 16.7   Net changes in cross-sections at mouth over study period: (A) Lovers' Lane stream; (B) Ma Oya

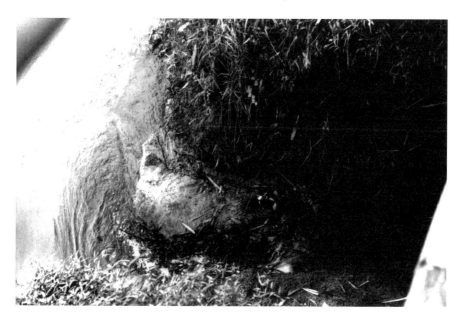

FIGURE 16.8 Bar/fan at Lovers' Lane outlet on 10 November 1984. Mahaveli discharge 43 m³ s⁻¹, tributary flow *c.* 0.02 m³ s⁻¹. Note well marked scour channel and knickpoint, to left of tributary channel

and the development of a small fan or delta below the outlet, which was left hanging above the Mahaveli thalweg as the stage dropped (cf. Kennedy, 1984). This corresponds to the "tributary bar with avalanche face" described by many workers (cf. Best, 1986) as extending from a tributary into the confluence zone. In this case, the direct equivalence (and, indeed, sequential conversion) between submerged avalanche face (cf. Bristow et al., 1993) and subaerial tributary fan (cf. Rachocki and Church, 1989) is worth stressing. It was also observed that variations in the relative stage of the Mahaveli vis-à-vis the Lovers' Lane outlet led to the development of small (less than 0.25 m) knickpoints which propagated up-tributary (Figure 16.8). In the case of the Ma Oya, the net change in cross-sectional area and form was minimal (Figure 16.7B), but made up of a wide range of erosional and depositional episodes (cf. Biron et al., 1993). Bed elevation changes between survey periods ranged up to 120 mm of net change and it became evident that sediment movement at the Ma Oya mouth was both frequent and on differing scales, ranging from wholesale scour or deposition right across the channel, to a patchwork of minor scour-and-fill episodes. The almost continuous changes in bed forms in the lowest reach of the Ma Oya, taken together with the evidence of the silt clouds, would seem a priori to match the view of Bruun and Gerritson (quoted by Battjes, 1967, p.186) that sediment moves back and forth through tidal inlets like "a rolling carpet". Certainly, that appears to describe the motion of the fine sand and silt in the lower Ma Oya.

It is when one looks at the nature of the Ma Oya mouth as a whole, however (Figures 16.9 and 16.10), that the analogy between this tributary mouth and a tidal estuary/tidal pass becomes most striking (cf. Carter, 1988, figure 91; Hume and Bell, 1993, figure 2.11: see also the figures in Biron et al., 1993). Figure 16.9 is based on observations made

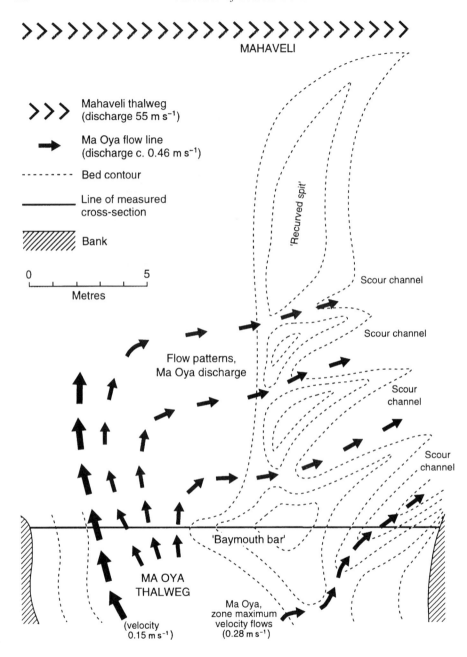

FIGURE 16.9 Bar complex at Ma Oya mouth, as observed on 2 November 1984. Ma Oya flow lines observed from surface floats

FIGURE 16.10   Sediments and topography of Ma Oya bar complex (units on staff, 0.1 m). (A)
Surface features, 6.5 m out into Mahaveli channel, 26 October 1984; Mahaveli discharge 55 $m^3 s^{-1}$.
(B) Junction of spit and tributary bar, 8 November 1984, showing scour channel. Mahaveli
discharge 56 $m^3 s^{-1}$. (C) Coarse deposits on bar surface, in Mahaveli channel, 10 November 1984,
Mahaveli discharge 42 $m^3 s^{-1}$

on a number of occasions, although the flow lines shown were those of 2 November
1984. The "baymouth bar" is again analogous to the tributary bar with avalanche face; it
was composed of coarse sand and developed low-relief (*c.* 20 mm) ripples on the top
(Figure 16.10B). The mixing interface between Ma Oya and Mahaveli waters (cf. Rhoads
and Kenworthy, 1995) was not infrequently observed to be located along the Mahaveli

side of this structure (e.g. on 14 November 1984, Mahaveli discharge was 63 m^3 s^{-1}, and Ma Oya 0.69 m^3 s^{-1}). Whilst there was some suggestion of a bank-attached bar in the separation zone along the right-hand outlet of the Ma Oya, a far more dramatic feature, which does not seem to have been described elsewhere in the fluvial literature, was an apparently classic example of a recurved spit extending along the "avalanche face"/ "bank-attached bar" some 15 m or more out into the mainstream channel. This feature was associated with divergent outflow from the Ma Oya, as shown in Figure 16.9. The spit itself (Figure 16.10) was apparently constructed of cobbles (exposed in scour episodes; Figure 16.10A) but topped with low sand ripples, some of them lunate, but mostly irregular. Important features appear to be: the tributary's "bay mouth bar"; the recurved spit with its scour channels (Figure 16.10B); and the maximum velocity outlet channel located at the right-bank exit from the Ma Oya (i.e. at the downstream point, vis-à-vis the Mahaveli flow). Most of these features would seem to be more readily recognized by coastal, rather than fluvial geomorphologists. They suggest some significant interactions between main channel and tributary with respect to both flows and sediment movements over a very wide range of discharges (in this case, *c.* 1400–40 m^3 s^{-1} for the Mahaveli) and, as such, deserve more detailed attention.

THE MA OYA MOUTH AND MESOTIDAL INLETS

The key elements, it appears, of tidal inlets are not only the reversal of flow with tidal cycle, but also the balance which exists between the in and out fluxes of water and sediment generated by tidal currents and the across-inlet ("by-passing") movements primarily linked to waves and longshore currents. Battjes (1967), who summarizes the general situation, also distinguishes between stable and unstable inlets: the former are dominated by in–out movements; the latter by longshore by-passing which may, in the limit, lead to choking of the inlet. Boothroyd (1978) places particular emphasis on those "mesotidal" estuaries where both tidal and longshore forces are significant. Such locations tend to develop characteristic and important erosional and depositional structures as a result of the interplay of "outlet jets" and longshore drift. Carter (1988, p.174, following Hubbard et al. (1979)) provides a useful summary of what are termed "tide-dominated passes".

It seems valuable to try to identify the key elements of these tidal systems. They appear to be:

1. The development of a barrier or sill across the inlet.
2. The existence of different channels for the flood and ebb flows. Characteristically, the ebb channel is relatively narrow and deep and flow may develop high velocities; the flood flow, in contrast, tends to fill a bigger channel with a higher width/depth ratio. Velocities on the flood may be highly variable, depending upon inlet morphology and the development of bores (Chen et al., 1994).
3. The development of distinct sedimentary structures on flood and ebb tides. The likely existence of a still-water lens up-inlet from the tidal pass (Guilcher, 1967; Van Rijn, 1993) has already been mentioned, together with its probable impact on bedload movement; there may also (cf. Carter, 1988, figure 91) be wholesale deposition of material upstream of the tidal pass in flood flow. On the ebb tide (i.e.

when outflow conditions dominate) there will be important deposition outside the inlet which will, in particular (Carter, 1988), give rise to spits and recurves offshore. According to the balance between tidal currents and longshore sediment movement, the barriers deposited across the inlet mouth, either side of the tidal pass, and the pass itself, may migrate up- or downstream (Lynch-Blosse and Kumar, 1976; Aubrey and Speer, 1984).

If we make the assumptions (1) that the tributary acts as the ebb-flow system and (2) that the mainstream, especially if relatively large in relation to the tributary, may contribute *both* flood flow effects (cf. Kochel and Baker 1988) *and* "longshore drift" of material across the tributary mouth at lower flow stages, then we can begin to see how the mouth regions of small tributary streams like the Ma Oya may behave as the trunk river stage, discharge and sediment load vary.

1. At flood stages on the trunk stream, there is no doubt that there is likely to be wholesale inflow of water and sediment up the tributary (Kochel and Baker, 1988). In addition:
   (a) there will almost certainly be extensive and persistent disruption of stage/ discharge relations in the tributary, well upstream of the water influx (cf. Postma (1967) for coastal cases and Grover (1938) for a fluvial example);
   (b) it may be possible for fine fluvial sediment to be diffused upstream from the zone of maximum turbidity at the mainstream's influx *without any necessary water flux involved* (cf. Postma, 1967, p.173); it is not possible, however, to state how much upstream deposition might arise from such a mechanism;
   (c) if there is a well developed sill at the tributary mouth, which is also funnel-shaped, with an erodible substrate, there is a strong likelihood that a hydraulic jump will form at the sill and then propagate up-tributary, as a bore (Hume and Bell, 1993; Chen et al., 1994).
2. Well below flood peaks on the main channel, there may still be residual "flood tide" effects on the small tributaries. There will also be important sediment movements across the tributary mouth (cf. Figure 16.6A) which will – depending on the balance between "longshore drift" volume and the strength of the "outlet jet" of the tributary's flow, as well as the relative density of sediment concentrations – cause the build-up and erosion of the upstream and downstream ("avalanche face") sides of the barrier at the tributary mouth (cf. Biron et al., 1993). Where there are very large sediment volumes moving down the main channel, the tributary outlet may migrate "upstream" (Aubrey and Speer, 1984). In extreme cases, the tributary outlet may even become choked or barred, with consequent creation of temporary lakes/ billabongs (Kennedy, 1984; Nanson et al., 1993).
3. At low stage on the trunk stream, when tributary outlet flows are dominant, maximum discharges and flow velocities are likely to develop through smallish, relatively deep scour channels cut across or through the barrier. There are likely to be extensive, "deltaic" deposits built in the main stream's channel (cf. Biron et al., 1993). The complexity and form of these deposits will depend upon: the momentum ratio of the flows (cf. Rhoads and Kenworthy, 1995); the volume and calibre of material in the tributary; and the gradient/elevation differences between tributary and

trunk stream channels (recall the difference between the simple fan at the Lovers' Lane outlet and the Ma Oya bar and spit: Figures 16.9 and 16.10). Warner (pers. comm., 1995) notes the case of a major delta, with some recurve, built by Glenbrook Creek into the Nepean River, New South Wales, in May 1994.

4. Finally, there is evidence that the trunk stream may contribute sediment to the tributary over a wide range of flows. Any interpretation of the provenance of sedimentary material in the tributary mouth zone must, therefore, be made with extreme caution (cf. Carter, 1988, p.178; Langhorne et al., 1985, p.12; Nanson et al., 1993, figure 7; Postma, 1967, p.173; Wright, 1978, p.33). Unless there are clear differences in sediment calibre and/or mineralogy, it may be difficult or impossible to trace sediment provenance unambiguously.

5. It is possible (Miller, pers.comm., 1997) that these findings may apply to "deeply incised tributary channels with relatively low gradients immediately upstream of the confluence", but further investigations are clearly needed to confirm or refute that suggestion.

CONCLUSION

A limited set of observations at the junctions of two small, tropical tributary streams with a major channel have indicated that the stage of the trunk river, even at relatively low flows, exerts a perceptible influence to disrupt the smooth down-network transfer of both water and sediment. The sedimentary structures observed, as well as the flow reversals and sediment incursions, suggest that these small, tropical tributary mouths experience flow regimes which are highly analogous to estuaries in general and mesotidal tidal inlets in particular. The probably rather limited role of density differences in the purely fluvial case, as against the estuarine, and the far more irregular pattern of tributary/mainstream floods, as compared with tidal cycles and longshore currents, are both reasons for suspecting that a simple homology between tributary mouths and estuaries will not generally occur. Nevertheless, it is suggested that the analogies adumbrated here are sufficiently striking to merit further investigation since, a priori, they should prove widespread throughout fluvial networks. It further seems profitable to consider developing a uniform typology of confluence flow patterns and sedimentary structures which would go well beyond Bristow et al. (1993) and embrace subaerial alluvial fans at one extreme and tidal estuarine features at the other. At the least, the data presented here appear to suggest that there is some real continuity between "fluvial" and "coastal" cases.

ACKNOWLEDGEMENTS

The research in Sri Lanka was supported by a Royal Society Study Visit Grant and by the Faculty of Anthropology and Geography, University of Oxford. The hospitality and assistance of the Departments of Geography, University of Peradeniya and University of Sydney, are gratefully acknowledged. Thanks are also due to Ms A. Allen, Mr M. Barfoot and Mr P. Hayward, School of Geography, University of Oxford, for the illustrations. Finally I am indebted to Dr G.C. Nanson and Dr R.F. Warner for their comments on the draft of this paper; the constructive advice of Dr A. Miller and Dr B.L. Rhoads, as well as of an anonymous referee, was also of enormous assistance.

REFERENCES

Ashmore, P.E. and Parker, G. 1983. Confluence scour in coarse braided streams. *Water Resources Research*, **19**, 392–402.

Aubrey, D.G. and Speer, P.E. 1984. Updrift migration of tidal inlets. *Journal of Geology*, **92**, 531–546.

Baker, V.R. 1973. *Palaeohydrology and sedimentology of Lake Missoula flooding in eastern Washington*. Special Paper, Geological Society of America, 144, 1–79.

Battjes, J.A. 1967. Quantitative research on littoral drift and tidal inlets. In G.H. Lauff (Ed.), *Estuaries*, AAAS, Washington, 185–190.

Best, J.L. 1986. The morphology of river channel confluences. *Progress in Physical Geography*, **10**, 157–174.

Best, J.L. 1988. Sediment transport and bed morphology at river channel confluences. *Sedimentology*, **35**, 481–498.

Best, J.L. and Reid, I. 1984. Separation zone at open-channel junctions. *Journal of Hydraulic Engineering, ASCE*, **110**, 1588–1594.

Biron, P., Roy, A.G., Best, J.L. and Boyer, C.J. 1993. Bed morphology and sedimentology at the confluence of unequal depth channels. *Geomorphology*, **8**, 115–129.

Boothroyd, J.C. 1978. Mesotidal inlets and estuaries. In R.A. Davis (Ed.), *Coastal Sedimentary Environments*, Springer Verlag, New York, 287–360.

Bretz, J.H. 1929. Valley deposits immediately east of the Channelled Scablands of Washington, *Journal of Geology*, **37**, 393–427.

Bridges, P.H. and Leeder M.R., 1976. Sedimentary model for intertidal mudflat channels, with examples from the Solway Firth, Scotland. *Sedimentology*, **23**, 533–552.

Bristow, C.S., Best, J.L. and Roy, A.G. 1993. *Morphology and facies models of channel confluences*. International Association of Sedimentologists, Special Publication, 17, 91–100.

Carter, R.W.G. 1988. *Coastal Environments*. Academic Press, London, 617 pp.

Chen, P., Ingram, R.G. and Gan, J. 1994. A numerical study of hydraulic jump and mixing in a stratified channel with a sill. In M.L. Spaulding et al. (Eds), *Estuarine and Coastal Modelling*, III. ASCE, 119-133.

Grover, N.C. 1938. Floods of Ohio and Mississippi Rivers, January–February 1937. *US Geological Survey Water-Supply Paper*, 838.

Guilcher, A. 1967. Origin of sediments in estuaries. In G.H. Lauff (Ed.), *Estuaries*. Washington, AAAS, 149–157.

Hubbard, D.K., Oertel, G. and Nummedal, D. 1979. The role of waves and tidal currents in the development of tidal-inlet sedimentary structures and sand body geometry: examples from North Carolina, South Carolina, and Georgia. *Journal of Sedimentary Petrology*, **49**, 1073–1092.

Hume, T.M. and Bell, R.G. 1993. *Methods for determining tidal flows and material fluxes in estuarine cross-sections*. DSIR, Water Quality Centre, Hamilton, NZ, 43 pp.

Kennedy, B.A. 1984. On Playfair's Law of Accordant Junctions. *Earth Surfaces Processes and Landforms*, **9**, 153–173.

Kenworthy, S.T. and Rhoads, B.L. 1995. Hydrologic control of spatial patterns of suspended sediment concentration at a stream confluence. *Journal of Hydrology*, **168**, 251–263.

Knighton, A.D., Woodroffe, C.D. and Mills, K., 1992. The evolution of tidal creek networks, Mary River, Northern Australia. *Earth Surfaces Processes and Landforms*, **17**, 167–190.

Kochel, R.C. and Baker, V.R. 1982. Palaeoflood hydrology. *Science*, **215**, 353–361.

Kochel, R.C. and Baker, V.R. 1988. Palaeoflood analysis using slackwater deposits. In V.R. Baker, R.C. Kochel and P.C. Patton (Eds), *Flood Geomorphology*. Wiley, New York, 357–376.

Langhorne, D.N., Malcolm, J.O. and Reed, A.A. 1985. *Observations of the changes of inter-tidal bedforms over a neap-spring tidal cycle*. Institute of Oceanographic Sciences Report, No. 203, 101 pp.

Lessa, G. and Masselink, G. 1995. Morphodynamic evolution of a macrotidal barrier estuary. *Marine Geology*, **129**, 25–46.

Lynch-Blosse, M.A. and Kumar, N. 1976. Evolution of downdrift-offset tidal inlets: a model based on the Brigantine Inlet system of New Jersey. *Journal of Geology*, **84**, 165–178.

Mosley, M.P. 1976. An experimental study of channel confluences. *Journal of Geology*, **84**, 535–562.

Myrick, R.M. and Leopold, L.B. 1963. Hydraulic geometry of a small tidal estuary. *US Geological Survey Professional Paper*, 422–B.

Nanson, G.C., East, T.J. and Roberts, R.G. 1993. Quaternary stratigraphy, geochronology and evolution of the Magela Creek catchment in the monsoon tropics of northern Australia. *Sedimentary Geology*, **83**, 277–302.

Pethick, J. 1984. *An Introduction to Coastal Geomorphology*, Edward Arnold, London, 260 pp.

Postma, H. 1967. Sediment transport and sedimentation in the estuarine environment. In G.H. Lauff (Ed.), *Estuaries*. AAAS, Washington, 158–179.

Rachocki, A.H. and Church, M.A. (Eds)1989. *Alluvial Fans: a Field Approach*. Wiley, New York.

Reid, I., Best, J.L. and Frostick, L.E. 1989. Floods and flood sediments at river confluences. In K. Bevin and P. Carling (Eds), *Floods: Hydrological, Sedimentological and Geomorphological Implications*. Wiley, Chichester, 135–150.

Rhoads, B.L. and Kenworthy, S.T. 1995. Flow structure at an asymmetrical stream confluence. *Geomorphology*, **11**, 273–293.

Roy, A.G. and Roy, R. 1988. Changes in channel size at river confluences with coarse bed material. *Earth Surface Processes and Landforms*, **13**, 77–84.

Van Rijn, L.C. 1993. *Principles of Sediment Transport in Rivers, Estuaries and Coastal Seas*. Aqua Publications, Amsterdam.

Wright, L.D. 1978. River deltas. In R.A. Davis (Ed.), *Coastal Sedimentary Environments*. Springer Verlag, New York, 5–68.

17

Geomorphology and Coastline Change of the Lower Yangtze Delta Plain, China

ZHONGYUAN CHEN

Estuarine and Coastal State Key Laboratory, East China Normal University, Shanghai, People's Republic of China

ABSTRACT

This study examines the suite of geomorphological features of the lower Yangtze delta plain. Detailed examination of surficial morphology allows an understanding of the interplay among eustasy, isostasy, tectonic displacement, fluvial transport and coastal hydrodynamics for forming the delta plain.

Holocene sea-level fluctuation is the key factor in controlling the geomorphic framework of the study area. Changes in topographic features with time, including the chenier ridge system, lake depression, tidal flat and estuary, occurred in direct response to sea-level rise. Deceleration of sea-level rise *c.* 8000–7000 years BP initiated deltaic build-up, which in turn made available abundant natural resources, primarily fresh water and fertile soil. As a result, the earliest Neolithic settlement of eastern China developed on the delta plain.

Since the late Pleistocene, the Yangtze delta depocentre has progressively migrated over a distance of about 150 km, from the northern to the southern delta plain. This long-term shift is explained by rapid tectonic subsidence during the Quaternary of the southern delta plain as compared to the northern delta plain. The directions of Coriolis and longshore currents are also southward in the estuarine region; this further promotes the migration of the delta towards the southeast.

The historical changes of the Yangtze coastline are closely linked with river-mouth shifts and anthropogenic activity. The spectacular southward displacement of the upper region of the northern delta coast was due to avulsion of the former Yellow River and to embankment construction for flood prevention. The lower region of the southern delta coast changed significantly because of the Yangtze River's migration through time and because of the effects of reclamation projects. The processes of formation of the lower Yangtze delta plain can be compared with other major deltas of the world in many respects. These scenarios reflect the influence of global and regional change events during the Holocene.

INTRODUCTION

The lower Yangtze delta plain supports around 12 million people engaged in agriculture and industry. The rapidly increasing development of the economy along this coast is now

Varieties of Fluvial Form. Edited by A.J. Miller and A. Gupta
© 1999 John Wiley & Sons Ltd

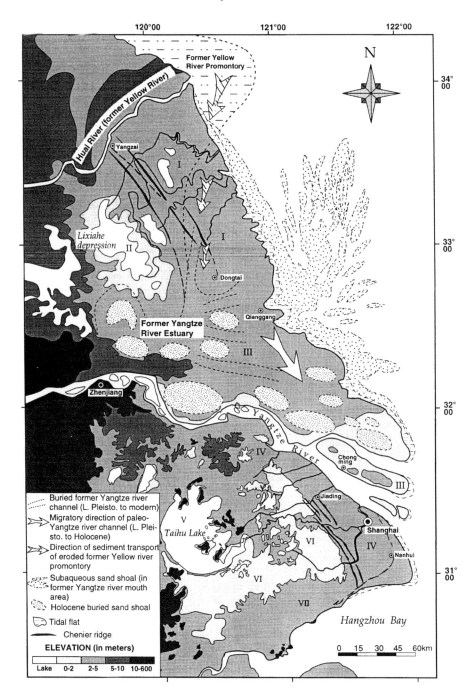

FIGURE 17.1   Morphological features of the lower Yangtze delta plain, Eastern China

greatly intensifying the need for land. Management of land use for various purposes and long-term planning for reclamation, channelization and harbour construction are critical for resolving conflicts among people, natural resources and environment. Thus, a better understanding of the history of the development of the lower Yangtze delta plain is essential for planning and managing future land-use patterns in the delta.

The Yangtze is the third longest of the world's rivers. Originating from the Himalayan plateau, it flows from west to east, crossing central China and eventually entering the East China Sea. Its drainage basin covers 1.81×10^6 km^2 and the total area of the lower delta plain is about 50 000 km^2. The lobe-like Yangtze coastal morphology, characterized by a series of elongated islands and sandy shoals distributed in the river mouth area (Figure 17.1), demonstrates the fluvial and tidal-dominated pattern (Coleman, 1981) in contrast to the fluvial process-dominated Mississippi delta or the wave-dominated delta of the Nile River.

The lower Yangtze delta plain is a natural Quaternary laboratory, in which evidence of the interaction of geomorphic phenomena with climate, sea-level fluctuation, isostasy, neotectonism, and fluvial and coastal transport can be observed. A large amount of existing data are available for the study area from many previous geological and archaeological investigations (Yi and Zhao, 1963; Xu et al., 1981; Yan and Xu, 1987; Yan and Hong, 1988; Yan and Shao, 1989; Sun and Huang, 1993; Chen and Stanley, 1995; Stanley and Chen, 1996). Numerous seismic profiles and radiocarbon-dated sediment cores have been collected and petrologically analysed; an account of the migration of the Neolithic settlements in response to Holocene sea-level rise is also a product of this series of studies. The entire database provides a background for studying the formation of the lower Yangtze delta plain, and also for studying human impact on the natural environment.

PHYSICAL BACKGROUND

The Yangtze delta plain is an area of low relief with ground elevations generally 3 to 5 m above mean sea level (m.s.l.). The present main river channel divides this vast low-lying delta plain into two parts, the northern and southern plains (Figures 17.1 and 17.2). The deltaic coastline is about 300 km long and there is a distance of about 200 km from the delta's apex (Zhengjiang City, Figure 17.1) to the river mouth. As it flows out of its rocky valley near Zhengjiang city, the river channel begins to release its tremendous discharge in the estuarine region, where topography obviously tends to be more gentle.

The Yangtze delta is formed primarily by fluvial and tidal processes (Xu et al., 1981; Xu and Chen, 1995). The volume of fresh water runoff is 995×10^9 m^3 a^{-1} and sediment input is approximately 0.5×10^9 ton a^{-1}. Irregular semi-diurnal tides occur in the estuary, with an average tidal range of 2.6 m and a maximum of about 5.0 m. The mean velocity of ebb tidal currents in the river mouth area is 1.13 m s^{-1}, which is greater than that of flood currents (0.96 m s^{-1}). The duration ratio of ebb to flood is 1.73:1.

Sediment dispersal from the Yangtze has been influenced by runoff, tides, longshore currents and deep ocean currents (Figure 17.3). The coarser materials (Shen et al., 1992) settles in the estuarine depocentre where long-term subsidence prevails, apparently as a result of neotectonic lowering of the underlying bedrock and compaction of unconsolidated Quaternary sediments (Chen and Stanley, 1993a, 1995; Stanley and Chen, 1993). In contrast, the remaining fine material, suspended on the salt wedge as a

FIGURE 17.2   Satellite image showing configuration of the lower Yangtze delta plain. The northern and southern plains divided by the main Yangtze river channel. Taihu lake is located on the southern plain. Source: Remote Sensing Laboratory of Urban and Environmental Archaeology, ECNU

fresh plume, is driven primarily southeastward by ebb and longshore currents (Chen et al., 1986; Chen and Zhang, 1989).

Directions of incoming and outgoing tidal currents are not quite diametrically opposed in the river mouth area. The flood azimuth is approximately 305° and ebb trends to SSE. Therefore, between the two tidal currents slackwater regions occur, where large amounts of coarse sediment settle rapidly to form patchy shoals. These shoals accrete seaward to connect and merge, especially at low tide, and eventually become the bases of river-mouth islands.

Longshore currents flow on both sides of the Yangtze River mouth (Figure 17.3). The Subei longshore current occurs along the northern coast and flows constantly southward. This current acts as a reworking agent on the northern subaqueous delta area, where surficial sediment mixed with shell fragments is relatively coarser than along the southern

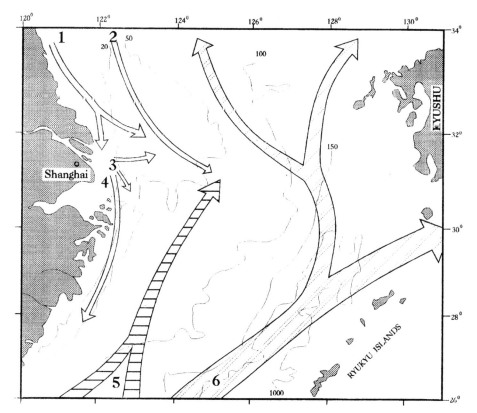

FIGURE 17.3   Hydrodynamic background in and off the Yangtze River mouth. 1, Subei longshore current; 2, Yellow sea longshore current; 3, Yangtze fresh plume; 4, Yangtze longshore current; 5, Taiwan warm current; 6, Kuroshio current (data modified after Chen, 1987)

coast (Chen et al., 1986, 1991). The Yangtze longshore current occurs along the southern coast, carrying most Yangtze suspended sediment southward to the Zhejiang coast and approaching the Taiwan strait. In addition, the Taiwan oceanic warm current flowing northward is located at varying distances from the Yangtze coastline at different seasons. In summer, southeast-prevailing winds push the current nearer to the estuary, whereas in winter, northwest-prevailing winds have the opposite effect.

Typhoons from the southeast Pacific Ocean intrude the Yangtze coast one to three times per year, on average. During this climatic hazard, sea-level set-up of 2–3 m is often observed (Shao et al., 1991; Xu et al., 1990). The situation worsens, especially when the typhoon coincides with astronomical high tides. The maximum water level of 8.40 m above m.s.l. was recorded during the 1990 typhoon season.

GEOMORPHOLOGY OF THE LOWER YANGTZE DELTA PLAIN

Four major types of morphology of the lower Yangtze delta plain are recognizable, including chenier ridge, estuary, lake and tidal flat. The characteristics of each are described briefly.

Chenier ridge

Three to five linear chenier ridges are conspicuous on the northern and southern delta plains (Figure 17.1). These are 4–5 m in elevation above m.s.l., almost parallel, and discretely distributed along the coastline. The chenier ridges on the northern delta plain are about 100 km long and 3–10 km wide, extending southeastward from the Yangzai by the southern Huai River bank to the Dongtai of the central northern plain (Figure 17.1). The chenier ridges on the southern delta plain are about 80 km long and 4–8 km wide, starting from Jiading on the southern bank of the Yangtze River channel, southeastward to the Fengxian near the northern part of Hangzhou Bay.

The chenier ridges consist largely of shells (whole and fragments) and muds. An average composition is *c.* 60% shell, *c.* 30% silt and *c.* 10% fine sand (with 1–5% heavy minerals). The thickness of the ridge sequences generally varies from <0.5 to 4 m and the inner sedimentary structures are characterized primarily by large-scale tabular and wedge cross-stratification, obviously resulting from strong tides and typhoon processes (Gu et al., 1983; Yan et al., 1989; Zhang et al., 1982). Coring and trenching profiles reveal that the chenier ridge system rests directly upon a yellow brown tenacious mud unit (stiff mud). This mud contains abundant reed roots, and is characterized by Fe and Mn oxides indicative of an upper tidal-flat environment (Zhang et al., 1982). The basal strata of the chenier ridge are therefore inferred to have been at mean high sea level at the time of deposition (Gu et al., 1983; Zhang et al., 1982).

In fact, the chenier ridges delineate former Yangtze coastlines. Based on a number of radiocarbon dates, the chronology of the ridges spans at least 3000 years, from *c.* 7000 to 4000 years BP (Liu et al., 1985; Zhang et al., 1982). The origin of the chenier ridges on both delta plains is apparently related to regional river mouth shift.

The evolution of the northern chenier ridges was closely associated with the shifting of the former Yellow River channel during the mid- to late Holocene (Lin, 1988). Coring and petrological analyses reveal that the modern Huai River mouth area at the upper northern plain (Figure 17.1) was periodically captured in earlier times by avulsion of the Yellow River (Chen, 1995; Lin, 1989), which has the highest measured sediment concentration ($33.2\,\mathrm{kg\,m}^{-3}$) in the world. The coastline remained almost stable from 7000 to 4000 years BP, during the period when the Yellow River incompletely took over the course of the Huai River. Waves, longshore currents and typhoons brought a large amount of shells from the nearshore to rest on the coast, piling up a series of chenier ridges. Meanwhile, aeolian processes formed magnificent sand dunes along the coast (Chen, 1995; Zhao et al., 1991).

The chenier ridges on the southern plain are also closely related to the position of the Yangtze River channel. During 7000 to 4000 years BP, the Yangtze River channel lay further north than at present, crossing the northern delta plain into the southern Yellow Sea. At that time, the southern delta coast did not receive much sediment input from the Yangtze, and the progradation rate was very low (*c.* $0.5–1.3\,\mathrm{m\,a}^{-1}$). Tides, longshore currents and typhoon processes reworked the tidal flat, piling up the chenier ridges, mixed with sand, mud and shell brought from nearshore (Zhang et al., 1982). As the Yangtze main river channel migrated gradually southward, the coast of the southern delta plain began to prograde seaward after 4000 years BP.

Estuary

The present Yangtze River mouth is primarily characterized by three large-scale elongated islands, separated by distributaries that are about 10 m deep. The Chongming island, the biggest one, is about 100 km in length and 30 km in width (Figure 17.1), and has been formed by connection of shoals since 2000 years BP (Chen et al., 1979). It is noteworthy that these islands, and many shoals in front of the islands, are oriented en echelon southeastward. Moreover, Chongming island gradually merges with the northern delta plain, due to siltation of the northern Yangtze distributary. This tendency has kept pace with the southeastward shift of the delta depocentre, at least since the late Quaternary (Chen and Stanley, 1995; Chen and Yang, 1991).

Numerous coring investigations indicate that Holocene sedimentary sequences in the estuary, accumulating directly on marine-influenced fluvial sand deposits of late Pleistocene age, are thickest (50–60 m thick) in the lower Yangtze delta plain. During early to late Holocene, the sedimentary facies evolved upward from estuarine muds to sandy shoals (Figure 17.4).

The former Yangtze estuarine region (Figure 17.1) is located between the two chenier ridges resting upon the northern and southern plains, and extends landward from the modern Yangtze estuary to the northwest. Numerous sediment cores, together with interpretation of satellite images (Xu et al., 1981; Zhong et al; 1983), indicate that within this former funnel-shaped estuary, Yangtze River channels of early to mid-Holocene age were deeply (40–60 m) incised into the subsurface, cutting through the northern delta plain into the southern Yellow Sea. These channels migrated progressively with time, from north to south. No Yangtze main river channel has been found south of the present river channel position (Chen and Stanley, 1995).

Cores and lithological analyses reveal many buried sandy shoals at different sites within the former estuarine region (Figures 17.1 and 17.3). They were also oriented en echelon southeastward, showing the same migratory trend as the shift of the present river mouth (Xu et al., 1981).

Our geological investigation proposes that the progressive north to south migration of the delta depocentre has been affected by tectonic subsidence of the study area (Chen and Stanley, 1995). The rate of subsidence of the southern plain, due to rapid lowering of the basement rock, is greater than that of the northern plain and has increased since the early Quaternary. This geological mechanism is inducing the long-term migration of the Yangtze River channel from north to south.

Lake

Of special interest is a cluster of freshwater lakes landward of the chenier ridges on both delta plains. The lakes on the northern delta plain form the Lixiahe depression (Figure 17.1), and the lakes on the southern plain are associated with the Taihu depression.

Taihu is the biggest of the lakes, with a water surface area of about 2400 km^2 and average water depth of about 2 m. The Taihu area is a saucer-like depression surrounded by hills (100–500 m in elevation) to the west of the lake, a chenier ridge system (4–5.5 m above m.s.l.) to the east of the lake, and deltaic fluvial plains (3–5 m above m.s.l) to the north and south of the lake (Figure 17.1). Coring and seismic profiling reveal that the

434

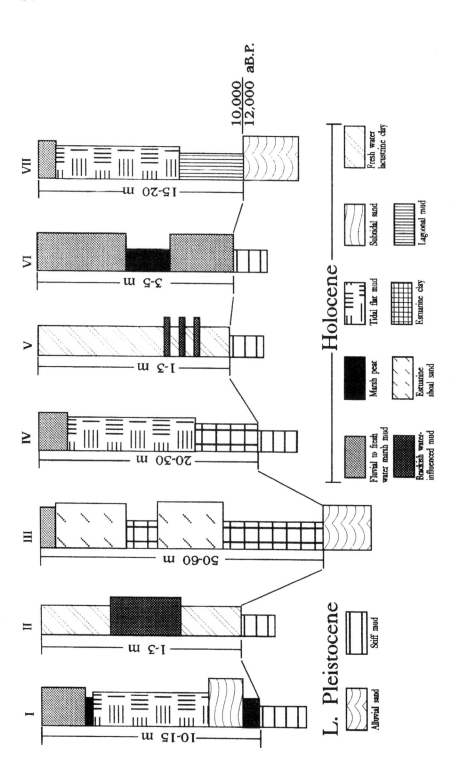

FIGURE 17.4   Generalized late Quaternary stratigraphic sequences of the lower Yangtze delta plain (sequences I to VII are marked in Figure 17.1)

Holocene sediment sequences in the lake, as well as the depression area to the east and southeast of the lake are generally *c.* 1–5 m thick (Figure 17.4V and VI), and are composed in large part of freshwater lacustrine clay, peat and fluvial mud deposits. Thin marine-influenced strata are sporadically interbedded (Sun and Huang, 1993). The lacustrine sediments are underlain directly by a dark green to brown stiff mud unit of late Pleistocene age which exists worldwide in delta regions, and is characterized primarily by many Fe and Mn oxides and by some calcareous and phosphate nodules (Chen and Stanley, 1993b; Xu and Chen, 1995). The stiff mud occurs as floodplain facies on the delta plain, and was formed during the low sea-level stand. This mud tends to be buried at greater depths (about 20–30 m) approaching the coastline.

Numerous Neolithic sites (more than 100), mostly 1–3 m under the present delta plain surface, occur in the saucer-like depression. A combined geological and archaeological investigation demonstrated that the origin of the Taihu lake, the chenier ridges, the delta plain in the periphery of the Taihu lake and the Neolithic migration in the study area have been closely associated with Holocene eustatic sea-level rise (Stanley and Chen, 1996).

Between 10 000 and 7 000 years BP, the Taihu region was subaerially exposed, and no evidence of settlement of this age has been found. This inference is supported by a lack of organic content in soil, which therefore would not have been optimum to support agricultural activity. By about 7000 years BP, sea level approached much closer to its present level and the groundwater table had moved upwards. Organic-rich sediment brought by the Yangtze River began to construct the fertile delta plain. These conditions, as well as a warming climate (Wang et al., 1984) during the late early Holocene, induced the settlements and rice cultivation in the Yangtze delta area. A continuous rise in sea level and sediment accumulation influenced by runoff, tides, waves, longshore currents and typhoon processes have led to aggradation of the delta plain margin surrounding the Taihu settlement area. This change in topography with time in response to sea-level rise has gradually turned the former habitat base into a saucer-like depression.

Limnolization of the Taihu depression due to rising groundwater apparently drove the Neolithic settlements to migrate eastward, both to the back of and onto the chenier ridge, which acted as a coastal barrier and partially protected settlements from sea invasion. However, the enlarged Taihu lake, which at one time was two to three times larger than the present lake (Jing, 1989), drowned most of the sites and aggravated living conditions in the settlement region after 4000 years BP. After this time the palaeoculture of the Taihu depression declined.

In comparison, though the Lixiahe depression of the northern plain was also formed by aggradation of the outer delta plain in response to sea-level rise after 7000 years BP, the southern part of the depression was basically open to the former Yangtze estuary (Figure 17.1), and many large-scale tidal creeks crossed the chenier ridges to the sea. Saline water intruded the depression through openings and tidal inlets, and therefore a lagoonal environment prevailed. Holocene sediment sequences in the Lixiahe depression generally range from 1 to 5 m in thickness (Figure 17.4), and accumulated directly on the late Pleistocene stiff mud deposits, along with abundant marine fossils, including molluscs and foraminifera (Lin, 1989, 1993).

The early Holocene transgression extended 50–80 km landward to the west of the Lixiahe depression. After about 7000 years BP, formation of the chenier ridge gradually separated the sea from the depression. The Lixiahe depression began to shrink due to (1) high siltation from Yellow River floods from the north after 2000 years BP and (2) intensifying anthropogenic activity, such as reclamation (Lin, 1989). Moreover, geological and archaeological investigations show that Neolithic sites, so far found around the Lixiahe depression, apparently were less numerous than those in the Taihu region. The sites, usually overlain by thicker (3–5 m) terrigenous sediments from the Yellow River floods, occur more along the northern margin of the depression (Yi and Zhao, 1963). This evidence strongly supports the interpretation of a sea-influenced environment in the depression area, which was unsuitable for habitation during the early to mid-Holocene.

Tidal flat

Tidal flats, one of the major geomorphic components of the lower Yangtze delta plain, occur widely along the estuarine coast, and are indicative of this delta being tide-dominated (Reading, 1978; Coleman, 1981). The vast delta plains east of the chenier ridge on both plains are actually of tidal-flat origin (Yan and Shao, 1989). Many modern major channels, as wide as 50-80 m and perpendicular to the coast, have been derived from former tidal creeks (Yan and Shao, 1989; Stanley and Chen, 1996). The width of the tidal flat varies in places along the coast. The widest part is located (*c.* 10 km) near the Qianggang coast (Figure 17.1) in the northern plain, where the former Yangtze River mouth was located. This section of the coast was abandoned as the Yangtze River channel shifted towards its present position. Tides and waves have reworked this area, forming a gigantic submerged shoal system (Figure 17.1), which extends about 100 km to the sea and emerges at low tide (Li and Li, 1981). Tidal flats normally tend to be narrower towards the south, diminishing to only a few hundred metres in width. However, in front of Chongming island and along the Nanhui coast of the southern plain, tidal flats of 3–7 km in width can still be observed.

Hundreds of sediment cores reveal that the tidal flat prevailed during the mid- to late Holocene (Figures 17.4I, IV and VII) and built up on the early Holocene sediments, as the Yangtze delta began to prograde seaward about 7000 years BP. The sediment sequences of early to late Holocene in the northern delta plain are usually 10–15 m thick and change primarily from subtidal sand and tidal-flat mud to fluvial silt and clay on the top (Figure 17.4I). In the southern delta plain, the Holocene sediment sequences change upward from lagoonal, estuarine silt to tidal-flat mud, and to fluvial silt, thickening from 20–30 m (Figure 17.4IV and VII).

Typhoons from the southeast Pacific Ocean strike the Yangtze coast periodically, averaging one to three times during every August and September. During such events, sea-level set-up of 2–3 m often occurs. Hummocky cross-stratification, which is generally believed to be preserved between normal and storm wave bases (Chen, 1987; Xu et al., 1990), is present due to the high sedimentation rate influenced by the large fluvial sediment input from the Yangtze River source.

HISTORICAL CHANGE OF COASTLINE

In recent decades, extensive investigations utilizing the techniques of sedimentology, geomorphology, palaeontology, climatology and archaeology have been conducted to demonstrate that the Yangtze coastline has changed greatly since about 7000 year BP

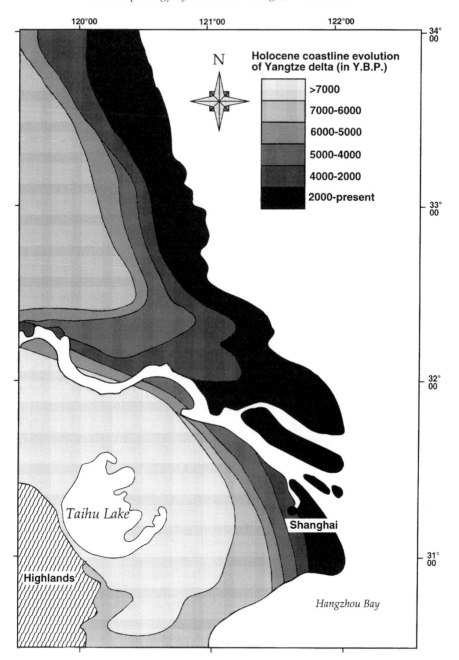

FIGURE 17.5   Historical changes of the Yangtze coastline (data modified after Wang et al., 1988)

(Figure 17.5). This change has occurred in response to the Holocene sea-level rise in competition with the sedimentation rate around the estuarine region, and in association with regional river mouth shift.

At the end of the Pleistocene and the beginning of the early Holocene, the level of the East China Sea was about 150–100 m lower than at present (Zhao et al., 1994). At that time, the older delta plain was subaerially exposed, and the river systems, including both the Yangtze and the river channels originating from the western highlands to the southern plain (Figure 17.1), cut deeply through the plain, entering the East China Sea.

Sea level rose rapidly during the early Holocene, and was about 7 m lower than at present during the period from 7500 to 7000 year BP (Zhao et al., 1979; Yan and Hong, 1988; Yan and Shao, 1989). The Yangtze coastline at that time was positioned about 150–200 km landward away from the present coast. The sea inundated the lowest delta plain, except the Taihu region where the early settlements occurred. With the deceleration of sea-level rise after 7000 year BP (Zhao et al., 1979), the modern Yangtze estuary began to form, as fluvial sediments started to be deposited in its estuarine region.

The coastline of the Yangtze delta plain began to prograde rapidly seaward after 4000 year BP, in close association with river-channel migration in the estuarine region. The geologic record shows that the northern coast was strongly influenced by the migration of the Yellow River. In AD 1128 the Yellow River totally engulfed the Huai River by avulsion. A tremendous amount of sediment derived from the Yellow River heavily aggraded the Huai estuarine region. Consequently, the river bed was about 3 m higher than the adjacent floodplain, and it has been described as a "suspended river" (Chen, 1995). To prevent frequent flooding and siltation, the river was straightened and embankments constructed along the river channel. However, this anthropogenic activity certainly accelerated the seaward movement of the river mouth. Between AD 1128 and 1855, the former Yellow River mouth had moved as far as about 80 km seaward from its present position (Figure 17.1). The return of the Yellow River to north China in AD 1855 caused this promontory to suffer from considerable erosion under reworking of coastal hydrodynamics, resulting in substantial recession of the coastline (Li and Li, 1980). The rate of retreat at the beginning of this episode of coastal destruction is estimated at between 300 and $1000 \, \text{m a}^{-1}$ (Lin, 1988).

The section of the northern coastline immediately south of the promontory accepted a large amount of sediment both from the eroded promontory by south-directed longshore currents and from the former Yangtze River mouth near the Qianggong. The progradation rate reached about $90–200 \, \text{m a}^{-1}$ (Lin, 1988), especially after c. 2000 years BP. The coastline of the southern plain also moved rapidly seaward by accumulation from the Yangtze sources, with progradation rates calculated to be about $8–25 \, \text{m a}^{-1}$ after 4000 years BP (Zhang et al., 1982). Maximum progradation rates of $60–100 \, \text{m a}^{-1}$ are recorded for coastline displacement of the southern plain over the past 2000 years, largely because of anthropogenic activity such as reclamation (Shao and Yan, 1982).

COMPARISON WITH OTHER MAJOR DELTAS WORLDWIDE

Radiocarbon-dated deltaic sequences of Holocene age show that the lower Yangtze delta plain began to form around 8000–7000 years BP (Stanley and Chen, 1993). The deceleration in sea-level rise during the mid-Holocene is the key factor in controlling the onset of the Yangtze delta morphology (Chen et al., 1991). Deltaic sediments prograded seaward, as the rate of fluvial sediment input overtook the declining rate of sea-level rise along the coast. This sequence can be compared with the findings of Stanley and Warne

(1994) from 35 major world deltas. Global change relative to climate variation during the early and mid-Holocene may have influenced sea-level deceleration, explaining the near-coincidental initiation of deltas worldwide during the Holocene

Topographic features of the lower Yangtze delta plain were related to the tidal range (about 2.6 m, on average) and typhoon processes (about 3 m sea-level set-up). The low relief of the delta plain (3–5 m elevation above m.s.l.) brings it within the range of normal tidal processes and typhoons every August to September. Few locations in the lower Yangtze delta plain have elevations higher than 6 m above m.s.l.; a similar situation occurs in most of the world's major lower delta plains.

Sediments overlying the late Quaternary substrata in the Yangtze estuary have been affected by subsidence resulting from compaction and dewatering processes. Coring and seismic profiling demonstrate that the accumulation of sediment in the river mouth area has pushed the base of the Holocene sediment pile down into the underlying late Pleistocene soft sediment horizon (Figure 17.6). This vertical gravitation in the delta depocentre has tilted substrata seaward, and formed mud diapirs immediately seaward of the depocentre. The mud diapirs have penetrated upward into the Holocene sediment and even through the sea bottom, with protrusions of up to 20 m (Chen and Stanley, 1993a). Meanwhile, water-saturated clay deposits of Holocene age have been mobilized horizontally, and subsidence has triggered mass movement seaward from the river mouth depocentre.

Highly deformed topography and substrata are also reported in the Mississippi and Niger depocentres (Morgan et al., 1968; Damuth, 1994), where mud diapirs and reverse faults are common. The deformation is caused largely under the influence of pressure imposed by the quickly accumulating overlying sediment. The underlying clay section, rapidly deposited with high water content, undergoes compaction. These deposits in the depocentre are potentially unstable, and start to mobilize slowly seaward as mass movements. This is similar to the deformation process of the Yangtze, where in addition a gravitational subsidence pattern prevails in a large-scale estuarine area dominated by fine-grained sediments (largely clay and silt). The altered topography and flowage of substrata in the river mouth areas are of great economic concern, especially for large-scale geo-engineering projects.

The historical change of the Yangtze coastline is closely related to shifts in location of the river mouth. The migration of the Yellow and the Yangtze River mouths is the primary cause of coastal progradation and erosion. An analogue is found in the Nile delta. From 8000 to 2000 years BP, the Nile coast changed its position dramatically by displacing the main river promontory (Sebennitic) from the central Nile to the eastern Damietta and the western Rosetta promontories (Stanley and Warne, 1993). The Nile coastline varied significantly as river mouths alternately prograded and retreated. For example, the progradation rate in the Rosetta promontory was only about $0.5\,\mathrm{m\,a^{-1}}$, when this river channel was a minor distributary before 2000 years BP. After that time, with the waning of the Sebennitic distributary largely due to intensifying anthropogenic excavation, a substantial sediment discharge switched to the Rosetta distributary. In consequence, its coast moved rapidly seaward with an average progradation rate estimated at about $6\,\mathrm{m\,a^{-1}}$. This promontory, however, has been eroded following the enormous reduction of sediment input into the estuary due to the closure of Aswan high dam in 1964. The maximum recorded coastline retreat rate is $275\,\mathrm{m\,a^{-1}}$ (Chen et al., 1993).

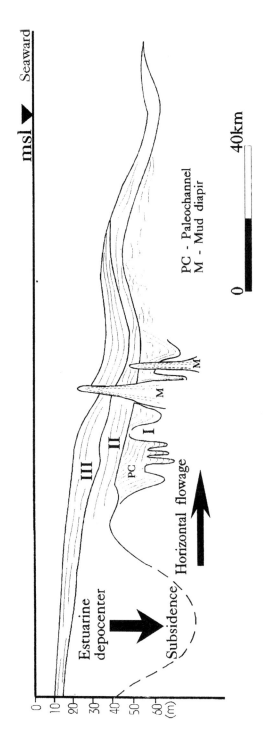

FIGURE 17.6 Deformed topography in association with tilted substrata and mud diapirs. Seismic profile extending seaward southeasterly from the Yangtze estuary (II+III = Holocene; I = late Pleistocene)

CONCLUSIONS

The lower Yangtze delta plain presents a suite of morphological characteristics, resulting from a combination of eustasy, isostasy, tectonic displacement, fluvial transport and coastal hydrodynamics.

Holocene sea-level rise, with variable rates over time and a cumulative rise of up to 100 m or more, obviously played the most important role in controlling the geomorphic framework of the study area. Topographic change occurred in response to sea-level rise since the mid-Holocene in the Yangtze delta plain, especially in the vast area east of the chenier ridge. This consequently led to the formation of the Taihu and Lixiahe depressions, where abundant natural resources, primarily the availability of fresh water and fertile soil, brought about the earliest Neolithic settlement of eastern China. The deceleration of sea-level rise at about 7000 years BP initiated the deltaic build-up. Deposition of a large amount of sediment within the estuary caused sedimentation rates that progressively exceeded the rate of sea-level rise, resulting in seaward progradation of the coastline. The Yangtze delta system, including present and ancient river channels and islands, constantly migrates southeastward. Tectonic subsidence of the southern delta plain since the Quaternary explains this shifting, and has induced long-term migration of the entire delta depocentre, progressing from the northern to the southern plain, across approximately 200 km distance (Chen and Stanley, 1995). The direction of the Coriolis force and longshore currents of the estuarine region are also southward. This further promotes the migration of the delta towards the southeast.

Chenier ridge formation on both the northern and the southern delta plains occurred almost contemporaneously from about 7000 to 4000 years BP. The spectacular change of the coastline in the study area during the past 4000 years is associated with shifts in the positions of the river mouth, as well as anthropogenic activity. Geomorphic changes along the upper and lower parts of the northern coast have been influenced by the former Yellow River and by Yangtze River channel migration, respectively.

Although the lower Yangtze delta plain is characterized by its own geomorphological and sedimentological processes, many aspects of its formation can be compared with other major deltas. These aspects include the initiation of Holocene delta construction, tidal range-controlled deltaic elevation, deformed topography and changed coastline owing to the river mouth shifting. These scenarios reflect global and regional change during the Holocene.

A better understanding of the geomorphological evolution and coastline change of the lower Yangtze delta plain is of great value at present, especially given the increasing pressure on land. Appropriate assessment of the processes of delta formation will prove invaluable towards the comprehension of various issues generated by nature and people, such as the Three Georges Dam.

ACKNOWLEDGEMENTS

Special thanks are given to Drs Andrew J. Miller, Avijit Gupta and Robert B. Jacobson, who kindly reviewed the manuscript and provided helpful suggestions. The project is financially funded by FEYUT.SEDC-China and National State Education Commission TCTPF–China to Z.C.

REFERENCES

Chen, J., Yun, C., Xu, H. and Dong, Y. 1979. The development of the Chang Jiang River Estuary during last 2000 years. *Acta Oceanologica Sinica*, **1**(1), 103–111 (in Chinese, with English summary).

Chen, Z. 1987. Characteristics of storm-generated sediments of the subaqueous Yangtze Delta. In Yan, Qingshang and Xu, Shiyuan (Eds), *Recent Yangtze Delta Deposits*. East China Normal University Press, Shanghai, 246–257 (in Chinese, with English summary).

Chen, Z. 1995. Sedimentary Characteristics of Subei Coastal Plain, Northern Yangtze Delta, China. *Journal of East China Normal University (Natural Science)*, **2**, 86–92 (in Chinese, with English summary).

Chen, Z. and Stanley, D.J. 1993a. Yangtze delta, eastern China: 2: Late Quaternary subsidence and deformation. *Marine Geology*, **112**, 13–21.

Chen, Z. and Stanley, D.J. 1993b. Alluvial stiff muds (Late Pleistocene) underlying the Lower Nile Delta Plain, Egypt: petrology, stratigraphy and origin. *Journal of Coastal Research*, **9**(2), 539–576.

Chen, Z. and Stanley, D.J. 1995. Quaternary subsidence and river channel migration in the Yangtze Delta Plain, Eastern China. *Journal of Coastal Research*, **11**(3), 927–945.

Chen, Z. and Yang, W. 1991. Quaternary Evolution of Palaeogeography and Paleoenvironment in the Yangtze Estuarine Region. *Acta Geographica Sinica*, **46**(4), 436–447 (in Chinese, with English summary).

Chen, Z. and Zhang, S. 1989. Sedimentary features and evolutionary trend of the subaqueous Yangtze Delta. *Journal of East China Normal University (Natural Science)*, **1**, 103–112 (in Chinese, with English summary).

Chen, Z., Yang, W., Zhou, C. and Wu, Z. 1986. Subaqueous topography and sediments off the modern Yangtze Estuary. *Journal of East China Sea*, **4**, 36–45 (in Chinese, with English summary).

Chen, Z., Xu, S. and Yan, Q. 1991. Sedimentary environment and facies of the Holocene subaqueous Yangtze Delta. *Oceanologic et Limnologica Sinica*, **22**(1), 29–37 (in Chinese, with English summary).

Chen, Z., Warne A.G. and Stanley, D.J. 1993. Late Quaternary evolution of the northwestern Nile Delta between the Rosetta Promontory and Alexandria, Egypt. *Journal of Coastal Research*, **8**(3), 527–561.

Coleman, J.M. 1981. *Deltas: Processes of Deposition and Models for Exploration* (2nd edition). International Human Resources Development Corp., Boston, 124 pp.

Damuth, J.E. 1994. Neogene gravity tectonics and depositional processes on the deep Niger Delta continental margin. *Marine and Petroleum Geology*, **11**(3), 320–346.

Gu, J., Yan, Q. and Yu, Z. 1983. The cheniers of the Northern Coastal Plain of Jiangsu Province. *Sedimentologica Sinica*, **1**(2), 47–59 (in Chinese, with English summary).

Jing, C. 1989. Formation and evolution of Taihu Lake. *Scientia Geographica Sinica*, **9**(4), 378–385 (in Chinese, with English summary).

Li, B. and Li, C. 1980. Geomorphologic features and coastal sediment transport, Subei Plain, Northern Yangtze Delta. *Marine Science*, **3**, 12–17 (In Chinese).

Li, C. and Li, B. 1981. Studies on the formation of Subei Sand Cays. *Oceanologica et Limnologica Sinica*, **12**(4), 321–331 (in Chinese, with English summary).

Lin, S. 1988. Southward migration of the Yellow River and coastal change of Northern Yangtze Plain. *Marine Science*, **5**, 54–58 (in Chinese).

Lin, S. 1989. Change of land and sea in Yancheng area since Holocene. *Donghai Marine Science*, **7**(3), 12–19 (in Chinese, with English summary).

Lin, S. 1993. Study on changes of Sheyang Lake in historical periods. *Journal of Lake Sciences*, **5**(3), 225–233 (in Chinese, with English summary).

Liu, C., Wu, L. and Cao, M. 1985. Sedimentary feature, origin and chronology of Chenier Plain (Gangshen), Southern Yangtze Delta. *Acta Oceanologica Sinica*, **7**(1), 55–66 (in Chinese).

Morgan, J.P., Coleman, J.M. and Gagiano, S.M. 1968. Mud slumps: diapiric structures in Mississippi delta sediments. In J. Braanstein and G.D. O'Brien (Eds), *Diapirism and Diapirs*.

American Association of Petroleum Geologists, Memoir 8, 145–161.

Reading, H.G. 1978. *Sedimentary Environments and Facies*. Blackwell Scientific Publications, Oxford, 138–142.

Shao, X. and Yan, Q. 1982. Intertidal flat sediments in Shanghai coastal region. *Acta Geographica Sinica*, **37**(3), 241–251 (in Chinese, with English summary).

Shao, X., Xu, S., Yan, Q. and Chen. Z. 1991. Storm deposits in the coastal region of Shanghai the Yangtze Delta, China. *Geologie en Mijnbouw*, **70**, 45–58.

Shen, H., He, S., Pan, D. and Li, J. 1992. A Study of Turbidity Maximum in the Changjiang Estuary. *Acta Geographica Sinica*, **47**(5), 472–479 (in Chinese, with English summary).

Stanley, D.J. and Chen, Z. 1993. Yangtze delta, eastern China: 1. Geometry and subsidence of Holocene depocenter, *Marine Geology*, **112**, 1–11.

Stanley, D.J. and Chen, Z. 1996. Neolithic settlement distributions as a function of sea-level controlled topography in the Yangtze delta, China. *Geology*, **24**(12), 1083–1086.

Stanley, D.J. and Warne, A.G. 1993. Nile Delta: recent geological evolution and human impact. *Science*, **260**, 628–634.

Stanley, D.J. and Warne, A.G. 1994. Worldwide initiation of Holocene marine deltas by deceleration of sea-level rise. *Science*, **265**, 228–231.

Sun, S.C. and Huang, Y.P. 1993. *Taihu Lake*. Ocean Press, Beijing, 271 pp.

Wang, D., Chen, Y. and Gu, Z. 1988. Map of Quaternary Isopach, Archaeological Sites and Paleo Channel, and Coastline changes in the Yangtze Delta Region, 1:1,000,000. Shanghai. Unpublished map.

Wang, K.F., Zhang, Y.L., Jiang, H. and Han, X.B. 1984. Quaternary pollen spore assemblage and its significance of stratigraphy and paleogeographic change in the Yangtze delta. *Oceanologia* **6**, 28–38 (in Chinese, with English summary).

Xu, S. and Chen, Z. 1995. Evolutional similarity and discrepancy of Late Quaternary deltas, China. *Acta Geographica Sinica*, **50**(6), 12–22 (in Chinese, with English summary).

Xu, S., Li, P. and Wang, J. 1981. Sedimentary model of Yangtze Delta. *Quaternary Research*, 55–62 (in Chinese, with English summary).

Xu, S., Shao, X., Chen, Z. and Yan, Q. 1990. Storm deposits in the Changjiang delta. *Science in China, Series B*, **33**(10), 18–27.

Yan, Q. and Hong, X. 1988. Issues of the Holocene transgression on the Southern Plain of the Changjiang Delta. *Acta Oceanologica Sinica*, **7**(4), 578–590.

Yan, Q. and Shao, X. 1989. Evolution of shorelines along the North Bank of Hangzhou Bay during the Late Stage of the Holocene Transgression. *Science in China (Series B)*, **32**(3), 347–360.

Yan, Q. and Xu. 1987. *Recent Yangtze Delta Deposits*. East China Normal University Press, Shanghai, 438 pp.

Yan, Q., Xu, S. and Shao, X. 1989. Holocene cheniers in the Yangtze Delta, China. *Marine Geology*, Special Issue: Cheniers and Chenier Plains, **90**, 337–343.

Yi, H. and Zhao, Q. 1963. Archaeological investigation of Huai Yin region. *Archaeology*, **1**, 1–8 (in Chinese).

Zhang, S., Yan, Q. and Guo, X. 1982. Characteristics and development of cheniers, southern part of Yangtze Delta Plain. *Journal of East China Normal University (Natural Science)*, **3**, 81–94 (in Chinese, with English summary).

Zhao, X., Geng, X. and Zhang, J. 1979. Sea Level Variation Since 20,000 Years, Eastern China. *Acta Oceanologica Sinica*, **1**(2), 269–280 (in Chinese, with English summary).

Zhao, X., Li, B., Tang, L. and Wu, S.G, 1991. Discovery of coastal dune and its significance of mid-Subei Plain. *Ke Xue Tong Bao*, **36**(22), 1727–1730 (in Chinese),

Zhao, X., Tang, L., Shen, C. and Wang, S. 1994. Holocene climate variation and sea level change of Qingfeng Profile, Jianghu, Jiangsu Province. *Acta Oceanologica Sinica*, **16**(1), 78–88 (in Chinese, with English summary).

Zhong, D., Shen, X., Xia, D. and Liu, Z. 1983. Explanation of satellite photography of the ancient Changjiang Delta region in Early Holocene. *Marine Science*, **2**, 16–17 (in Chinese).

18

In-channel Benches: the Role of Floods in their Formation and Destruction on Bedrock-confined Rivers

WAYNE D. ERSKINE[1] AND ELIZABETH A. LIVINGSTONE[2]

[1] *Forest Research and Development Division, State Forests of New South Wales, Beecroft, NSW, Australia*
[2] *School of Geography, University of New South Wales, Sydney, Australia*

ABSTRACT

In-channel benches are discontinuous, sometimes paired, elongate, tabular, often vegetated, usually bank-attached sediment bodies which occur at intermediate elevations between the river bed and the main valley flat on many rivers worldwide. In southeastern Australia, up to six, but usually three benches are present at a site. They are most often located on the inside of bends as bench-like point bars and in straight reaches as parallel benches. Compound channels exhibiting a number of benches were investigated on bedrock-confined rivers in the Hunter Valley, New South Wales (NSW). These channels are characterized by very high flood variability and have essentially straight channel patterns. There is no overlap in the return periods for each bench allowing for plus or minus two standard errors of estimate of the mean. This proves that vertically adjacent benches are separate in terms of their return periods. Furthermore, each of the four benches is associated with a constant but different flood frequency. Only the second lowest bench (bench 2) has a mean return period which lies within the range commonly quoted for bankfull discharge, i.e. mean return periods of 1.93 years and 2.05 years on the annual maximum series for the front edge and mean height, respectively. Bench 3 has a mean return period which lies within the range found for the upper floodplain on other NSW rivers, i.e. mean return periods of 4.5 and 4.8 years on the annual maximum series for the front edge and mean height, respectively. However, in-channel benches are usually too small to be floodplains but are ephemeral channel sediment storages which are episodically subjected to cut and fill. Three types of bench sediments were classified: stratic sediments of low benches which consist of interstratified silts, sands and gravels; massive sediments of low benches which consist of multiple, texturally uniform beds between 0.5 and 1.0 m thick; and cumulic sediments of high benches which consist of deep, uniform, finer-grained deposits. Each bench is separated from vertically adjacent benches by an erosional, usually inset contact. Catastrophic floods erode benches which are reconstructed by subsequent smaller events within the erosional void. Benches are not stable over time, with many exhibiting significant variations in height over 30 year time spans and

Varieties of Fluvial Form. Edited by A.J. Miller and A. Gupta
© 1999 John Wiley & Sons Ltd

with many containing a number of European artefacts. Benches are found in a wide range of hydrological settings, many of which are different from northern hemisphere conditions.

INTRODUCTION

Benches, berms and bars

Compound channel cross-sections are characterized by one or more benches, berms or bars below the main valley flat level (Gregory and Park, 1974) and have been recognized on a number of rivers worldwide (e.g. Kilpatrick 1961; Kilpatrick and Barnes, 1964; Hedman, 1970; Miller et al., 1971; Hedman and Kastner, 1977; Petts, 1977; Wahl, 1984). However, these channels are especially common in eastern Australia (e.g. Woodyer, 1968, 1970; Hickin, 1968; Riley, 1972; Warner, 1972, 1993, 1994; Warner et al., 1975; Abrahams and Cull, 1979; Woodyer et al., 1979; Erskine and Melville, 1983; McDermott and Pilgrim, 1983; Sherrard and Erskine, 1991; Erskine, 1994a; 1996). In-channel benches are depositional landforms which are essentially tabular, often vegetated, elongate, discontinuous, sometimes paired, usually bank-attached sediment bodies and which occur at intermediate elevations between the river bed and the main valley flat (Figure 18.1). They are *not* slump blocks (Taylor and Woodyer, 1978; Abrahams and Cull, 1979; Woodyer et al., 1979; Erskine, 1994a, 1996), as suggested by Costa (1974), although slump blocks occasionally act as the nucleus for subsequent bench formation (Woodyer et al., 1979). Benches on bedload and suspended load channels are usually composed of two lithofacies: a coarse basal unit of sand and/or gravel similar in grain size to the adjacent bed material, and a fine surficial unit of thinly interbedded sediments ranging from sand to clay (Taylor and

FIGURE 18.1   Parallel benches on Foy Brook at the original Ravensworth gauging station (station 2 in Table 18.2)

PLANFORM

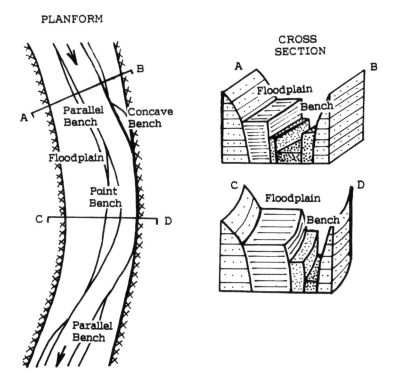

FIGURE 18.2 Planform and cross-sections of a typical bedrock-confined river valley in coastal New South Wales showing the distribution of in-channel benches

Woodyer, 1978; Woodyer et al., 1979; Erskine, 1994a). Woodyer (1966, 1968) recognized up to three benches, which he called "low", "middle" and "high", with increasing elevation above the thalweg. The high bench was present only at some sites believed to be recently incised and was *assumed* to be equivalent to the floodplain level. Subsequent work by Wright (1969), Cull (1975), Warner (1972, 1993) Warner et al. (1975) and Erskine (1986a) found at least four and occasionally five benches at gauging stations on coastal rivers in New South Wales. Benches develop at various locations within the channel, most commonly in straight reaches as parallel benches, on the inside of bends as crescentic point benches (bench-like point bars) which often extend away from the bend to join parallel benches and, less commonly, along the outside of bends as the concave benches of Woodyer (1975), Taylor and Woodyer (1978) and Page and Nanson (1982) (Figure 18.2). None of the benches discussed in the remainder of this paper is a concave bench.

Natural benches were first identified on streams in the Piedmont Province of the southeastern United States (Kilpatrick, 1961; Kilpatrick and Barnes, 1964) and similar landforms elsewhere have been variously called berms or bars (Hedman, 1970; Miller et al., 1971; Hedman and Kastner, 1977). The term "bench" is used here for the following reasons:

(i) "berm" was originally used for very small sediment bodies in irrigation canals (Blench, 1957);

(ii) "bar" has been used indiscriminately for many different large-scale bedforms (Smith, 1978); and

(iii) "bench" was used first for these landforms.

Miller et al. (1971), among others, maintained that rivers only construct one contemporary alluvial surface and that this landform is a floodplain formed dominantly by lateral accretion and is flooded at a fixed recurrence interval of approximately 1.5 years on the annual series (the Wolman and Leopold (1957) lateral accretion model). However, Nanson and Croke (1992) concluded that this model is but one of a number of potential models of floodplain formation. Bedrock-confined channels do not migrate, forming vertically accreted floodplains, which, when well developed, are flooded rarely (Nanson, 1986; Nanson and Erskine, 1988; Nanson and Croke, 1992). Benches are often found on floodplain rivers, indicating that rivers can construct more than one contemporary alluvial surface (Woodyer, 1968; Walker and Coventry, 1976; Woodyer et al., 1979; Erskine and Melville, 1983; Warner, 1993, 1994). However, the main valley flat level is not always a contemporary floodplain and may be a river terrace (Williams, 1978). For this reason, Woodyer (1968) called the main valley flat on rivers in southeastern Australia, the *apparent floodplain*.

Formative processes

Woodyer (1968) and Page (1972) believed that benches formed by recent channel incision, stranding former parts of the bed above the thalweg. However, Cull (1975) and Abrahams and Cull (1979) concluded that benches are formed by massive deposition and reworking during high magnitude floods. Erskine (1994a, 1996) found that catastrophic floods clean out sand-bed channels which recover by bench formation within the void excavated by the superflood. Schick (1974) also proposed that desert stream terraces are obliterated by superfloods and subsequently reform by quasi-continuous aggradation. On the other hand, many authors have documented the formation of benches during persistent periods of small floods and their subsequent reworking during extended periods of large floods (e.g. Erskine and Bell, 1982; Storrie, 1990; Erskine, 1992; Warner, 1994). Although Woodyer (1966, 1968) concluded that bench levels are probably stable over time, he noted that there is an absence of data on temporal changes. Warner et al. (1975) recommended that channel changes at particular sites were required to increase the understanding of the geomorphological origins of benches. Hedman (1970) presented limited information which indicated that berms at two stations in California retained their relative position and size following floods. This paper presents data on the long-term stability of benches.

Flood frequency

Woodyer (1968) developed new statistical tests to demonstrate that first, the expected recurrence intervals of vertically adjacent bench levels are separate and second, the middle bench and high bench/floodplain categories are flooded with a constant frequency (McGilchrist and Woodyer, 1968; Woodyer, 1968). Design flood estimation methods were then developed from these relations (Woodyer and Fleming, 1968, McDermott and

Pilgrim, 1983). However, neither Wright (1969) nor Warner et al. (1975) found a common frequency of flooding for any bench. This is hardly surprising because a range of flows and flow sequences are effective in determining channel geometry (Pickup and Warner, 1976; Wolman, 1977; Pickup and Rieger, 1979; Erskine and Melville, 1983; Erskine, 1994a, 1996). There is also a large literature which has estimated various indices of river discharge from channel dimensions (e.g. Osterkamp et al., 1983; Wharton et al., 1989). However, this line of investigation is not pursued here.

Flood variability

Many streams in central eastern Australia exhibit a distinctive hydrology characterized by high variability and significant secular changes. McMahon et al. (1992) demonstrated, from their detailed study of comparative global hydrology, that the variability of both annual streamflows and flood peak discharges for Australian and southern African rivers is much higher than elsewhere in the world. Flood variability is usually expressed as the standard deviation of the $\log_{10}$ of the annual maximum series flood frequency curve (Baker, 1977). The terms flash flood magnitude index and index of variability have been used by Baker (1977) and McMahon et al. (1992), respectively, for this measure of flood variability.

The results of McMahon et al's (1992) analysis of the annual maximum flood series for 931 river gauging stations throughout the world are summarized in the first three lines of Table 18.1. Rivers in Australia and southern Africa have flash flood magnitude indices which are at least twice as large as those in other parts of the world. However, some regions of Australia, such as the Hunter Valley in central eastern New South Wales and the Genoa and Cann Rivers in East Gippsland, Victoria, exhibit flash flood magnitude indices which are at least three times as large as those for the rest of the world (Table 18.1). For geomorphic purposes, large flood variability refers to flash flood magnitude indices greater than 0.6 (Erskine and Saynor, 1996; Erskine and White, 1996). Rivers with high flood variability experience large floods relatively frequently because they have such steep flood frequency curves (Baker, 1977; Erskine, 1993, 1994a, 1996; Erskine and White, 1996).

TABLE 18.1 Flood variability indices for selected areas (data from McMahon et al., 1992; Erskine, 1986b; 1993; 1994a; 1996; Erskine and White, 1996)

Area	Number of gauging stations	Mean flash flood magnitude index*
World rivers	931	0.28
Australian and southern African rivers	280	0.45
Rest of the world rivers	651	0.21
Hunter Valley, NSW (this study)	24	0.65
Genoa and Cann Rivers, Victoria	4	0.62
Ovens and King Rivers, Victoria	8	0.40

*This is the standard deviation of the $\log_{10}$ of the annual maximum flood series (Baker, 1977)

Catastrophic floods

Catastrophic floods are large perturbations of the magnitude–frequency distribution of flood flows. They must be relatively infrequent events which are large enough to exceed equilibrium thresholds of channel stability (Baker, 1977). For geomorphic purposes, catastrophic floods should be defined solely on the basis of magnitude because flood frequency varies systematically with flood variability (Abrahams and Cull, 1979; Erskine, 1993, 1994a, 1996). Empirical results indicate that floods with a peak discharge at least ten times greater than the mean annual flood totally destroy the pre-flood channel whereas smaller floods do not cause such extensive, long-term geomorphic changes (Stevens et al., 1975, 1977; Erskine, 1993, 1994a, 1996). Erskine (1994a, 1996) found that in-channel benches are recovery landforms formed by moderate floods within the void eroded by catastrophic floods.

Shifts in flood regime

Rivers in central eastern Australia are often characterized by rainfall-driven alternating periods of low and high flood activity (Hall, 1927; Bell and Erskine, 1981; Erskine and Bell, 1982; Erskine, 1986a,b; Sammut and Erskine, 1995). Erskine and Warner (1988) defined flood-dominated regimes (FDRs) as time periods of several decades during which there is a marked upward shift of the whole flood frequency curve. They do not consist solely of flood years but of periods of episodic and persistent flood activity. Drought-dominated regimes (DDRs) are time periods of several decades during which there is a marked downward shift of the whole flood frequency curve from the previous FDR. They do not consist solely of drought years but relatively long periods of low flood activity. FDRs have been documented between 1857 and 1900, and between 1947 and 1990, and DDRs have been documented between 1821 and 1856, and again between 1901 and 1946 on many coastal rivers in central eastern Australia (Hall, 1927; Erskine and Warner, 1988; Sammut and Erskine, 1995; Warner, 1994; Smith and Greenaway, 1982). While many researchers have documented the wholesale destruction of benches after 1947 (see Henry, 1977; Erskine, 1986b, 1992; Warner, 1993), only Warner (1994) has had a sufficiently detailed data set to demonstrate that the volume of sediment going into bench construction during DDRs is essentially equal to the volume eroded during FDRs. Attempts by Raine and Gardiner (1994) to explain such changes solely by loss of riparian vegetation ignore the fact that the density of riparian vegetation increased during the FDRs (Erskine and Bell, 1982; Erskine, 1996).

The purpose of this paper is fourfold:

(i)   to determine whether there is a constant frequency of surcharging of individual benches;
(ii) to assess the stability of in-channel benches over time;
(iii) to determine the stratigraphy of individual benches and the stratigraphic relationships between benches; and
(iv) to evaluate the role of large floods as agents of bench formation and/or destruction.

Most of the channels investigated here are vertically and/or laterally confined by bedrock to various degrees and are *not* gravel-bed, laterally migrating, meandering streams. The

Hunter Valley in central eastern New South Wales (NSW) was chosen for this study because both Woodyer (1966, 1968) and Warner et al. (1975) worked there and because the river gauging network has the highest density of stations in NSW.

STUDY AREA

For this study, 24 gauging stations (Table 18.2) on unregulated rivers in the Hunter Valley of central eastern NSW (Figure 18.3) were chosen for detailed investigation. The Hunter River drains a basin area of about 22 000 km^2 which is characterized by eight physiographic regions (Galloway, 1963) (Figure 18.3). Gauging stations were located in the following four physiographic regions. Extending inland from the coastal zone is a corridor of lowlands (*Central Lowlands*) developed on weak Permian sedimentary rocks. This undulating to hilly terrain contrasts with the steep country on either side. Six gauging stations (1, 2, 11, 20, 21, 23 in Table 18.2) are located in this physiographic region. The *Southern Mountains* are a dissected Triassic sandstone plateau which cover the southern one-third of the Hunter Valley. Three gauging stations (17, 18, 19 in Table 18.2) are located in this physiographic region. To the north of the Central Lowlands lies a mountainous tract of country called the *North-Eastern Mountains* which are composed of Devonian marine sediments and Carboniferous terrestrial rocks. Eight gauging stations (3, 4, 5, 6, 7, 8, 22, 24 in Table 18.2) are located in this physiographic region. The *Central Goulburn Valley* is a belt of high plateaus and ridges of Triassic sandstone dissected by steep-sided valleys cut into Permian sedimentary rocks which extend the complete length of the Goulburn River. Seven gauging stations (9, 10, 12, 13, 14, 15, 16 in Table 18.2) are located in this physiographic region.

Rainfall decreases with distance from the coast and increases with elevation (McMahon, 1964). The highest precipitation occurs near the coast (1100 mm a^{-1}) and at Barrington Tops (1500 mm a^{-1}). Although the lowest rainfall is in the Goulburn Valley (500 mm a^{-1}), rainfall increases to 900 mm a^{-1} to both the north and south of the Goulburn River due to orographic effects. There has been a statistically significant increase of up to 32% in mean annual rainfall since 1946, up to 51% in summer rainfall, up to 39% in rainfall frequency and up to 40% in rainfall intensity (Bell and Erskine, 1981; Erskine and Bell, 1982; Erskine, 1986a,b). The spatial distribution of runoff is similar to that for rainfall and secular changes in runoff are synchronous with catchment rainfall (McMahon, 1964; Bell and Erskine, 1981). McMahon (1968; 1969a,b) previously defined subjectively five hydrological areas in the Hunter Valley on the basis of mean flow, base flow, low-flow frequency, storage-draft characteristics and a catchment-deficiency index. These regions are not discussed here because they were not defined on the basis of flood indices which are more relevant to channel morphology. Erskine (1986a) used cluster analysis of the flood indices adopted by Mosley (1981) (i.e. mean annual flood per unit basin area and flash flood magnitude index) to define the following hydrologic areas (Figure 18.4).

The *Hunter* hydrologic area is characterized by a low specific mean annual flood (0.035–0.170 m^3 s^{-1} km^{-2}) and moderate to large flood variability (flash flood magnitude indices of 0.48–0.64). There are only two gauging stations (13, 22 in Table 18.2) in this area because the majority of stations are on the regulated Hunter River downstream of Glenbawn Dam (Erskine, 1985). The *Western Barrington* hydrologic area exhibits a

TABLE 18.2 Hydrological and geomorphological characteristics of the gauging stations used in this study. For location of study sites, see Figure 18.3.

River	Gauging station	Hydrologic area (see Figure 18.3)	Basin area (km²)	Mean annual runoff (ML)	Mean annual runoff (mm)	Mean annual flood (m³ s⁻¹)	Flash flood magnitude index	Bed slope (m km⁻¹)	Mean bed material size (mm)	Sinuosity (km km⁻¹)
1. Glennies Creek*	Middle Falbrook	Central Hunter	466	82285	177	147.1	0.699	1.50	1.7	1.19
2. Foy Brook	Ravensworth	Western Barrington	170	16196	95.2	40.5	0.473	1.90	3.0	1.23
3. Omadale Brook	Roma	Western Barrington	104	35648	342	24.9	0.359	16.0	125‡	1.05
4. Moonan Brook	Moonan Brook	Western Barrington	103	33311	323	20.0	0.365	16.0	83‡	1.04
5. Hunter River†	Moonan Dam Site	Western Barrington	764	121385	158	137.3	0.424	4.44	51‡	1.36
6. West Brook	Glendon Brook	Southern Barrington?	80	15038	187	112.6	0.663	4.00	Bedrock‡	1.18
7. Allyn River	Halton	Southern Barrington	205	96300	469	153.3	0.376	2.44	30‡	1.10
8. Williams River	Tilligra	Southern Barrington?	194	90033	464	216.0	0.414	2.77	84	1.27
9. Baerami Creek	Baerami	Wollombi-Goulburn	384	11615	30.2	7.57	1.025	2.70	0.44	1.13
10. Martindale Creek	Martindale	Wollombi-Goulburn	352	16969	48.2	10.8	0.690	2.00	0.67	1.15
11. Doyles Creek	Doyles Creek	Wollombi Goulburn	202	9331	46.1	16.1	0.773	2.96	0.74	1.08
12. Halls Creek	Gungal	Wollombi-Goulburn	242	4423	18.2	18.5	0.901	3.40	Bedrock‡	1.32
13. Wybong Creek	Wybong	Hunter River	676	33844	50.0	83.5	0.639	1.43	Bedrock‡	1.56
14. Goulburn River	Coggan	Wollombi-Goulburn	3340	74932	22.4	145.4	0.692	0.82	60	1.26
15. Goulburn River	Kerrabee	Wollombi-Goulburn	4950	140041	28.2	154.0	0.749	0.88	3.9	1.28
16. Goulburn River	Sandy Hollow	Wollombi-Goulburn	6810	188029	27.6	170.4	0.682	1.20	0.49	1.14
17. Congewai Creek	Eglinford	Central Hunter	83	34979	421	53.5	0.638	1.30	0.37	1.48
18. Congewai Creek	Hanging Rock	Wollombi-Goulburn	395	36351	92	66.9	0.757	0.83	0.60	1.18
19. Wollombi Creek	Paynes Crossing	Wollombi-Goulburn	1064	116882	109	90.1	0.783	0.42	0.34	1.10
20. Wollombi Brook	Bulga	Wollombi-Goulburn	1672	108823	65	127.8	0.865	0.72	0.30	1.20
21. Wollombi Brook	Warkworth	Wollombi-Goulburn	1848	162381	87.8	162.9	0.834	0.62	0.25‡	1.12
22. Pages River	Gundy Recorder	Hunter River	1050	79372	75.5	180.5	0.581	2.39	5.4‡	1.27
23. Pages River	Blandford	Central Hunter	302	44148	146	123.2	0.562	2.29	18†	1.44
24. Rouchel Brook	Rouchel Brook	Western Barrington	395	59939	151	84.4	0.590	2.96	108‡	1.22
Range of values			80-6810	4423-188029	18.2-469	7.57-216.0	0.359-1.025	0.42-16.0	0.25-125	1.04-1.56

* All work reported here was completed before the commencement of flow regulation in June 1983
† All work reported here was completed before the commencement of interbasin water transfers from the Barnard River
‡ All of these stations have bedrock bars as the gauge control

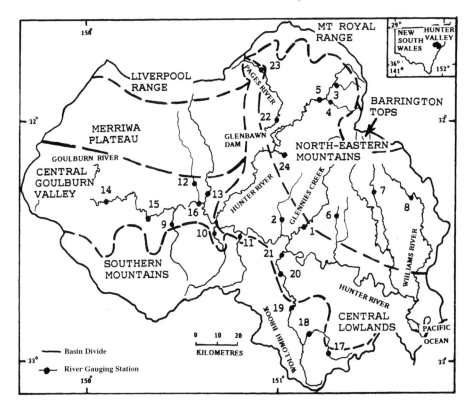

FIGURE 18.3   Hunter Valley in central eastern New South Wales showing the location of the 24 gauging stations (Table 18.2) used in this study and the distribution of physiographic regions

moderate specific mean annual flood (0.18–$0.53\,\mathrm{m^3\,s^{-1}\,km^{-2}}$) and moderate to large flood variability (flash flood magnitude indices of 0.32–0.64). Five gauging stations (2, 3, 4, 5, 24 in Table 18.2) are in this area. The *Wollombi-Goulburn* hydrologic area is characterized by low to moderate specific mean annual flood (0.02–$0.22\,\mathrm{m^3\,s^{-1}\,km^{-2}}$) and high to very high flood variability (flash flood magnitude indices >0.68). There are 11 gauging stations (9, 10, 11, 12, 14, 15, 16, 18, 19, 20, 21 in Table 18.2) in this area because these rivers are not regulated. The *Central Hunter* hydrologic area is characterized by a moderate specific mean annual flood (0.32–$0.65\,\mathrm{m^3\,s^{-1}\,km^{-2}}$) and high flood variability (flash flood magnitude indices of 0.56–0.70). There are three gauging stations (1, 17, 23 in Table 18.2) in this area. The *Southern Barrington* hydrologic area is characterized by a high specific mean annual flood (0.72–$0.80\,\mathrm{m^3\,s^{-1}\,km^{-2}}$) and low flood variability (flash flood magnitude index <0.38). West Brook and Tillegra do not cluster with the Southern Barrington stations because West Brook has high flood variability (flash flood magnitude index of 0.66) and Tillegra has an exceptionally high specific mean annual flood ($1.30\,\mathrm{m^3\,s^{-1}\,km^{-2}}$). For ease of discussion, both stations are tentatively assigned to the Southern Barrington hydrologic area, pending further work.

Hydrological and geomorphological characteristics of the study sites are summarized in Table 18.2. It should be emphasized that these rivers have straight to slightly sinuous

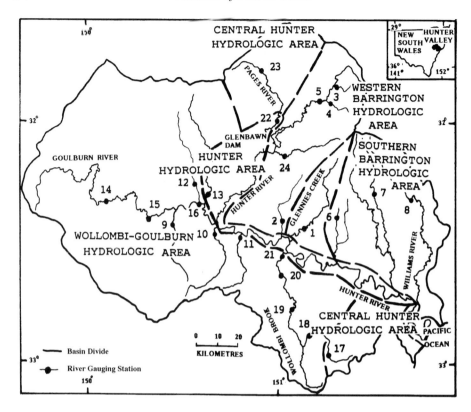

FIGURE 18.4   Hydrologic areas in the Hunter Valley. Each area is defined in the text. Gauging station numbers refer to those listed in Table 18.2

patterns (only one river has a sinuosity greater than 1.5), that bed material is highly variable, ranging from medium sand to cobble gravel and bedrock, and that flood variability varies directly with basin area, a result consistent with that of other workers in Australia (Erskine, 1986b; McMahon et al., 1992; Erskine and Saynor, 1996).

METHODS

Fieldwork commenced at the sites previously investigated by Woodyer (1966, 1968) and Warner et al. (1975) so that benches could be easily identified using the criteria listed by these authors. None of the benches used in this study is similar to those outlined by Melville and Erskine (1986) which backfill the trenches of small upland valleys produced by post-European settlement channel incision. At each gauging station, complete cross-sections from the main valley flat on one bank to the same surface, if present, on the other were surveyed in close proximity to the recorder. Supplementary cross-sections were also surveyed upstream and downstream (Figure 18.5). The extensive flats bordering the stream channel may be river terraces and not floodplains. For this reason, the highest, most extensive surface at a site is called the apparent floodplain, following the proposed terminology of Woodyer (1968). Longitudinal profiles of each bench were also surveyed to help define more accurately the bench levels at the recorder. Previous workers have fixated on the level of the front edge of a bench or the streamward shoulder of the bar as being the most representative (Woodyer, 1968; Hedman and Kastner, 1977). However, the front edge often is not indicative of the mean height of the bench (Cull, 1975). To determine a mean bench level, a large two-dimensional grid of three equally spaced rows of six sampling sites was paced across every bench, each row being parallel to the direction of stream flow for one channel width upstream and one channel width downstream of the recorder. In the following analyses, both a mean bench level, determined from the grid survey, and the level of the front edge of each bench near the recorder, determined from the long profile and cross-section surveys, were used. All surveys were levelled into the gauging station bench mark and reduced to gauge heights by subtracting the level for gauge zero from the reduced level for each point. Gauge zero is the level corresponding to a gauge height of 0 m.

The bench levels were converted to discharge by the relevant rating table for the time of survey. Return periods corresponding to these discharges were determined by fitting a log Pearson type III distribution to the annual maximum flood series for the period since 1946 according to the procedure of Pilgrim and Doran (1987). Where there were zero events, the method of Jennings and Benson (1969) was used. Geomorphologists have tended to avoid using this flood frequency procedure (Hedman and Kastner (1977) is a significant exception) but it is recommended for general use in both the USA (Benson, 1968) and Australia (McMahon and Srikanthan, 1981; Pilgrim and Doran, 1987). The period since 1946 was used because of the statistically significant increase in flood frequency reported by Bell and Erskine (1981), Erskine and Bell (1982), Erskine (1986a) and Erskine and Warner (1988). Only the annual maximum series could be used for flood frequency analysis because most stations had missing records. It is easy to estimate one or two missing annual maxima by regression with a neighbouring station or by using the method of Jennings and Benson (1969). If the partial series was used, too many floods would have been estimated. The mean record length for flood frequency analysis was 34 years, the shortest being 20 years and the longest, 47 years.

The stability of bench levels over time was assessed by comparing the results of this study with those of Woodyer (1966, 1968) and Warner et al. (1975). Woodyer's (1966) surveys at five stations were completed in 1966 (some of the 1966 surveys were undertaken earlier but the date is unknown (K.D. Woodyer, 1979, written comm.)). Warner et al. (1975) resurveyed four of Woodyer's sites plus five of our sites in 1971 (R.F. Warner, 1979, pers. comm.). Our surveys were undertaken between 1979 and 1982, and some were repeated again in 1996. All bench levels used for flood frequency analysis refer to the 1979–1982 surveys. Changes in gauge zero since the early 1960s necessitated adjustments to the Woodyer and Warner data so that the levels for all surveys are expressed in terms of current gauge heights. Furthermore, the station referred to as Bowmans Creek at Glenora by Woodyer and Bowmans Creek at Ravensworth by Warner is now known as Foy Brook at Ravensworth (2 in Table 18.2).

Bench stability was also assessed by comparing repeat surveys of the same site (Figure 18.5). These surveys were undertaken by the authors, Woodyer, Warner and the Department of Land and Water Conservation. Changes in bed level over time were reconstructed from gauging station rating tables (Figure 18.5). The role of floods in

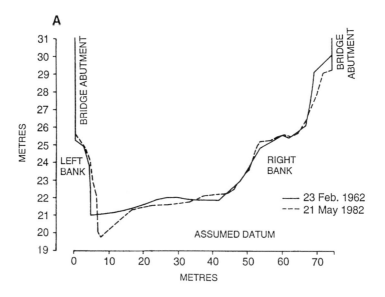

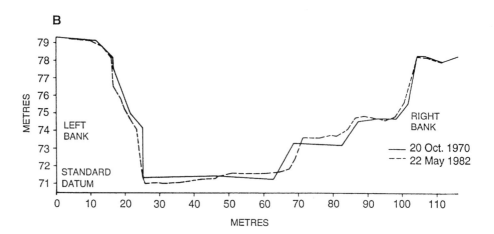

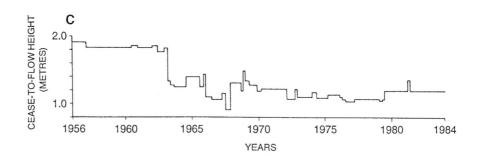

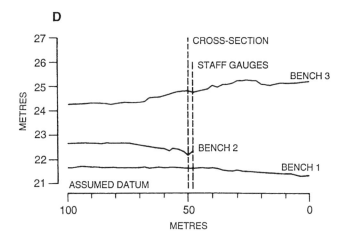

FIGURE 18.5   Glennies Creek at Middle Falbrook (Station 1 in Table 18.2). This site has been regulated by Glennies Creek Dam since June 1983. (A) Cross sections at the recorder site. (B) Cross-sections 106 m downstream of the gauge. (C) Cease-to-flow height variations over time. (D) Long profiles along the front edge of benches 1, 2 and 3

forming and destroying benches was determined by comparing benches before and after various sized floods. Pits, trenches, exposures and auger holes were dug in each of the 68 benches and 24 apparent floodplains at the 24 gauging stations to determine bench stratigraphy and in an attempt to find European artefacts.

FREQUENCY OF BENCH SURCHARGING

Bench levels

The channels exhibited large capacities when measured at the apparent floodplain (Figures 18.5 and 18.6). Up to six, but usually four benches were present at most sites. The mean number of benches at the sites is 3, supporting Woodyer's (1968) results. Table 18.3 compares the mean bench height with the level for the front edge of each bench at each site for the four lowest benches. Benches are numbered 1 to 4 from lowest to highest. The level for the front edge of 53% of benches was outside the range of ± 2 standard errors of estimate of the mean level. In particular, the front edge was significantly lower for 41.2% of benches and it was significantly higher for 11.8% of benches. This indicates that many benches are not flat-topped and that the level of the front edge is *not* representative of the general bench height. Furthermore, where the front edge is significantly higher than the mean height, there is a well developed levee present, as has been noted for benches in the Piedmont province of the USA (Kilpatrick and Barnes, 1964).

Return periods

Tables 18.4 and 18.5 list the return periods for the front edge and for the mean level of each bench, respectively at each gauging station. To compute the mean return periods, it

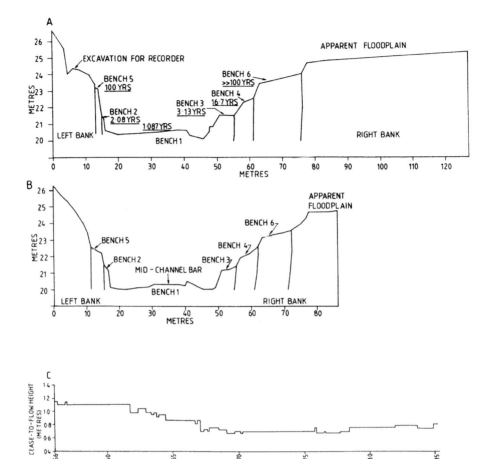

FIGURE 18.6   Foy Brook at Ravensworth (Station 2 in Table 18.2). (A) Cross-section at the original recorder site. (B) Cross-section 20 m upstream of the original recorder. (C) Cease-to-flow height variations over time at the original gauge site

has been necessary to ignore both the "greater' and "less than" signs. No statistics have been calculated for the apparent floodplain because the return periods are approximate.

The return periods for both the front edge and mean level for each bench are significantly different. There is no overlap in the return periods for each bench allowing for ± 2 standard errors of estimate of the mean. This indicates that the vertically adjacent bench levels at a gauge are separate in terms of return period. The mean return period for the front edge of each of the three lowest benches is slightly, but non-significantly less than the mean return period for the mean level of each bench. Therefore, for practical purposes, it does not matter whether the front edge or mean level of a bench is used to determine the frequency of bench surcharging (cf. Cull, 1975).

Figure 18.7 shows the frequency distributions of return periods for each bench level and the apparent floodplain. These frequency distributions are strongly clustered around a well defined modal range, unlike Williams' (1978) results for his active floodplain and

TABLE 18.3 Bench levels for the current gauge zero at each gauging station

Gauging station	Gauge height (m)									
	Bench 1		Bench 2		Bench 3		Bench 4		Apparent floodplain	
	Front edge	Mean	Front edge	Mean	Front edge	Mean	Front edge	Mean	Front edge	Mean
1. Glennies Creek at Middle Falbrook	1.39*	1.49	2.15	2.18	4.55	4.71	-	-	8.68*	8.92
2. Foy Brook at Ravensworth	1.28†	1.04	1.93*	2.11	2.19†	2.02	2.95	2.91	5.33	5.40
3. Omadale Brook at Roma	1.13*	1.37	1.66†	1.58	1.96	1.98	2.15†	2.11	2.50*	2.63
4. Moonan Brook at Moonan Brook	0.59*	0.83	1.20	1.22	2.15*	2.46	-	-	3.25*	3.98
5. Hunter River at Moonan Dam Site	-	-	1.69*	2.41	3.15*	3.42	-	-	5.05	5.06
6. West Brook upstream of Glendon Brook	-	-	0.94*	1.21	1.56*	1.77	-	-	3.96†	3.83
7. Allyn River at Halton	-	-	3.10	3.20	3.93*	4.05	-	-	6.13*	6.37
8. Williams River at Tillegra	0.94*	1.17	-	-	4.76	4.78	-	-	6.65	6.7
9. Baerami Creek at Baerami	0.97*	1.05	1.37*	1.62	2.09	2.11	2.44	2.65	2.91?	-
10. Martindale Creek at Martindale	-	-	1.12	1.18	1.17†	1.59	-	-	4.02	4.04
11. Doyles Creek at Doyles Creek	0.38*	0.46	1.37*	1.54	1.67†	1.46	2.25*	2.51	3.24*	3.29
12. Halls Creek at Gunga	-	-	1.40	1.44	3.22	3.01	3.87	3.78	5.07†	4.76
13. Wybong Creek at Wybong	-	-	2.28	2.29	3.75	3.68	-	-	10.50*	11.27
14. Goulburn River at Coggan	-	-	3.31	3.31	5.35	5.35	-	-	12.03*	12.19
15. Goulburn River at Kerrabee	1.36	1.35	3.08	3.07	3.96	3.95	7.19*	7.51	14.65*	16.37
16. Goulburn River at Sandy Hollow	-	-	3.36†	3.33	4.87	4.93	6.66†	6.51	10.41*	10.69
17. Congewai Creek at Eglinford	0.75*	1.29	-	-	4.38*	4.48	5.19*	5.32	5.45	5.46
18. Congewai Creek at Hanging Rock	-	-	-	-	-	-	7.67	7.65	9.01†	8.92
19. Wollombi Brook at Paynes Crossing	1.40	1.46	3.09	3.03	3.79*	4.08	6.37	6.33	12.6?	-
20. Wollombi Brook at Bulga	-	-	5.11	5.13	-	-	-	-	5.94†	5.83
21. Wollombi Brook at Warkworth	2.21†	2.13	2.27	2.53	2.82*	3.09	-	-	7.11	7.20
22. Pages River at Gundy Recorder	1.26*	1.58	1.70	1.79	2.73†	3.43	-	-	7.19*	7.32
23. Pages River at Blandford	0.80	0.80	2.10	2.24	4.27	4.29	5.91*	6.92	8.38*	8.50
24. Rouchel Brook at Rouchel Brook	0.29*	0.57	1.48*	1.69	2.51	2.49	-	-	6.15†	5.89

* Height less than two standard errors of estimate below the mean   † Height greater than two standard errors of estimate above the mean

TABLE 18.4 Return periods corresponding to the front edge of each bench at each gauging

Gauging station	Return period (years on the annual maximum series)				
	Bench 1	Bench 2	Bench 3	Bench 4	Apparent flood-plain
1. Glennies Creek at Middle Falbrook	<1.0101	1.136	3.13	-	>25
2. Foy Brook at Ravensworth	1.087	2.08	3.13	16.7	>>100
3. Omadale Brook at Roma	1.43	5.26	13.3	22	53
4. Moonan Brook at Moonan Brook	<1.0101	1.52	11.1	-	~100
5. Hunter River at Moonan Dam Site	-	1.111	2.78	-	22
6. West Brook upstream of Glendon Brook	-	1.111	1.58	-	>100
7. Allyn River at Halton	-	1.72	3.33	-	29
8. Williams River at Tilligra	1.023	-	2.78	-	7.1
9. Baerami Creek at Baerami	1.16	2.27	5.9	9.1	12.5
10. Martindale Creek at Martindale	-	2.17	2.22	-	>100
11. Doyles Creek at Doyles Creek	1.11	2.17	3.47	>100	>>100
12. Halls Creek at Gungal	-	2.00	14.3	40	>>40
13. Wybong Creek at Wybong	-	1.89	4.55	-	>100
14. Goulburn River at Coggan	-	1.52	3.45	-	33
15. Goulburn River at Kerrabee	1.03	2.00	2.86	11.1	>100
16. Goulburn River at Sandy Hollow	-	1.72	3.57	10.0	18
17. Congewai Creek at Eglinford	1.042	-	2.38	3.33	3.85
18. Congewai Creek at Hanging Rock	-	-	-	6.1	8.7
19. Wollombi Brook at Paynes Crossing	1.07	3.7	5.3	33	>50
20. Wollombi Brook at Bulga	-	1.60	-	-	71
21. Wollombi Brook at Warkworth	1.28	1.33	1.72	-	11.1
22. Pages River at Gundy Recorder	1.149	1.39	2.47	-	14.3
23. Pages River at Blandford	<1.0101	1.32	3.12	7.7	>25
24. Rouchel Brook at Rouchel Brook	1.014	1.49	3.33	-	>>100
Mean	1.102	1.93	4.54	24	
± Two standard errors of estimate ($\bar{x}$)	1.039–1.165	1.73–2.13	3.05–6.03	8–40	
Range	<1.0101–1.43	1.111–5.26	1.72–14.3	3.33–>100	

valley flat. However, the return periods for the apparent floodplain are much greater than those usually associated with active floodplains. This means that they are either river terraces or extremely well developed vertically accreted floodplains. Unlike the high floodplains described by Nanson (1986), these surfaces usually have well developed soils with texture, colour, structure or consistence B horizons, bleached A_2 horizons and abundant ferromanganiferous nodules and concretions similar to Walker and Coventry's (1976) high contrast solum soils. Therefore, most of these surfaces are truly terraces. While each of the four benches is associated with a constant but different flood frequency, it is *not* appropriate to regard any of these features as a floodplain because each is a small-scale temporary sediment storage, as outlined below. Furthermore, benches are usually much narrower than the adjacent channel (Figures 18.5 and 18.6). Nevertheless, the mean return periods of 1.93 years and 2.05 years for the front edge and mean height of bench 2, respectively, lie within the range commonly quoted for bankfull discharge (Wolman and Leopold, 1957; Leopold et al. 1964; Woodyer, 1968; Warner, 1993). The mean return periods of 4.5 and 4.8 years for the front edge and mean height of

TABLE 18.5   Return periods corresponding to the mean level of each bench at each gauging station.

Gauging station	Return period (years on the annual maximum series)				
	Bench 1	Bench 2	Bench 3	Bench 4	Apparent flood-plain
1. Glennies Creek at Middle Falbrook	1.016	1.149	3.33	-	>25
2. Foy Brook at Ravensworth	1.0101	2.63	2.27	12.5	>>100
3. Omadale Brook at Roma	2.63	4.55	13.3	20	71
4. Moonan Brook at Moonan Brook	1.026	1.56	16.1	-	>100
5. Hunter River at Moonan Dam Site	-	1.64	4.0	-	22
6. West Brook upstream of Glendon Brook	-	1.30	1.85	-	100
7. Allyn River at Halton	-	1.79	3.85	-	4.0
8. Williams River at Tilligra	1.042	-	2.63	-	7.7
9. Baerami Creek at Baerami	1.33	3.33	6.3	11.1	15
10. Martindale Creek at Martindale	-	2.27	3.57	-	>100
11. Doyles Creek at Doyles Creek	1.16	2.63	2.94	>100	>>100
12. Halls Creek at Gungal	-	2.04	10.0	33	>>40
13. Wybong Creek at Wybong	-	1.89	4.55	-	>100
14. Goulburn River at Coggan	-	1.52	3.45	-	56
15. Goulburn River at Kerrabee	1.025	1.92	2.86	12.5	>100
16. Goulburn River at Sandy Hollow	-	1.72	3.7	8.7	59
17. Congewai Creek at Eglinford	1.105	-	2.5	3.45	3.85
18. Congewai Creek at Hanging Rock	-	-	-	6.1	7.7
19. Wollombi Brook at Paynes Crossing	1.08	3.33	6.7	32	>50
20. Wollombi Brook at Bulga	-	1.61	-	-	50
21. Wollombi Brook at Warkworth	1.24	1.52	1.89	-	11.1
22. Pages River at Gundy Recorder	1.32	1.45	3.8	-	14.3
23. Pages River at Blandford	<1.0101	1.41	3.13	>10.0	>25
24. Rouchel Brook at Rouchel Brook	1.064	1.70	3.18	-	>>100
Mean	1.22	2.05	4.8	23	
± Two standard errors of estimate ($\bar{x}$)	1.00–1.44	1.70–2.40	3.3–6.3	7–39	
Range	<1.0101–2.63	1.149–4.55	1.85–16.1	3.45–>100	

bench 3, respectively, closely agree with Warner's (1993) results for his upper floodplain on the Bellinger River, NSW, and with Pickup and Warner's (1976) results for bankfull discharge in the Cumberland Basin, NSW. As benches 1, 2 and 3 are often modified by floods (see below), there is likely to be a range of flows effective in shaping these channels.

TEMPORAL CHANGES IN BENCH LEVELS

Woodyer (1968) excluded four of his five Hunter Valley gauging stations from his analysis of the common frequency of bankfull discharge because of shifts in rating curves and control instability. Furthermore, McGilchrist and Woodyer (1968) deleted all five Hunter Valley gauging stations from their analysis. This suggests that the benches in the Hunter Valley are unstable. The 17 gauging stations listed in Table 18.6 were surveyed on more than one occasion. All levels (Table 18.6) refer to a point at the front of the

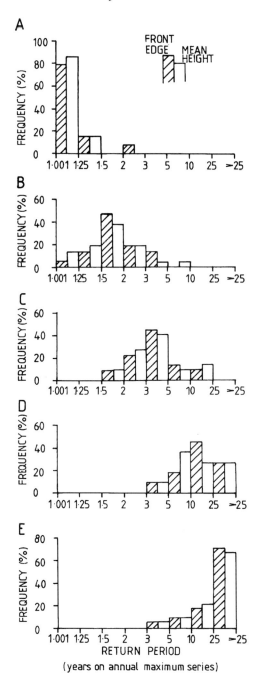

FIGURE 18.7   Frequency distributions of return periods corresponding to the front edge and mean height of each bench: (A) Bench 1; (B) Bench 2; (C) Bench 3; (D) Bench 4; (E) Apparent Floodplain

TABLE 18.6   Changes in bench levels over time. All gauge heights are reported for current gauge zero. The 1966 data are from Woodyer (1966, 1968, pers. comm. 1979), the 1971 data are from Warner et al. (1975) and the 1979–1996 data are the authors'

Gauging station	Year	Bench gauge heights (m)			
		B1	B2	B3	B4
1. Glennies Creek at Middle Falbrook*	1966	2.7	3.2	4.8	–
	1979	1.4	2.2	4.6	-
	1982	1.4	1.9	4.6	-
2. Foy Brook at Ravensworth†	1966	1.5	1.6	2.4	-
	1971	1.9	2.5	-	3.5
	1981	1.3	1.9	2.2	3.0
	1996	1.3	2.6	2.5	3.1
3. Omadale Brook at Roma	1966	0.7	1.0	1.3	-
	1971	1.0	1.3	-	1.8
	1981	1.1	1.7	2.0	2.1
4. Moonan Brook at Moonan Brook	1966	0.9	1.2	1.7	-
	1971	-	1.3	2.0	2.5
	1981	0.6	1.2	2.2	2.8
	1996	0.6	1.2	2.2	2.8
7. Allyn River at Halton	1971	1.7	3.6	3.9	6.4
	1981	-	3.1	3.9	6.1
	1996	2.0	3.3	-	6.1
8. Williams River at Tillegra	1981	0.9	-	4.8	-
	1996	0.7	-	4.9	-
9. Baerami Creek at Baerami	1971	-	0.6	-	1.8
	1980	1.0	1.4	2.1	2.4
	1996	0.3	1.4	2.1	2.5
10. Martindale Creek at Martindale	1981	-	1.1	1.2	4.0
	1996	-	1.3	1.8	4.0
11. Doyles Creek at Doyles Creek	1981	0.4	1.4	1.7	2.3
	1996	0.8	1.4	1.7	2.3
12. Halls Creek at Gungal	1980	-	1.4	3.2	3.9
	1996	-	0.7	2.0	2.6
13. Wybong Creek at Wybong	1980	-	2.3	3.8	-
	1996	-	2.2	3.8	-
15. Goulburn River at Kerrabee‡	1966	1.6	2.4	4.0	-§
	1971	1.4–1.8	2.3	3.8	4.8
	1979	1.4	3.1	4.0	7.2
17. Congewai Creek at Eglinford	1971	1.2	1.8	4.3	5.2+
	1980	0.8	-	4.4	5.2
19. Wollombi Brook at Paynes Crossing	1980	1.4	3.1	3.8	6.4
	1996	1.3	1.7	-	6.4
22. Pages River at Gundy Recorder	1971	1.6	1.8	2.4	3.3
	1981	1.3	1.7	2.7	-¶
23. Pages River at Blandford	1981	0.8	2.1	4.3	6.0
	1996	0.7	2.5	4.2	6.0
24. Rouchel Brook at Rouchel Brook	1971	1.0	1.6	2.9	3.7
	1981	0.3	1.5	2.5	-¶
	1996	0.2	1.5	2.6	-¶

* Due to a change in gauge zero, 1.5 m must be added to the values published by Woodyer (1968)
† Due to a change in gauge zero, 0.9 m must be added to the values published by Woodyer (1968) and Warner et al. (1975)
‡ Due to a change in gauge zero, 1.0 m must be added to the values published by Woodyer (1968) and Warner et al. (1975)
§ Woodyer (1968) noted the presence of a fourth bench here which coalesced with the high bench both upstream and downstream of the gauge
¶ There is a bench present at these stations but it is not present in close vicinity to the recorder

bench in close proximity to the gauge (following Woodyer, 1966, 1968; Warner et al., 1975). It is unlikely that the levels for all surveys refer exactly to the same point on all features. Unfortunately the bench levels for the earliest two surveys were reported only to one decimal place. Therefore, the results for the more recent surveys are reported also to one decimal place so that they are consistent with the earlier data.

It is not proposed to undertake a detailed analysis of temporal changes in bench levels because the accuracy of the data does not justify it. Nevertheless, there is a strong indication from the data in Table 18.6 that some benches are ephemeral and that others have been actively modified. Bench 1 at both Middle Falbrook and Ravensworth decreased in level due to bed degradation, as shown by the bed level changes in Figures 18.5 and 18.6. Large floods in March 1977 and March 1978 eroded most of bench 2 at Middle Falbrook and account for the significant reduction in bench level. Substantial degradation between 1980 and 1996 also resulted in the marked reduction in the level of bench 1 at Baerami. Similar rapid changes in bench levels occurred at Roma, Moonan Brook, Halton, Tillegra, Gungal, Eglinford, Paynes Crossing and Rouchel Brook.

Some benches have aggraded substantially over time. Bench 2 at Baerami and Kerrabee aggraded by 0.8 m between successive surveys ($>89\,mm\,a^{-1}$) while bench 2 at Roma aggraded by 0.7 m over 15 years ($47\,mm\,a^{-1}$). Less rapid deposition was recorded on bench 1 at Roma, on bench 3 at Moonan Brook, on benches 2 and 3 at Martindale and on bench 2 at Blandford.

While some benches are certainly stable (for example, bench 1 at Paynes Crossing and Blandford; bench 2 at Moonan Brook, Doyles Creek, Gundy Recorder and Rouchel Brook; bench 3 at Middle Falbrook, Halton, Doyles Creek, Kerrabee, Wybong and Eglinford; bench 4 at Martindale, Doyles Creek, Eglinford, Paynes Crossing and Blandford), there are many more examples of rapid changes in bench levels. This strongly suggests that Woodyer's (1968) and Hedman's (1970) assumption of fixed bench levels during the complete length of gauging records is inappropriate. There have been many cycles of cut and fill of benches at some gauging stations.

BENCH STRATIGRAPHY

Internal bench stratigraphy

The internal bench stratigraphy has been categorized into three classes. *Stratic sediments* are similar to Walker and Coventry's (1976) stratic stage soils developed in alluvial deposits throughout New South Wales. They consist of thinly bedded, interstratified silts and sands (Figure 18.8), with occasional pebble lenses often overlying a thick basal sand or gravel sheet. The sands are usually horizontally laminated although trough-laminated cosets are occasionally present. Convoluted laminations are often present in the silty sediments. Generally benches 1, 2 and 3 exhibit stratic sediments. Pedogenesis is interrupted by frequent sediment accessions, and little post-depositional modification of the sediments is apparent (Walker and Coventry, 1976; Erskine, 1994b). Large-scale cut-and-fill structures are occasionally present.

Massive sediments consist of multiple, thick, uniform beds ranging from cobble and pebble gravels to sands to silty sands. The individual beds are usually 0.5 m thick and occasionally exceed 1 m. The boundary between individual beds is abrupt to clear. There

FIGURE 18.8   Spanner recovered at base of bench 3 on Martindale Creek at Martindale (station 10 in Table 18.2). Exposure shows stratic sediments in bench 3

is usually no vertical grain size variation within massive beds although graded beds are occasionally present. Benches 1, 2 and 3 exhibit massive sediments and there is no significant difference in the mean return period of benches with stratic and massive sediments.

Cumulic sediments are similar to Walker and Coventry's (1976) cumulic stage soils. They consist of deep, relatively uniform, finer grained sediments with strong organic coloration and numerous faunal channels to depth. Biological mixing or homogenization is active and keeps pace with the rate of surface sediment accessions (Walker and Coventry, 1976; Walker and Green, 1976; Erskine, 1994b). Cumulic sediments are occasionally found in bench 3 and the apparent floodplain, and are common in bench 4. The mean return period of benches with cumulic sediments is significantly higher than for both stratic and massive sediments.

Buried cumulic soils were rarely found in benches 3 and 4, and demonstrate that episodic large sediment successions occur on the higher benches. Walker and Green (1976) have also recorded a buried soil in the bench sediments of Gooromon Ponds near Canberra.

Many apparent floodplains exhibited well developed soils with bleached A_2 horizons and with colour, texture, fabric, structure and/or consistence B horizons. The difference in soil properties between the A and B horizons ranged from slight colour and texture differences (low contrast solum stage of Walker and Coventry (1976)) to marked texture, fabric, structure and consistence contrast (high contrast solum stage of Walker and Coventry (1976)). These apparent floodplains are true river terraces which have been abandoned long enough for pedogenesis to alter the overbank sediments into distinct, well defined soil horizons.

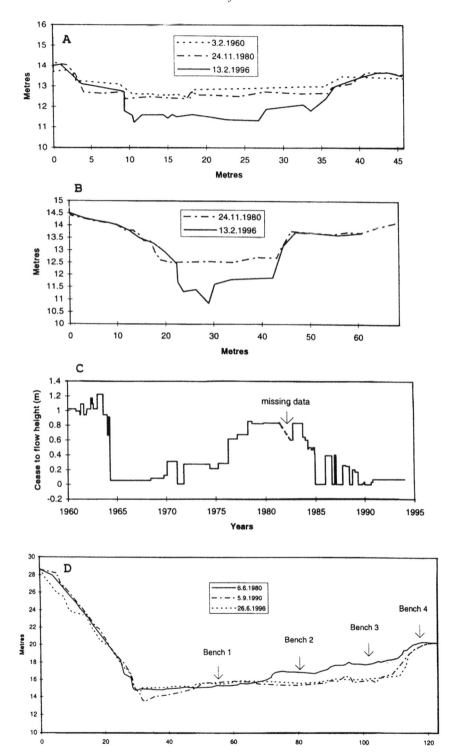

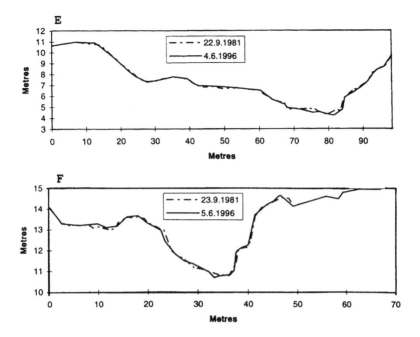

FIGURE 18.9 Channel changes at selected gauging stations: (A) Baerami Creek gauge; (B) Baerami Creek gauge (30 m upstream of gauge); (C) Cease-to-flow height variation at Baerami Creek; (D) Paynes Crossing gauge; (E) Rouchel gauge; (F) Moonan Brook gauge

Stratigraphic relationships between benches

Using the terminology proposed by Warner (1972) for river terraces, the in-channel benches in the Hunter Valley are usually *inset* into adjacent benches and/or the apparent floodplain. *Incised* and *overlapped* relationships are occasionally found but are very rare. The stratigraphic relationships shown in Figure 18.6 for Foy Brook at Ravensworth have been determined from detailed trenching and augering, and are repeated at most of the sites investigated here (Table 18.2). Abrupt erosional contacts separate each bench and a simple stratigraphy such as that found by Jacobson and Coleman (1986) for floodplains in the Maryland Piedmont is absent. Channel widening by floods partially or totally removes benches which reform subsequently in the eroded void. The large-scale cut-and-fill structures, outlined above, represent the backfill of a channel eroded into a bench. Although these channels or chutes dissect a bench, they do not totally erode it.

European artefacts

European artefacts were recovered from various depths in nine of the 68 benches. These artefacts included a spanner (Figure 18.8), beer bottle, Coca Cola bottle, rusted can, fencing wire, barbed wire, a bridge spike, an iron rod and rusted metal. All artefacts were found in benches 1 to 3 inclusive, but *not* in bench 4. The depth of the artefacts ranged from near the surface (0.1 m) to the base of the bench (1.1 m). The recovery of such a relatively large number of European artefacts from benches at gauging stations indicates that some benches have been subject to widespread cut-and-fill episodes since European settlement. Furthermore, most of the artefacts are very recent (<30 years old) and provide further evidence of frequent fluvial reworking of benches.

FORMATION AND DESTRUCTION OF BENCHES

Repeat surveys of the same cross-section at a number of sites and sequential air photographs demonstrate that benches are episodically cut and filled (Erskine, 1994a, 1996; Erskine and Melville, 1983). Additional data presented below reinforce this conclusion.

Baerami Creek exhibits the highest flood variability of any station in the Hunter Valley (Table 18.2). Cross-sectional changes at the gauge (Figure 18.9A, B) show over 1 m of degradation between 1980 and 1996. However, a plot of continuous bed level changes over time (Figure 18.9C) reveals a cyclic pattern of alternating aggradation and degradation in association with the passage of bedload slugs (Erskine and Melville, 1983; Erskine, 1993, 1994a, 1996). Over the last 35 years, two cycles of alternating aggradation and degradation have occurred. The high bed levels of the early 1960s (Figure 18.9A, C) were caused by massive bed aggradation following channel widening during the floods of the early 1950s (Ellis, undated). However, a major flood in 1964 caused about 0.9 m of degradation and the February 1971 flood also initiated renewed degradation. Nevertheless, the series of floods between 1976 and 1978 produced substantial aggradation which persisted for about six years after which rapid degradation occurred. Benches are very unstable landforms on this river with bench construction occurring during aggradational phases and with both bench construction and destruction occurring during degradational phases.

Cross-sectional changes at Paynes Crossing between 1980 and 1996 are shown in Figure 18.9D. Benches 2 and 3 were completely destroyed by a flood in June 1989 which had a peak discharge of approximately 10 times the mean annual flood. Since 1989 there has been no tendency for these benches to reform. At Doyles Creek, no catastrophic floods were recorded during our monitoring period between 1981 and 1996. As a result, the lowest benches have been extended into the channel, narrowing the bed width as the bed degraded. Data for other sites (for example, Moonan Brook and Rouchel Brook) also demonstrate that the channel and benches were stable during periods (1981 to 1996) without catastrophic floods. Although the floods which did occur were competent to transport the gravelly bed material, the benches remained very stable (Figure 18.9E and F).

The above sequence of bench destruction by catastrophic floods and subsequent stability or bench construction by smaller events was found in most of the five hydrologic areas in the Hunter Valley (Erskine, 1994a, 1996). However, the number of cut-and-fill sequences in the Wollombi–Goulburn hydrologic area was far greater than in the other areas. This is to be expected given the exceptionally high flash flood magnitude indices of the streams in this area (Erskine, 1994a; Erskine and Saynor, 1996; Erskine and White, 1996).

DISCUSSION AND CONCLUSIONS

In-channel benches form simultaneously but at different rates at various levels between the river bed and the main valley flat. The height of each bench is separate from vertically adjacent benches. Furthermore, the distributions of return periods associated with each bench level are different. The mean return period for the front edge and the mean level of bench 2 conforms to those usually associated with bankfull discharge, i.e. less than three years on the annual maximum series. However, analogous data for bench 3 conform to the results of Pickup and Warner (1976) and Warner (1993), who concluded that many NSW rivers have vertically accreted floodplains corresponding to return periods in excess of four years on the annual maximum series. This is particularly the case on rivers which do not migrate laterally and which form vertically accreted floodplains with natural levees (Nanson, 1986). However, in-channel benches are not floodplains for the following six reasons:

(i)   benches, as shown in Figures 18.1, 18.5, 18.6 and 18.9, are narrow features which only rarely approach channel width in lateral extent;
(ii) benches are very fragmentary with a discontinuous longitudinal distribution;
(iii) multiple benches are present at most sites (Woodyer, 1968; Warner, 1972; Warner et al., 1975; Abrahams and Cull, 1979);
(iv) benches *and* floodplains coexist;
(v)   benches are ephemeral landforms which are fluvially reworked on decadal time scales; and
(vi) benches are temporary sediment storages which are episodically destroyed by catastrophic floods which release the sediment as a bedload slug or wave (Erskine and Melville, 1983; Erskine, 1994a, 1996; Erskine and Saynor, 1996).

Page (1988) and Erskine and Keshavarzy (1996) found that meandering Australian rivers with broad floodplains have bankfull return periods within the range reported by Wolman and Leopold (1957), Leopold et al. (1964) and Woodyer (1968). Much geomorphic theory has been developed for such meandering rivers. However, the rivers investigated here are *not* meandering streams. The mean sinuosity of the 24 sites is 1.22 and most rivers frequently impinge against the bedrock valley sides: the second and third degree lateral confinement of Lewin and Brindle (1977). Furthermore, the streams are steep, having a mean bed slope of $3.17 \, \text{m km}^{-1}$. Therefore, stream power is high and substantial channel enlargement is recorded during catastrophic floods (Erskine, 1992, 1994a, 1996; Erskine and Saynor, 1996). Most of the channel adjustments on these streams occur as large-scale changes in width and depth (Erskine and Melville, 1983; Erskine, 1994a, 1996) because the meander wavelength is largely controlled by the bedrock valley. Regulated rivers which experience significant flood suppression often respond by channel contraction via bench construction (Gregory and Park, 1974; Petts, 1977; Sherrard and Erskine, 1991). Furthermore, flood variability is very high by world standards (Table 18.1), with the flash flood magnitude index for Baerami (1.025) being one of the highest recorded. Flood peak discharge on most flood frequency curves extends over three and rarely over six orders of magnitude. The potential for generating floods with peak discharges greater than 10 times the mean annual flood is high. These events have caused widespread erosion on many streams (Erskine, 1986a, 1993, 1994a, 1996; Erskine and

A. PRE-CATASTROPHIC FLOOD

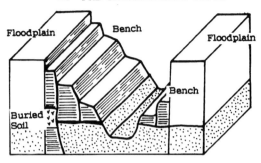

B. IMMEDIATELY POST-CATASTROPHIC FLOOD

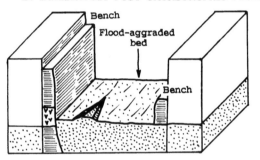

C. A DECADE POST-CATASTROPHIC FLOOD

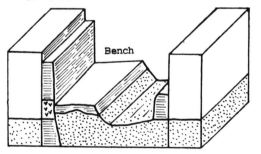

FIGURE 18.10   Model of bench formation and destruction by different sized floods on bedrock-confined channels.

Saynor, 1996). Furthermore, the largest floods recorded at long-term Hunter Valley gauging stations have all occurred since 1945 (Bell and Erskine, 1981; Erskine and Bell, 1982; Erskine 1986a) during the current FDR. Therefore, the imprint of such floods is still evident.

While it has been possible to identify five hydrologic areas in the Hunter Valley by cluster analysis of selected flood indices, these areas are *not* relevant to the formation of in-channel benches. Benches are present at every station, irrespective of the flood hydrology. However, this means that either inappropriate hydrologic indices have been used or that the hydrologic conditions throughout the Hunter Valley are similar in terms of the formative conditions for in-channel benches. The latter is more likely because flood variability in the Hunter Valley is much greater than world norms (Table 18.1). The lowest flash flood magnitude index (0.359) is 71% greater than the average of McMahon et al's. (1992) world rivers, excluding Australia and southern Africa (Tables 18.1 and 18.2). This high variability means that floods much greater than the mean annual flood occur more frequently than elsewhere. These large events generate such high stream powers that the channel boundary is extensively eroded (Erskine, 1993, 1994a; Erskine and Saynor, 1996). Large floods episodically exceed the resistance threshold of the channel boundary to erode and clean out the channel. The resultant large-capacity channel is not obliterated by subsequent sedimentation because of the episodic recurrence of large floods. Benches are temporary sediment storages and ephemeral landforms constructed by small to moderate floods within the void eroded by the large event. Therefore, the compound channel so common in eastern Australia is a morphological response to the differential impacts of catastrophic and smaller floods. As noted by Pickup and Warner (1976) and Neller (1980), two ranges of floods are important in forming different aspects of channel morphology.

Figure 18.10 presents a model of bench destruction and formation by different sized floods. Catastrophic floods are the disturbing agents which extensively erode the channel boundary of bedrock-confined channels. The liberated sediment is temporarily stored in the channel bed producing a bedload slug. Subsequent lateral redistribution of the slug sediment by moderate floods rebuilds benches in the void excavated by the catastrophic flood. While European settlement has decreased channel boundary resistance by removing riparian vegetation (Erskine, 1993), it is *not* the primary disturbance producing compound channels (Erskine, 1993). Rather, the highly flood-variable character of Australian hydrology also produces a distinctive channel morphology and sediment transfer pattern on rivers that are often sediment supply-limited (Nanson, 1986). As noted by Tooth and Nanson (1995), overseas findings have limited application to Australian fluvial systems.

ACKNOWLEDGEMENTS

This work was funded by an Australian Postgraduate Research Award and a Faculty of Applied Science, University of New South Wales Staff Research Grant. Keith Woodyer, Graham Taylor and Rob Warner kindly provided some of their own material for us to use. The fieldwork was completed with the assistance of C.C. Erskine, D.O. Erskine, G. Campbell, M.D. Melville, M.J. Saynor, P. Greenwood, C. Peacock, J.J. Sherrard, D. Edwards, B. Smart and M. Eather. Professor J. Dodson, Associate Professor M.D. Melville, Mr J. Sammut and the referees constructively criticized a draft manuscript. Ms G. Campbell drafted the figures.

REFERENCES

Abrahams, A.D. and Cull, R.F. 1979. The formation of alluvial landforms along New South Wales coastal streams. *Search*, **10**, 187–188.

Baker, V.R. 1977. Stream-channel response to floods, with examples from central Texas. *Geological Society of America, Bulletin*, **88**, 1057–1071.

Bell, F.C. and Erskine, W.D. 1981. Effects of recent increases in rainfall on floods and runoff in the upper Hunter Valley. *Search*, **12**, 82–83.

Benson, M.A. 1968. Uniform flood-frequency estimating methods for federal agencies. *Water Resources Research*, **4**, 891–908.

Blench, T. 1957. *Regime Behaviour of Canals and Rivers*. Butterworths, London.

Costa, J.E. 1974. Stratigraphic, morphologic and pedologic evidence of large floods in humid environments. *Geology*, **2**, 301–303.

Cull, R.F. 1975. *Alluvial benches in the drainage basins of the Macleay and Shoalhaven Rivers, NSW*. BSc Thesis, University of New South Wales.

Ellis, I. (undated). *A History of Baerami Creek Valley*. Muswellbrook Chronicle Print, Muswellbrook.

Erskine, W.D. 1985. Downstream geomorphic impacts of large dams: the case of Glenbawn Dam, NSW. *Applied Geography*, **5**, 195–210.

Erskine, W.D. 1986a. *River metamorphosis and environmental change in the Hunter Valley, New South Wales*. PhD Thesis, University of New South Wales.

Erskine, W.D. 1986b. River metamorphosis and environmental change in the Macdonald Valley, New South Wales since 1949. *Australian Geographical Studies*, **24**, 88–107.

Erskine, W.D. 1992. Channel response to large-scale river training works: Hunter River, Australia. *Regulated Rivers: Research and Management*, **7**, 261–278.

Erskine, W.D. 1993. Erosion and deposition produced by a catastrophic flood on the Genoa River, Victoria. *Australian Journal of Soil and Water Conservation*, **6**(4), 35–43.

Erskine, W.D. 1994a. *Sand slugs generated by catastrophic floods on the Goulburn River, New South Wales*. International Association of Hydrological Sciences, Publication No. 224, 143–151.

Erskine, W.D. 1994b. Late Quaternary alluvial history of Nowlands Creek, Hunter Valley, NSW. *Australian Geographer*, **25**(1), 50–60.

Erskine, W.D. 1996. Response and recovery of a sand–bed stream to a catastrophic flood. *Zeitsfriche für Geomorphologie*, **40**, 359–383.

Erskine, W.D. and Bell, F.C. 1982. Rainfall, floods and river channel changes in the upper Hunter. *Australian Geographical Studies*, **20**, 183–196.

Erskine, W.D. and Keshavarzy, A. 1996. Frequency of bankfull discharge on South and Eastern Creeks, NSW, Australia. *Water and the Environment, 23rd Hydrology and Water Resources Symposium*, 21–24 May, 1996, Hobart, Tasmania, Volume 2, Institution of Engineers, Australia, National Conference Publication 96–05, 381–387.

Erskine, W.D. and Melville, M.D. 1983. Impact of the 1978 floods on the channel and floodplain of the lower Macdonald River, NSW. *Australian Geographer*, **15**, 284–292.

Erskine, W.D. and Saynor, M.J. 1996. *Effects of catastrophic floods on sediment yields in southeastern Australia*. International Association of Hydrological Sciences, Publication No. 236, 381–388.

Erskine, W.D. and Warner, R.F. 1988. Geomorphic effects of alternating flood- and drought-dominated regimes on NSW coastal rivers. In R.F. Warner (Ed.), *Fluvial Geomorphology of Australia*. Academic Press, Sydney, 223–244.

Erskine, W.D. and White, L.J. 1996. Historical river metamorphosis of the Cann River, East Gippsland, Victoria. *Proceedings of First National Conference on Stream Management in Australia*, Merrijig, 19–23 February, 1996, 277–282.

Galloway, R.W. 1963. *Geomorphology of the Hunter Valley*. CSIRO (Aust.) Land Research Series No. 8, 90–102.

Gregory, K.J. and Park, C.C. 1974. Adjustment of river channel capacity downstream from a reservoir. *Water Resources Research*, **10**, 870–873.

Hall, L.D. 1927. The physiographic and climatic factors controlling the flooding of the Hawkesbury River at Windsor. *Proceedings of the Linnean Society of New South Wales,* **52,** 133–152.

Hedman, E.R. 1970. Mean annual runoff as related to channel geometry of selected streams in California. *US Geological Survey Water-Supply Paper* 1999E.

Hedman, E.R. and Kastner, W.M. 1977. Streamflow characteristics related to channel geometry in the Missouri River basin. *Journal of Research, US Geological Survey,* **5,** 285–300.

Henry, H.M. 1977. Catastrophic channel changes in the Macdonald Valley, New South Wales, 1949–1955. *Journal and Proceedings of Royal Society of New South Wales,* **110,** 1–16.

Hickin, E.J. 1968. Channel morphology, bankfull stage, and bankfull discharge of streams near Sydney. *Australian Journal of Science,* **30,** 274–275.

Jacobson, R.B. and Coleman, D.J. 1986. Stratigraphy and recent evolution of Maryland Piedmont flood plains. *American Journal of Science,* **286,** 617–637.

Jennings, M.E. and Benson, M.A. 1969. Frequency curves for annual flood series with some zero events or incomplete data. *Water Resources Research,* **5,** 276–280.

Kilpatrick, F.A. 1961. Bankfull depth and depth of flow for mean annual flood, Piedmont Province. *US Geological Survey Professional Paper* 424C, C49–C50.

Kilpatrick, F.A. and Barnes, H.H. Jr 1964. Channel geometry of Piedmont streams as related to frequency of floods. *US Geological Survey Professional Paper* 422E.

Leopold, L.B., Wolman, M.G. and Miller, J.P. 1964. *Fluvial Processes in Geomorphology.* W.H. Freeman, San Francisco.

Lewin, J. and Brindle, B.J. 1977. Confined meanders. In K.J. Gregory (Ed.), *River Channel Changes.* Wiley, Chichester, 221–233.

McDermott, G.E. and Pilgrim, D.H. 1983. A design flood method for arid western New South Wales based on bankfull estimates. *Transactions, Institution of Engineers Australia, Civil Engineering,* **CE25,** 114–120.

McGilchrist, C.A. and Woodyer, K.D. 1968. Statistical tests for common bankfull frequency in rivers. *Water Resources Research,* **4,** 331–334.

McMahon, T.A. 1964. *Hydrologic features of the Hunter Valley, NSW.* Hunter Valley Research Foundation Monograph No. 20.

McMahon, T.A. 1968. Geographical interpretation of hydrologic characteristics in the Hunter Valley. *Australian Geographer,* **10,** 404–407.

McMahon, T.A. 1969a. Water resources research: aspects of a regional study in the Hunter Valley, New South Wales. *Journal of Hydrology,* **7,** 14–38.

McMahon, T.A. 1969b. Water yield and physical characteristics of catchments. *Transactions, Institution of Engineers Australia, Civil Engineering,* **CE11,** 74–81.

McMahon, T.A. and Srikanthan, R. 1981. Log Pearson III distribution – is it applicable to flood frequency analysis of Australian streams? *Journal of Hydrology,* **52,** 139–147.

McMahon, T.A., Finlayson, B.L., Haines, A.T. and Srikanthan, R. 1992. *Global Runoff.* Catena Paperback, Cremlingen, 166pp.

Melville, M.D. and Erskine, W.D. 1986. *Sediment remobilization and storage by discontinuous gullying in humid southeastern Australia.* International Association of Hydrological Sciences, Publication No. 159, 277–286.

Miller, R.A., Troxell, J. and Leopold, L.B. 1971. Hydrology of two small river basins in Pennsylvania before urbanisation. *US Geological Survey Professional Paper* 701A.

Mosley, M.P. 1981. Delimitation of New Zealand hydrologic regions. *Journal of Hydrology,* **49,** 173–192.

Nanson, G.C. 1986 Episodes of vertical accretion and catastrophic stripping: a model of flood–plain development. *Geological Society of America, Bulletin,* **97,** 1467–1475.

Nanson, G.C. and Croke, J.C. 1992. A genetic classification of floodplains *Geomorphology,* **4,** 459–486.

Nanson, G.C. and Erskine, W.D. 1988. Episodic changes of channels and floodplains on coastal rivers in New South Wales. In R.F. Warner (Ed.), *Fluvial Geomorphology of Australia.* Academic Press, Sydney, 201–221.

Neller, R.J. 1980. Channel changes on the Macquarie Rivulet, New South Wales. *Zeitschrift für Geomorphologie,* **24,** 168–179.

Osterkamp, W.R., Lane, L.J. and Foster, G.R. 1983. An analytical treatment of channel-morphology relations. *US Geological Survey Professional Paper* 1288.

Page, K.J. 1972. *A field study of the bankfull discharge concept in the Wollombi Brook drainage basin, New South Wales.* MA (Hons) Thesis, University of Sydney.

Page, K.J. 1988. Bankfull discharge frequency for the Murrumbidgee River, NSW. In R.F. Warner (Ed.), *Fluvial Geomorphology of Australia.* Academic Press, Sydney, 267–281.

Page, K.J. and Nanson, G.C. 1982. Concave–bank benches and associated floodplain formation. *Earth Surface Processes and Landforms*, 7, 529–543.

Petts, G.E. 1977. Channel response to flow regulation: the case of the River Derwent, Derbyshire. In K.J. Gregory (Ed.), *River Channel Changes,* Wiley, Chichester, 145–164.

Pickup, G. and Rieger, W.A. 1979. A conceptual model of the relationship between channel characteristics and discharge. *Earth Surface Processes*, 4, 37–42.

Pickup, G. and Warner, R.F. 1976. Effects of hydrologic regime on magnitude and frequency of dominant discharge. *Journal of Hydrology*, 29, 51–75.

Pilgrim, D.H. and Doran, D.G. 1987. Flood frequency analysis. In D.H. Pilgrim (Ed.), *Australian Rainfall and Runoff Vol. 1.* Institution of Engineers Australia, Barton, 195–236.

Raine, A. and Gardiner, J. 1994. *Use and Management of native vegetation for River Bank Stabilisation and Ecological Sustainability.* Department of Water Resources, Parramatta.

Riley, S.J. 1972. A comparison of morphometric measures of bankfull. *Journal of Hydrology*, 17, 23–31.

Sammut, J. and Erskine, W.D. 1995. Hydrological impacts of flow regulation associated with the Upper Nepean Water Supply Scheme, NSW. *Australian Geographer*, 26, 71–86.

Schick, A.P. 1974. Formation and obliteration of desert stream terraces a conceptual analysis. *Zeitschrift für Geomorphologie,* Suppl. Bd 21, 88–105.

Sherrard, J.J. and Erskine, W.D. 1991. Complex response of a sand–bed stream to upstream impoundment. *Regulated Rivers: Research and Management*, 6, 53–70.

Smith, D.I. and Greenaway, M.A. 1982. Flood probabilities and urban flood damage in coastal northern New South Wales. *Search*, 13, 312–314.

Smith, N.D. 1978. Some comments on terminology for bars in shallow rivers. In A.D. Miall (Ed.), *Fluvial Sedimentology.* Canadian Society of Petroleum Geologists Memoir No. 5, 85–88.

Stevens, M.A., Simons, D.B. and Richardson, E.V. 1975. Nonequilibrium river form. *Journal of Hydraulics Division, Proceedings, American Society of Civil Engineers*, 101, 557–566.

Stevens, M.A., Simons, D.B. and Richardson, E.V. 1977. Closure–nonequilibrium river form. *Journal of Hydraulics Division, Proceedings, American Society of Civil Engineers*, 103, 197–198.

Storrie, G. 1990. *The geomorphic impact of alternating rainfall regimes on Webbs Creek and Macdonald River.* BSc Thesis, University of New South Wales.

Taylor, G. and Woodyer, K.D. 1978. Bank deposition in suspended-load streams. In A.D. Miall (Ed.), *Fluvial Sedimentology.* Canadian Society of Petroleum Geologists, Memoir No. 5, 257–275.

Tooth, S. and Nanson, G.C. 1995. The geomorphology of Australia's fluvial systems: retrospect, perspect and prospect. *Progress in Physical Geography*, 19(1):35–60.

Wahl, K.L. 1984. Evolution of the use of channel cross-section characteristics for estimating streamflow characteristics. *US Geological Survey Water-Supply Paper* 2262, 53–66.

Walker, P.H. and Coventry, R.J. 1976. Soil profile development in some alluvial deposits of eastern New South Wales. *Australian Journal of Soil Research*, 14, 305–317.

Walker, P.H. and Green, M.P. 1976. Soil trends in two valley fill sequences. *Australian Journal of Soil Research,* 14, 291–303.

Warner, R.F. 1972. River terrace types in the coastal valleys of New South Wales. *Australian Geographer*, 12, 1–22.

Warner, R.F. 1993. Downstream changes in channel morphology and capacity in the Bellinger Valleys, NSW, Australia. *Zeitschrift für Geomorphologie,* Suppl. Bd 88, 29–47.

Warner, R.F. 1994. Temporal and spatial variations in erosion and sedimentation in alternating hydrological regimes in southeastern Australian rivers. In L.J. Olive, R.J. Loughran and J.A. Kesby (Eds), *Variability in Stream Erosion and Sediment Transport.* International Association of Hydrological Sciences, Publication No. 224, 211–219.

Warner, R.F., Sinclair, D. and Ewing, J. 1975. A comparative study of relations between bedrock meander wavelengths, benchfull discharges and channel perimeter sediments for the Sydney Basin and the North East of New South Wales. In I. Douglas, J.E. Hobbs and J.J. Pilgram (Eds), *Geographical Essays in Honour of Gilbert J. Butland*. Department of Geography, University of New England, 159–212.

Wharton, G., Arnell, N.W., Gregory, K.J. and Gurnell, A.M. 1989. River discharge estimated from channel dimensions. *Journal of Hydrology*, **106**, 365–376.

Williams, G.P. 1978. Bank-full discharge of rivers. *Water Resources Research*, **14**, 1141–1154.

Wolman, M.G. 1977. Changing needs and opportunities in the sediment field. *Water Resources Research*, **13**, 50–54.

Wolman, M.G. and Leopold, L.B. 1957. River flood plains: some observations on their formation. *U.S. Geological Survey Professional Paper* 282C.

Woodyer, K.D. 1966. *Frequency of bankfull flow*. MSc Thesis, University of New South Wales.

Woodyer, K.D. 1968. Bankfull frequency in rivers. *Journal of Hydrology*, **6**, 114–142.

Woodyer, K.D. 1970. The terraces of the lower Colo River, New South Wales. *Search*, **1**, 164–165.

Woodyer, K.D. 1975. Concave-bank benches on the Barwon River, NSW. *Australian Geographer*, **13**, 36–40.

Woodyer, K.D. and Fleming, P.M. 1968. Reconnaissance estimation of stream discharge–frequency relationships. In G. Al Stewart (Ed.), *Land Evaluation*. Macmillan, Australia, 287–298.

Woodyer, K.D., Taylor, G. and Crook, K.A.W. 1979. Depositional processes along a very low-gradient, suspended-load stream: the Barwon River, New South Wales. *Sedimentary Geology*, **22**, 97–120.

Wright, J. 1969. *Some relationships between elements of geometry and morphology of stream channels*. BA (Hons) Thesis, University of New England.

19

Anabranching Rivers: Divided Efficiency Leading to Fluvial Diversity

GERALD C. NANSON AND H. Q. HUANG

School of Geosciences, University of Wollongong, NSW, Australia

ABSTRACT

In order to maintain an efficient conveyance of water and sediment, most rivers flow in single channels with adjustable slopes. Less common are anabranching rivers consisting of multiple channels separated by vegetated, semi-permanent alluvial islands or ridges. They occur across a wide range of environments and form a variety of river types, yet they remain the last major category of alluvial system to be described and explained. Anabranching rivers develop multiple channels that are, in combination, usually substantially narrower and deeper than if the same discharge were to occupy a single channel. An analysis of basic hydraulic relationships for alluvial channels that include flow continuity, roughness and several sediment transport functions, reveals that in situations where there is little or no opportunity to increase channel slope, conversion from a wide, single channel to a semi-permanent system of multiple channels will reduce total width and increase average flow depth, hydraulic radius and velocity. This will maintain or enhance water and sediment throughput, even overriding moderate increases in channel roughness. Anabranching rivers appear to be closer to exhibiting the most efficient sections for the conveyance of both water and sediment than are equivalent wide, single channels at the same slope. However, not all anabranching systems are finely tuned to hydraulic efficiency. Once formed, some channels may, due to inertia, continue to operate despite increasing inefficiency and channel atrophication with time. Others may form as distributary systems for dispersing and storing water and sediment across extensive low-gradient floodplains.

Australia provides a continental setting conducive to the formation of anabranching rivers. Very low relief, intensely weathered and cohesive fine-grained sediment in an arid environment with declining flow discharges and increasing sedimentation downstream encourage sediment storage and the development of anabranches. Tree-lined rivers with muddy, cohesive banks permit narrow channels to form that maintain water and sediment conveyance for the maximum distance downstream. Conditions leading to the development of alluvial anabranching systems are additional to those required for other types of river pattern and, as a consequence, anabranching rivers form a highly diverse group that occurs inclusively with stable or laterally active straight, sinuous, meandering or braided systems.

Varieties of Fluvial Form. Edited by A.J. Miller and A. Gupta

INTRODUCTION

Alluvial rivers respond to changes in environmental conditions by adjusting their hydraulic geometry, planform and boundary roughness, with the result that river morphology can vary across a wide spectrum. Anabranching streams remain the last major category of alluvial river to be described and explained, for until recently they received no mention in geomorphology text books except in confusion with braided systems. They are a diverse style of river ranging from straight to highly sinuous, fine-grained and low-energy to coarse-grained and high-energy, and two-channelled to many-channelled (Nanson and Knighton, 1996). In certain regions, such as parts of arid and semi-arid Australia, they are by far the most dominant type of river.

In the first general analysis of the conditions that may lead to the development of multichannel systems, Knighton and Nanson (1993) described anastomosing rivers (low energy, multiple-channel rivers) as part of a continuum of channel pattern defined by three dimensions: flow strength, bank erodibility and local sediment budget. They determined that such anastomosing reaches are characterized by low-energy flow, erosion-resistant banks and often by a slight surplus in their sediment budget that can result in aggradation. In a subsequent paper, Nanson and Knighton (1996) examined rivers over a much wider range of energy conditions, adopting the term *anabranching* as the generic term for all multichannelled rivers and limiting the term *anastomosing*, by common usage, to those at the low-energy, fine-sediment end of the range. They showed that anabranching is by no means restricted to low-energy systems and that it can be superimposed on any of the four basic types of single channel pattern: straight and sinuous laterally stable systems, and meandering and braiding laterally active systems. Using the criteria of stream power, sediment texture and a number of physiographic characteristics, they classified anabranching rivers into six types (while accepting that other types probably exist, as yet unrecognized) and found that anabranching rivers are usually characterized by flood-dominated flow regimes, banks that are resistant to erosion relative to stream energy, and mechanisms that lead to channel damming and avulsion. Furthermore, and of particular relevance here, is their assertion that a fundamental advantage anabranching rivers have over their single-channel counterparts is that, in situations where it is not possible to increase channel slope, the division of a single channel into two or more anabranches concentrates shear stress and stream power and enables the system to maintain or enhance the transport of water and sediment. Restrictions on slope adjustment can result from valley obstructions, tectonism, basin infilling and subsidence, and delta formation, but anabranching rivers are not restricted to these situations only.

In some cases, anabranching is a chance event, a single channel dividing around some temporary or more permanent obstruction. However, in other cases it is clearly a general condition. The objectives of this paper are to explain why rivers, most of which flow in single, well defined channels, may become permanently anabranched over considerable distances, and why anabranching is the dominant style of river over large areas of Australia.

WHY DO RIVERS USUALLY FLOW IN SINGLE CHANNELS?

To explain why rivers in some situations flow in more than one channel first requires clarification as to why most rivers flow in a single channel. One compelling reason for rivers to occupy a single channel is that a river is forced to follow the thalweg of what is commonly a narrow valley. Yet, this begs the question of why, on very extensive floodplains, rivers usually occupy a single channel often well away from any influence of the valley sides.

The explanation proposed here is one of flow efficiency and minimum energy expenditure (Yang, 1971; Chang, 1979), consistent with what appears to occur in many natural systems. In open channels, the smallest ratio of skin resistance to flow discharge will result from a single, semi-circular cross-section. However, because the boundary of an alluvial channel usually consists of unconsolidated sediment, a semi-circular or other relatively deep form of cross-section cannot be maintained. As demonstrated by Wolman and Brush (1961), the alluvial banks will collapse and the channel will widen and shallow as sediment moves to fill the centre of the channel, resulting in a shallow, parabolic cross-section (also see Lane, 1953; Henderson, 1966). Pickup (1976) noted that the most efficient channel for the transmission of bedload has a broad rectangular cross-section and, as a consequence, the form of an alluvial channel becomes a compromise between an efficient conduit for water and one able to transport the bed material supplied to it. In some cases the widening of the channel results in braiding with the deposition of within-channel bars (Leopold and Wolman, 1957). Owing to its much larger boundary area and an increase in roughness, such a wide, shallow system is hydraulically less efficient than a narrow, deep channel, a situation that is usually overcome by an increase in gradient. The energy compromise from narrow and deep to wide and shallow channels involves an increase in gradient in order to maintain or increase the conveyance of water and sediment. An increase in gradient can often be achieved by simply reducing sinuosity but this is not always an option. An alternative strategy may be, where boundary material permits, for a river to reduce its boundary resistance by reverting to a narrower and deeper cross-section or, as will be shown to be more likely, several such sections in the form of an anabranching system. Understanding such a change becomes a problem of channel geometry.

Because of their complex behaviour, no general theory has been determined to predict the hydraulic characteristics of alluvial channels. However, a great deal of empirical research on this topic is now referred to as regime theory or hydraulic geometry. The basis of this research is that a reach of alluvial river (or self-adjusting canal) will exhibit hydraulic characteristics (channel width, depth, slope, flow velocity and roughness) that are optimum for the transport of just that amount of water and sediment supplied to it from upstream (Lacey, 1929, 1933, 1946; Lane, 1953, 1955; Leopold and Maddock, 1953; Blench, 1957, 1969; Simons and Albertson, 1960). A river in such a state of balance is said to be "at grade" or "in equilibrium" (Mackin, 1948; Hack, 1957; Chorley, 1962), slight adjustments for long-term landscape denudation or basin infilling notwithstanding. The analysis of a very large number of alluvial river cross-sections shows that this is most commonly achieved by a single channel with a shallow trapezoid or parabolic cross-section. However, until now the channel geometries of anabranching reaches of river channels have not been compared with their single-channel counterparts.

CHANNEL GEOMETRY OF ANABRANCHING RIVERS

As identified by Knighton and Nanson (1993) and Nanson and Knighton (1996), anabranching rivers commonly occur where the conveyance of water and sediment must be maintained without a significant increase in stream gradient and indeed where gradients may have been slightly reduced. On this basis, an analysis of the channel geometry of anabranching rivers is conducted here.

Most efficient hydraulic sections of stream channels

Channel flow follows the laws of flow continuity and flow resistance with the former expressed as:

$$Q = AV \qquad (19.1)$$

where Q, A and V are flow discharge, channel cross-sectional area and average flow velocity, respectively.

The law of flow resistance is illustrated by the Darcy–Weisbach flow relation:

$$V = \sqrt{\frac{8}{f} gRS} \qquad (19.2a)$$

where f, g, R and S are Darcy–Weisbach resistance factor, acceleration due to gravity, hydraulic radius ($= A/P$, P is channel wetted perimeter) and channel slope, respectively. However, f varies with channel boundary condition and needs to be replaced with f_p, f_l or f_u, which are defined by the following relations:

$$f_p = 8g\left(\frac{n^6}{R}\right)^{1/3} ; f_l = 2.03\left(\frac{S^4 d}{R}\right)^{1/17} ; f_u = 0.19\left(\frac{Sd}{R}\right)^{1/9} \qquad (19.2b)$$

where n and d are Manning's roughness coefficient and a measure of sediment size, respectively. In Equation 19.2b, f_p represents the well known Manning's flow resistance equation in which roughness coefficient n can be readily determined, particularly in the cases of fixed-bed or gravel-bed channels where sediment transport is not very active. In contrast, f_l and f_u represent the Darcy–Weisbach resistance factors for the lower and upper flow regimes, respectively, where bed forms exist due to active transport of sediment. There are many procedures in determining f_l and f_u, but most of them involve complicated trial-and-error techniques. Only the empirical relations developed by Brownlie (1983) lead to the explicit expressions of f_l and f_u in Equation 19.2b[1].

Equations 19.2a and 19.2b show clearly that when channel roughness or sediment size and channel slope remain unchanged, flow discharge is determined only by section size (cross-sectional area) and a shape factor (hydraulic radius or wetted perimeter). From a hydraulic point of view, therefore, the channel section having the least wetted perimeter for a given area has the maximum flow discharge; in open-channel hydraulics, this is known as the best or the most economic section (Chow, 1959). However, it is found that the best hydraulic section also provides a maximum flow velocity (that is, it has a minimum sectional size) when flow discharge is given[2]. In river studies, it might be better to call it the most *efficient* hydraulic section, for it should provide the maximum capacity for sediment transport (sediment transport capacity is normally in direct proportion to velocity with a power of four or more, as shown by Laursen (1956) and Colby (1964)).

Provided that anabranching rivers possess rectangular cross-sections that have equal width and depth in all channels, and provided that the sediment over the whole channel boundary is homogeneous, the following geometric relations remain:

$$A = WD; \ P = W + 2mD; \ R = \frac{WD}{W + 2mD} \tag{19.2c}$$

where W, D and m are aggregated channel surface width, averaged channel depth and number of channels, respectively. For single-channel streams $m = 1$ and for multichannel streams $m > 1$.

In terms of the definition of the most efficient hydraulic section given above, it is found that any of the expressions of f presented in Equation 19.2b (i.e. f_p, f_l and f_u) can lead to a maximum flow velocity for anabranching rivers when the following condition is satisfied:

$$\frac{W}{D} = 2m \tag{19.2d}$$

Substituting Equation 19.2d into Equation 19.1, 19.2a and 19.2b yields the following maximum velocity expressions for different flow resistance factors as defined in Equation 19.2b:[3]

$$V_{\text{max p}} = \left(\frac{S^{3/2}Q}{8n^3 m} \right)^{1/4}; \ V_{\text{max l}} = 1.11 \left(\frac{g^{17}S^{13}Q^9}{m^9 d} \right)^{1/43}; \ \dot{V}_{\text{max u}} = 2.75 \left(\frac{g^9 S^8 Q^5}{m^5 d} \right)^{1/23}$$
$$\tag{19.2e}$$

As stated in Equation 19.2d, for a single channel the most efficient hydraulic section would be where $W/D = 2$, a highly unlikely geometry. However, for an anabranching system, this condition is met by wider (aggregate width) and shallower system ($W/D = 2m$). Alluvial rivers that are required to transport bedload and have erodible banks tend to have wide, shallow channels compared to those that do not (Pickup, 1976). In other words, anabranching rivers are closer to exhibiting the most efficient section for the conveyance of both water and sediment than are rivers with a single channel, for anabranching rivers commonly have width–depth ratios (W/D) in the range of 10–30 (Knighton and Nanson, 1993), getting larger in systems with coarser or more abundant sediment (Nanson and Knighton, 1996).

Adjustments of channel form to sediment transport

A natural alluvial river adjusts channel geometry to accommodate both water discharge and sediment load from its drainage basin. However, this link between sediment transport and channel form has been poorly understood and results in conflicting interpretations. For example, Henderson (1966) and Nevins (1969) demonstrated that the increase in either bedload or suspended load results in reduction in channel width, while Bagnold (1977, 1980) and Parker (1979) argued that a direct proportional relationship between channel width and bedload exists. Others have suggested that there is a one- or two-peak maximum in sediment transport capacity at some intermediate width (Gilbert, 1914; Chang, 1979; White et al., 1982). As a consequence, both direct and inverse relationships between sediment transport and channel width have been proposed. Carson and Griffiths

(1987) made a detailed investigation of this dichotomy and found that it can be explained by the use of different sediment transport formulas. Therefore, the following analysis of the adjustment of channel form to changes in sediment transport is conducted initially without reference to any sediment transport formulas, but rather on the understanding that sediment transport is strongly related to channel average velocity (e.g. Laursen, 1956; Colby, 1964; Karim and Kennedy, 1990).

When channel roughness or sediment size and channel slope can be regarded as constants in Equations 19.2a, 19.2b and 19.2c and stream channels as rectangular in cross-section, channel flow velocity can be generally illustrated as:

$$V \propto R^i \propto \left(\frac{WD}{W + 2mD} \right)^i \propto \left(\frac{1}{D} + 2m\frac{1}{W} \right)^{-i} \tag{19.3a}$$

where $i = 2/3, 9/17$ and $5/9$, respectively, for the expressions of f_p, f_l and f_u in Equation 19.2b, and W is the aggregated width.

Assuming an increase in velocity (differential variation $dV > 0$), but no change in water discharge Q, channel gradient S, Manning's roughness n (this assumption is considered in detail below), sediment size d or the number of channels m, a decrease in both (WD) and $D^{-1} + 2mW^{-1}$ in Equation 19.3a is required if flow continuity is to be maintained $(Q = WDV)$. Indeed, an increase in both (WD) and $D^{-1} + 2mW^{-1}$ is necessary for a decrease in velocity (differential variation $dV < 0$). There are many possible adjustments in both channel width and depth in order to accommodate a variation in velocity. To provide a prediction of the tendency of the adjustments of channel geometry, the following functions are assumed here:

$$\zeta_1 = WD \tag{19.3b}$$

$$\zeta_2 = \frac{1}{D} + 2m\frac{1}{W} \tag{19.3c}$$

When there is an increase in velocity $(dV > 0)$ then, according to the principles of differential calculus, both the following conditions must be met:

$$d\zeta_1 < 0 \quad \text{and} \quad d\zeta_2 < 0 \tag{19.3d}$$

To satisfy Equation 19.3d, the following conditions are derived from Equations 19.3b and 19.3c:

$$D.dW + W.dD < 0 \tag{19.3e}$$

$$-D^{-2}dD - 2mW^{-2}dW < 0 \tag{19.3f}$$

Further solving Equations 19.3e and 19.3f produces:

$$-\frac{W}{D} \left(\frac{W}{2mD} \right) dD < dW < -\frac{W}{D} dD \tag{19.3g}$$

Because $W/(2mD)$ is positive and therefore so is W/D, if $dD > 0$, $dW < 0$ when $W/D > 2m$. Similarly, if $dD < 0$, $dW > 0$ when $W/D < 2m$. It is apparent that the variations of dW and dD are in opposite directions, increasing depth accompanying a decrease in width, and vice versa. For channels having smaller width/depth ratios

$(W/D < 2m)$, flow velocity can be readily increased by decreasing depth and increasing width, while for channels having larger width/depth ratios $(W/D > 2m)$, the increase in flow velocity can be achieved most readily by increasing depth and reducing width. As mentioned earlier, when $W/D = 2m$, flow velocity achieves a maximum and thus $dV < 0$.

When a similar analytical procedure is followed for the case of decreasing flow velocity $dV < 0$, it is found that for channels having smaller width/depth ratios $(W/D < 2m)$, flow velocity can be readily increased by reducing depth and increasing width. For channels having larger width/depth ratios $(W/D > 2m)$, however, an increase in flow velocity can be achieved by increasing depth and reducing width.

Since the above analysis suggests the existence of a relationship between velocity and the ratio of width and depth, this study introduces the ratio of width and depth (W/D) as an independent variable and consequently derives the following relationships for velocity according to different flow resistance factors defined in Equation 19.2b[4]:

$$V_p = \left(\frac{S^{1/2}}{n}\right)^{3/4} Q^{1/4} \left[\frac{W/D}{(W/D + 2m)^2}\right]^{1/4} \tag{19.3h}$$

$$V_1 = \left(\frac{8gS^{13/17}}{2.03d^{1/17}}\right)^{17/43} Q^{9/43} \left[\frac{W/D}{(W/D + 2m)^2}\right]^{9/43} \tag{19.3i}$$

$$V_p = \left(\frac{8gS^{8/9}}{0.19d^{1/9}}\right)^{9/23} Q^{5/23} \left[\frac{W/D}{(W/D + 2m)^2}\right]^{5/23} \tag{19.3j}$$

Figure 19.1 shows the relationship between flow velocity and width/depth ratios in single and multichannel systems. It is plotted only with Equation 19.3h because Equations 19.3h to 19.3j are very similar in illustrating the effect of independent variables and they exhibit very similar exponents relevant to W/D (1/4, 9/43 and 5/23, respectively). When n, S and Q are assumed to take constant values of 0.03, 0.0006 and 500 m^3s^{-1}, respectively (this is the channel modelled in the following sections based on several anabranching rivers in northern and central Australia), it is seen in Figure 19.1 that channels need to adjust their form (either width/depth ratio or number of channels or both) to accommodate the requirements of velocity, but the magnitude of adjustments of velocity is restricted by channel maximum conveyance defined by Equation 19.2d. To maintain the same velocity when $W/D > 2m$ requires either a larger W/D for a smaller number of channels or a smaller W/D for a larger number of channels. Therefore, an inverse relationship exists between W/D and the number of channels. However, when $W/D < 2m$, Figure 19.1 illustrates a directly proportional relationship between W/D and the number of channels. In other words, to maintain the same velocity requires either both a smaller W/D and a smaller number of channels, or larger W/D and a larger number of channels.

It should be emphasized, however, that the above analysis is based entirely on mathematical logic. In reality, it is virtually unknown for natural alluvial streams to have width/depth ratios of less than two. By excluding the case of $W/D < 2m$, it can then be found from Equations 19.3h to 19.3j that in natural alluvial streams, particularly in single-channel systems, channel flow velocity is directly proportional to depth but inversely proportional to aggregated width:

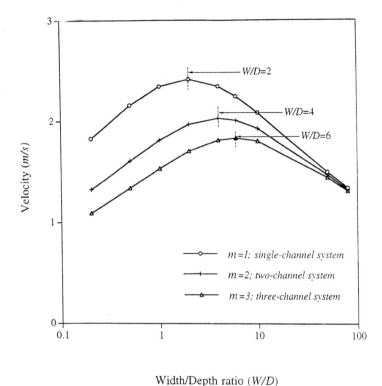

Width/Depth ratio (*W/D*)

FIGURE 19.1   The relationship between flow velocity and width/depth ratio in single and multiple channel systems (lines are drawn in terms of Equation 19.3h in which $n = 0.03$, $S = 0.0006$ and $Q = 500$ m^3 s^{-1}). *W* is aggregate channel width

$$V \sim \left(\frac{1}{W}; \, D \right)$$ (19.3k)

When bed sediment is readily available, sediment concentration is directly related to flow velocity (e.g. Yang, 1973, 1976; Colby, 1964; Karim and Kennedy, 1990). Yang's formulas are believed to provide reasonably accurate predictions by way of relating sediment concentration to unit stream power (the product of velocity and channel slope) (ASCE, 1982; Chang, 1988; Yang and Wan, 1990). Hence, when channel slope and flow discharge remain unchanged, the increase in sediment discharge or in sediment concentration is entirely the product of an increase in velocity which itself is the product of a reduction in aggregated width or an increase in depth (Equation 19.3k).

There are many sediment formulas in which channel velocity is directly related to sediment discharge per channel width (e.g. Colby, 1964; Karim and Kennedy, 1990). An increase in velocity will increase sediment discharge per unit width, but a reduction in width may also reduce the sediment discharge for the channel as a whole (the product of width and sediment concentration). The result is a complex relationship between sediment concentration and channel width. A detailed analysis of this relationship was conducted by Carson and Griffiths (1987) who found that there is a maximum sediment discharge at some optimum channel width. Before reaching the peak, sediment discharge is in direct proportion to width and after the peak an inverse relationship appears. The position of this peak relative to channel width varies with different sediment transport formulas and therefore cannot be readily identified (Carson and Griffiths, 1987). However, the notion that channel form (width/depth ratio), velocity and sediment transport are interacting in the manner described above is very important for understanding the existence and function of anabranching rivers.

ILLUSTRATING THE ADVANTAGES OF ANABRANCHING

Should a river be required to increase or maintain velocity in order to increase or maintain the conveyance of sediment, a single-channel system ($m = 1$) with constant discharge and roughness (based on sediment size) broadly speaking has two degrees of freedom: it can adjust slope and/or channel form (Equations 19.3h to 19.3j). However, a system with the potential to anabranch has a third option; it can adjust its number of channels (Equations 19.3h to 19.3j). In other words, there is mutual interdependence amongst flow velocity, slope, channel form and the number of channels (the latter two being closely related). However, in a situation where it is not possible to increase slope, a single channel is restricted to adjusting only its cross-sectional form in order to alter flow velocity. As discussed above, due to the limitation of bank strength, it is virtually impossible for a single channel to approach the optimum form value for flow efficiency where there is a W/D ratio of 2 (Equation 19.2d); furthermore, such a narrow channel would be a highly inefficient conduit for bed-material load. The most efficient section, or something approaching it, is more attainable for a system of multiple channels where each individual channel is wider and not so extraordinarily deep (Figure 19.1).

As an example of the hydraulic advantages of an anabranching system over a single-channelled equivalent, the following illustrations are given, broadly modelled on several rivers in northern and central Australia. The model channel starts as a wide, single-thread

TABLE 19.1 Adjustments (as a percentage) of channel depth, width/depth ratio and velocity for a 25% reduction in channel width

m	Width	Depth			Width/depth ratio			Hydraulic radius			Velocity		
		(a)	(b)	(c)	(a)	(b)	(c)	(a)	(b)	(c)	(a)	(b)	(c)
2	−25	+20	+27	+17	−37	−41	−36	+17	+24	+12	+11	+5	+14
3	−25	+21	+29	+19	−38	−42	−37	+15	+22	+9	+10	+4	+12
4	−25	+22	+30	+20	−39	−42	−38	+14	+20	+7	+9	+3	+11
5	−25	+23	+31	+22	−39	−43	−38	+12	+19	+4	+8	+2	+9

m = number of channels; flow discharge $Q = 500.0$ m³/s; channel gradient $S = 0.0006$

Case (a) for Manning's flow resistance equation ($f = f_p$ in Equations 19.2a and 19.2b) where roughness coefficient $n = 0.03$

Case (b) for Manning's flow resistance equation ($f = f_p$ in Equation 19.2b) where roughness coefficient $n = 0.033$

Case (c) for Brownlie's (1983) lower flow regime equation ($f = f_l$ in Equations 19.2a and 19.2b and sediment size $d = 0.35$ mm)

Variations in channel depth are determined from an integrated analysis of Equations 19.1, 19.2a, 19.2b and 19.2c using trial and error technique. Width is the aggregate width for m channels

TABLE 19.2   Adjustments (as a percentage) of channel geometry with 10% increase in velocity

m	Velocity	Width			Depth			Width/depth ratio			Hydraulic radius		
		(a)	(b)	(c)	(a)	(b)	(c)	(a)	(b)	(c)	(a)	(b)	(c)
2	+10	−23	−34	−28	+18	+39	+27	−35	−53	−43	+15	+33	+20
3	+10	−25	−36	−32	+21	+43	+33	−38	−56	−49	+15	+33	+20
4	+10	−26	−39	−36	+24	+48	+42	−41	−59	−55	+15	+33	+20
5	+10	−28	−41	−41	+28	+55	+53	−44	−62	−61	+15	+33	+20

m = number of channels; flow conditions are the same as defined in the Table 19.1 footnote
Variations in channel width and depth are determined by solving a quadratic equation resulting from Equations 19.1, 19.2a, 19.2b and 19.2c

sandy reach of river with a bankfull discharge of 500 m³ s⁻¹, a slope of 0.0006, a width of 250 m, an average depth of 1.72 m, and a width/depth ratio of 145. This then divides, based on field observation, into an anabranching reach with the same discharge and slope but with a 25% reduction in total channel width due to the formation of islands or ridges. With this change, three possible roughness conditions are imposed: (a) roughness remains unchanged (Manning's roughness $n = 0.03$); (b) roughness is increased by 10% ($n = 0.033$); (c) roughness varies due to sediment transport. In the last case Brownlie's (1983) lower flow regime relation ($f = f_l$ in Equations 19.2a and 19.2b) is applied. Table 19.1 shows that the multiple channels develop a marked increase in channel depth, a commensurate decline in the aggreagate width/depth ratio, and an increase in average flow velocity, depending on the number of channels formed (up to five). It is interesting to note in Table 19.1 that the changes in mean depth, aggregate width/depth ratio and velocity with unchanged roughness (column a) are significantly different from those where roughness is increased by 10% (column b), but are very close to Brownlie's low-flow regime relation (column c). Table 19.2 shows how aggregate width, depth and

TABLE 19.3   Total sediment discharge estimations (as a percentage) based on a 10% increase in velocity and the resulting channel adjustments presented in Table 19.2

m	Rubey equation			Einstein–Brown equation			Karim–Kennedy equation			Colby's relationship		
	(a)	(b)	(c)	(a)	(b)	(c)	(a)	(b)	(c)	(a)	(b)	(c)
2	+24	+46	+32	+17	+55	+24	+19	+2	+12	+36	+17	+28
3	+27	+51	+40	+14	+51	+18	+16	0	+5	+33	+13	+21
4	+30	+56	+49	+13	+44	+11	+15	−1	0	+31	+8	+13
5	+34	+62	+60	+10	+39	+2	+12	−9	−9	+27	+5	+5

m = number of channels; the same flow conditions as defined in Tables 19.1 and 19.2 are used
Rubey equation: $Q_s \propto (W/D)^{-1/2}$ (Rubey, 1952); Einstein–Brown equation: $q_s \propto \tau^{3.0}$ (Brown, 1950); Karim–Kennedy equation: $V \propto q_s^{0.216}$ (i.e. $q_s \propto V^{4.6}$) (Karim and Kennedy, 1990); Colby's relationship: $q_s \propto V^{6.0}$ (Colby, 1964)
Symbols: q_s = sediment discharge per unit channel width; Q_s = total sediment discharge (= $q_s W$); W = aggregate channel width; D = channel average depth; V = channel average velocity; τ = channel two-dimensional shear stress

aggregate width/depth ratio respond to an imposed 10% increase in velocity (the approximate average increase in velocity for unchanged roughness and the Brownlie low-flow regime relation in Table 19.1) as the model river changes from single-thread to anabranching (up to five channels).

While there is no universally accepted sediment transport formula, four examples are applied here to evaluate the relative effect of anabranching on sediment concentration and total sediment discharge (Table 19.3). Rubey (1952) developed a function derived from energy considerations for "large adjusted streams" which is appropriate for use here, for it is based on changing channel form (with slope here being held constant). The Einstein–Brown equation (Brown, 1950) is linked to changing shear stress and has been widely used. Both these equations relate sediment discharge indirectly to velocity and depth. The Colby (1964) relationship and Karim and Kennedy (1990) equation are employed for they are based strictly on velocity changes and were developed from a large set of field data. By taking as an example the single channel outlined above and dividing it into five anabranches with a velocity increase of 10% (Table 19.2), a fixed slope and the changes in aggregate width, depth, hydraulic radius and aggregate width/depth ratio from Table 19.2, all four transport functions indicate that sediment concentration and total sediment load have the potential to increase substantially (Table 19.3).

In reality it is unlikely that the increase in sediment load of a single-channel system that changes to anabranching is as great as the results suggested in Table 19.3, for without an additional supply of bed material, such an increase would cause the channels to scour and become unstable. In natural systems, hydraulic adjustments more complex than those proposed in this simple model (for example, adjustments in roughness; see below), may restrain the system's capacity to move additional sediment. What is significant from the results in Table 19.3 is that anabranching appears to provide the *potential* to increase or at least maintain sediment flux without an increase in slope, thereby illustrating why rivers may develop an anabranching form.

THE PROBLEM OF ROUGHNESS

Immediately apparent from the above analysis is that any variation in roughness is crucial to determining the extent to which there will be an increase in flow efficiency with anabranching, or if in fact there will be a decrease. As illustrated, even a modest increase in roughness of say 10% could negate the apparent hydraulic advantages of anabranching, although the response appears complex, with sediment transport equations based on form and hydraulic radius predicting a substantial increase in transport efficiency. While changes in depth, width and slope between adjacent anabranching and single-thread reaches can be measured in the field, without flow discharge measurements, roughness cannot; it can only be estimated approximately. However, there is evidence from the ridge-forming anabranching rivers in arid central and monsoonal northern Australia indicating that roughness does not increase significantly from single-thread to ridged reaches (Tooth, 1997; Wende and Nanson, 1998). First, the sandy-floored channels in the ridge sections appear to be equally clear of obstructions as the unridged reaches upstream, and in some cases even more so as the wider channels support trees growing on the bed that obstruct even relatively low flows. The narrower anabranches appear to be self-cleansing of growing trees and large woody debris (Wende and Nanson, 1998).

Second, the depth of flow for a given total discharge is greater in the anabranched system (by as much as 18–28% from Tables 19.1 and 19.2) and roughness is known to decline considerably with increased flow depth (Barnes, 1967; Hicks and Mason, 1991), although there can be exceptions where dense bank vegetation is involved. Third, a 25% reduction in width (based on selected field observations in central Australia and used in the calculations in Tables 19.1 to 19.3) results in a marked decline in the total wetted perimeter and an increase in hydraulic radius (R). This is true even where the flow is divided into five channels (Tables 19.1 and 19.2), although there is an increase in boundary area and a reduction in R with an increasing number of anabranches. Finally, detailed measurements show no increase in slope from single-thread to anabranched reaches of channel in central Australia despite a significant drop in total width (Tooth, 1997). With no increase in slope, a 25% drop in width, and an increase in roughness, flow continuity could only be maintained with a very substantial increase in flow depth, a variable already shown to be negatively associated with roughness. In summary, it appears unlikely that there would be a substantial increase in total roughness in the central or northern Australian anabranching systems considered here. However, complex roughness changes are very likely to be part of an overall adjustment of flow variables that act to restrain the very considerable apparent increase in flow velocity and transport efficiency demonstrated in the majority of computations from this simple model (Tables 19.1 to 19.3).

WHY DO MOST RIVERS NOT ANABRANCH?

Nanson and Knighton (1996) proposed that anabranching rivers are formed by two sets of fundamentally different processes: avulsion (involving erosion) and accretion. They illustrated the existence of six anabranching types and argued that these occur across a wide range of environments due to a number of mechanisms including highly variable flow regimes, resistant banks, factors leading to flow displacement (e.g. channel sedimentation or the formation of vegetation or ice jams), and tectonism. While suggesting that each of these may be important in certain situations, they concluded with the proposal examined in detail here that the primary advantage of anabranching is that such rivers can maintain or increase water and sediment throughput without increasing gradient. If this is indeed such an advantage, then why do most rivers not anabranch? Yet, as Nanson and Knighton (1996) observed, anabranching rivers are sufficiently uncommon worldwide to suggest that they result from an unusual combination of flow and sediment conditions.

It appears that rivers can most readily adjust sediment transport conditions, and thereby maintain grade or equilibrium, by making minor adjustments to channel form (by altering width) and/or slope (by altering sinuosity). Only where such changes cannot be made to enhance the efficiency of a single channel will the system anabranch, for this requires a substantial change in channel form (width/depth ratio and numbers of channels) that can only be achieved where stream banks are reinforced by cohesive sediment or suitable riparian vegetation. The actual mechanisms involved in making more than one channel will be either by flow avulsing from an inefficient and partially clogged single channel to form a multiple-channel system across the floodplain (Schumann, 1989; Smith et al., 1989; Brizga and Finlayson, 1990), or by the deposition of sediment in the form of islands or ridges in a relatively inefficient single channel, thereby concentrating flow energy (Tooth, 1997; Wende and Nanson, 1998). Once formed, these multiple channels will continue to operate unless they become inefficient and atrophy. In some cases, multiple channel systems have clearly been operating for thousands of years (Nanson et al., 1986) whereas others are short-lived and resort to a single channel again within a few years (Brizga and Finlayson, 1990; Schumm et al., 1996).

AUSTRALIA'S ANABRANCHING RIVERS

Uncommon as they are worldwide, anabranching rivers are widely represented in Australia thereby providing a key to a better understanding of why they exist at all. The continent's two largest drainage basins, the Murray–Darling and Lake Eyre basins, exhibit many anabranching rivers, as do numerous smaller basins elsewhere on the continent. The explanation appears to be as follows. First, Australia is an old and unusually flat landmass with an average elevation of only *c.* 340 m above sea level, less than half the world average. As a consequence, away from their headwaters, Australian rivers have little opportunity to increase their gradients if required to. Second, with the exception of the globe's ice-covered landmasses, Australia is the driest and many of its rivers exhibit declining discharges with increasing sediment accumulation and storage downstream. Third, owing to low gradients and prolonged geochemical weathering, much alluvium is fine-grained and cohesive, particularly in the Channel Country of western Queensland and parts of the Murray–Darling basin where anabranching rivers abound. Finally, even in the most arid regions, a variety of riparian-adapted trees preferentially locate along stream banks and, in association with cohesive mud, act to stabilize the channels. As these rivers commonly have very low flows or are dry, trees tend to establish well down the banks, protecting them even at the base. This is in contrast to large rivers in humid regions where the lower banks, subaqueous for prolonged periods, are often relatively unprotected by vegetation. On a continent where streams must attempt to maintain even limited sediment export over low-gradient basins with extensive sediment stores and with discharges that decline downstream, it appears that anabranching is commonly the most effective channel type, but only where cohesive alluvium and/or suitable riparian vegetation permit stable, low width/depth ratio channels to form. Even in the sand-dominated areas of central and northern Australia, a type of anabranching system termed "ridge-forming" occurs where the ridges are stabilized by riparian vegetation (Nanson and Knighton, 1996; Wende and Nanson, 1998).

This interpretation of Australia's anabranching rivers refines our understanding of anabranching in general. While Nanson and Knighton (1996) proposed up to four causative mechanisms, these appear to be mostly subsidiary to the primary requirement to maintain water and sediment throughput without recourse to increasing gradient but under conditions where channel width/depth ratios can be reduced due to the formation of erosion-resistant banks.

The above explanation notwithstanding, it appears that not all anabranches are adjusted to move water and sediment efficiently downstream. In very large systems, such as occur in the Channel Country of Australia (Nanson et al., 1986, 1988), some channels appear to be formed to relocate and store water and sediment on quite diverse parts of the floodplain. As a consequence, the distal portions of such networks will atrophy and

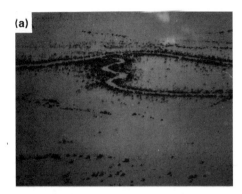

FIGURE 19.2   Examples of anabranching channels in the Channel Country of western Queensland, Australia. (a) Two converging anabranches at the upper end of Meringhina Waterhole near Durham Downs on Cooper Creek. The photo was taken during the 1990 flood with only the trees defining the channel positions. (b) Secondary anabranching channels starting to form reticulate channels on Cooper Creek near South Galway. (c) Fully developed reticulate channels near Davenport Downs on the Diamantina River

become buried in a gradually accreting floodplain, only to be replaced by subsequent anabranching systems. Rivers like the Cooper and Diamantina, which characterize the Channel Country of western Queensland, consist of complex multiple-channel systems that exhibit several different "levels" of anabranching. There are usually two or three dominant, commonly low-sinuosity anabranches (Figure 19.2a), but there is also a secondary system of smaller, often sinuous anabranching channels inset within the floodplain; these, like the primary channels, carry flows up to bankfull (Figure 19.2b) (Rust and Nanson, 1986; Knighton and Nanson, 1994). In addition, there is a complex array of shallow floodplain channels that have their "take off points" and subsequent channel beds elevated well above the floor of the main anabranches. Some of these elevated channels are braid-like in planform and possess large width/depth ratios, while others are narrower with a reticulate planform (Figure 19.2c). During overbank flow, these act as distributary channels that relocate and store both water and sediment on disparate parts of floodplains that total many thousands of square kilometres in area. However, their stability and longevity imply that these flood-channels are more efficient than unchannelized overbank flow. Clearly, anabranching can occur across a wide variety of fluvial environments and, even in one river system, can exist at more than one scale and have more than one function.

CONCLUSION

The persistence of anabranching systems both spatially and temporally indicates that, in certain situations, they must have hydraulic advantages over equivalent single-thread rivers. Under conditions where gradients cannot be increased, or indeed where they may have been locally slightly reduced, or where additional sediment load is added to the system or water removed, the sediment flux can be maintained or increased by an increase in the rate of work being done per unit area of the bed. From flow continuity and flow resistance determinations for both single-channel and multiple-channel systems, it is demonstrated that for a given flow discharge, channel slope, sediment size (roughness) and number of channels, channel form can adjust in a curvilinear relationship to accommodate the requirements of water and sediment conveyance (Figure 19.1). Because natural alluvial channels have width/depth ratios great than two, then where velocity must be controlled largely or entirely by channel form, Figure 19.1 indicates, first, that velocity responds in direct proportion to depth yet in inverse proportion to aggregate width, and second, that the most hydraulically efficient channel form for the transport of water and sediment will be achieved by an anabranching rather than an equivalent, wide, single-thread system. From application of the continuity and flow resistance equations, Table 19.1 shows that, without a significant change in flow resistance, velocity increases when a river anabranches. Even if flow resistance increases by 10%, it appears that velocity will still increase in the anabranching section. The application of sediment transport functions in Table 19.3 shows that, despite a loss in aggregate width, an anabranching system can increase or at least maintain its sediment transport. This analysis suggests that, in rivers where there is little opportunity to adjust channel slope, maintenance of sediment flux can be achieved through the mutual adjustment of the number of channels and channel form (width/depth ratio).

The above analyses show how anabranching rivers can represent a balance between channel form, and sediment and water transport; the persistent and yet dynamic nature of some anabranching systems (Knighton and Nanson, 1993; Nanson and Knighton, 1996) show them to be one of a variety of long-lived equilibrium river conditions. However, not all anabranching systems may be so. Once formed some may, due to inertia, continue to operate despite increasing inefficiency and channel atrophication with time. Others may form as temporary distributaries for dispersing and storing water and sediment across extensive low-gradient floodplains, with channels switching to other locations as sediment builds up.

Australia provides a continental setting conducive to the formation of anabranching rivers. Very limited relief and declining flows that increase sedimentation downstream require anabranching systems for the maximum conveyance of water and sediment, and erosion-resistant banks enable such stable systems to form. Multiple channels are often associated with sediment deposition and storage and, where this consists of readily erodible material, vegetation is usually an important stabilizer of the boundary.

ACKNOWLEDGEMENTS

We are particularly grateful to Michael Church who made numerous helpful suggestions for revising the analyses and interpretations that underlie this paper, and to Rainer Wende and Stephen Tooth, who willingly discussed their ideas and reviewed drafts of the manuscript.

ENDNOTES

(1) Brownlie (1983) established explicit relations for easily determining the influence of bed forms on flow resistance from a regression analysis of numerous observations. For the lower flow regime, the regression relation is:

$$\frac{R}{d} = 0.3724 \left(\frac{q}{\sqrt{gd^3}} \right)^{0.6539} S^{-0.2542} \sigma_g^{0.1050}$$

and for the upper flow regime, it is:

$$\frac{R}{d} = 0.2836 \left(\frac{q}{\sqrt{gd^3}} \right)^{0.6248} S^{-0.2877} \sigma_g^{0.10801}$$

where q, d and σ_g are flow discharge per unit width ($= Q/W$, W is channel surface width), median sediment size (i.e. d_{50}) and sediment geometric standard deviation, respectively. When σ_g is dropped by integrating its average values into the constants in the above two equations for R/d with the knowledge that $1 < \sigma_g < 5$, that is $1 < \sigma_g^{0.0801} < 1.1375$ and $1 < \sigma_g^{0.1050} < 1.1840$, Bruschin (1985) found that the above equations for R/d can be written in the form of the Darcy–Weisbach flow relation (Equation 19.2a), leading f_l and f_u to have the expressions presented in Equation 19.2b.

(2) Because $V \propto R^i \propto (A/P)^i$ and $Q = AV \propto A^{1+i}/P^i$ ($i = 2/3$, $9/17$ and $5/9$, respectively for the expressions of f_p, f_l and f_u in Equation 19.2b), $P \rightarrow minimum$ for constant A leading to $Q \rightarrow maximum$. On the other hand, $Q = AV = PRV$ leading to $V = Q/(PR) \propto Q/(PV^{1/i})$, thus $V \propto (Q/P)^{i/(1+i)}$ and $P \rightarrow minimum$ for constant Q resulting in $V \rightarrow maximum$.

(3) Firstly deriving $Q \sim D$ or $Q \sim W$ relationship by combining Equations 19.1, 19.2a, 19.2b and 19.2c together to remove V, and then back to find V using Equations 19.1 and 19.2c.

(4) Firstly deriving $R \sim R(Q; W/D)$ relationship by combining Equations 19.1, 19.2a, 19.2b and 19.2c together to remove V, and then back to find V using Equations 19.1 and 19.2c.

REFERENCES

ASCE Task Committee. 1982. Relationships between morphology of small streams and sediment yield. *Journal of the Hydraulics Division, ASCE*, **108**, 1329–1365.

Bagnold, R.A. 1977. Bedload transport by natural rivers. *Water Resources Research*, **13**, 303–312.

Bagnold, R.A. 1980. An empirical correlation of bedload transport rate in flumes and natural rivers. *Proceedings of the Royal Society*, London, England, Series A, **372**, 453–473.

Barnes, H.H. 1967. Roughness characteristics of natural channels. *US Geological Survey Water-Supply Paper* 1849, 213 pp.

Blench, T. 1957. *Regime Behaviour of Canals and Rivers*. Butterworths Scientific Publications, London.

Blench, T. 1969. *Mobile-Bed Fluviology*, 2nd revised edition. University of Alberta Press, Canada.

Brizga, S.O. and Finlayson, B.L. 1990. Channel avulsion and river metamorphosis: the case of the Thompson River, Victoria, Australia. *Earth Surface Processes and Landforms*, **15**, 391–404.

Brown, C.B. 1950. Sediment transportation. In H. Rouse (Ed.), *Engineering Hydraulics*, Wiley, New York, Chapter 12.

Brownlie, W.R. 1983. Flow depth in sand–bed channels. *Journal of Hydraulic Engineering*, **109**, 959–990.

Bruschin, J. 1985. Flow depth in sand-bed channels – Discussion. *Journal of Hydraulic Engineering*, **111**, 736–739.

Carson, M.A. and Griffiths, G.A. 1987. Influence of channel width on bed load transport capacity. *Journal of Hydraulic Engineering*, **113**, 1489–1509.

Chang, H.H. 1979. Geometry of rivers in regime. *Journal of the Hydraulics Division, ASCE*, **105**, 691–706.

Chang, H.H. 1988. *Fluvial Processes in River Engineering*, Wiley, New York.

Chorley, R.J. 1962. Geomorphology and general systems theory. *US Geological Survey Professional Paper* 500–B.

Chow, V.T. 1959. *Open Channel Hydraulics*. McGraw-Hill, New York.

Colby, B.R. 1964. Sand discharge and mean-velocity relationships in sandbed streams. *US Geological Survey Professional Paper* 462A, 47 pp.

Gilbert, G.K. 1914. The transportation of debris by running water. *US Geological Survey Professional Paper* 86.

Hack, J.T. 1957. Studies of longitudinal stream profiles in Virginia and Maryland. *US Geological Survey Professional Paper* 294B, 97 pp.

Henderson, F.M. 1966. *Open Channel Flow*. MacMillan Publishing Co., New York.

Hicks, D.M. and Mason, P.D. 1991. *Roughness Characteristics of New Zealand Rivers*. Water Resources Survey, DSIR, Marine and Fresh Water, Wellington, 327 pp.

Karim, M.F. and Kennedy, J.F. 1990. Menu of coupled velocity and sediment–discharge relations for rivers. *Journal of Hydraulic Engineering*, **116**, 978–996.

Knighton, A.D. and Nanson, G.C. 1993. Anastomosis and the continuum of channel pattern. *Earth Surface Processes and Landforms*, **18**, 613–625.

Knighton, A.D. and Nanson, G.C. 1994. Waterholes and their significance in the anastomosing channel system of Cooper Creek, Australia. *Geomorphology*, **9**, 311–324.

Lacey, G. 1929. Stable channels in alluvium. *Proceedings of the Institute of Civil Engineers*, London, **229**(1), 259–292.

Lacey, G. 1933. Uniform flow in alluvial rivers and canals. *Proceedings of the Institute of Civil Engineers*, London, **237**, 421–453.

Lacey, G. 1946. A general theory of flow in alluvium. *Proceedings of the Institute of Civil Engineers*, London, **27**, 16–47.

Lane, E.W. 1953. Design of stable channels. *Proceedings of the American Society of Civil Engineers*, **79**, 280.1–280.31.

Lane, E.W. 1955. The importance of fluvial morphology in hydraulic engineering. *Proceedings of the American Society of Civil Engineers*, **81**, 1–17.

Laursen, E.M. 1956. Application of sediment-transport mechanics to stable-channel design. *Journal of the Hydraulics Division, ASCE*, **82**, 1034.1–1034.11.

Leopold, L.B. and Maddock, T. Jr 1953. The hydraulic geometry of stream channels and some physiographic implications. *US Geological Survey Professional Paper* 252.

Leopold, L.B. and Wolman, M.G. 1957. River channel patterns – braided, meandering and straight. *US Geological Survey Professional Paper* 282B, 39–85.

Mackin, J.H. 1948. Concept of the graded river. *Bulletin of the Geological Society of America*, **59**, 463–512.

Nanson, G.C. and Knighton, A.D. 1996. Anabranching rivers: their cause, character and classification. *Earth Surface Processes and Landforms*, **21**, 217–239.

Nanson, G.C., Rust, B.R. and Taylor, G. 1986. Coexistent mud braids and anastomosing channels in an arid–zone river: Cooper Creek, central Australia. *Geology,* **14**, 175–178.

Nanson, G.C., Young, R.W., Price, D.M. and Rust, B.R. 1988. Stratigraphy, sedimentology and Late Quaternary chronology of the Channel Country of western Queensland. In R.F. Warner (Ed.), *Fluvial Geomorphology of Australia*. Academic Press, Sydney, 151–175.

Nevins, T.H.F. 1969. River–training: the single–thread channel. *New Zealand Engineering*, **15**, 367–373.

Parker, G. 1979. Hydraulic geometry of active gravel rivers. *Journal of the Hydraulics Division, ASCE*, **105**, 1185–1201.

Pickup, G. 1976. Alternative measures of river channel shape and their significance. *Journal of Hydrology (New Zealand)*, **15, 9–16.**

Rubey, W.W. 1952. Geology and mineral resources of the Hardin and Brussels quadrangles (in Illinois). *US Geological Survey Professional Paper* 218, 175pp.

Rust, B. R. and Nanson, G. C. 1986. Contemporary and palaeochannel patterns and the Late Quaternary stratigraphy of Cooper Creek, southwest Queensland, Australia. *Earth Surface Processes and Landforms*, **11**, 581–590.

Schumann, R. 1989. The morphology of Red Creek, Wyoming, an arid–region anastomosing channel system. *Earth Surface Processes and Landforms*, **14**, 277–288

Schumm, S.A., Erskine, W.D. and Tileard, J.W. 1996. Morphology, hydrology and evolution of the anastomosing Ovens and King Rivers, Victoria, Australia. *Geological Society of America, Bulletin*, **108**, 1212–1224.

Simons, D.B. and Albertson, M.L. 1960. Uniform water conveyance channels in alluvial materials. *Journal of the Hydraulics Division, ASCE,* **86**, 33–71.

Smith, N.D., Cross, T.A., Dufficy, J.P. and Clough, S.R. 1989. Anatomy of an avulsion. *Sedimentology*, **36**, 1–23.

Tooth, S. 1997. *The morphology, dynamics and late Quaternary sedimentary history of ephemeral drainage systems on the Northern Plains of central Australia.* Unpublished PhD thesis, University of Wollongong, Australia.

Wende, R. and Nanson, G.C. 1998. Anabranching rivers: ridge–forming alluvial channels in tropical northern Australia. *Geomorphology*, **22**, 205–224.

White, W.R., Bettes, R. and Paris, E. 1982. Analytical approach to river regime. *Journal of the Hydraulics Division, ASCE*, **108**, 1179–1193.

Wolman, M.G. and Brush, L.M. 1961. Factors controlling the size and shape of stream channels in coarse uncohesive sands. *US Geological Survey Professional Paper* 282G, 183–210.

Yang, C.T. 1971. On river meanders. *Journal of Hydrology*, 13, 231–253.

Yang, C.T. 1973. Incipient motion and sediment transport. *Journal of the Hydraulics Division, ASCE*, **99**, 1679–1704.

Yang, C.T. 1976. Minimum unit stream power and fluvial hydraulics. *Journal of the Hydraulics Division, ASCE*, **102**, 919–934.

Yang, C.T. and Wan, S. 1990. Comparisons of selected bed-material load formulas. *Journal of Hydraulic Engineering*, **117**, 973–989.

Part 5

Review Papers

20

Varieties of Fluvial Form: the Relevance to Geologists of an Expanded Reality

CHRISTOPHER R. FIELDING

Department of Earth Sciences, University of Queensland, Australia

INTRODUCTION

The geomorphological and sedimentological literature on fluvial environments is strongly biased towards studies of perennial streams in humid, temperate climate areas, notably in North America and Europe. Fluvial environments representative of more extreme climates, despite covering significant proportions of the present Earth's landmasses, are comparatively poorly understood. If the present disposition of the continents is any guide, then such systems are likely to be well represented in the geological record, yet comparatively few data are available to assist geologists in recognizing their deposits. The contributions in this volume emphasize some of the more extreme varieties of fluvial form that are under-represented in the research literature. By drawing attention to such systems, I hope that the volume will stimulate research activity in this direction.

In this review article, I will address the relevance of the papers presented herein to the geological community, specifically their role in improving the veracity of facies models where appropriate, and in contributing to a better understanding of alluvial stratigraphy. I will then make some more general remarks concerning the directions of present and future research, and on the interrelationships between sedimentological and geomorphological research activities.

In entertaining the papers presented within the volume, I will categorize them somewhat arbitrarily on the basis of climate, as the role of climate seems to be a recurrent theme among the fluvial systems described. There are papers concerning systems on every continent (and beyond!), although the greatest number of papers is from Australia, reflecting the large number of submissions from that source at the Singapore Conference. Together, all these contributions make a powerful statement about the varieties of fluvial form that are less prominent in the minds of scientists worldwide.

Varieties of Fluvial Form. Edited by A.J. Miller and A. Gupta
© 1999 John Wiley & Sons Ltd

HYPER-ARID CLIMATE SYSTEMS

Three papers in the volume concern fluvial and quasi-fluvial environments developed in the most extreme climatic conditions. Perhaps the most spectacular of these is the account of extra-terrestrial fluvial landforms by Baker and Komatsu. In reviewing NASA-derived remote sensing data from the Moon, Mars and Venus, the authors pose some important philosophical questions about the restricted scope of investigations into natural environments. Baker and Komatsu suggest quite reasonably that the study of such extra-terrestrial environments allows us to expand reality in a way that might improve our understanding of cataclysmic processes on Earth. Geologists are still seeking ways of understanding "Non-Uniformitarian" processes and their deposits; that is, those formed in conditions that for reasons of scale are unlikely to be experienced within the timespan of human historical records (e.g. Ager, 1993). In the paper, analogies are drawn between extra-terrestrial fluvial forms and the record of large-scale glacial lake outbursts such as in the western USA (Baker, 1978, 1996) and Siberia (Rudoy and Baker, 1993; Carling, 1996), and with those of submarine channels. The exact physical process of formation may differ between these features, but Baker and Komatsu argue that the scale is comparable.

Similar problems are posed in understanding the deposits of the hyper-arid Uniab River of the Skeleton Coast, Namibia, as described in the paper by Scheepers and Rust. At present, this system is only intermittently capable of breaking through an aeolian dunefield to reach the coast, although it is evident that large-scale flooding occurred within the recent geological past to form substantial sand and gravel deposits. Similar deposits in a nearby area (Kuiseb Silts) were previously described by Smith et al. (1993), and attributed to backflow into tributaries and embayments during Pleistocene flash floods. In both the Kuiseb and Uniab cases, the scale of contemporary processes cannot explain deposits stored within the present alluvial topography. Such inconsistencies have also been recorded by workers in central Australia, where the deposits of Pleistocene and early Holocene mega-floods have been recognized (Baker et al., 1983; Pickup et al., 1988; Bourke, 1995; see also the paper in this volume by Bourke and Pickup).

In a further variation on the theme, a series of alluvial fans that developed in response to seasonal snowmelt in coastal East Antarctica is described by Webb and Fielding. Both debris flow-dominated and sheetflood-dominated types are recognized and related to differences in physiographic context and sediment/water supply. The authors conclude that there are few features diagnostic of polar climates preserved within the observed deposits. The question of whether alluvial fan deposits preserve a record of the climate in which they form is controversial, with arguments both in the affirmative (e.g. Ritter et al., 1995) and negative (e.g. Blair and McPherson, 1994).

ARID TO SUBHUMID CLIMATE SYSTEMS

A series of papers in this volume addresses aspects of fluvial systems in arid to subhumid areas, where rainfall is modest in amount, intermittent in occurrence and unpredictable in all respects. Comprising studies of rivers in central Australia, India, southern Africa, central America and western USA, the group of papers provides what is probably a representative view of this genre.

Three papers examine aspects of canyon-confined rivers, in the USA, Mexico and South Africa. Grams and Schmidt examine the relationships between riverbed level geology, tributary sediment delivery and channel character within the Green River as it passes through part of the Colorado Plateau in western USA. They conclude that there is a strong relationship between the three parameters, and draw attention to the depositional environments created by the confluences of tributary fans with the main channel. Hattingh and Rust describe a similar geomorphic setting in which the post-Late Miocene Sundays River of South Africa has incised a spectacular flight of alluvial terraces in response to variations in sea level, climate and tectonic activity. In this study, variations in alluvial sediment calibre from coarse- to fine-grained through time are related qualitatively to climatically driven eustatic overprinting of a long-term tectonic uplift. Studies such as this provide useful tests of as yet largely theoretical models that seek to predict the geological consequences of changes in various extrinsic factors such as tectonic uplift/subsidence and sea-level fluctuation. Palacios and Chávez examine the geomorphology of a canyon-confined system in Mexico that has experienced both tectonic and volcanic activity, and interpret the long-term history of this valley in terms of periodic tectonic upheaval and volcanism.

Rivers that are bedrock-floored or partly bedrock-floored but not as confined as those discussed above are also represented in papers by Wende, Van Niekirk et al., Heritage et al., Gupta et al. and Deodhar and Kale. Wende describes boulder bedforms developed in bedrock-floored, flash-discharge streams in northwestern Australia, where large platy clasts accumulate as cluster bedforms (cf. Brayshaw, 1984) against upstream-facing rock steps formed by erosion of strongly jointed bedrock. Van Niekirk et al. and Heritage et al. describe patterns of initial sediment accumulation in bedrock-floored, anabranching channels of the Sabie River, South Africa. The authors note a feedback mechanism operating between sediment accumulation and channel-bed vegetation, a relationship that has also been recognized in several streams in subhumid to semi-arid parts of Australia (Fielding and Alexander, 1996; Fielding et al., 1997; see also papers in this volume by Bourke and Pickup, and Tooth). Heritage et al. propose a model for progressive transformation of the multithread river system by aggradation of sediment, via a space-for-time substitution. Studies such as this provide new insights into the initial patterns of sediment accumulation in incised valleys such as might be formed during tectonic uplifts or drops in sea level.

Large, partly bedrock-floored rivers in subhumid parts of India are described by Gupta et al. and Deodhar and Kale. Gupta et al. provide a descriptive analysis of the Narmada River, a system which is affected by both occasional, high-magnitude floods and by seasonal, monsoonal rainfall. The river has developed an inset channel geometry, with a small misfit channel nested within a larger channel that is filled regularly by monsoon-related discharge, in turn set within a broader valley that is filled only during high-magnitude floods. Recent studies of palaeoflood deposits by Baker et al. (1995) and Ely et al. (1996) have provided a longer-term view of this regime of punctuated, high-magnitude events. Gupta et al. suggest that the fluvial style of the Narmada is distinctive of Indian rivers in general. The characteristics exhibited by these systems are, however, well developed in numerous rivers in northern Australia, southern Africa, central Asia and perhaps parts of western USA; these are all areas where rainfall is strongly seasonal giving rise to pronounced variations in discharge ("tropical, variable-discharge systems" of Fielding and Alexander, 1996). I would suggest, therefore, that this distinctive class of river owes its character to its climatic context rather than geographical location. It is, furthermore, likely that facies models for this type of river, currently lacking in the literature, will be developed in the near future as more data come to light on their deposits and internal architecture.

Several other papers describe alluvial or partly alluvial systems within this climatic range. Some of these systems are examples of what Deodhar and Kale term "allochthonous rivers", whereby the discharge of the system reflects the climate of a distant headwater region rather than the (downstream) area under investigation. Other examples of similar situations are abundant across the drier continents (e.g. Grimes and Doutch, 1978; McCarthy et al., 1991); indeed, it seems inevitable that such situations will arise where mountain ranges fringe semi-arid to subhumid landscapes. There are important implications here for geological investigations of reddened alluvial successions, where there is often a paradox of large, channelized sediment bodies formed apparently by sediment accumulation in fast-flowing big rivers found interbedded with interfluve strata indicative of slow and episodic accumulation under hot, dry climatic conditions (e.g. Allen, 1979; Allen and Williams, 1979).

Aspects of Australian semi-arid zone fluvial systems are detailed in papers by Nanson and Huang, Bourke and Pickup, and Tooth. Nanson and Huang provide a mechanical analysis of the group of multi-thread rivers termed "anabranching" by Nanson and Knighton (1996). They conclude that anabranching occurs as a means of reducing boundary resistance and hence maintaining water and sediment throughput in fluvial systems where there is little opportunity to increase channel slope. Certainly, anabranching systems of various types are very common across the low-gradient landscapes of central Australia, and elsewhere (see, for example, papers in this volume by Heritage et al. and Van Niekirk et al.). Are the deposits of such multichannel rivers therefore likely to be disproportionately well represented in the geological record, given that most preserved fluvial deposits reflect systems that were active in low-gradient, basinal settings?

Bourke and Pickup provide an overview of the fluvial geomorphology of the Todd River in central Australia. This study illustrates the futility of applying concepts of dynamic equilibrium to rivers in arid to subhumid climatic zones, and the authors appropriately describe the Todd River as a "non-equilibrium" system (cf. Graf, 1988). Spatial and temporal variability in flow patterns, sediment deposition and channel planform are characteristics of this and other such systems.

Another aspect of arid-zone systems, that of "floodouts", is described by Tooth. Such features occur where channel capacity is reduced abruptly, causing flows to become unconfined and to spread across the alluvial surface. Some such floodouts are replaced downstream by a return to normal channel activity, whereas others are terminal, marking the downstream terminus of a stream. Tooth investigates the causes of this phenomenon, and concludes that various forms of natural barrier are responsible. Although Tooth regards floodouts as distinct from the "terminal fans" of Friend (1978) and Kelly and Olsen (1993), at least partial analogy is suggested. Terminal river systems are evidently common in arid-zone landscapes, some displaying a distributary channel form at their downstream end (e.g. Abdullatif, 1989). It is this property that has evidently led to use of the term "fan", a term that is quite inappropriate for many systems such as the ones described by Tooth that do not show such a fan morphology. Tooth's work illustrates yet more variability in nature that has not been properly documented hitherto!

HUMID CLIMATE SYSTEMS

A smaller group of papers address aspects of fluvial systems developed in humid climate situations. These range from studies of inland rivers to coastal plain systems, and cover a variety of geographical locations.

Three papers describe geomorphological aspects of rivers in Australia and Papua New Guinea. Erskine and Livingstone provide an analysis of the in-channel benches that are common in many eastern Australian rivers, and conclude that their development and consistent elevation relative to channel floor are related to flows of discrete return periods. This phenomenon may again be related to the considerable variability in discharge experienced by almost all Australian rivers, and may indeed be recognizable in other areas experiencing similar climatic conditions. Ferguson and Brierley investigate the relationships between floodplain development and valley confinement in the Tuross River of southern New South Wales, and find that a strong control evident in inland reaches breaks down in the less confined coastal plain. They also provide some interesting documentation of the surface arrangement of geomorphic elements in the Tuross River, which will be of interest to those studying the architecture of alluvial strata. Dietrich et al. provide some useful new insights into the factors controlling the recent history of the Fly River of Papua New Guinea. In contrast to some previous studies, they conclude that the well developed scroll topography preserved on the Fly River floodplain is unlikely to be related to rapid channel migration. This is particularly interesting when the river is considered in its tectonic context, as an axial drainage system into a rapidly subsiding foreland basin.

The dynamics of water and sediment movement at a channel confluence in Sri Lanka are the topic of a paper by Kennedy. This study provides new data on a topic that is of considerable relevance to sedimentologists (e.g. Bristow et al., 1993). In her analysis, Kennedy has chosen a situation where the flow in a tributary is much smaller than that of a trunk stream, and concludes that there is considerable similarity in process to that experienced by mesotidal estuaries. Another paper with a coastal flavour, by Chen, discusses the geomorphology of the Yangtze River delta plain in eastern China. This research, aspects of which are also documented elsewhere (e.g. Chen and Stanley, 1995; Stanley and Chen, 1996), provides a detailed geomorphological and sedimentological database on one of the modern world's major delta systems. In particular, the arrangement, sedimentation and migration patterns of tidally influenced rivers are of great relevance to understanding the facies architecture in ancient deltas.

DISCUSSION

The papers in this volume draw attention to the styles of fluvial system that occupy arid to subhumid climatic landscapes, both in mountainous terrain and in lowland settings. As mentioned previously, comparably little is known of the deposits of such rivers relative to those of streams in humid, temperate climate situations. There have been recent developments in the understanding of some big tropical rivers such as the Fly (Dietrich et al.), Ganges/Brahmaputra (Bristow, 1987, 1993; Thorne et al., 1993), Kosi (Singh et al., 1993) and Amazon (Mertes and Dunne, 1988; Mertes et al., 1996). Our understanding of anabranching rivers is improving rapidly thanks largely to the work of Nanson and co-workers, and some others (e.g. Schumm et al., 1996), in Australia. Yet, there are very few data on what is probably an equally important fluvial milieu, the group of large, variable-discharge or extreme-discharge, tropical rivers fed by cyclonic rain. There are no facies models as yet that can account for the deposits of rivers such as the Narmada in India (Gupta et al.), Sabie and Sundays in South Africa (Heritage et al., Van Niekirk et al., Hattingh and Rust), or others like them. Yet such rivers will characterize any tropical climate area subject to strongly seasonal rainfall where there is an elevated hinterland that experiences elevated levels of precipitation.

Tropical, extreme-discharge rivers such as the Burdekin (Fielding and Alexander, 1996) and Pioneer (Gourlay and Hacker, 1986) of northeast Australia are generally single-channel systems, straight to moderately sinuous (<1.7), locally bedrock-floored and evidently prone to rapid stripping of alluvium. These rivers experience up to two to four orders of magnitude daily discharge variations. Major flow events arrive as physical floodwaves, and floodwaters may travel at velocities in excess of 10 m s^{-1}. Such flows are capable of entraining coarse gravel, and can transport huge volumes of silt and sand in possibly quite dense suspensions. River beds are sand and gravel-dominated, and preserve high-energy bedforms such as gravel antidunes (Alexander and Fielding, 1997) as a consequence of the rapid falls in stage following major flows. Hopefully, the studies presented in this volume and the others quoted will stimulate further research on this important but neglected class of river.

It is also worth considering the broader context of facies and stratigraphic modelling of fluvial deposits in general. The early fluvial facies models based purely on vertical sequences have been shown to be hopelessly inadequate and counter-productive, in that they caused unreasonable categorization of fluvial deposits, in many cases where there was insufficient information to make a scientifically valid interpretation. In the past decade, it has become clearer just how much is not known about the activities of rivers and their resulting deposits. Concurrently, there has been a trend towards reconstructing ancient fluvial systems from an understanding of their internal deposit geometry and facies arrangements. In a recent paper, Brierley (1996) has coined the term "constructivist" for this approach, although it is clear to the present writer that sedimentologists have been using such an approach at least since the early 1980s. The major stumbling block for sedimentologists in adopting such a methodology, noted by Brierley (1996), is the paucity of detailed facies studies of modern rivers. How can we hope to diagnose the deposits of ancient fluvial environments as arising from one or other planform type when we do not yet know what those deposits look like in the modern environment?

One recent development that has the potential to rapidly improve knowledge of facies distribution and arrangement (or architecture) in surface environments is the increased use of ground-penetrating radar (GPR). This technique, which allows the rapid acquisition of subsurface imagery at a scale suitable for sedimentological study, has allowed possibly the first rigorous test of facies models for point bars (Bridge et al., 1995) and braided rivers (Bridge et al., 1998), as well as providing other, very useful data (e.g. Gawthorpe et al., 1993; Huggenberger, 1993; Van Overmeeren, 1997). When combined with core drilling (which provides actual lithological data, as opposed to geophysical imagery), this technique allows three-dimensional coverage of a modern system to a level of detail not attainable by pure drilling or other means. Only by use of techniques such as this, which allow rapid acquisition of data over significant areas/volumes, can we hope to adequately characterize the deposits of modern rivers in a way meaningful to future geological investigations.

Equally, in order for studies of surface environments to be relevant to geologists, they must evaluate the preservation potential of different fluvial deposits and systems. What we see at the surface today reflects only one set of climatic, physiographic and, importantly, tectonic conditions. Computer modelling studies of alluvial stratigraphy (Allen, 1978; Leeder, 1978; Bridge and Leeder, 1979; Bridge and Mackey, 1993; Mackey and Bridge, 1995) are therefore critically important to geologists as they incorporate the third dimension, controlled by the rate of subsidence or accommodation. Geologists are now beginning to find settings in which to test these models against real data (Leeder et al., 1996). As geologists, we must ask what an ancient fluvial deposit will look like given a subsidence rate of, for example, 100 mm per 1000 years (typical of slowly subsiding cratonic areas), as opposed to one formed during a subsidence regime of 1000 mm per 1000 years (such as might be encountered in a tectonically active rift, foreland or transtensional basin). Clearly, in slowly subsiding areas there is the strong possibility of preserving within a discrete stratigraphic interval the record of several phases of fluvial activity, each recording different climatic or other conditions (see Page and Nanson (1996) for an excellent example).

Brierley (1996, p.263) writes "The challenge to fluvial sedimentologists is to generate a suite of facies models which covers the spectrum of river styles, while each model must retain specific characteristics which define a particular river type". I would suggest, rather, that the challenge is for all of us, sedimentologists and geomorphologists alike, to work towards a fuller understanding of fluvial processes and products in different settings, which may then be used as a foundation for the next generation of fluvial facies models.

REFERENCES

Abdullatif, O.M. 1989. Channel–fill and sheet-flood facies sequences in the ephemeral terminal River Gash, Kassala, Sudan. *Sedimentary Geology*, **63**, 171–184.

Ager, D. 1993. *The New Catastrophism: the importance of the rare event in geological history.* Cambridge University Press, Cambridge, 231 pp.

Alexander, J. and Fielding, C.R. 1997. Gravel antidunes in the tropical Burdekin River, Queensland, Australia. *Sedimentology*, **44**, 327–338.

Allen, J.R.L. 1978. Studies in fluvial sedimentation: an exploratory quantitative model for the architecture of avulsion–controlled alluvial suites. *Sedimentary Geology*, **21**, 129–147.

Allen, J.R.L. 1979. Old Red Sandstone facies in external basins, with particular reference to southern Britain. *Special Papers in Palaeontology*, **23**, 65–80 (Palaeontological Association, London).

Allen, J.R.L. and Williams, B.P.J. 1979. Interfluvial drainage on Siluro–Devonian alluvial plains in Wales and the Welsh Borders. *Journal of the Geological Society*, **136**, 361–366.

Baker, V.R. 1978. Large–scale erosional and depositional features of the Channeled Scabland. In V.R. Baker and D. Nummedal (Eds), *The Channeled Scabland*. National Aeronautics and Space Administration, Washington, DC, 81–115.

Baker, V.R. 1996. Megafloods and glaciation. In I.P. Martini (Ed.), *Late Glacial and Postglacial Changes: Quaternary, Carboniferous-Permian and Proterozoic.* Oxford University Press, Oxford, 98–108.

Baker, V.R., Pickup, G. and Polach, H.A. 1983. Desert palaeofloods in central Australia. *Nature,* **301**, 502–504.

Baker, V.R., Ely, L.L., Enzel, Y. and Kale, V.S. 1995. Understanding India's rivers: Late Quaternary palaeofloods, hazard assessment and global change. In S. Wadia, R. Korisettar and V.S. Kale (Eds), *Quaternary Environment and Geoarchaeology of India.* Geological Society of India, Memoir **32**, 61–77.

Blair, T.C. and McPherson, J.G. 1994. Alluvial fans and their natural distinction from rivers based on morphology, hydraulic processes, sedimentary processes, and facies assemblages. *Journal of Sedimentary Research,* **A64**, 450–489.

Bourke, M.C. 1995. Geomorphic effects of Holocene super-floods on the Todd River, semi–arid central Australia. In *XIV INQUA Congress, Proceedings,* Berlin (unpaginated).

Brayshaw, A.C. 1984. Characteristics and origin of cluster bedforms in coarse-grained alluvial channels. In E.H. Koster and R.J. Steel (Eds), *Sedimentology of Gravels and Conglomerates.* Canadian Society of Petroleum Geologists, Memoir **10**, 77–85.

Bridge, J.S. and Leeder, M.R. 1979. A simulation model of alluvial stratigraphy. *Sedimentology,* **26**, 617–644.

Bridge, J.S. and Mackey, S.D. 1993. A revised alluvial stratigraphy model. In M. Marzo and C. Puigdefabregas (Eds), *Alluvial Sedimentation.* International Association of Sedimentologists, Special Publication **17**, 319–336.

Bridge, J.S., Alexander, J., Collier, R.E. Ll., Gawthorpe, R.L. and Jarvis, J. 1995. Ground-penetrating radar and coring used to study the large-scale structures of point-bar deposits in three dimensions. *Sedimentology,* **42**, 839–852.

Bridge, J.S., Collier, R.E.Ll. and Alexander, J. 1998. Large-scale structure of Calamus River deposits revealed using ground-penetrating radar. *Sedimentology,* **45** (in press).

Brierley, G. 1996. Channel morphology and element assemblages: a constructivist approach to facies modelling. In P.A. Carling and M.R. Dawson (Eds), *Advances in Fluvial Dynamics and Stratigraphy.* Wiley, Chichester, 263–298.

Bristow, C.S. 1987. Brahmaputra River: channel migration and deposition. In F.G. Ethridge, R.M. Flores and M.D. Harvey (Eds), *Recent Developments in Fluvial Sedimentology.* Society of Economic Paleontologists and Mineralogists, Special Publication 39, 63–74.

Bristow, C.S. 1993. Sedimentary structures exposed in bar tops in the Brahmaputra River, Bangladesh. In J.L. Best and C.S. Bristow (Eds), *Braided Rivers.* Geological Society of London, Special Publication 75, 277–289.

Bristow, C.S., Best, J.L. and Roy, A.G. 1993. Morphology and facies models of channel confluences. In M. Marzo and C. Puigdefabregas (Eds), *Alluvial Sedimentation.* International Association of Sedimentologists, Special Publication, 17, 91–100.

Carling, P.A. 1996. Morphology, sedimentology and palaeohydraulic significance of large gravel dunes, Altai Mountains, Siberia. *Sedimentology,* **43**, 647–664.

Chen, Z. and Stanley, D.J. 1995. Quaternary subsidence and river channel migration in the Yangtze Delta plain, eastern China. *Journal of Coastal Research,* **11**, 927–945.

Ely, L.L., Enzel, Y., Baker, V.R., Kale, V.S. and Mishra, S. 1996. Changes in the magnitude and frequency of late Holocene monsoon floods on the Narmada River, central India. *Geological Society of America, Bulletin,* **108**, 1134–1148.

Fielding, C.R. and Alexander, J. 1996. Sedimentology of the upper Burdekin River of north Queensland, Australia – an example of a tropical, variable discharge river. *Terra Nova,* **8**, 447–457.

Fielding, C.R., Alexander, J. and Newman-Sutherland, E. 1997. Preservation of *in situ* arborescent vegetation and fluvial bar construction in the Burdekin River of north Queensland, Australia. *Palaeogeography, Palaeoclimatology, Palaeoecology,* **135**, 123–144.

Friend, P.F. 1978. Distinctive features of some ancient river systems. In A.D. Miall (Ed.), *Fluvial Sedimentology.* Canadian Society of Petroleum Geologists, Memoir 5, 531–542.

Gawthorpe, R.L., Collier, R.E.Ll., Alexander, J., Bridge, J.S. and Leeder, M.R. 1993. Ground-penetrating radar: application to sandbody geometry and heterogeneity studies. In C.P. North and D.J. Prosser (Eds), *Characterisation of Fluvial and Aeolian Reservoirs*. Geological Society of London, Special Publication 73, 421–432.

Gourlay, M.R. and Hacker, J.L.F. 1986. *Pioneer River Estuary – Sedimentation Studies*. Department of Civil Engineering, University of Queensland, Brisbane.

Graf, W.L. 1988. *Fluvial Processes in Dryland Rivers*. Springer-Verlag, New York, 343 pp.

Grimes, K.G. and Doutch, H.F. 1978. The late Cainozoic evolution of the Carpentaria Plains, North Queensland. *Bureau of Mineral Resources Australia, Journal of Geology and Geophysics*, 3, 101–112.

Huggenberger, P. 1993. Radar facies: recognition of facies patterns and heterogeneities within Pleistocene Rhine gravels. NE Switzerland. In J.L. Best and C.S. Bristow (Eds), *Braided Rivers*. Geological Society of London, Special Publication 75, 163–176.

Kelly, S.B. and Olsen, H. 1993. Terminal fans – a review with reference to Devonian examples. In C.R. Fielding (Ed.), *Current Research in Fluvial Sedimentology, Sedimentary Geology*, **85**, 339–374.

Leeder, M.R. 1978. A quantitative stratigraphic model for alluvium, with special reference to channel body density and interconnectedness. In A.D. Miall (Ed.), *Fluvial Sedimentology*. Canadian Society of Petroleum Geologists, Memoir, 5, 587–596.

Leeder, M.R., Mack, G.H., Peakall, J. and Salyards, S.L. 1996. First quantitative test of alluvial stratigraphic models: southern Rio Grande rift, New Mexico. *Geology*, 24, 87–90.

McCarthy, T.S., Stanistreet, I.G. and Cairncross, B. 1991. The sedimentary dynamics of active fluvial channels on the Okavango Fan, Botswana. *Sedimentology*, 38, 471–487.

Mackey, S.D. and Bridge, J.S. 1995. Three–dimensional model of alluvial stratigraphy: theory and application. *Journal of Sedimentary Research*, 65, 7–31.

Mertes, L.A.K. and Dunne, T. 1988. Morphology and construction of the Solimoes–Amazon River floodplain. In *On the fate of particulate and dissolved components within the Amazon dispersal system: river and ocean*. Chapman Conference Proceedings, American Geophysical Union, 82–86.

Mertes, L.A.K., Dunne, T. and Martinelli, L.A. 1996. Channel–floodplain geomorphology along the Solimoes–Amazon River, Brazil. *Geological Society of America, Bulletin*, 108, 1089–1107.

Nanson, G.C. and Knighton, A.D. 1996. Anabranching rivers: their cause, character and classification. *Earth Surface Processes and Landforms*, 21, 217–239.

Page, K.J. and Nanson, G.C. 1996. Stratigraphic architecture resulting from Late Quaternary evolution of the Riverine Plain, south–eastern Australia. *Sedimentology*, 43, 927–945.

Pickup, G., Allan, G. and Baker, V.R. 1988. History, palaeochannels and palaeofloods of the Finke River, central Australia. In R.F. Warner (Ed.), *Fluvial Geomorphology of Australia*. Academic Press, Sydney, 105–127.

Ritter, J.B., Miller, J.R., Enzel, Y. and Wells, S.G. 1995. Reconciling the roles of tectonism and climate in Quaternary alluvial fan evolution. *Geology*, 23, 245–248.

Rudoy, A.N. and Baker, V.R. 1993. Sedimentary effects of cataclysmic late Pleistocene glacial outburst flooding, Altay Mountains, Siberia. In C.R. Fielding (Ed.), *Current Research in Fluvial Sedimentology, Sedimentary Geology*, **85**, 53–62.

Schumm, S.A., Erskine, W.D. and Tilleard, J.W. 1996. Morphology, hydrology, and evolution of the anastomosing Ovens and King Rivers, Victoria, Australia. *Geological Society of America, Bulletin*, 108, 1212–1224.

Singh, H., Parkash, B. and Gohain, K. 1993. Facies analysis of the Kosi megafan deposits. In C.R. Fielding (Ed.), *Current Research in Fluvial Sedimentology, Sedimentary Geology*, **85**, 87–113.

Smith, R.M.H., Mason, T.R. and Ward, J.D. 1993. Flash-flood sediments and ichnofacies of the Late Pleistocene Homeb Silts, Kuiseb River, Namibia. In C.R. Fielding (Ed.), *Current Research in Fluvial Sedimentology, Sedimentary Geology*, **85**, 579–599.

Stanley, D.J. and Chen, Z. 1996. Neolithic settlement distributions as a function of sea level–controlled topography in the Yangtze delta, China. *Geology*, 24, 1083–1086.

Thorne, C.R., Russell, A.P.G. and Alam, M.K. 1993. Planform pattern and channel evolution of the Brahmaputra River, Bangladesh. In J.L. Best and C.S. Bristow (Eds), *Braided Rivers*, Geological Society of London, Special Publication 75, 257–276.

Van Overmeeren, R.A. 1997. Radar facies of unconsolidated sediments in the Netherlands. *Journal of Applied Geophysics* (in press).

21

Towards an Understanding of Varieties of Fluvial Form

M.J. Kirkby

School of Geography, University of Leeds, UK

INTRODUCTION

Case studies of fluvial form reveal how widely actual rivers differ from the typical case, generally identified with temperate alluvial rivers. Significant interest also focuses on gravel bed rivers, but only individual scientists have previously recognized the widespread occurrence of rivers which deviate strongly from these modal types.

Although there is an increasing level of knowledge and theory about detailed flow hydraulics in complex channels and floodplains, represented at one extreme by computational fluid dynamics (CFD), there are still severe shortcomings in dynamic coupling with sediment transport to generate evolving bed and channel forms. There are also substantial difficulties in extending simulations over time to the periods required for channel self-adjustment or change. For this longer term, of perhaps thousands or tens of thousands of years, our level of understanding and associated theoretical concepts are generally much simpler than the specification of CFD. For example, quasi-equilibrium is often assumed although it is known to be, at best, a first-order approximation. The coarse-scale models which do exist are also necessarily cruder in their hydraulic approximations, due to the need, in the time domain, for formal integration over the frequency distribution of discharge conditions. Although this is certain to raise some doubts about their validity, they appear to offer the only practicable way forward towards an understanding of the way in which water flows create channelways in the wide variety observed in nature. Perhaps it is necessary to go back to Schumm's (1977) book, *The Fluvial System*, to find a connected attempt to address the issues at this landscape scale.

The variety of channel forms described here also raises many questions about how channelway morphology is related to the distribution of flows over time, to their spatial geography and hydrology and to the interactions within the fluvial system between erosion and deposition. First, it is clear that many rivers have magnitude–frequency distributions which are very far from those of normal rivers, and that these distributions

Varieties of Fluvial Form. Edited by A.J. Miller and A. Gupta
© 1999 John Wiley & Sons Ltd

may not be stationary over time. Thus the notion of an equilibrium form may not even be relevant, let alone well determined. Second, many rivers show spatial patterns which depart from assumptions of normality. At both catchment and finer scales, rivers flow between or across very diverse regions which differ in climate, lithology and/or tectonic regime, so that channel forms may be out of phase with the driving processes. Third, rivers generally erode upstream and deposit downstream, have gravel beds upstream and sand beds downstream, and these transitions result from a dynamic interplay between hydrology and sediment characteristics.

The case studies show how incompletely these relationships are understood, and provide significant signposts for future research. Some possible approaches are sketched out in general terms below, not as a definitive road-map, but to stimulate work at the important landscape scale. The viewpoint expressed is that only a mesoscale theory can adequately explain the great variety of channel forms in their proper hydrological and geological context.

TIME DISTRIBUTIONS AND CHANNEL FORM

The channelway is the whole set of cut-and-fill features formed by the river in the valley bottom. It is the relevant unit of study in taking a long-term and coarse-scale view of the river system. Channelway form is generally seen as an expression of the magnitude and frequency distribution in the long term, but must clearly be built up from the impact of individual floods. The most extreme examples of rapid adjustment are seen in the catastrophic floods of the scablands, Mars or the Skeleton Coast, but these seem to form part of a progression towards some of the rivers described for Australia and the monsoonal Narmada River, where there are well defined inner and outer channels. It may be hypothesized that, for any given discharge level, there is an equilibrium channel form, given the availability of sediment, the gradient and the lithology. At a given discharge, the channel form moves towards the appropriate equilibrium, at a rate which is higher for larger discharges. Equilibrium depth, h, might, for example adjust according to a relationship of the form:

$$\frac{dh}{dt} \propto q(q - uh) \tag{21.1}$$

where q is the discharge per unit width and u is a constant with the dimensions of velocity. This type of expression has the property that catastrophic floods are able to create something approaching an established channel very quickly, and also that large extreme events have a strong impact, from which it takes many years of normal floods to recover. The relationship is also asymmetric in that small channels do not immediately remove traces of larger channels, whereas enlargement of channels can more quickly destroy pre-existing smaller channels.

Equation 21.1 is a very simple – almost certainly over-simplified – representation of a class of models where the cross-section is described by a statistical or fractal distribution of bed elevations, which is modified by the passage of a given total discharge and sediment load. Such a representation will lose the detail of secondary flow structures in an ensemble average, but may be sufficient to indicate overall channel form. It forms one extreme on a continuum of channel models stretching to CFD at the other extreme. There is inevitably a trade-off between a more detailed understanding and the practicality of representing the development of whole rivers or substantial reaches over a long period. It is only by exploring coarse-scale models of reduced complexity that the optimum scale can be chosen for any particular purpose.

For a well behaved extremal frequency distribution, say of exponential form, there is a fairly clear association of equilibrium depth, h_∞ with the dominant sediment transporting event $(h_\infty = \sum q^2 / (u \sum q))$. Many rainfall and discharge records can be well approximated as an additive mixture of two exponential distributions. Figure 21.1 illustrates the cumulative effect of sediment transport for flows of each size, assuming that sediment transport is roughly proportional to discharge squared. It can be seen that once the coefficient of variation reaches a value of 2.0 or more, there are clearly two dominant sediment transport events. It is inferred that bimodal channel-in-channel forms are associated with channels in quasi-equilibrium with each of the two modes. In contrast, temperate rainfall records have a coefficient of variation of about 1.3. associated with the "normal" single channel.

Where valleys are confined, for example in bedrock gorges resulting from tectonic uplift, channel-in-channel forms are less apparent because there is limited scope to increase channel width. However, in unconfined situations, even in bedrock, outer channels may be very extensive, as in some of the Australian channels.

Most classic work on extremal distributions has assumed that they are asymptotically exponential in form. However, there is also at least the possibility that they are less well behaved. For example, if the cumulative frequency behaves, for large storms, like (storm size)$^{-m}$, then the sediment transport curve comparable to Figure 21.1 rises indefinitely if $m < 2$. In this case, there is no dominant discharge, and any actual sequence is essentially dominated by the largest historical event. Over long time periods there is also likely to be non-stationarity in the record due to climate change, so that the extremal distribution

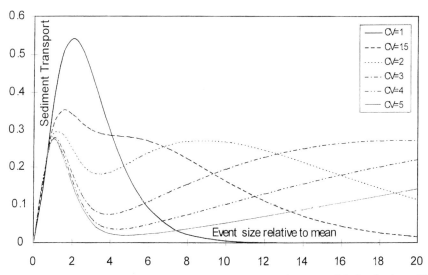

FIGURE 21.1 Multiple dominant sediment transports for mixed exponential distributions. CV is the coefficient of variation for each example

becomes hard even to define. In these cases, which cannot generally be excluded on the basis of existing data, it may be necessary to abandon all notions of quasi-equilibrium channels or dominant flows, and revert to a more idiographic and historical approach.

The long-term response to climate change may not only change the frequency distribution, but may also change the nature of the sediment supply through exhaustion. In the long term of 10^5–10^6 years, there may be a balance between sediment generation through weathering and sediment removal through the fluvial system. If rates of sediment removal increase, the source soil materials may be irreversibly stripped. In extreme cases, fluvial sediments will also be removed, exhuming bedrock landscapes without any change in climate. This kind of behaviour leads to landscape response times which are not solely dependent on the fluvial system, but on responses and thresholds in the landscape as a whole. The impact of the most extreme floods may be to accelerate hillslope erosion, and ultimately to strip the soil irreversibly. As this occurs, following either an exceptional storm and/or a significant change in climate or land use, the channelway is progressively starved of coarser and coarser sediment, generally leading to increased incision.

DISTRIBUTIONS OVER SPACE

Fluvial systems commonly pass through a range of environments. Some transitions between zones produce forms which are only observed in large systems, while others occur on more local scales. One common transition is from humid source areas upstream to arid areas downstream, although the reverse is less usual. Discharge in such systems tends to decrease downstream for substantial distances, leading to conditions which violate some of the normal assumptions of hydraulic geometry and similar relationships.

Long profiles are convex in parts of some of the river systems described, most strikingly for the US Canyonlands, but also parts of the Sabie and Narmada Rivers. Much of this work has echoes of Hack's (1957) classic work on long profiles, but perhaps the problem is best re-posed in terms of the downstream sediment budget for the channelway, taking account of tributary and side-slope inputs. Formally, the mass budget for sediment along the channelway may be expressed in the form:

$$\frac{\partial S}{\partial x} = \frac{\partial S}{\partial q}\frac{\partial q}{\partial x} + \frac{\partial S}{\partial \Lambda}\frac{\partial \Lambda}{\partial x} = T + \frac{S_T}{w} \qquad (21.2)$$

where s is the downstream sediment transport per unit width of channelway, q is the water discharge per unit width, Λ is the channelway gradient, x is horizontal distance in the downstream direction, T is the local rate of incision, S_T is the tributary and side-slope sediment contribution and w is the width of the channelway.

Equation 21.2 can be re-arranged algebraically to give the degree of convexity or concavity of the profile:

$$\frac{\partial \Lambda}{\partial x} = \frac{T + \dfrac{S_T}{w} - \dfrac{\partial S}{\partial q}\dfrac{\partial q}{\partial x}}{\dfrac{\partial S}{\partial \Lambda}} \qquad (21.3)$$

with positive signs for convexity. A number of these terms are necessarily positive,

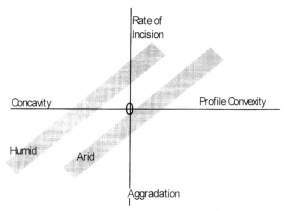

particularly S_T/w, $\partial S/\partial q$ and $\partial S/\partial \Lambda$, but the other two terms can take either sign. Clearly the strongest conditions for convexity occur where there is strong incision $(T > 0)$, discharge is decreasing $(\partial q/\partial x < 0)$ and there is a large tributary or side-slope influx. Figure 21.2 sketches the relationships between profile convexity or concavity and local rates of aggradation or deposition.

For arid rivers, the term $\partial q/\partial x$ in Equations 21.2 and 21.3 is zero or negative, which moves the relationship between the profile convexity (or minus concavity), $\partial \Lambda/\partial x$, and the rate of incision (or minus aggradation), T, to the right compared with humid rivers $(\partial q/\partial x > 0)$, leading to reduced concavity or actual convexity of the long profile. The Green River canyon is a good example of a system which meets these conditions for strong profile convexity. Arid areas in general tend towards conditions of steady or decreasing discharge downstream, so that convex channels are most common in dry areas. Karst drainage can also create these conditions, but may do so only after a period of surface flow has created and enlarged sink holes in the channel bed sufficiently to allow drainage at flood stages. Tectonic uplift of antecedent channels drives strong downcutting which also favours convexity, and there is a strong positive interaction between high incision rates and high lateral influxes from steep canyon walls and tributaries. Similarly subsidence encourages deposition and tends to reduce lateral inflows. Both downcutting and deposition may be further enhanced by isostatic compensation.

The sediment budget relationship expressed in Equation 21.2 is true at all times. However, gradient only slowly adjusts to the prevailing discharge, weathering and tectonic regimes. In the short term, therefore, erosion and deposition respond to the changes in gradient imposed by tectonics and other factors, only slowly approaching a quasi-equilibrium in which gradient sensitively reflects the local tectonic environment. The response time depends on the capacity for sediment transport and the ease or difficulty of changing the channelway cross-section. Thus low-gradient bedrock channelways may have very long response time, and so reflect some mixture of former conditions, whereas narrow alluvial floodplains may respond within perhaps 100 years.

Many catchments show an alternation of alluvial reaches and confined reaches, a "string of pearls" morphology, at many possible scales. At the coarse scale, the differences are most commonly based on tectonic basins, for example where a river is running across a series of half-grabens, with repeated faulting and back-rotation of the blocks. At finer scales, confined zones are related to differences in sediment supply due to variations in material properties (in, say, a glacial till) or the location of tributary inputs. Confined sections act as transmission zones, with little long-term preservation of sediments, although slackwater deposits may occur near tributary junctions and in other favoured niches. Alluvial zones typically preserve evidence of environmental history in both terraces and flood deposits, and tend to buffer the sedimentary response to individual floods and short-term climatic fluctuations. In many arid and semi-arid environments, moderate floods are locally generated so that not all tributaries "fire" at the same time or in the same storm, and tributary fans back up mainstream alluviation; but in major storms, the whole catchment fires at once, and there is general sediment transport.

EROSION AND DEPRESSION IN TIME AND SPACE

In the short term, each flood erodes and redistributes sediment within a river system. A floodwave generated in part of the catchment creates a sediment wave or "slug" which rises in the area of storm generation and gradually declines downstream outside the storm area, as gradient declines and the flood moves into larger channelways downstream. This is illustrated schematically in Figure 21.3. The sediment budget Equation 21.2 gives erosion or deposition as the slope of the transport curve, so that there is erosion upstream within the source area, and deposition downstream. If the flow travels as a debris flow, then sediment may be very poorly sorted and the slug can have an abrupt head, but for fluvial flows, or flows which become fluvial downstream, there is strong selection for grain size.

In principle, there are superimposed waves of sediment associated with each grain size so that, in combination, the slug of deposited material has a coarser tail and a finer head.

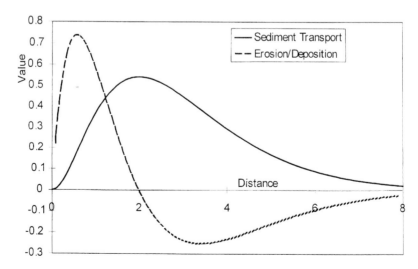

FIGURE 21.3 Schematic relationship between sediment transport and erosion

The form is also complicated by incision of the deposit both during the recession of the initiating hydrograph and in subsequent, smaller floods.

Sediment transport need not occur at the transporting capacity if there is a scarcity of sediment. There is a complete spectrum from conditions of completely flux-limited to completely supply-limited transport. The behaviour of each grain-size fraction in a flood can be approximately represented by its mean travel distance, h, which decreases as grain size increases. For large boulders, where this distance is small compared to the diameter of the eroded source area, H, transport is essentially limited by the flux capacity. Where $h \gg H$, for small sizes, transport is at a fraction H/h of capacity, approaching a condition of complete supply limitation for the finest material. Thus the channelway deposits are coarser than those from soil material on the source hillslopes, and most of the fine fraction moves far downstream and/or accumulates in alluvial zones. Channel formation is then partly controlled by the dominant grain size carried in the fluvial sediment, which is coarser in larger floods, adding another factor which influences cross-sectional form and channel pattern.

In a downstream direction, rivers generally decrease in gradient (although with important exceptions) and lose contact with sediment coming directly from their side slopes as sediment transport becomes dominated by sand rather than gravel, and substantial floodplains develop. In a large, tectonically controlled river, this changeover may occur more than once in a transition zone, within which gravel dominates in confined sections and sand in alluvial sections.

CONCLUSION: TOWARDS RELEVANT MODELS AND UNDERSTANDING

Three physically based approaches have been outlined here, none for the first time. The first is concerned with the evolution of channelway cross-sections through the frequency distribution of formative sediment transport. The physical basis is not fully established, but is closely related to an appraisal of sediment transport equations, applied at an appropriate scale to be determined. The second approach examines the evolution of long profiles in relation to rates of incision and lateral sediment contributions, and the third is concerned with the movement and size differentiation associated with sediment slugs. Both of these are linked to considerations of the fluvial sediment budget, summed across the whole channelway. It is argued that approaches at this scale have greater relevance to an appropriate understanding of the variety of case studies presented than either detailed CFD or derivatives of hydraulic geometry (Leopold and Maddock, 1953).

There are, of course, other relevant conceptual approaches, for example the concept of maximum efficiency described by Nanson and Huang (see Chapter 19). Cellular automaton models also provide a significant approach at an intermediate scale. These may refer to the channelway only (Murray and Paola, 1994), or to the whole catchment (Coulthard et al, 1999). It may also be important to incorporate feedbacks with vegetation growth and destruction (Hooke et al., 1997) in establishing channel morphology. These methods generally make simplifying assumptions to retain enough computational simplicity to deal with long periods and large areas, and it is still a matter for debate whether they are sufficiently compatible with CFD and other detailed models.

The present trend in fluvial geomorphology is towards increasingly detailed understanding of smaller and smaller features – more and more about less and less.

The variety of fluvial forms described in this book, and the generally light level of analysis that has been possible within these case studies, illustrate the very limited way in which detailed fluvial research has been able to contribute to a broad understanding of rivers and channelways. It is time to draw conclusions from all that has been learned in the last 50 years, and apply them, in a suitably simplified way and at relevant scales, to entire fluvial systems, and to studying how such systems have evolved over time.

REFERENCES

Coulthard, T.J. Kirkby, M.J. and Macklin. M.G. 1999. Modelling the impacts of Holocene environmental change in an upland river catchment, using a cellular automaton approach. In A.G. Brown and T.M. Quine (Eds), *Fluvial Processes and Environmental Change*. Wiley, Chichester, 31–46.

Hack, J.T. 1957. *Studies of longitudinal stream profiles in Virginia and Maryland*. USGS Professional Paper 294-B, 45–97.

Hooke, J. M., Brookes, C. and Mant, J., 1997. Modelling desertification and land use impacts on ephemerally-flowing channels of the Mediterranean Region. Paper presented in *Symposium of Fluvial Processes and Environmental Change*, RGS-IBG Conference, Exeter, Jan. 1997.

Leopold, L.B and Maddock, T. Jr., 1953. *The hydraulic geometry of stream channels and some phyisographic implications*. USGS Professional Paper 242, 57 pp.

Murray, A.B, and Paola, C. 1994. A cellular model of braided rivers. *Nature*, **371**, 54–57.

Schumm, S.A., 1977. *The Fluvial System*. Wiley, New York, 338 pp.

Index